I0049660

Sheet Metal 2025

The 21st SheMet Conference hosted by the Laboratory for Material and Joining Technology at the Paderborn University, Germany from April 1st to April 3rd, 2025.

Editor

G. Meschut[1], M. Bobbert[1], J. Duflou[2], L. Fratini[3], H. Hagenah[4], P. Martins[5], M. Merklein[4], F. Micari[1]

[1]Universität Paderborn, Germany
[2]Katholieke Universiteit Leuven, Belgium
[3]Università degli Studi di Palermo, Italy
[4]Friedrich-Alexander-Universität Erlangen-Nürnberg, Germany
[5]Universidade de Lisboa, Portugal

Peer review statement

All papers published in this volume of "Materials Research Proceedings" have been peer reviewed. The process of peer review was initiated and overseen by the above proceedings editors. All reviews were conducted by expert referees in accordance to Materials Research Forum LLC high standards.

Copyright © 2025 by authors

[cc] BY Content from this work may be used under the terms of the Creative Commons Attribution 3.0 license. Any further distribution of this work must maintain attribution to the author(s) and the title of the work, journal citation and DOI.

Published under License by **Materials Research Forum LLC**
Millersville, PA 17551, USA

Published as part of the proceedings series
Materials Research Proceedings
Volume 52 (2025)

ISSN 2474-3941 (Print)
ISSN 2474-395X (Online)

ISBN 978-1-64490-354-4 (Print)
ISBN 978-1-64490-355-1 (eBook)

This book contains information obtained from authentic and highly regarded sources. Reasonable efforts have been made to publish reliable data and information, but the author and publisher cannot assume responsibility for the validity of all materials or the consequences of their use. The authors and publishers have attempted to trace the copyright holders of all material reproduced in this publication and apologize to copyright holders if permission to publish in this form has not been obtained. If any copyright material has not been acknowledged please write and let us know so we may rectify in any future reprint.

Distributed worldwide by

Materials Research Forum LLC
105 Springdale Lane
Millersville, PA 17551
USA
https://mrforum.com

Manufactured in the United State of America
10 9 8 7 6 5 4 3 2 1

Table of Contents

Simulation

Characterization

Polymers and composites

Machine learning

Sustainability

Welding and additive manufacturing

Keyword Index

Committees

Comission

Prof. G. Meschut, Universität Paderborn, Germany

Dr.-Ing. M. Bobbert, Universität Paderborn, Germany

Prof. J. Duflou, Katholieke Universiteit Leuven, Belgium

Prof. L. Fratini, Università degli Studi di Palermo, Italy

Prof. H. Hagenah, Friedrich-Alexander-Universität Erlangen-Nürnberg, Germany

Prof. P. Martins, Universidade de Lisboa, Portugal

Prof. M. Merklein, Friedrich-Alexander-Universität Erlangen-Nürnberg, Germany

Prof. F. Micari, Università degli Studi di Palermo, Italy

Scientific Advisory Board

Prof. J. Allwood, University of Cambridge, Great Britain

Prof. D. Banabic, Technical University from Cluj-Napoca, Romania

Prof. B.-A. Behrens, Leibniz Universität Hannover, Germany

Prof. T. Bergs, Rheinisch-Westfälische Technisch Hochschule Aachen, Germany

Prof. A. Brosius, Technische Universität Dresden, Germany

Prof. S. Bruschi, Università degli Studi di Padova, Italy

Prof. G. Buffa, Università degli Studi di Palermo, Italy

Prof. A. H. van den Boogaard, University of Twente, The Netherlands

Prof. D. Drummer, Friedrich-Alexander-Universität Erlangen-Nürnberg, Germany

Prof. J. Duflou, Katholieke Universiteit Leuven, Belgium

Prof. L. Filice, Università della Calabria, Italy

Prof. W. Flügge, Universität Rostock, Germany

Prof. L. Fratini, Università degli Studi di Palermo, Italy

Prof. A. Ghiotti, Università degli Studi di Padova, Italy

Prof. C. Giardini, Università degli Studi die Bergamo, Italy

Prof. M. Gude, Technische Universität Dresden, Germany

Prof. H. Hagenah, Friedrich-Alexander-Universität Erlangen-Nürnberg, Germany

Prof. G. Ingarao, Università degli Studi di Palermo, Italy

Prof. D. R. Kumar, IIT Delhi, India

Prof. M. Liewald, Universität Stuttgart, Germany

Prof. L. Madej, AGH University of Science and Technology, Poland

Prof. M. Merklein, Friedrich-Alexander-Universität Erlangen-Nürnberg, Germany

Prof. G. Meschut, Universität Paderborn, Germany

Prof. F. Micari, Università degli Studi di Palermo, Italy

Prof. B. Rolfe, Deakin University, Australia

Prof. H. C. Schmale, Technische Universität Dresden, Germany

Prof. M. Schmidt, Friedrich-Alexander-Universität Erlangen-Nürnberg, Germany

Prof. J. Yanagimoto, The University of Tokyo, Japan

Prof. Z. Zhao, Shanghai Jiao Tong University, PR China

Forming

SheMet 2025

Sheet Metal 2025
Materials Research Proceedings 52 (2025) 2-9

Materials Research Forum LLC
https://doi.org/10.21741/9781644903551-1

Folding pre-shaped blanks

David Evans[1,a] * and Julian M. Allwood[1,b]

[1]Department of Engineering, University of Cambridge, Trumpington Street, Cambridge, CB2 1PZ, UK

[a]de314@cam.ac.uk, [b]jma42@cam.ac.uk

Keywords: Deep Drawing, Folding, Geometry

Abstract. Deep drawing creates top-hat-shaped components. Some geometries are difficult to form using deep drawing but can be achieved by folding a top hat into another shape. Previous literature has explained the rules governing the folding of sheets but has not applied these rules to the folding of square top-hat-shaped components. This paper derives the equations governing the geometry of a quarter-top-hat and categorises the family of shapes which can be folded from it by changing the orientation of its creases. Any geometry within this family can be described by a point in a three-dimensional configuration space; this paper identifies the feasible region within this space and shows visually how the geometry of the quarter-top-hat changes as it is folded along three distinct trajectories in the feasible region.

Introduction

Deep Drawing is a rapid and highly repeatable method of forming sheet metal components, in which the blank is drawn radially into a die using a punch. The edges of the sheet are constrained by a blankholder. Deep drawing typically creates a "*top hat shape*", with a crown, brim, and vertical sidewalls. A square punch produces a square top hat (Fig. 1), which is a common benchmark part for analysing the performance of deep drawing processes [1].

Paper models of square top hats show that they can be folded into other shapes without significant in-plane deformation (e.g. Fig. 1). This phenomenon has not previously been investigated, but it provides a way to create components which would be difficult to produce using existing forming methods. Deep drawing or folding-shearing [2] can be used to create an intermediate shape or "*pre-shaped blank*" which can then be folded into the desired shape.

This paper will focus on folding the square top hat, which is made from four identical quarters with the following properties: four flat faces, five creases and two vertices. In this paper, pre-shaped blanks with these properties are known as "*the reference case*", so the quarter-top-hat is just one member of a family of reference case geometries (Fig. 2). Since the reference case is the fundamental building block of all top-hat shapes, the folding behaviour of this simple geometry is of significant interest to understand how metal shells can be formed into new shapes after an initial deep drawing stage. This paper applies origami theory to the quarter-top-hat, to examine the rules that govern the folding of the square top hat as a whole (assuming thin inextensible material with tight folds).

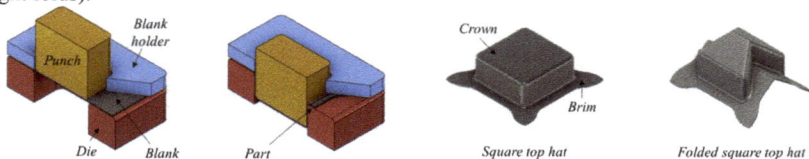

Fig. 1: Illustration of the deep drawing process, the resulting square top hat, and a flat-sided shape which can be folded from the square top hat.

Content from this work may be used under the terms of the Creative Commons Attribution 3.0 license. Any further distribution of this work must maintain attribution to the author(s) and the title of the work, journal citation and DOI. Published under license by Materials Research Forum LLC.

Sheet Metal 2025
Materials Research Proceedings 52 (2025) 2-9

Materials Research Forum LLC
https://doi.org/10.21741/9781644903551-1

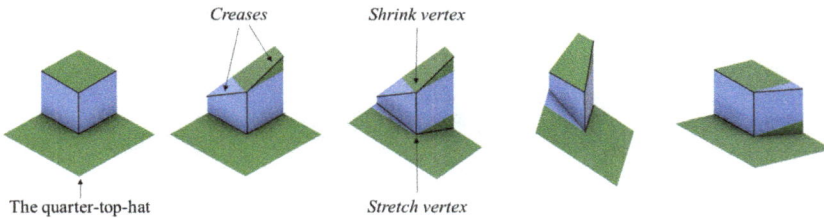

Fig. 2: Selected members of the quarter-top-hat's reference case family. The brim and crown of the original quarter-top-hat have been coloured green to show how material moves between the four flat faces as the creases rotate.

Literature Review

Folding is studied by mathematicians in a field known as Origami. The art of origami has uncertain origins, but paper-folding was certainly popular in Japan by the 1600s [3]. Origami began to be associated with mathematics in the 1800s, but only as an explanatory tool [4]. The first rigorous treatment of the mathematics of origami itself was in 1936 by Beloch [5], who analysed its use for geometric constructions.

In this paper, the folding of the square top hat is analysed using techniques from the subfield of *rigid origami*, in which all faces remain flat without bending or stretching. In rigid origami, folded shapes are described by networks of creases. Early origami theorems by Kawasaki [6] and Maekawa [7] proved that the connectivity and relative position of these creases have an important impact on folding behaviour. Since these theorems, there has been an attempt to create a standard set of terms to describe the geometry of folded shapes. This paper will use the rigid origami nomenclature adopted by Demaine and O'Rourke in their textbook *Geometric Folding Algorithms* [4]:

- A surface is *developable* if it can be folded from a flat sheet, and *non-developable* if it cannot be flattened without tearing, stretching, or material overlap.
- A *crease* is a straight line on a surface.
- A *crease pattern* is a network of creases.
- A *degree-n vertex* is a common endpoint of n creases.
- Such a vertex is defined by n *face angles* θ_i between consecutive creases,
- and n *fold angles* ϕ_i between the normals of consecutive faces. This paper will use *dihedral angles* $\varphi_i = 180 - \phi_i$, the angle between the faces themselves.
- The *angle sum* $\alpha = \sum \theta_i$ at a vertex determines whether it is a *developable vertex* ($\alpha = 360$) or *non-developable vertex* ($\alpha \neq 360$).

In metal forming literature (e.g. [2]), non-developable vertices are divided into *shrink vertices* ($\alpha < 360$) and *stretch vertices* ($\alpha > 360$). Developable vertices can be folded from a flat sheet. Shrink vertices cannot be flattened without tearing or stretching; stretch vertices cannot be flattened without material overlap.

Single-Vertex Patterns. Any vertex – shrink, stretch or developable – is uniquely described by its face angles θ and dihedral angles φ. Face angles are normally labelled $\theta_1, \theta_2, ..., \theta_n$ going cyclically around the vertex, and the dihedral angle φ_i corresponds to the crease between faces defined by θ_i and θ_{i+1}. For a single vertex, the shape of the sheet is irrelevant, so it is convenient to view it as a unit disk centred on the vertex [4].

The face angles θ cannot be chosen arbitrarily; if the creases are not positioned carefully, it may not be possible to find a folded configuration with flat faces [8]. The dihedral angles φ cannot be chosen arbitrarily either; they must obey the continuity condition derived by belcastro and Hull [9]:

$$A_n A_{n-1} \dots A_2 A_1 = \begin{pmatrix} 1 & 0 & 0 \\ 0 & 1 & 0 \\ 0 & 0 & 1 \end{pmatrix}, \tag{1}$$

where

$$A_i = \begin{pmatrix} 1 & 0 & 0 \\ 0 & -\cos\varphi_i & -\sin\varphi_i \\ 0 & \sin\varphi_i & -\cos\varphi_i \end{pmatrix} \begin{pmatrix} \cos\theta_i & -\sin\theta_i & 0 \\ \sin\theta_i & \cos\theta_i & 0 \\ 0 & 0 & 1 \end{pmatrix}. \tag{2}$$

This matrix equation forms a necessary condition for any vertex whose faces fit together with no gaps. It is not a sufficient condition because it does not rule out self-intersection of the vertex [4]. Writing this condition in scalar form results in nine complicated equations. For degree-3 vertices, there are simpler relationships between the dihedral angles and face angles based on the spherical cosine rule, as observed by He and Guest [10]:

$$\cos\varphi_1 = \frac{\cos\theta_3 - \cos\theta_1 \cos\theta_2}{\sin\theta_1 \sin\theta_2}, \tag{3}$$

$$\cos\varphi_2 = \frac{\cos\theta_1 - \cos\theta_2 \cos\theta_3}{\sin\theta_2 \sin\theta_3}, \tag{4}$$

$$\cos\varphi_3 = \frac{\cos\theta_2 - \cos\theta_1 \cos\theta_3}{\sin\theta_1 \sin\theta_3}. \tag{5}$$

Multi-Vertex Patterns. More complex crease patterns are built up from individual vertices which are connected by common crease lines. If a crease pattern can fold with rigid faces, each of the vertices must satisfy Eq. 1. However, the folded states of connected vertices are interdependent, and for a given crease pattern it is not always possible to find a set of dihedral angles which satisfies Eq. 1 at every vertex [4].

The simplest multi-vertex patterns are periodic, made from repeating unit cells which are easy to analyse using trigonometry. The earliest non-trivial example of this type of crease pattern is Miura-Ori [11]: this is a developable crease pattern which folds with one degree of freedom. Similar structures can be created based on the analysis of non-developable unit cells, such as the eggbox pattern [12]. Hybrid patterns containing a combination of developable and non-developable cells have also been investigated [13].

More complex, aperiodic patterns have been developed by Tachi [14], who generalised Miura-Ori to allow rigid-foldable origami without trivial repeating symmetry. He and Guest [15] built on this research by including non-developable vertices.

The multi-vertex work covered so far examined the folding of crease patterns made solely from degree-4 vertices. Akitaya et al. [16] analysed developable crease patterns with full generality: if an arbitrary crease pattern is drawn on a flat sheet, can the sheet be folded rigidly using those creases? They proved that answering this question for a given fold pattern is "NP-hard", which means it is probably impossible to create an efficient algorithm to verify rigid foldability of a multi-vertex fold pattern.

No literature was found which describes the folding of non-developable shell structures, except for very simple single-vertex shapes, periodic patterns, and non-developable variations on Miura-

Sheet Metal 2025   Materials Research Forum LLC
Materials Research Proceedings 52 (2025) 2-9   https://doi.org/10.21741/9781644903551-1

Ori. The folding of the reference case and top-hat geometries has not yet been investigated, and this is the goal of this paper.

Analysis of the folding of a Quarter-Top-Hat

A quarter-top-hat can be folded into a continuum of other geometries by moving its five creases. This section derives the relationships between face angles and dihedral angles which ensure all four faces of the reference case remain flat. By solving this system of equations, this section categorises all possible reference case geometries which can be folded from a quarter-top-hat using incremental motion of its creases.

The reference case system is made from two linked systems: a degree-3 shrink vertex, and a degree-3 stretch vertex. This analysis begins by deriving the equations governing the face angles and dihedral angles of a vertex with three flat faces.

Analysis of a Degree-3 Vertex. This analysis assumes that vertices are deformed by folding only, with no stretching of faces. Changing the position of creases can transfer material from one face to another, but the vertex angle sum α will always be the same. If two face angles are known, the third can be calculated easily:

$$\theta_1 + \theta_2 + \theta_3 = \alpha. \tag{6}$$

The face angles θ and dihedral angles φ of degree-3 vertices must satisfy Eq. 3-5. A degree-3 vertex is thus described by six variables $(\theta_1, \theta_2, \theta_3, \varphi_1, \varphi_2, \varphi_3)$, and four equations (Eq. 3-6). If the system were linear, this would imply that there were two independent parameters which could be used to solve for all the others. In practice, this is not the case, because this system of equations has three ambiguities.

The first ambiguity arises because Eq. 3-5 do not constrain the faces of the vertex to join up at all three creases; this can be resolved by picking solutions which also satisfy the continuity condition (Eq. 1). The second ambiguity comes from the periodicity of the sine and cosine functions; this can be resolved by restricting all angles to a single period: $0 \le \theta < 360$ and $0 \le \varphi < 360$. The third ambiguity occurs because there are two ways to assemble a degree-3 vertex from sectors with given face angles: one solution with dihedral angles $\varphi_1, \varphi_2, \varphi_3$, and an "inside-out" solution with dihedral angles $360 - \varphi_1, 360 - \varphi_2, 360 - \varphi_3$. This ambiguity can be resolved by restricting one of the dihedral angles to the range $0 \le \varphi_1 < 180$ to discard the inside-out solution.

When all three ambiguities are resolved as above, Eq. 3-6 form a system with two independent parameters; setting these provides a unique solution for the remaining four. In this paper, one face angle θ_1 and one dihedral angle φ_1 will be selected as the independent parameters and Eq. 3-5 will be solved numerically.

Fig. 3 shows the configuration space for the degree-3 vertices in the quarter-top-hat: the shrink corner ($\alpha = 270$) and the stretch corner ($\alpha = 450$). The coloured rectangles show the feasible regions within θ_1, φ_1 space; the colour designates the value of θ_2 at each point. There are twelve points where example vertices have been plotted. Changing φ_1 opens or closes the angle between the coloured faces, while changing θ_1 tilts the grey face.

The edges of the feasible regions represent configurations where at least two of the faces are coplanar. Points outside of the feasible region cannot be reached because they require faces to flex, break apart, or move through one another. When the input parameters lie in the feasible region and the output parameters are constrained as described above, each point in the two-dimensional configuration space represents a unique folding of the vertex.

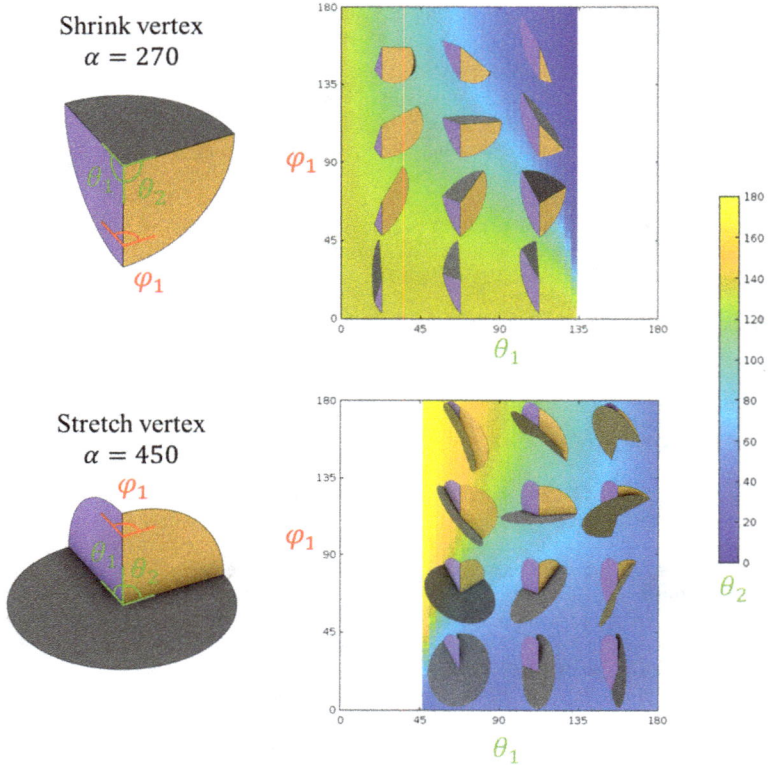

Fig. 3: The configuration space for the shrink and stretch vertices of the quarter-top-hat's reference case family.

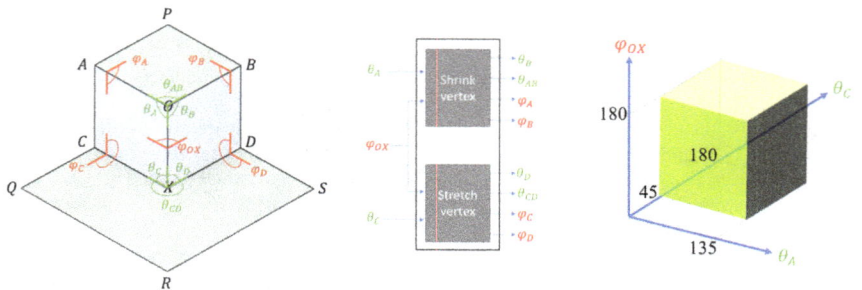

Fig. 4: The labelled reference case, a graphical representation of the analysis, and the feasible region in $(\varphi_{OX}, \theta_A, \theta_C)$ space.

Analysis of the Full Reference Case. A reference case like the quarter-top-hat is described by 11 parameters (Fig. 4). The reference case consists of two degree-3 vertices, so the relationship

Sheet Metal 2025
Materials Research Forum LLC
Materials Research Proceedings 52 (2025) 2-9
https://doi.org/10.21741/9781644903551-1

between the parameters can be described by two sets of four equations (which have the same form as Eq. 3-6). If the three ambiguities are resolved as before, the system has three independent parameters, as shown in the flow chart (Fig. 4).

All members of the quarter top hat's reference case family can therefore be plotted in a three-dimensional parameter space. Fig. 4 shows the feasible region in $(\varphi_{OX}, \theta_A, \theta_C)$ space. The position in the ϕ_{OX}, θ_A plane describes the shape of the shrink vertex, and the position in the φ_{OX}, θ_C plane describes the shape of the stretch vertex.

Results

Having shown that any member of the quarter-top-hat's reference case family can be uniquely described by a point in $(\varphi_{OX}, \theta_A, \theta_C)$ parameter space, this section will show what the geometries look like when navigating through this space.

Fig. 4 shows the initial quarter-top-hat, which is the reference case family member with coordinates $(\varphi_{OX}, \theta_A, \theta_C) = (90,90,90)$. The following figures show other examples from the reference case family. The analysis above has been used to calculate the face and dihedral angles, which have been fed into CAD software to produce the images. The CAD software automatically produces flat faces, and the geometries were validated using the software's angle inspection tool. The green regions show how material moves from one face to another as the creases rotate.

These results show visually how the geometry of the reference case changes as it is folded along three trajectories in $(\varphi_{OX}, \theta_A, \theta_C)$ space. Firstly, changing φ_{OX} independently (Fig. 5); next, changing θ_A independently (Fig. 6); finally, changing θ_C independently (Fig. 7).

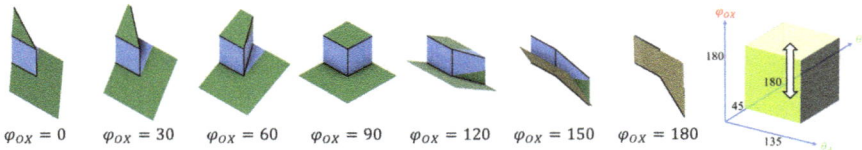

$\varphi_{OX} = 0$ $\varphi_{OX} = 30$ $\varphi_{OX} = 60$ $\varphi_{OX} = 90$ $\varphi_{OX} = 120$ $\varphi_{OX} = 150$ $\varphi_{OX} = 180$

Fig. 5: Changing φ_{OX} ($\theta_A = \theta_C = 90$).

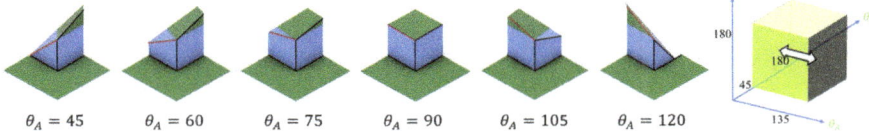

$\theta_A = 45$ $\theta_A = 60$ $\theta_A = 75$ $\theta_A = 90$ $\theta_A = 105$ $\theta_A = 120$

Fig. 6: Changing θ_A ($\varphi_{OX} = \theta_C = 90$).

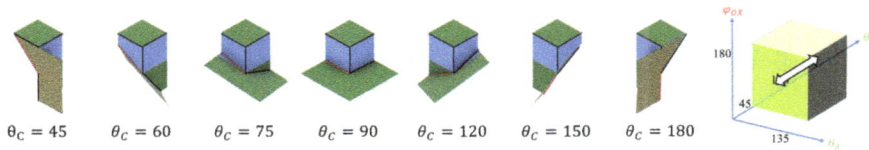

$\theta_C = 45$ $\theta_C = 60$ $\theta_C = 75$ $\theta_C = 90$ $\theta_C = 120$ $\theta_C = 150$ $\theta_C = 180$

Fig. 7: Changing θ_C ($\varphi_{OX} = \theta_A = 90$).

Discussion

Each of the three axes in the configuration space represents a different transformation of the reference case. Changing θ_A tilts the top face; changing θ_C tilts the bottom face; and changing φ_{OX} opens or closes the vertical faces, thereby tilting both top and bottom faces.

Sheet Metal 2025   Materials Research Forum LLC
Materials Research Proceedings 52 (2025) 2-9   https://doi.org/10.21741/9781644903551-1

When $\theta_A = \theta_C = 90$, as in Fig. 5, lines PA and CQ lie in the same plane, while PB and DS lie in another plane, orthogonal to the first. If the reference case is reflected in both planes, the resulting geometry can be folded into a full square top hat, as shown in Fig. 8. Note the additional creases and developable vertices created at the boundary between the four reference cases. Creating these new vertices and moving them requires continuous motion of creases.

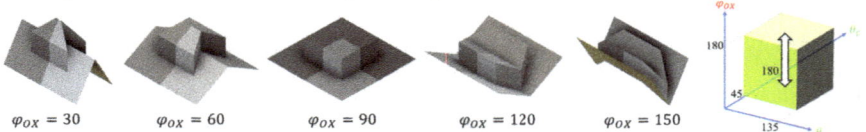

$\varphi_{OX} = 30$ $\varphi_{OX} = 60$ $\varphi_{OX} = 90$ $\varphi_{OX} = 120$ $\varphi_{OX} = 150$

Fig. 8: Folding of a full square top hat. Each quarter has the same shape as a reference case from Fig. 5.

There is one impossible geometry among the examples in the previous section: Fig. 6, θ_A=120. This geometry has crossed creases, so one of the blue faces is split into two subfaces. This is permitted by the mathematics, but it is not physical, because one of the subfaces has negative area. If the aspect ratio of the reference case were changed to lengthen line OX, lines OB and XD would no longer intersect on the face, and this geometry would be permissible. Future work could refine the feasible region to disallow all geometries with crossed creases for a given aspect ratio.

If crossed creases are avoided, a quarter-top-hat may be folded into any of the shapes within the feasible region shown in Fig. 5. The folding process can be completed by continuous rotation of the creases around the vertices according to a trajectory in $(\varphi_{OX}, \theta_A, \theta_C)$ space. There are other reference case geometries belonging to a second feasible region which is shown in Fig. 9. The feasible regions are not connected, so the new geometries cannot be folded from a quarter-top-hat without temporarily relaxing the requirement to retain four flat faces.

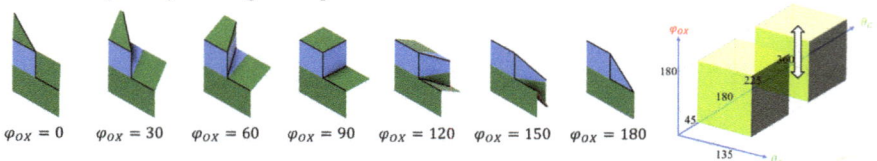

$\varphi_{OX} = 0$ $\varphi_{OX} = 30$ $\varphi_{OX} = 60$ $\varphi_{OX} = 90$ $\varphi_{OX} = 120$ $\varphi_{OX} = 150$ $\varphi_{OX} = 180$

Fig. 9: Changing θ_C within the additional feasible region ($\varphi_{OX} = \theta_A = 90$).

There are other families of reference cases with different values for angle sum α at the shrink vertex, stretch vertex, or both. The analysis in the previous sections is directly applicable to any of these. The analysis can also be extended beyond the reference case to two-vertex systems with higher-order vertices, but each additional crease in the pattern introduces two more degrees of freedom (one each of θ, φ), so two more parameters are required every time a new crease is added.

Conclusion

The reference case is the fundamental building block of the top hat shapes created by deep drawing. Now that the folding behaviour of the reference case is known, a designer can use this information to predict the ways in which a top hat shape can be folded to create a new component without further in-plane deformation. This is an example from a new class of processes, in which deep drawing or folding-shearing is used to produce a pre-shaped blank, followed by a folding stage to form the final component.

References

[1] Danckert J. Experimental investigation of a square-cup deep-drawing process. Journal of Materials Processing Technology. 1995 Mar 1;50(1):375–84. https://doi.org/10.1016/0924-0136(94)01399-L

[2] Cleaver CJ, Arora R, Loukaides EG, Allwood JM. Producing isolated shrink corners by folding-shearing. CIRP Annals. 2022;71(1):217–20. https://doi.org/10.1016/j.cirp.2022.03.036

[3] Hull T. Origami. In: Selin H, editor. Encyclopaedia of the History of Science, Technology, and Medicine in Non-Western Cultures. Dordrecht: Springer Netherlands; 2016. p. 3457–60. https://doi.org/10.1007/978-94-007-7747-7_8818

[4] Demaine ED, O'Rourke J. Geometric folding algorithms: linkages, origami, polyhedra. Cambridge; New York: Cambridge University Press; 2007. 472 p. https://doi.org/10.1017/CBO9780511735172

[5] Beloch MP. Sulla risoluzione dei problemi di terzo e quarto grado col metodo del ripiegamento della carta. Scritti matematici offerti a Luigi Berzolari. 1936.

[6] Kawasaki T. On the relation between mountain-creases and valley-creases of a flat origami. In: Proceedings of the First International Meeting of Origami Science and Technology, 1991.

[7] Kasahara K, Takahama T. Origami for the Connoisseur. Japan Publications; 1998.

[8] Abel Z, Cantarella J, Demaine ED, Eppstein D, Hull TC, Ku JS, et al. Rigid Origami Vertices: Conditions and Forcing Sets. 2015.

[9] Belcastro SM, Hull TC. Modelling the folding of paper into three dimensions using affine transformations. Linear Algebra and its Applications. 2002 Jun;348(1–3):273–82. https://doi.org/10.1016/S0024-3795(01)00608-5

[10] He Z, Guest SD. On rigid origami I: piecewise-planar paper with straight-line creases. Proceedings of the Royal Society A: Mathematical, Physical and Engineering Sciences. 2019 Dec 11;475(2232):20190215. https://doi.org/10.1098/rspa.2019.0215

[11] Miura K. A note on intrinsic geometry of origami. Research of pattern formation. 1989;91–102.

[12] Schenk M, Guest SD. Origami folding: A structural engineering approach. Origami. 2011;5:291–304.

[13] Liu KT, Paulino GH. Geometric mechanics of hybrid origami assemblies combining developable and non-developable patterns. Proceedings of the Royal Society A: Mathematical, Physical and Engineering Sciences. 2024 Jan 24;480(2282):20230716. https://doi.org/10.1098/rspa.2023.0716

[14] Tachi T. Generalization of rigid foldable quadrilateral mesh origami. Journal Of The International Association For Shell And Spatial Structures. 2009 Dec 1;50.

[15] He Z, Guest SD. On rigid origami II: quadrilateral creased papers. Proc R Soc A. 2020 May;476(2237):20200020. https://doi.org/10.1098/rspa.2020.0020

[16] Akitaya HA, Demaine ED, Horiyama T, Hull TC, Ku JS, Tachi T. Rigid Foldability is NP-Hard. 2020.

Sheet Metal 2025
Materials Research Proceedings 52 (2025) 10-18

Materials Research Forum LLC
https://doi.org/10.21741/9781644903551-2

A first approach towards in-line shape monitoring and control in flexible roll forming automotive components

Abdelrahman Essa[1,a] *, Buddhika Abeyrathna[1,b] *, Bernard Rolfe[1,c] *,
Li Yu[2,d] and Matthias Weiss[1,e]

[1]Deakin University, Geelong, VIC 3216, Australia

[2]Baoshan Iron & Steel Co. Ltd., Wuhan Branch of Baosteel Central Research Institute, Wuhan 430080, China

[a]a.essa@deakin.edu.au, [b]buddhika.abeyrathna@deakin.edu.au, [c]bernard.rolfe@deakin.edu.au, [d]nimon821@163.com, [e]matthias.weiss@deakin.edu.au

Keywords: Sheet Metal, Springback, Shape Monitoring

Abstract. Flexible Roll Forming (FRF) enables the manufacture of long and complex parts for automotive applications. In this paper, a frame rail component is successfully flexible roll formed from Advanced-High-Strength-Steel (AHSS). The component has a variable width and depth contour and shows variable springback over the length. Laser sensors were implemented to monitor the springback angle over the length of the part between forming steps. This was followed by a variable overbending pass to compensate for springback. Based on this, a flexible in-line shape control was developed and validated experimentally. in trials of three approaches to compensating for springback in a single geometry made in a single material, the springback of one approach was lower, but we cannot yet predict how this applies in other parts, or whether our compensation approach has other unwanted effects on other component features. The results of this study suggest that FRF in combination with inline shape control can achieve complex component shape quality standards that suit automotive applications. However, this study was limited to one single material condition and part shape and further studies are needed to validate the general applicability of the procedure to other workpiece geometries and material conditions.

Introduction

Advanced-High-Strength-Steels (AHSS) are increasingly used in the automotive industry for structural and crash components due to their high strength-to-weight ratio [1, 2]. Stamping these steels is difficult and costly. Roll forming where the sheet metal is shaped in consecutive roll stands to long parts represents a cost effective alternative [3]. However, conventional roll forming cannot produce variable cross sections along the length that are required for weight optimisation in automotive applications [4].

Flexible Roll Forming can produce a non-uniform cross section along the length [5] but shape defects such as web warping, wrinkling and springback currently reduce part quality and limit industrial commercialization. Previous work has produced solutions for wrinkling compensation [6, 7] and some of them include the forming of a top-hat flange to stiffen the flange [8, 9]. However, springback compensation is still a major issue especially when flexible roll forming high strength steels [10].

Digitized machines are strongly emerging in many industries to increase productivity and reduce quality issues. Many studies have applied smart sensors to digitize mechanical presses in metal stamping, and have developed operator decision-making assistance systems. These enable the identification of the component state to provide recommendations to the operator regarding the optimum machine settings to improve material flow and prevent material failure [11, 12].

Content from this work may be used under the terms of the Creative Commons Attribution 3.0 license. Any further distribution of this work must maintain attribution to the author(s) and the title of the work, journal citation and DOI. Published under license by Materials Research Forum LLC.

Sheet Metal 2025
Materials Research Proceedings 52 (2025) 10-18

Materials Research Forum LLC
https://doi.org/10.21741/9781644903551-2

However, only a limited amount of studies have applied the digitalization concept to conventional roll forming [13, 14]. Some of these studies implemented torque sensors to control and adjust the acceleration and deceleration of the forming rolls to increase the energy efficiency. Another study [15] successfully implemented a close-loop system for inline springback measurement and compensation in roll forming a U-channel profile.

In flexible roll forming a discontinuous profile shape is formed and this leads to a variable distribution of springback over the length of the component. How to monitor and compensate for this variable springback is still an open field of research.

In this paper, a frame rail profile with variable depth and width is produced. Extensive springback is observed that changes over the length of the profile. A first approach towards a procedure for inline quality control in flexible roll forming is developed that uses a 2D laser sensor in combination with a flexible overbending roll tool.

Profile Geometry and Material

Fig. 1 shows the frame rail component that is flexible roll formed from DP780 material with 2 mm thickness. The part has a profile shape variation in width (Fig. 1b) and depth (Fig. 1c) and a top-hat flange. As can be seen in Fig. 1a, the right and left flanges are not symmetric.

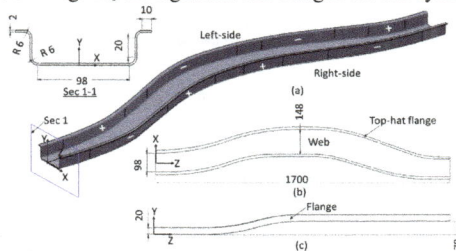

Fig. 1: The simplified frame rail component (a) the 3D view, (b) the top view, and (c) the front view (dimensions in mm)

The profile is flexible roll formed with a prototype flexible roll forming facility (Fig. 2) developed by DataM Sheet Metal Solutions GmbH [16]. It consists of a single forming stand (carriage) and two opposing robotic arms connected to forming rolls which gives the forming tool six degrees of freedom. The pre-cut blank is clamped between the top and the bottom dies by six hydraulic cylinders, and the forming stand moves back and forth, while the forming tool changes its angle and position to follow the contour of the top die to bend the flange into a complex profile shape.

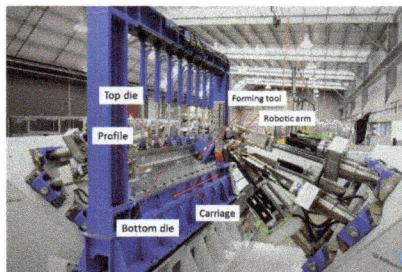

Fig. 2: The Flexible Roll Forming facility with critical features indicated

Sheet Metal 2025   Materials Research Forum LLC
Materials Research Proceedings 52 (2025) 10-18   https://doi.org/10.21741/9781644903551-2

Forming Approach

The forming sequence shown in Fig. 3a was used. Here the flange is formed up to 45° in 3 passes (passes 1-3), and then a top hat is formed in the next 2 passes (passes 4-5). Finally, the flange with the top hat is formed to 90° in the final four forming passes (passes 6-9).

A roll cluster tool is used to form the flange in the first 3 passes as shown in Fig. 3b while the next 2 passes (see Fig. 3c) apply a pair of top hat-forming rolls. After that, the flange and the top hat are formed to the final component shape in the last 4 passes with the roll tool shown in Fig. 3d.

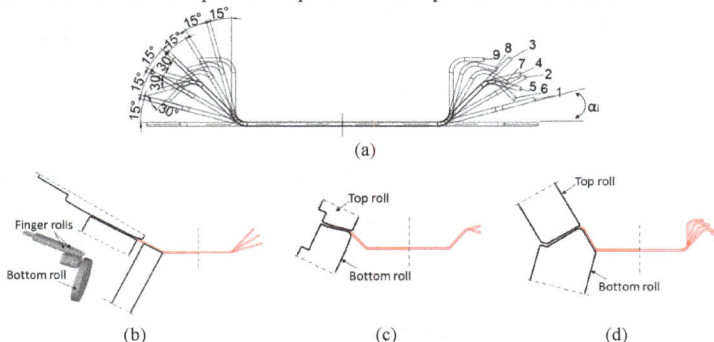

(a)

(b)   (c)   (d)

Fig. 3: (a) bending sequence (Flower design) and tool design used for (b) pass 1-3 (c) pass 4-5 (d) pass 6-9

Shape Monitoring

To monitor the variable springback angle during the forming process, a scanCONTROL 2D laser scanner was used. The scanner comes with support software that applies two reference lines to measure the profile angle over time that can be converted to angle vs profile's length using the speed of the carriage.

The 2D laser scanner was mounted on the forming stand (Fig. 4), and the angle between the flange and the side wall of the bottom die (θ_i) measured. The variable springback angle (β_i) was be calculated over the profile length using Eq. 1 with (α_i) being the ideally formed angle in each individual forming pass *(i)* based on the flower (see Fig. 3a).

$$\beta_i{}^\circ = \alpha_i - \gamma_i{}^\circ \tag{1}$$

$$\gamma_i{}^\circ = (\theta_i - 90°) \tag{2}$$

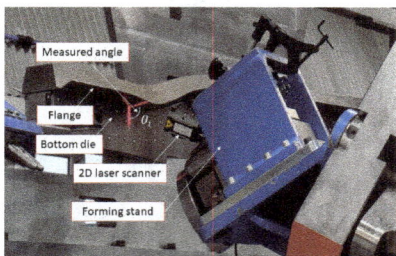

Fig. 4: The 2D laser assembly used to measure the flange angle θ_i between the forming passes

Sheet Metal 2025
Materials Research Proceedings 52 (2025) 10-18

Materials Research Forum LLC
https://doi.org/10.21741/9781644903551-2

The Initial Springback Measurement

Fig. 5 shows the springback angle (β_i) over the length of roll formed profile measured by the 2D laser after the final and 9th forming pass. Note that given that the right and left flanges showed the same trends for springback, only the results for the right flange are shown in this study. Fig. 5 shows that the distribution of springback is non-uniform over the length of the part. This is combined with significant outwards end flare at both component's ends. The results suggests that the compensation of springback in flexible troll forming will require a variable overbending approach.

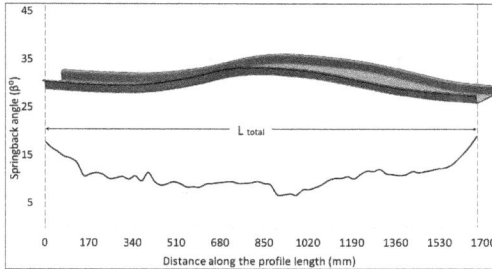

Fig. 5: The distribution of the springback angle along the profile length after the final forming pass 9

The Variable Overbending Approach

The next part of this work will implement a solution for variable overbending. For this, the magnitude and distribution of the variable overbending angle θ_b is assumed to be equal to the springback angle β_i° measured with the 2D laser in the last forming step (i) before the overbending pass $(i+1)$ using Eq. 3. The resulting overbending function is smoothened to enable implementation in the control program of the flexible forming facility. Fig 6 shows a schematic for the distribution of the springback angle after a forming step i (β_i) and the smoothed variable overbending pass ($\theta_b + 1$) that is calculated using Eq. 3 and formed after the forming step i to compensate for springback.

$$\theta_{b_{i+1}} = \left(\beta_i^\circ\right)_{L=0}^{L=L_{total.}} \tag{3}$$

Where $L_{total.}$ is the total profile length.

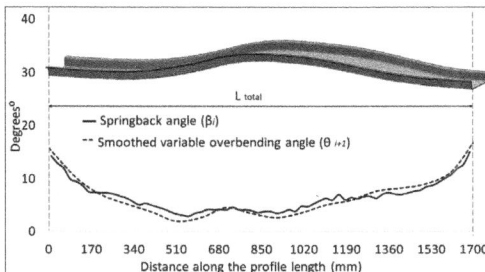

Fig. 6: A schematic shows the distribution of the springback angle after forming pass i ($\boldsymbol{\beta_i}$) and the smoothed variable overbending pass for the next overbending forming pass ($\boldsymbol{\theta_b}+1$)

13

Sheet Metal 2025   Materials Research Forum LLC
Materials Research Proceedings 52 (2025) 10-18   https://doi.org/10.21741/9781644903551-2

Three approaches for overbending were analysed:

Approach 1: Incremental overbending starting after forming pass 1. In this approach, overbending is performed at the start of forming after each individual forming pass. This promises to reduce the amount of plastic deformation that must be introduced in each individual overbending pass and in this way promises to decrease the risk for undesired shape defects that may result from excessive material deformation. This approach was tested by overbending after forming passes 1 and 2.

Approach 2: Incremental overbending starting after forming pass 6. At this forming stage, the top hat shape is fully formed. This provides stiffness in the flange and prevents potential wrinkling of the flange due to the overbending deformation. Like in Approach 1 the incremental overbending in multiple passes reduces the level of plastic deformation that is introduced. This approach was tested by overbending after passes 6, 7 and 8.

Approach 3: Full variable overbending after the final forming pass 9. In this approach, the full amount of overbending is applied after the final forming pass.

Results and Discussion

Approach 1: Incremental overbending starting after forming pass 1:

Fig. 7a shows the springback angle $\beta°$ after forming passes 1 and 2 and after overbending forming pass 1. The 2D laser measurement suggests a large shape deviation in the critical forming zones between profile length distances L =100 mm to 500 mm and L= 1100 mm and 1600 mm for both forming stages. Overbending does not improve the shape accuracy but on the contrary leads to a more severe shape deviation. Investigating the 3D part shape in Fig. 7b suggests that the shape deviation in this initial forming stage is due to the buckling of the flange and not a result of springback. Applying additional plastic deformation by overbending at this forming stage increases the compressive stress in and flange and leads to wrinkle development.

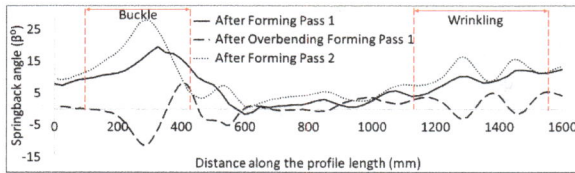

(a)

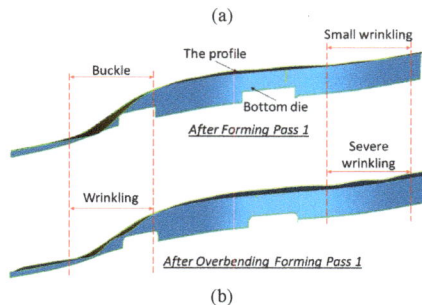

(b)

Fig. 7: (a) 2D laser measurement after forming passes 1 and 2 and after incremental overbending after pass 1 (before pass 2), and (b) images of flange shape after forming pass 1 and after overbending forming pass 1 suggesting an increasing wrinkling severity

Sheet Metal 2025

Materials Research Forum LLC

Materials Research Proceedings 52 (2025) 10-18

https://doi.org/10.21741/9781644903551-2

Approach 2: Incremental overbending starting after forming pass 6:

Compared to the initial forming stage (Fig. 7) the distribution of springback after forming passes 6 and 7 does not suggest any severe shape deviation in form of flange wrinkling, Fig. 8a and this is confirmed by Fig. 8b. After forming pass 5 the top hat shape is fully formed; this stiffens the flange and prevents buckling and wrinkling issues [8]. Variable overbending after forming pass 6 successfully eliminates springback. However, when forming the flange further in pass 7, springback reoccurs and gives a higher level of springback compared to forming pass 6, see Fig. 8b.

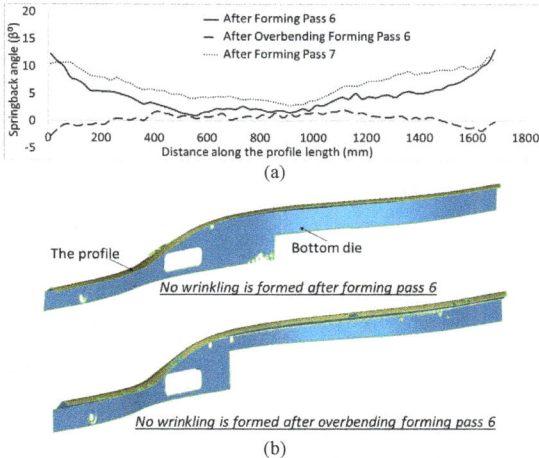

(a)

(b)

Fig. 8: (a) 2D laser measurement after forming passes 6 and 7 and after subsequent incremental overbending after pass 6 (before pass 7) and (b) images of flange shape after forming pass 6 and after overbending pass 6 suggesting an intact flange with no wrinkling

Fig. 9 shows the distribution of the springback angle after forming pass 9 with and without incremental overbending in passes 6, 7 and 8. The result suggests that the incremental overbending approach does not lead to a significant reduction of springback after the final forming pass 9.

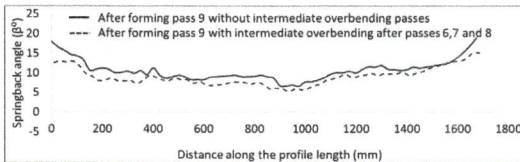

Fig. 9: Springback distribution after forming pass 9 with and without incremental overbending after forming passes 6, 7 and 8

Approach 3: Full variable overbending after the final forming pass 9:

In this approach, the full amount of variable overbending is applied after the final forming pass 9, i.e., without any incremental overbending performed in the previous passes. The result shown in Fig. 10a suggests that this leads to a complete elimination of springback and that the final part shape is formed without any further shape errors (Fig. 10b).

15

Sheet Metal 2025
Materials Research Proceedings 52 (2025) 10-18

Materials Research Forum LLC
https://doi.org/10.21741/9781644903551-2

(a)

No wrinkling is formed after forming pass 9 with full
amount of variable overbending applied

(b)

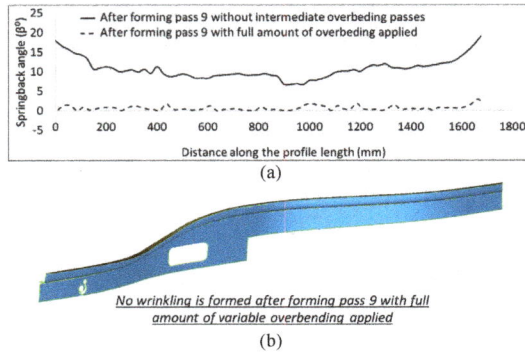

Fig. 10: Springback distribution after the last forming pass 9 without intermediate overbending and with full amount of variable overbending applied and (b) images of flange shape after forming pass 9 with the full amount of overbending applied showing that no further shape error occurred

Conclusion

In this study a complex-shape automotive rail component is successfully flexible roll formed from DP780 material but shows inhomogeneous springback over the length of the components. A 2D laser scanner is implemented into the forming facility to monitor springback after each forming pass and to apply this information to develop a variable overbending pass for springback reduction. Three different approaches for overbending are applied. In the first, overbending is applied at the start of the forming process after each forming pass. In the second, incremental overbending is applied after the top hat shape had been finalised. In the last approach the full level of overbending is applied after the final forming pass.

The following conclusion can be made.

- Incremental overbending before the top hat shape is finalised leads to an increase in the wrinkling severity. This is due to the flange not being stiff enough to prevent wrinkling initiation and growth when the additional flange deformation is applied in the overbending passes.

- Incremental overbending after the top hat is finalised prevents wrinkling initiation and eliminates of springback after the individual forming pass. However, when forming the flange in the subsequent forming passes springback re-occurs at a higher level. In this way the effect of incremental overbending is eliminated leading to a similar level of springback when the final component shape is finalised.

- Applying the full amount of overbending at the end of forming leads to a permanent compensation of springback without any initiation of wrinkling.

Overall, this study suggests that flexible roll forming combined with in line-shape monitoring and variable overbending at the end of forming allows the manufacture of complex components from high strength steel at tight part shape tolerances. However, the results shown in this paper are limited to one single part shape and material condition, and future work is required to prove the general applicability of this method.

Sheet Metal 2025
Materials Research Proceedings 52 (2025) 10-18

Materials Research Forum LLC
https://doi.org/10.21741/9781644903551-2

Acknowledgment
The authors would like to thank the Baosteel Australia Research and Development Centre (BAJC) for their financial support as part of BA19001. The authors are further grateful for the software support provided by data M Sheet Metal Solutions GmbH, Germany.

References

[1] N. Baluch, Z. M. Udin, and C. S. Abdullah, Advanced high strength steel in auto industry: an overview, (in English), Eng. Appl. Sci. Res. 4 (2014) 686-689. https://doi.org/10.48084/etasr.444

[2] G. Sun, M. Deng, G. Zheng, and Q. Li, Design for cost performance of crashworthy structures made of high strength steel, Thin-Walled Struct. 138 (2019) 458-472. https://doi.org/10.1016/j.tws.2018.07.014

[3] A. D. Deole, M. R. Barnett, and M. Weiss, The numerical prediction of ductile fracture of martensitic steel in roll forming, Int J Solids Struct. 144-145 (2018) 20-31. https://doi.org/10.1016/j.ijsolstr.2018.04.011

[4] C. Jiao-Jiao, C. Jian-Guo, Z. Qiu-Fang, L. Jiang, Y. Ning, and Z. Rong-guo, A novel approach to springback control of high-strength steel in cold roll forming, Int. J. Adv. Manuf. Technol. 107 (2020) 1793-1804. https://doi.org/10.1007/s00170-020-05154-8

[5] P. Groche, G. von Breitenbach, M. Jockel, and A. Zettler, New tooling concepts for future roll forming applications, Presented at the 4th, International Conference on Industrial Tools, Slovenia (2003).

[6] M. M. Kasaei, H. M. Naeini, G. H. Liaghat, C. M. A. Silva, M. B. Silva, and P. A. F. Martins, Revisiting the wrinkling limits in flexible roll forming, The J of Strain Anal for Eng. Des. 50 (2015) 528-541. https://doi.org/10.1177/0309324715590956

[7] M. M. Kasaei et al., Flange wrinkling in flexible roll forming Process, Procedia Eng. 81 (2014) 245-250. https://doi.org/10.1016/j.proeng.2014.09.158

[8] A. Essa, B. Abeyrathna, B. Rolfe, and M. Weiss, Incremental shape rolling of top-hat shaped automotive structural and crash components, J. Mater. Process. Technol. 321 (2023) 118162-118177. https://doi.org/10.1016/j.jmatprotec.2023.118162

[9] A. Sreenivas, B. Abeyrathna, B. Rolfe, and M. Weiss, Development of a reversible top-hat forming approach for reducing flange wrinkling in flexible roll forming, Int. J. Mech. Sci. 252 (2023) 108359-108370. https://doi.org/10.1016/j.ijmecsci.2023.108359

[10] B. Abeyrathna, S. Ghanei, B. Rolfe, R. Taube, and M. Weiss, Springback and end flare compensation in flexible roll forming, in International Deep Drawing Research Group Annual Conference, (2020). https://doi.org/10.1088/1757-899X/967/1/012048

[11] M. Kott, D. Echler, and P. Groche, Methodological approach for the development of an operator assistance system for the press shop, Int. J. Adv. Manuf. Technol. 119 (2021) 2409-2428. https://doi.org/10.1007/s00170-021-08199-5

[12] T. Gally et al., Identification of model uncertainty via optimal design of experiments applied to a mechanical press, Opti. and Eng. 23 (2021) 579-606. https://doi.org/10.1007/s11081-021-09600-8

[13] T. Traub, B. Güngör, and P. Groche, Measures towards roll forming at the physical limit of energy consumption, Int. J. Adv. Manuf. Technol. 104 (2019) 2233-2245. https://doi.org/10.1007/s00170-019-03992-9

[14] A. Sedlmaier, T. Dietl, and J. Harraßer, "Digitizing roll forming with smart sensors, Presented at the Proceedings of the 22nd International Esaform Conference on Material Forming, Spain (2019). https://doi.org/10.1063/1.5112733

[15] P. Groche, P. Beiter, and M. Henkelmann, Prediction and inline compensation of springback in roll forming of high and ultra-high strength steels, Produc. Eng. 2 (2008) 401-407. https://doi.org/10.1007/s11740-008-0131-3

[16] A. Sedlmaier and T. Dietl, 3D roll Forming centre for automotive applications, in 17th International Conference on Metal Forming, Toyohashi (2018). https://doi.org/10.1016/j.promfg.2018.07.319

Sheet Metal 2025
Materials Research Proceedings 52 (2025) 19-26

Materials Research Forum LLC
https://doi.org/10.21741/9781644903551-3

A Study of beak geometries for achieving pure shear deformation in folding-shearing

Rishabh Arora[1,a], Omer Music[2,b] and Julian M. Allwood[1,c *]

[1]Department of Engineering, Trumpington Street, University of Cambridge, CB2 1PZ, UK

[2]DeepForm, Allia Future Business Centre, Cambridge, CB4 2HY, UK

[a]ra632@cam.ac.uk, [b]omer@deepform.co.uk, [c]allwood-office@eng.cam.ac.uk

Keywords: Metal Forming, Design, Folding-Shearing

Abstract. The automotive industry produces significant material waste from the deep drawing process. The Folding-shearing process was developed as a solution that involves folding a blank while collecting the excess material in a region called the 'beak'. The beak is then sheared in-plane to form the part without any thickness changes. Previous work used folding-shearing to form a quarter of a square based cup and half of a U-channel part; however, these parts suffered from high levels of thickening in the sheared region. This study, for the first time, explores the underlying design principles to improve the thickness distribution by introducing curvature to the beak perimeter to induce compressive strains in one direction and tensile strains in the other. Numerical simulations are validated using physical trials and show a reduction in thickening from 17.5% to 6%.

Introduction

The classical deep drawing technique is widely used in automotive press lines. In this process, a flat metal blank is clamped along the perimeter in the blankholder region, and a male punch deforms the metal blank to form the target part [1]. As shown in metal forming books, this process can result in substantial material waste, primarily due to the blank being clamped in the blankholder but also from small trimmings generated from hole punching. It is necessary to clamp the blank as it helps control the balance between thinning and wrinkling [2]. A detailed analysis of material waste in the automotive industry showed that up to 45% of the purchased sheet metal is scrapped [3]. Despite this considerable amount of waste, much research to date has focused only on improving the recycling potential of automotive scrap [4]. A case study undertaken to improve material utilization found that a more effective approach is to prevent the waste in the first place, thus yielding both financial and environmental advantages [5]. To address the issue of low material utilization, the folding-shearing process was developed by [6], where the parts are formed through a pure shear deformation mode. The mechanics of the folding-shearing process is inspired by the spinning process, where a shape change occurs without thickness variation [7]. The process involves two stages: folding and shearing. Firstly, a blank is folded over a radius, while collecting the excess material in regions of incompatibility to form an intermediate shape called the 'beak'. Secondly, the material in the beak is deformed through an in-plane shearing process to form the target geometry. In preliminary studies, [6] validated the shearing stage through experimental and numerical simulations and subsequent work by [8] explored the full process. More recently, [9] demonstrated the process using a single set of tools in a single forming direction, acting as a drop-in solution to replace the stamping dies in deep drawing. Therefore, in theory, folding-shearing could achieve stroke rates comparable to deep drawing, producing upto 16 parts per minute. A process operating window was mapped by [9] while considering three failure modes: springback, thinning and thickening. The studies conducted by [8] and [9] revealed that excess thickening in the part could result in the workpiece to lock up between the tools, potentially leading to a

Content from this work may be used under the terms of the Creative Commons Attribution 3.0 license. Any further distribution of this work must maintain attribution to the author(s) and the title of the work, journal citation and DOI. Published under license by Materials Research Forum LLC.

Sheet Metal 2025   Materials Research Forum LLC
Materials Research Proceedings 52 (2025) 19-26   https://doi.org/10.21741/9781644903551-3

premature tearing failure. A comparison between folding-shearing and deep drawing was conducted by [8], concluding that folding-shearing process resulted in a lower maximum thinning for the same part geometry and could form 2.5 times deeper parts, across a range of material thickness. Furthermore, it was demonstrated that the folding-shearing process can form 2.5 times deeper parts, across a range of material thickness. This raises a fundamental question: if the process had truly achieved pure shear, why was a non-uniform thickness distribution observed?

This paper develops a new design methodology to achieve a pure shear deformation in the folding-shearing process. Through both numerical and physical trials, this paper validates new tool designs and addresses the challenges of uneven thickness distribution. This work is a continuation of the work conducted by [9] and will be used as a reference throughout the paper. The study begins with an analysis of the current design methodology, followed by a critical discussion of the limitations, and proposes an improvement to the design methodology ultimately, leading to the development of a novel design space.

Current beak design and reference part

The previous study, [9], used folding-shearing to form half a U-channel part with a shrink corner. An origami model, with sharp edges acts as a useful tool to define the process parameters and Fig. 1, demonstrates the steps required to achieve the target geometry. The process begins by first crash forming a flat blank with a width of w [mm] by an angle of $\delta°$ to obtain Fig.1b. Subsequently, the part is folded by 90° to obtain Fig. 1c and the excess material is collected in the beak. The beak is shown in pink in Fig. 1c. In the shearing stage, the upper and lower shear tools grip the material within the beak and deform the material by in-plane shear to form the flat sidewall.

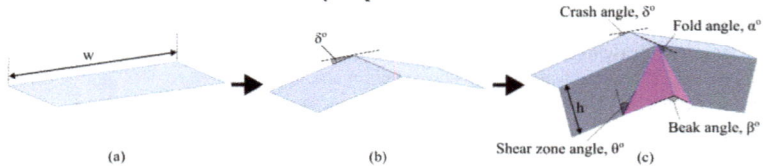

Figure 1: Process steps: (a) Flat rectangular blank is crash formed by $\delta°$ to form (b), which is then folded by **90°** to form (c)

Fig. 1c shows the input parameters that define the beak geometry: the crash angle (δ) and the shear zone angle (θ). The shear zone angle is the angle formed between the sidewall of the beak and the line parallel to the edge of the sheet. The resulting angles are the fold angle (α) and the beak angle (β) and help define the beak geometry. These are found to be a function of the crash angle and the shear zone angle as shown in Eq. 1 and Eq. 2, respectively.

$$\alpha = \sin^{-1}\left(\frac{\cos\theta}{\cos(\theta-\delta)}\right) \tag{1}$$

$$\beta = 2\sin^{-1}\left(\frac{\sin(\theta-\delta)}{\sin\theta}\right) \tag{2}$$

The shearing tool geometry follows the beak geometry shape found from origami folding. It was shown by [9] that a part can have various failure modes: it fails due to springback at a low crash angle ($\delta < 5°$), due to thinning at intermediate crash angle ($5° < \delta < 17.5°$) and due to thickening at high crash angles ($\delta > 17.5°$). The transition from thinning and thickening is seen around $\delta = 15°$ and at a part height of 50 mm. The part fails due to both thinning and thickening, making it a viable reference part for this study. This reference part is made using AA1050-H14 and has a part width of 200 mm and part height of 50 mm. Physical trials found a maximum thinning of 8% at the apex of the part and a maximum thickening of 17.5% at the edge of the

Sheet Metal 2025
Materials Research Proceedings 52 (2025) 19-26

Materials Research Forum LLC
https://doi.org/10.21741/9781644903551-3

formed region [9]. With the current design methodology and the reference geometry established, the next section will investigate the design of the beak to develop a new approach for modelling the corresponding tools.

Design Analysis
The thickness distribution from [9] showed that the reference part undergoes a pure shear deformation along the centre of the beak, while it experiences thickening along the sides. This can arise due to a varying curvature along the deformation zone. Fig. 2 shows, using arrows, that as the material deforms from the beak region, it passes through the feed radius and eventually forms the flat sidewall. As the material in the centre passes through the feed radius, it deforms along a surface with a curvature in two directions, while the material on the sides deforms due to curvature along one direction only. The study of curvature can be a useful mechanism to understand the impact of these curvatures on metal deformation.

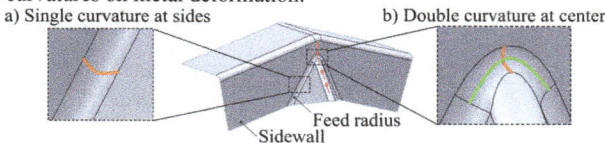

Figure 2: Current curvature along (a) side region (b) centre region

Existing metal forming literature focusses on the concept of Gaussian curvature, which is derived from the principal curvature theory from differential geometry. As explained by [10], the principal curvatures quantify surface bending in different directions, and [11] shows that the principal directions are always perpendicular on smooth surfaces. The principal curvatures at a point 'p' are denoted by κ_1 and κ_2, which represent the maximum and the minimum curvatures, respectively. The Gaussian curvature (K) is the product of the two principal curvatures ($K = \kappa_1 \kappa_2$) and it allows for a comprehensive description of surface curvature through 2D mapping on 3D surfaces. A surface can have a concave or a convex curvature, or a combination of the two. Fig. 3 shows three surface types with different Gaussian curvatures: zero, positive and negative. Fig. 3a resembles a cylindrical surface with a zero Gaussian curvature, as curvature occurs in only one of the orthogonal directions. Such surfaces are termed as 'developable' as they can be flattened without any thickness changes. Fig. 3b resembles a hemisphere with a positive Gaussian curvature, where both the principal curvatures are in the same direction. Fig. 3c resembles a saddle with a negative Gaussian curvature where one direction is positive, and the other direction is negative.

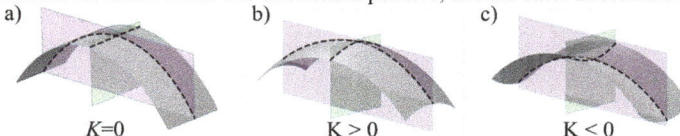

Figure 3: Gaussian curvature: (a) Zero (b) Positive (c) Negative

The influence of Gaussian curvature has been widely studied in metal forming processes. For instance, in stretch forming, a hemispherical punch with a positive Gaussian curvature induces tensile strains in perpendicular planes, thereby increases the risk of tearing [1]. Conversely, a saddle-shaped tool with a negative Gaussian curvature, can lead to a tensile strain in one direction and a compressive strain in the other [12]. Similar mechanics are observed in spinning where a working roller presses against an existing curvature, which results in circumferential compression and radial tension and when balanced, leads to a shape change without any thickness change [7]. These observations prove the importance of a Gaussian curvature in determining the deformation mechanics. The methodology discussed next seeks to optimize the current tool geometry by

Sheet Metal 2025

Materials Research Proceedings 52 (2025) 19-26

Materials Research Forum LLC

https://doi.org/10.21741/9781644903551-3

introducing a double curvature along the complete feed radius, thereby expanding the pure shear deformation region. Fig. 4 shows the beak geometry after the origami folding, which results in straight lines in the beak – lines AB, BC and AC. Line AB is formed from the extrusion of line BC along the line AC. Hence, it is only necessary to control the line AB and the line AC to define the overall shape of the beak. Henceforth, line AB is called the 'fold line' and line AC is called the 'ridge line'. Fig. 4a shows that the points A, B and M_1 lie on the same line and are in the same plane. M_1 is the midpoint of the fold line and can be offset by a perpendicular distance of λ_1, until it reaches a new point M_1'. Points $(AM_1'B)$ now sit on a circular arc and forms the new convex curved fold line. Similarly, points A, M_2 and C lie on the ridge line (Fig. 4b) and are on the same plane. M_2 can be offset by a perpendicular distance of λ_2 to reach a new point M_2'. Points $(AM_2'C)$ now sit on a circular arc and form the new concave ridge line. λ_1 and λ_2 can be offset in the opposite direction to get a concave or a convex curvature respectively.

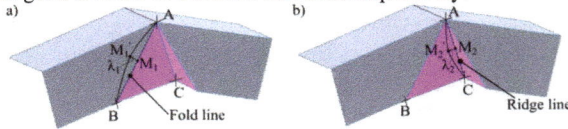

Figure 4: Adding curvature to (a) Fold line (AB) and to (b) Ridge line (AC) by using a perpendicular offset

The maximum allowable offsets (λ_1, λ_2) are defined by the tangency limits made by the circular arcs to points A, B or C, such that the arc does not go past the existing boundary of the part or should not end up penetrating the surface in any direction. Table 1 summarises the trigonometric function that defines the maximum offset for both the ridge or fold line at A,B or C as:

Table 1: Trigonometric functions to define maximum offset to satisfy tangency condition

Line	Point	Convex [mm]	Concave [mm]
Fold line - λ_1	A	$\dfrac{h(1-sin\theta)}{2\,cos^2\,\theta}$	$\dfrac{h(1-\cos(\theta-\delta))}{2\cos\theta\sin(\theta-\delta)}$
	B	$\dfrac{h(1-cos\theta)}{2cos\theta\,sin\theta}$	$\dfrac{h(1-sin\theta)}{2\,cos^2\,\theta}$
Ridge line - λ_2	A	$\dfrac{h(1-cos\theta)}{2\,sin^2\,\theta}$	$\dfrac{h(1-sin\theta)}{2cos\theta}$
	C	$\dfrac{h(1-sin\theta)}{2cos\theta}$	$\dfrac{h(1-cos\theta)}{2\,sin^2\,\theta}$

Caution must be taken as the design is based on the minimum value of the offsets for each point on the corresponding line. For example, if the fold line is intended to be convex, then the minimum λ_1 from points A or B must be considered, as shown in Eq. 3. If the offset exceeds this minimum value, then the tool geometry is no longer tangent to one of the axes at A or B and this will result in uneven tool geometries.

$$\lambda_{1_{convex,max}} = \min\left(\frac{h(1-sin\theta)}{2\,cos^2\,\theta}, \frac{h(1-cos\theta)}{2cos\theta\,sin\theta}\right) \qquad (3)$$

With the maximum displacements defined, a new design space is mapped and is shown in Fig. 5. The graph considers the fold line along the x axis and the ridge line along the y axis. Since a concave surface bends inwards, it is in the negative region of the axis and a convex offset is on the positive side of the axis. At the centre (0,0), the typical geometry from [9] is seen where both the ridge and fold are straight. Progressing along the x axis, from left to right, the fold line is initially

Sheet Metal 2025
Materials Research Proceedings 52 (2025) 19-26

Materials Research Forum LLC
https://doi.org/10.21741/9781644903551-3

concave and gradually becomes convex, while the ridge remains linear. Along the y axis, the ridge line is initially concave and gradually becomes convex from the top to the bottom, while the fold line remains linear. In each corresponding quadrant, four distinct tool combinations are identified, and a curvature is added along both the lines. With the tool designs defined, the next section presents the experimental design used to validate the design methodology to find a suitable tool that deforms the beak in pure shear.

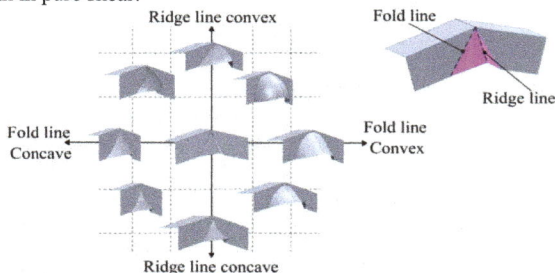

Figure 5: Design space found by adding curvature to the ridge and the fold lines

Design of experiments

Given the substantial costs and prolonged lead times associated with the physical prototyping, the initial evaluation was conducted using numerical trials. This approach allows for efficient screening of the tool designs before proceeding to manufacture and physically validate a design. The numerical setup used in this paper was previously validated by [9] and a thickness prediction was demonstrated to be within 5% for varying material and process parameters. Consequently, this numerical setup from [9] was considered sufficiently reliable for the preliminary investigation. The numerical model was developed using Autoform R10. The upper and lower shear tool was modelled as a rigid tool, while the blank was defined as deformable. A mesh sensitivity study was conducted until the differences in the stresses and the strains between subsequent mesh refinement was less than 5%. A Coulomb-friction model was used with the coefficient of friction set as 0.15. To achieve an acceptable level of resolution in each quadrant, five levels of curvature were selected for the fold and the ridge line: maximum concave offset, half of the maximum concave offset, zero offset, half of the maximum convex offset, and maximum convex offset. This resulted in a total of 25 tool designs across the design space. The key parameters are shown in table 2:

Table 2: Key parameters used in design of experiments

Total number of tool designs	25
Material	AA1050-H14
Crash angle (δ), Shear zone angle (θ) [deg]	$\delta = 15°, \theta = 45°$
Part height, Part thickness [mm]	h = 50, t = 1.5

Design space analysis

For a part to be successful, the thinning and thickening must be less than 10% and the springback must be lower than 1 mm. Fig. 6 evaluates the reference part (Fig. 6a) with a 15° crash angle (δ), 45° shear zone angle (θ), 55 mm part height and 1.5 mm part thickness across the 25 tool geometries in the design space.

Sheet Metal 2025   Materials Research Forum LLC
Materials Research Proceedings 52 (2025) 19-26   https://doi.org/10.21741/9781644903551-3

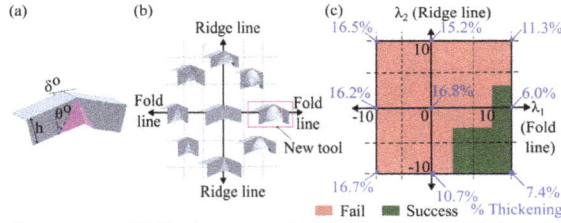

Figure 6: (a) Reference part (b) Design space (c) Window showing failure limits across the range of tool designs with thickening failure limits

Results from the design space are mapped in Fig. 6c, where if the part is successful it is shown in green and if the part has failed, it is shown in red. The only failure mode observed was thickening and Fig. 6c highlights the maximum thickening values from numerical trials in blue at the edges of the design space. As the curvature changes along the fold line, the thickening varies significantly from 16.2% to 6%. As the curvature changes along the ridge line, the thickening varies from 15.2% to 10.7%. Evidently, the curvature along the ridge line has a smaller effect compared to adding a curvature in the fold line. These trends align with existing Gaussian curvature literature. The primary deformation from the beak to the flat side wall is governed by the curvature of the fold line only. On the left side of the design space, the tool geometries exhibit an overall negative curvature in the orthogonal planes, resulting in compressive strains that lead to material accumulation and, hence thickening. On the right side of the design space, the convex curvature along the fold line results in an overall negative gaussian curvature, thereby resulting in compressive strains in one direction and tensile strains in the other. The tool with the maximum convex curvature along the fold line (boxed in Fig. 6b) has the biggest impact and was manufactured to validate the numerical trials. Henceforth, this tool is regarded as the 'new' tool and the tool at the origin, from [9], is regarded as the 'old' tool. Fig. 7a shows the physical setup.

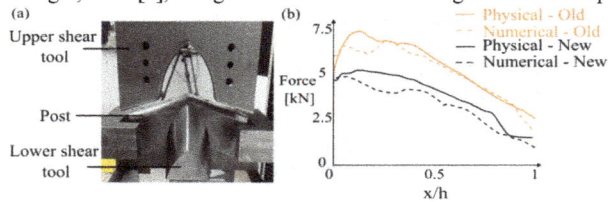

Figure 7: (a) Physical setup of new tool (b) Force evolution from physical and numerical setup

The upper shear tool was attached to the crosshead and the post was fixed at the base of the machine. The workpiece was first clamped to the post and is crash formed over the post, until it reached the target crash angle. The upper shear tool then moves vertically downwards at a constant speed of 1 mm/s until the upper shear tool closed onto the lower shear tool. In the second stage, the upper shear tool was bolted onto the lower shear tool and both the tools are moved vertically downwards at a constant speed of 1mm/s. The process continued until the end of the stroke. Two measurements were taken after each trial: the force and the thickness. The force was recorded by the loadcells connected to the compression testing machine and the thickness measurements were taken using an ultrasonic precision thickness gauge. Fig. 7b compares the force output from the old tools to the new tool. The solid lines represent the physical trials, and the dotted line represents the numerical simulation. The orange colour represents the old tools, and the black colour represents the new tools. As shown, the maximum shearing force in the old tools and the new tools is 7.5 kN and 5.2 kN, respectively.

Sheet Metal 2025
Materials Research Proceedings 52 (2025) 19-26

Materials Research Forum LLC
https://doi.org/10.21741/9781644903551-3

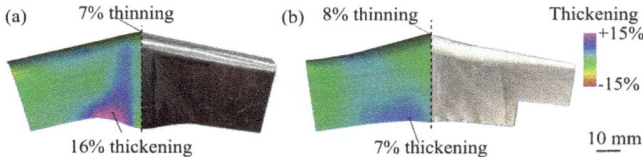

Figure 8: Results of the thickness distribution comparing linear tools and convex fold line

Fig. 8a, 8b shows the results from the old and new tools, respectively. As expected, the thinnest point is seen at the apex of the part and the thickest point is seen towards the edges of the beak. The thinning percentage remains relatively comparable (7% vs 8%) between the two tools, which prove that most of the thinning arises due to the first crash forming and the secondary folding stage. The thickening percentage at the edge of the beak has reduced from 16% to 7%. The reduction in force input from the upper shear tool and the improved thickness distribution can be attributed to the extension of a negative Gaussian curvature along the feed radius. Both these phenomena can be explained by plotting the major and minor principal strains. Fig 9. plots the major and the minor principal strain, of equal magnitude, in red and blue respectively. An equal strain magnitude indicates no thickness change, and a greater major strain leads to thinning, while a greater minor strain causes thickening.

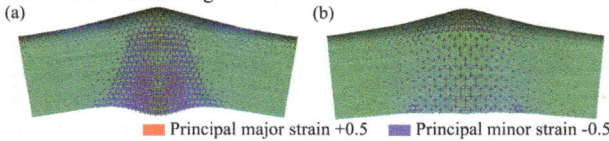

Figure 9: Strain during shearing using (a) Conventional tool design (b) Convex fold lines

Visually, the workpiece under the old tool experiences a complex strain history (Fig. 9a) with a high proportion of principal minor strains, indicating high level of thickening. In comparison, the new tools with a negative Gaussian curvature, induces a compressive and tensile strain, and when balanced leads to minimal thickness change. Generally, a large proportion of the major and minor strains are balanced when the new tool (Fig. 9b). This results in a more uniform strain stage and as explained by [13], this can reduce the hydrostatic forces in the workpieces, hence minimizing the tool forces. This minimises the energy input required to operate the stamping tool, but also enhances the tool life, thereby reducing operational costs in high volume manufacturing.

Conclusion

This study demonstrated the influence of curved tool geometries, to improve thickness distribution in parts made using folding-shearing. The introduction of a negative Gaussian curvature across the line of contact yielded significantly reduced material thickening from 16% to 7% and decreased the upper shear tool force input by 25%. The maximum value of thinning at the apex of the part remained relatively unchanged, suggesting that the thinning primarily occurs from the first crash forming and the secondary folding stages. The design space reveals that tools in quadrant 4 are suitable for shrink corners as it can help reduce thickening. However, the presented methodology can be extended to stretch corners, where tools from quadrant 2 can potentially improve areas that are prone to thinning. This symmetry in the design space underscores the potential for developing a broader range of tool configurations tailored to different corner types, hence enhancing the overall versatility of the folding-shear process. Future work will investigate the interaction between the part geometry and process performance across a range of materials, while focussing on adapting the process to high-volume production.

Materials Research Forum LLC

https://doi.org/10.21741/9781644903551-3

References

[1] Z. Marciniak, S. J. Hu and J. L. Duncan, Mechanics of Sheet Metal Forming, London: Elsevier, 2002.

[2] T. Altan and A. E. Tekkaya, Sheet Metal Forming: Fundamentals, Asm International, 2012. https://doi.org/10.31399/asm.tb.smff.9781627083164

[3] P. M. Horton and J. M. Allwood, "Yield improvement opportunities for manufacturing automotive sheet metal components," Journal of Material Processing Technology, vol. 249, pp. 78-88, 2017. https://doi.org/10.1016/j.jmatprotec.2017.05.037

[4] J. Atherton, "Declaration by the metals industry on recycling principles," The International Journal of Life Cycle Assessment , no. 12, pp. 59-60, 2012. https://doi.org/10.1065/lca2006.11.283

[5] P. Horton, J. M. Allwood, P. Cassell, C. Edwards and A. Tautscher, "Material Demand Reduction and Closed-Loop Recycling Automotive Aluminium," MRS Advances, no. 3, pp. 1393-1398, 2018. https://doi.org/10.1557/adv.2018.280

[6] J. M. Allwood, C. J. Cleaver, E. G. Loukaides, O. Music and A. Nagy-Sochacki, "Folding-shearing: Shrinking and stretching sheet metal with no thickness change," CIRP Annals - Manufacturing Technology, vol. 68, pp. 285-288, 2019. https://doi.org/10.1016/j.cirp.2019.04.045

[7] O. Music, J. M. Allwood and K. Kawai, "A review of the mechanics of metal spinning," Journal of Materials Processing Technology, no. 210, pp. 3-23, 2010. https://doi.org/10.1016/j.jmatprotec.2009.08.021

[8] C. J. Cleaver, R. Arora, E. G. Loukaides and J. M. Allwood, "Producing isolated shrink corners by folding-shearing," CIRP Annals, vol. 71, pp. 217-220, 2022. https://doi.org/10.1016/j.cirp.2022.03.036

[9] R. Arora, O. Music and J. M. Allwood, "Understanding the Process Limits of Folding-Shearing," Journal of Materials Processing Technology, 2024. https://doi.org/10.1016/j.jmatprotec.2024.118660

[10] H. W. Guggenheimer, Differential Geometry, New York: McGraw- Hill, 1977.

[11] K. A. Stevens, "The Visual Interpretation of Surface Countours," Artificial Intelligence, pp. 47-73, 1981. https://doi.org/10.1016/0004-3702(81)90020-5

[12] N. Wei, Y. Ding, J. Zhang, L. Li, M. Zeng and L. Fu, "Curvature geometry in 2D materials," National Science Review, no. 8, 2023. https://doi.org/10.1093/nsr/nwad145

[13] O. Richmond and K. Chung, "Ideal stretch forming for minimum weight axisymmetric shell structures," International Journal of Mechanical Sciences, vol. 42, pp. 2455-2468, 2000. https://doi.org/10.1016/S0020-7403(99)00006-5

[14] K. Isik, M. B. Silva, A. E. Tekkaya and P. A. Martins, "Formability limits by fracture in sheet metal forming," Journal of Materials Processing Technology, no. 214, pp. 1557-1565, 2014. https://doi.org/10.1016/j.jmatprotec.2014.02.026

[15] J. M. Allwood and D. R. Shouler, "Generalised forming limit diagrams showing increased forming limits with non-planar stress states.," International journal of Plasticity, no. 25.7, pp. 1207-1230, 2009. https://doi.org/10.1016/j.ijplas.2008.11.001

Sheet Metal 2025
Materials Research Proceedings 52 (2025) 27-34

Materials Research Forum LLC
https://doi.org/10.21741/9781644903551-4

Potential of part quality monitoring for deep drawing processes by integrating sensors into drawbeads

Papdo Tchasse[1,a] *, David Briesenick[1,b], Kim Rouven Riedmüller[1,c] and Mathias Liewald[1,d]

[1]Institute for Metal Forming Technology, University of Stuttgart, Holzgartenstraße 17, 70174 Stuttgart, Germany

[a]papdo.tchasse@ifu.uni-stuttgart.de, [b]david.briesenick@ifu.uni-stuttgart.de, [c]kim.riedmueller@ifu.uni-stuttgart.de, [d]mathias.liewald@ifu.uni-stuttgart.de

Keywords: Deep Drawing, In-Process Measurement, Drawbead

Abstract. The sheet metal material flow in deep drawing is the result of the prevailing blank restraining forces that occur due to frictional interactions between the workpiece and the active tool components. Therefore, monitoring the restraining forces during deep drawing is an indirect way of recording the material flow and thus the performance of the ongoing sheet metal forming operation. A common tool adjustment that is frequently implemented in industrial applications for controlling the material flow during deep drawing is the integration of drawbeads. While the effect of such drawbeads can be modelled numerically, it is still a great challenge to track and thus verify the acting restraining forces experimentally. Against this background, this paper deals with novel sensor concepts for drawbeads to online monitor the restraining forces acting on the sheet metal material and thus determine the quality of the deep drawn part. For this study, two monitoring methods were evaluated, considering two types of sensors that can be integrated into drawbeads, namely fibre Bragg grating and thin film sensors. For both types of sensors, the operating principle was numerically simulated. Furthermore, the application of FBG was experimentally investigated. In conclusion, the numerical results highlight the promising potential of integrating FBG and thin-film sensors. While the FBG setup showed a minor correlation with part defects, further work is needed to resolve uncertainties in sensor integration and bonding.

Introduction

The quality of deep drawn sheet metal parts strongly depends on the sheet restraining forces and the resulting material flow during the forming operation. For parts with a complex geometry, it is essential to specifically control or restrict the material flow in some areas of the blank. This is the case, for example, of many car body parts, for which drawbeads are integrated in the forming tool in order control material flow and enhance part quality [1]. In most cases, the effect of drawbeads is assessed by numerical simulations, which often do not accurately model the actual retention force and thus the material flow in the real production process. Validation of the simulations through sensory recording of the process forces on drawbeads, e.g. by using strain gauges or piezoelectric transducers, proves to be difficult due to the different shapes and sizes of drawbeads. A promising approach to overcome these limitations is the use of fibre Bragg grating (FBG) and thin-film sensors, which are small and flexible and can therefore be adapted to different measuring structures.

FBG sensors consist of a short section of distributed Bragg reflectors in the core of an optical fibre. These Bragg reflectors are dielectric layers with a periodical modulated refractive index that can reflect particular wavelength light depending on the spacing between the layers. Thus, the grating structure of FBG act as an optical filter that reflects the wavelength that fulfills the Bragg condition of the fibre core index modulation. This periodic refractive index modulation is often created using ultraviolet light and its spacing is sensitive to fibre elongations or compressions that

Content from this work may be used under the terms of the Creative Commons Attribution 3.0 license. Any further distribution of this work must maintain attribution to the author(s) and the title of the work, journal citation and DOI. Published under license by Materials Research Forum LLC.

Sheet Metal 2025

Materials Research Proceedings 52 (2025) 27-34

Materials Research Forum LLC

https://doi.org/10.21741/9781644903551-4

can be induced by thermal or structural variations in the fibre environment [2]. Therefore, FBG sensors are well suited for temperature and strain measurements.

Thin-film sensors, on the other hand, are transducers that can be directly applied to various types of bodies and tool components through a coating process. In this case, a diamond-like carbon sensor layer with piezoresistive properties is applied via a plasma-enhanced chemical vapor deposition (PECVD) process, resulting in a 6 μm thick layer. Photolithography and chemical etching are then used to create load-capturing structures. The sensor layer is electrically insulated by a 1 μm-thick layer of SICON®. Depending on the required data quality and local mechanical loads, these sensors can be strategically integrated into any process zone of interest [3]. In recent years, both sensor technologies have been used for structural strain and temperature monitoring in aerospace, energy, civil infrastructure, biomedical, and maritime applications [4].

Because of their versatility, FBG and thin film sensors are assumed to have a unique potential for the deep drawing industry, where accessibility to the die surfaces is limited and process conditions are difficult. For this reason, the objective of the present work was to investigate their applicability on drawbeads for the part quality monitoring during deep drawing processes.

Related works
Since restraining forces and material flow during forming operation are essential quality influencing variables of deep drawing processes, numerous research works in the past years have focused on measuring them using optical, resistive, inductive, piezoelectric or rolling ball tracking sensors [5]. Among these different sensor technologies, optical transducers such as laser triangulation sensors have gained a particular attention as they allow to measure the blank draw-in online without geometry modification of the active tool components [6]. However, since the beam spot of laser triangulation sensors is very small, the correct and exact positioning of these sensors can be challenging. For this reason, measuring the blank restraining forces indirectly using piezoelectric sensors integrated in the blankholder [7] or punch radii [8] proved to be a more reliable part quality measurement.

In contrast to deep drawing, some research activities in the field of metal casting have been focused on the measuring of process strains and temperatures using FBG. In one of these use cases, Bian et al. investigated the monitoring of high temperatures and large strains during a copper casting process using so called regenerated FBG sensors (RFBG) [9]. RFBG are a specific kind of FBG, obtained by a heat-treatment process, which can record high temperatures and strains. The same technology was also used by Stadler et al. in order to measure the part temperature and strain during an aluminum casting process [10] and by Bian et al. for monitoring the strain evolution and distribution in the casting process of an AlSi9Cu3 alloy [11]. Moreover, FBG have also been considered for monitoring cold forging processes. For this purpose, Deliktas et al. integrated FBG sensors in an additive manufactured die intended for a full forward extrusion process [12]. In this study, the authors investigated the application of FBG sensors for the die temperature and strain measurement.

Thin-film sensors were used by Rekowski et al. [3] in order to predict the concentricity of the final part in a cup backward extrusion process. In this use case, a piezoelectric thin-film disc was attached to the extrusion punch and used for detecting concentricity deviations based on the measurement of the eccentric load during the forming operation. Further applications of thin film sensors can be found in monitoring cutting forces of turning process [13] and the flow front movement and temperature of injection molding [14].

Methodical Approach
According to the state of the art, a research gap can be found for the application of FBG and thin-film sensors in deep drawing applications, although the flexibility and measuring range of these sensors represent a non-negligible aspect that would enable versatile and non-weakening

Sheet Metal 2025
Materials Research Proceedings 52 (2025) 27-34

Materials Research Forum LLC
https://doi.org/10.21741/9781644903551-4

integration into the forming tools. Due to this potential of the both sensor technologies, the present work focused on designing methods that allow FBG and thin-film sensors to be integrated into metal forming tools in order to track local material flow. Here, an exemplary application on drawbeads was examined, thus monitoring the part quality of deep drawing processes.

For this purpose, deep drawing with drawbeads of a rectangular cup was considered and numerical investigations were performed in order to evaluate the potential integration strategies for both FBG and thin-film sensors. Then, first experiments focusing on the FBG sensors were carried out in order the evaluate their performance considering the numerical analysis.

Numerical analysis

The active tool components used for the numerical analysis are shown in Fig. 1. These components were considered for a single acting forming operation with a fixed punch, a motion-controlled die and a force-controlled blankholder. Presented analysis focused on two common defects in deep drawing processes, namely cracks and wrinkles. For this reason, three different blankholder forces (BHF) 60 kN, 250 kN and 400 kN were considered. These forces correspond respectively to a part with wrinkles (60 kN), without defects (250 kN) and with cracks (400 kN). For all three simulations, a constant Coulomb friction coefficient μ of 0.125 was used.

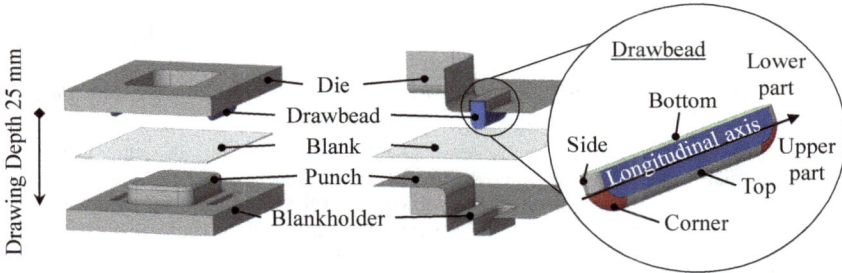

Fig. 1 Tool components for the numerical analysis and drawbead geometry

The numerical investigation was performed in the simulation environment of LS-Dyna. With the exception of the drawbead, which was modeled as an elastic steel body approximated by hexahedral elements (*MAT_001), all other active tool components were set to be rigid shells (*MAT_020). The blank material for this study was DC04, represented by fully integrated shell elements and anisotropic elastic-plastic material model (*MAT_133) with FLD-based failure criteria for crack propagation (*MAT_ADD_DAMAGE_DIEM). **Table 1** summarizes the properties of material used in this study.

Table 1: Material properties of DC04

Material	t [mm]	E [GPa]	YS [MPa]	UTS [MPa]	U.E. [%]	n	r_m	Material model	Damage model
DC04	0.7	210	167.9	303.5	25.3	0.211	1.65	Yld2000	DIEM

The evaluation of each simulation was conducted by correlating elastic strains and stresses within the drawbead body with predicted initiation points for cracks and wrinkles to identify optimal sensor integration locations. The analysis of the calculated strain distribution across the drawbead under varying BHF indicates that only its upper part (see Fig. 1) is suitable for the integration of a fibre with FBG sensors. Due to their small diameter, fibres with FBG sensors are

Sheet Metal 2025   Materials Research Forum LLC
Materials Research Proceedings 52 (2025) 27-34   https://doi.org/10.21741/9781644903551-4

highly sensitive to buckling, necessitating integration that respects their minimum bending radius. Accordingly, the upper part along the longitudinal axis of the drawbead was selected for further evaluation of elastic strains as related to FBG sensor deformation. Finite elements (FE) for strain evaluation were chosen approximately 2.5 mm below the drawbead's top surface, with three central FE beneath the active surface designated as measurement points for strain assessment.

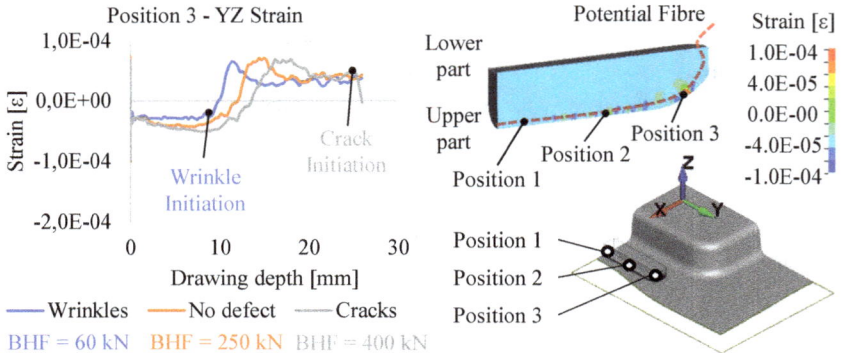

Fig. 2: Strain curve in the corner zone of the drawbead middle section in reference to the part quality (FBG sensor concept)

Fig. 2 shows strain curves of the most sensitive FBG sensor positions according to the three cases of part quality, namely wrinkles, no defect and cracks. A clear correlation between the part quality and the evaluated strain development, particularly at the position 3, was found. Increased contact pressure raises the load on the drawbead, reducing material flow. Under high BHF, the FE at the position 3 experiences significant compression at the start of the deep drawing process, followed by a delayed initiation of material flow, marked by a shift from compression to tension in the strain signals, and finally, an unloading as cracks form due to element deletion in the corner areas of the part. Conversely, lower BHF result in minimal fibre compression, with blankholder separation due to corner wrinkling aligning with the signal switch from compression to tension. These effects are most prominent at the drawbead's side, positioning it as the optimal integration area (Position 3) for sensor sensitivity.

For thin-film sensors, which are ideally applied on flat surfaces, the bottom of the drawbead is identified as the best integration site. This region is less impacted by dynamic frictional forces and experiences a distinctive pressure distribution during deep drawing, with higher compression on the inner side due to bending moments on the drawbead (see Fig. 3). For the evaluation, nine positions were selected along the drawbead's bottom for the thin-film sensor, with three positions each in the middle section, outer side, and inner side near the drawbead corner.

Sheet Metal 2025
Materials Research Proceedings 52 (2025) 27-34

Materials Research Forum LLC
https://doi.org/10.21741/9781644903551-4

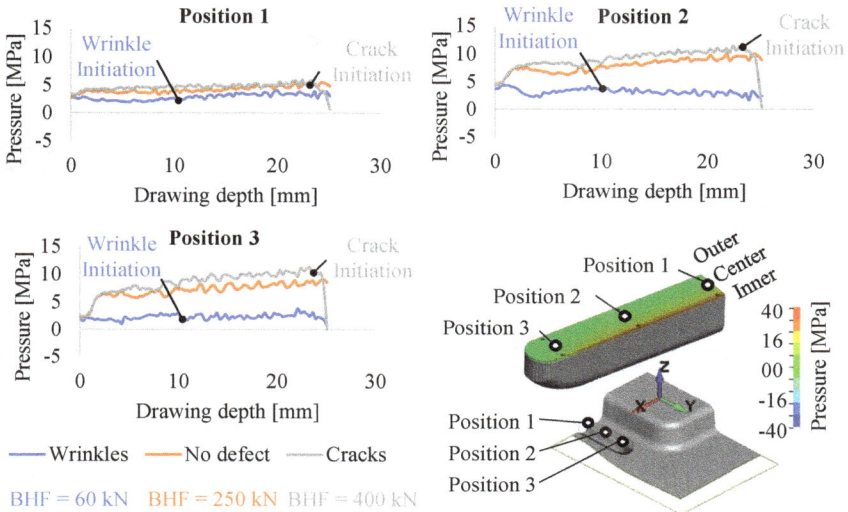

Fig. 3: Evaluation of pressure at three locations of the bottom zone of the drawbead (Thin-film sensor concept)

The most distinct pressure signals were recorded in the central area of the drawbead's bottom. Here, variations in BHF led to differing vertical loads on the drawbead, with responses varying by measurement position. Overall, signal sensitivity increases towards the side areas of the drawbead (Position 3), where a clear distinction between simulated BHF—and thus part quality—is observed. High BHF produce a stronger pressure signals, and crack initiation is identifiable through unloading in the signal due to element deletion. Conversely, the onset of wrinkle formation was not detectable in the pressure signals across all nine measurement points.

Experimental evaluation of the FBG sensor concept

Following the numerical study that highlighted the potential of both sensor concepts, an experimental study was conducted using commercially available FBG sensors in the deep drawing process of DC04 steel. Three locations were chosen based on the evaluation of symmetric elastic drawbead simulations. To facilitate testing, two fibers, each containing six FBGs spaced 20 mm apart over a total length of 120 mm, were sourced from Fisens GmbH. A new drawbead, designed with a split structure, allowed for the FBG sensors to be embedded just below the top surface in alignment with the simulation results, enabling attachment to the die as shown in Fig. 4. After manufacturing the modified drawbead, FBG sensors were affixed with epoxy resin adhesive for strain measurement, while an unbonded FBG was positioned for temperature compensation. The entire tool was then assembled and mounted in a servo-mechanical press (Aida NST-S2-6300(D)-305-150-SERVO).

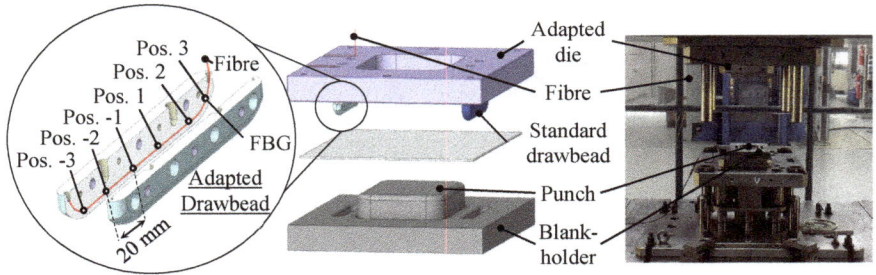

Fig. 4: Experimental setup for the evaluation of the FBG sensor concept

In the experimental tests, the BHF were set based on parameters established in the numerical study, while lubrication and drawing depth were kept constant at 2 g/m² of mineral oil (M100) and 25 mm, respectively. Multiple parts were drawn for each BHF setting to assess measurement repeatability. Fig. 5 presents all strain measurements recorded during deep drawing under varying BHF at the most sensitive measurement point related to part quality (Position 3).

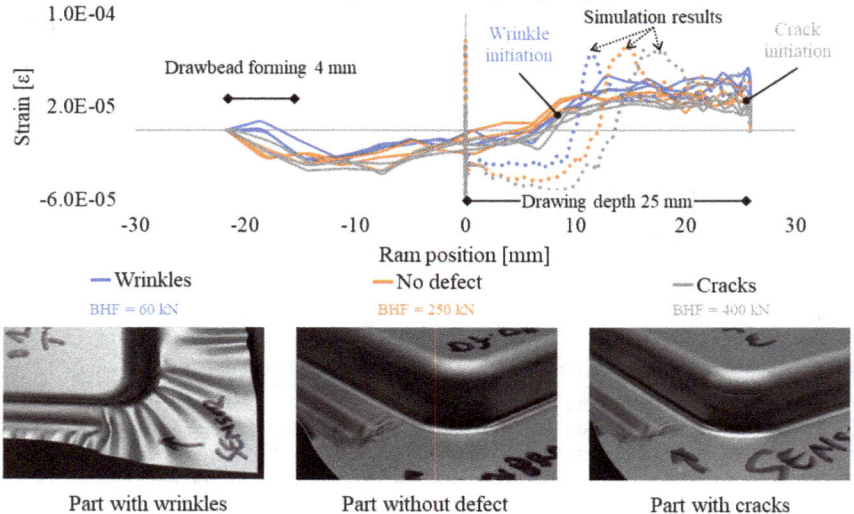

Fig. 5: Experimental strain measurements in the corner of the drawbead (Position 3) with FBG sensors for different settings of the blankholder force

Overall, the FBG sensor strain signals exhibited increased scattering and drift. After a compensation, the influence of different BHF became apparent, reflected in varied compressive strains during the drawbead formation and a switch to tensile strains in the subsequent deep drawing phase. However, Although the experimental results displayed the same trend as the simulation results, identifying part-quality-related signal features across all data was more challenging. These differences can be explained by the geometry of the drawbead and the process environment. The simulations assumed an idealized, full-body structure, while the experimental setup included a body with holes and grooves, subjected to a highly dynamic environment.

Sheet Metal 2025
Materials Research Proceedings 52 (2025) 27-34

Materials Research Forum LLC
https://doi.org/10.21741/9781644903551-4

Additionally, the manual bonding of small FBG sensors introduced variability in adhesion and positioning, adding to signal scattering and strain bonding uncertainties. Future fabrication of drawbead sensor prototypes will necessitate advanced manufacturing techniques, including 3D printing and coating, to achieve reliable integration and enhance measurement robustness[15].

Although the strain signals carry considerable uncertainties, the simultaneous temperature measurements consistently capture the heat generated during deep drawing. To evaluate temperature evolution in the drawbead zones, a batch of 20 defect-free parts was formed. Fig. 6 displays temperature signals across all measurement positions, alongside trends for average and maximum temperatures during the ramp-up phase of a deep drawing process with manual part transfer and an average stroke rate of 1.5 parts per minute. Overall, the average temperature across the drawbead increases by approximately 5 K. A closer look of each stroke and measurement point reveals significantly higher heating in the drawbead side area (Position 3) at the beginning of the forming process, with an average peak 6 K above that of the center areas (Positions 1 and 2). This elevated temperature is attributed to greater contact pressure and plastic strain concentrations in the respective region.

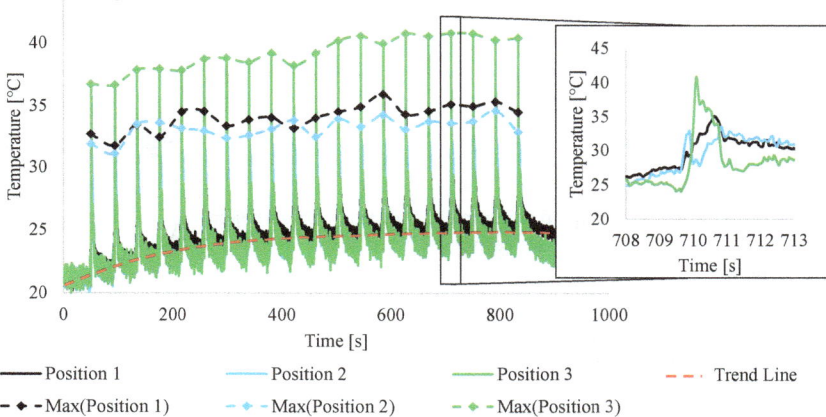

Fig. 6: Temperature evolution in a continuous production of 20 parts

Conclusion

The objective of this study was to evaluate the potential of FBG and thin-film sensors for monitoring part quality in deep drawing processes. Due to their compact size, geometric flexibility, and wide measuring range, these sensors can be integrated into various tool components. In this investigation, FEM simulations were conducted to assess the sensor applicability in drawbeads, showing that both FBG and thin-film sensors can detect part quality, particularly crack formation in corners. An experimental evaluation with commercially available FBG sensors however revealed challenges in identifying quality-related features, likely due to the drawbead geometry, the dynamic testing conditions, and the manual sensor bonding. Advanced fabrication techniques, such as 3D printing and coating, will be necessary for an improved integration and reliability. Temperature monitoring appears more suitable for the FBG sensor concept, while experimental verification of thin-film sensors might enable a robust part quality monitoring in the future.

References

[1] A. Birkert, S. Haage, and M. Straub, *Umformtechnische Herstellung komplexer Karosserieteile.* 2013. https//doi.org/10.1007/978-3-662-46038-2

[2] S. J. Mihailov, "Fiber bragg grating sensors for harsh environments," *Sensors*, vol. 12, no. 2, pp. 1898–1918, 2012. https//doi.org/10.3390/s120201898

[3] M. Rekowski, K. C. Grötzinger, A. Schott, and M. Liewald, "Thin-film sensors for data-driven concentricity prediction in cup backward extrusion," *CIRP Ann.*, vol. 73, no. 1, pp. 205–208, 2024. https//doi.org/10.1016/j.cirp.2024.04.035

[4] M. Yang and J. Dai, "Review on optical fiber sensors with sensitive thin films," *Photonic Sensors*, vol. 2, no. 1, pp. 14–28, 2012. https//doi.org/10.1007/s13320-011-0047-y.

[5] J. M. Allwood *et al.*, "Closed-loop control of product properties in metal forming," *CIRP Ann. - Manuf. Technol.*, vol. 65, no. 2, pp. 573–596, 2016. https//doi.org/10.1016/j.cirp.2016.06.002

[6] M. Kott, *Methodik zur Entwicklung eines Bedienerassistenzsystems für das Presswerk.* Darmstadt: Dr.-Ing. Dissertation, Technische Universität Darmstadt, 2022.

[7] J. Hengelhaupt, M. Vulcan, F. Darm, P. Ganz, and R. Schweizer, "Robust Deep Drawing Process of Extensive Car Body Panels," *Neuere Entwicklungen der Blechumformung*, pp. 277–304, 2006.

[8] C. Blaich, *Robuster Tiefziehprozess durch Erfassung und Optimierung der örtlichen Bauteilqualität.* Dissertation, Universität Stuttgart, 2012.

[9] Q. Bian *et al.*, "In-Situ High Temperature and Large Strain Monitoring during a Copper Casting Process Based on Regenerated Fiber Bragg Grating Sensors," *J. Light. Technol.*, vol. 39, no. 20, pp. 6660–6669, 2021. https//doi.org/10.1109/JLT.2021.3101524

[10] A. Stadler *et al.*, "Decoupled temperature and strain measurement with regenerated fiber Bragg gratings during an aluminum casting process," vol. 11591, p. 82, 2021. https//doi.org/10.1117/12.2588926

[11] Q. Bian *et al.*, "Monitoring strain evolution and distribution during the casting process of AlSi9Cu3 alloy with optical fiber sensors," *J. Alloys Compd.*, vol. 935, p. 168146, 2023. https//doi.org/10.1016/j.jallcom.2022.168146

[12] T. Deliktas, M. Liewald, and N. Nezic, "Contribution to process digitisation of cold forging processes using additive manufactured tools," *ESAFORM 2021 - 24th Int. Conf. Mater. Form.*, vol. 13, pp. 1–10, 2021. https://doi.org/10.25518/esaform21.1931

[13] M. Plogmeyer, G. González, C. Pongratz, A. Schott, V. Schulze, and G. Bräuer, "Tool-integrated thin-film sensor systems for measurement of cutting forces and temperatures during machining," *Prod. Eng.*, vol. 18, no. 2, pp. 207–217, 2024. https//doi.org/10.1007/s11740-023-01251-1

[14] A. Schott, M. Rekowski, F. Timmann, C. Herrmann, and K. Dröder, "Development of Thin-film Sensors for In-process Measurement during Injection Molding," *Procedia CIRP*, vol. 120, pp. 619–624, 2023. https//doi.org/10.1016/j.procir.2023.09.048

[15] F. Ahmed, M. S. Forhad, and M. H. Porag, "Spectral Behavior of Fiber Bragg Gratings during Embedding in 3D-Printed Metal Tensile Coupons and Cyclic Loading," *Sensors*, vol. 24, no. 12, 2024. https//doi.org/10.3390/s24123919

Sheet Metal 2025
Materials Research Proceedings 52 (2025) 35-42

Materials Research Forum LLC
https://doi.org/10.21741/9781644903551-5

Cost-effective repair solution for twin-roll-caster rollers

Martin Lauth[1,a] *, Kay-Peter Hoyer[1,b], Mirko Schaper[1,c], and Winfried Gräfen[2,d]

[1]Chair of Materials Science, Paderborn University, Warburger Strasse 100, 33098 Paderborn, Germany

[2]Benninghoff Oberflächentechnik, Richard-Löchel-Straße 14, 47441 Moers, Germany

[a]lauth@lwk.upb.de, [b]hoyer@lwk.upb.de, [c]schaper@lwk.upb.de, [d]graefen@benninghoff-gmbh.de

Keywords: Metal Forming, Manufacturing, Twin-Roll-Casting

Abstract. Twin-Roll-Casting (TRC) is an energy- and cost-efficient process to produce near-net-shape aluminum strips. Due to the high affinity of molten aluminum to steel surfaces, those rollers show signs of wear throughout the rolling campaign. This leads to the necessity of restoring the worn surfaces to suitable parameters. The easiest way is to grind the surface till all superficial defects are omitted. However, the thickness of the roller is not endless, therefore the rollers must be replaced after a certain amount of surface reconditioning. This ultimately leads to the non-usability of the roller. This research shows a route to recondition the surface including the possibility of renewing worn-down surfaces with an energy- and cost-efficient high-velocity oxygen fuel (HVOF) treatment with subsequent grinding to the desired initial surface parameters.

Introduction

Twin-Roll-Casting (TRC) is an energy-efficient way to produce thin metal strips. It is commonly used in the production route for technical pure aluminum but also steel. It is a near-net-shape process which allows to omit the hot-rolling step needed to produce thin strip material utilized by direct cast (DC). Therefore, it leads to a significantly lower primary energy usage than the DC process. [1, 2]

There are, however, disadvantages which this research tries to conquer, as the wear of the roller surface material is focused on. Fig. 1 shows the TRC process schematically. In principle, a light metal melt, not necessarily aluminum, is poured through a nozzle between two water-cooled rollers. The molten material solidifies under the influence of the water-cooled rollers and builds up a shell on each side.

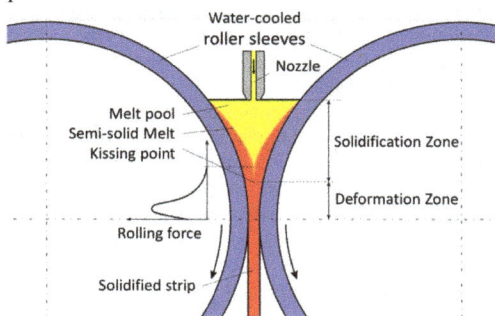

Fig. 1: Schematic of TRC process [3]

The rollers turn in opposing directions leading the shells to touch each other at the so-called kissing point. Up until the kissing point, only rapid solidification takes place. After reaching the kissing point, only deformation occurs. Therefore, TRC is a combination of solidification and formation in one process step, leading to high energy efficiency and, due to high cooling rates of 100 K/s, a specific fine microstructure in the cast strip. [1]

Due to certain circumstances enacted in TRC, like heat-shrinked water-cooled roller surfaces on the water-transferring driving shaft (Fig. 2), most of the industry standard surface treatment

Content from this work may be used under the terms of the Creative Commons Attribution 3.0 license. Any further distribution of this work must maintain attribution to the author(s) and the title of the work, journal citation and DOI. Published under license by Materials Research Forum LLC.

Sheet Metal 2025   Materials Research Forum LLC
Materials Research Proceedings 52 (2025) 35-42   https://doi.org/10.21741/9781644903551-5

methods are not well-suited. The high-velocity oxygen fuel (HVOF) process is a well-known process for applying different materials on iron-based surfaces for wear and corrosion resistance. [4, 5]

Fig. 2: *Roller driving shafts with circumferential water lines as cooling channels [6]*

State of the Art
TRC uses a combination of casting and forming to generate thin strip materials, which are near-net-shape. This allows for a cheap downstream transforming process without the necessity for further heat treatment of the aluminum strip. Due to the high affinity of the molten aluminum to the steel surface, a release agent is mandatory to prevent the sticking of the hot aluminum strip to the steel surface of the rollers. [7]
The release agent (RA), however, is not tight-bonded to the roller's surface, therefore merging itself with the produced thin strip, disregarding the product for higher-purpose applications like drug packaging or food-grade packaging. [7, 8] Due to this fact, the amount of release agent on the roller's surface is permanently diminished, leading to the intrusion of aluminum melt directly into the steel roller's surface. While reaching high exposure levels due to the permanent loss of RA on the roller's surface and the length of a rolling campaign of up to three weeks, a reconditioning of the surface is inevitable. [9]

The quasi-standard as of today, is to grind down the surface, to match "as-new" conditions, e.g. a surface roughness of Ra = 1 μm. Grinding is a necessity due to the desired surface hardness of at least 48 HRC. Standard dimensions of TRC rollers range from 1,100 mm to 2,000 mm in width and up to 1,000 mm in diameter with a shell thickness of 20 mm. [10] Therefore, the amount of recondition cycles is finite.

HVOF is a process in which a powder-based material is molten and accelerated through a nozzle in which a flammable gas is burnt with the assistance of oxygen. By the usage of different burning gases, the temperature of the flame can be chosen between 1,700 °C and 3,100 °C. By the pressure expelled by the rapid burning of oxygen with the chosen gas, the flame catapults the molten particles with supersonic velocity onto the surface (up to 1,060 m/s). Due to the high velocity and high temperature of the particles, a weld-like bonding between particles and the substrate surface can be achieved. [11, 12] Internal stresses derived from the high temperature of the molten powder particles and a vast temperature gradient need a tempering to be aligned. Furthermore, surface grinding is mandatory due to the irregular surface shape derived from the powder particles thrown on the surface. The process itself must be tailored to the circumstances of the desired surface roughness.

The fact that the material employed for the rollers as of today, a hot work tool steel (1.2344, X38CrMo5-3) is also usable in HVOF, led to the idea of reapplying substrate material to thicken the roller material, to be able to further use the rollers.

Materials and Methods
Typical materials for TRC-rollers are hot work tool steel such as 1.2367 (X38CrMoV5-3) or 1.2344 (X40CrMoV5-1). The rollers which were chosen for this research are made of 1.2367. Detailed information regarding the dimensions of the rollers used in the vertical laboratory caster

Materials Research Forum LLC
https://doi.org/10.21741/9781644903551-5

at *Paderborn University* is given in Fig. 3. As these rollers have already been used in several research projects, they show wear and tear in the form of revolving groves, surface cracks and adherend aluminum particles. Hence, these rollers exhibit a perfect starting surface and initial condition for the addressed reconditioning approach.

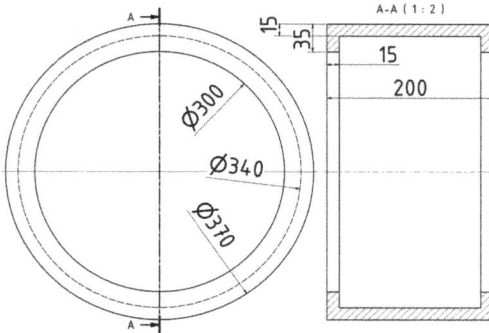

Fig. 3: *Schematic drawing of a laboratory caster roll*

In pre-trials, 1.2344 particles were sprayed via HVOF onto a strip material of 1.2367 which was preconditioned using peening with a particle size distribution between 425 μm to 600 μm The strip was then heated to 350 °C to model the standard circumstances of a TRC process The heated strip and a further strip of technical pure aluminum (EN-AW 1080) heated to 650 °C, which is close to the melting temperature, were fed into the rolling mill with a rolling reduction of 45 %. All pre-trials were enacted utilizing an industry-standard release agent (Pyrotek Nekote 35XL) which was applied with a sponge onto the cold surface of the renewed roller. The specimen showed no signs of cracks in the surface after several rolling trials. In additional light microscopy (LM) and scanning electron microscopy (SEM) investigations, no cracks in the surface layer were detected. Based on energy-dispersive X-ray spectroscopy (EDS), no residue of aluminum on the contact surface could be found.

These pre-trials were conducted utilizing the 4-column roller mill at the *Chair of Materials Science* (LWK) at *Paderborn University* (UPB).

The surface is, depending on the condition, ground to a suitable roundness (e.g. lower than 0.001 mm runout) and afterwards peened. The first application of an HVOF coating showed immediate adhesion loss while casting therefore the performed blasting was altered. In the first attempt (performed by the project partner *Benninghoff Oberflächentechnik*) peening was conducted with commercially available Al_2O_3 with a particle size distribution from 425 μm to 600 μm. In all further attempts preconditioning was performed with a blasting particle size distribution between 850 μm and 1180 μm. The HVOF process was enacted by propane and oxygen, a designated temperature of 2,350 °C, and 1.2344 as powder material with a particle size distribution from 15 μm to 45 μm. The application of 1.2344 instead of the substrate material 1.2367 was chosen as a compromise between applying the same material on the roller surface like the roller material and the availability of the material as powder material for HVOF. To diminish secondary martensite derived from the high cooling rate of the HVOF process, leading to crack susceptibility, tempering at 530 °C for 2 hours was applied. Pre-trials ensured to have a minimal porosity to not diminish the heat dissipation capabilities of the surface in comparison to the substrate material. Furthermore, a well-established connection layer between the surface and the substrate was used.

To resurface worn-down roller surfaces, up to 0.3 mm of material is removed to ensure all surface cracks and intermetallic phases are dislodged. A standard roller can therefore only be reconditioned a certain number of times to maintain sufficient residual stability for further TRC campaigns. To ensure the highest possible outcome of the resurfacing of the roller, a maximum amount of material must be sprayed onto the surface. While applying materials via HVOF, to ensure a suitable heat transfer due to any porosity, a coating of up to 2 mm is chosen, which is more than a typical industrial application. The tempering was performed by *Ferrum Edelstahlhärterei* in a vacuum chamber oven, ensuring a slack-free surface. After conducting the heat treatment, the rollers were installed in the laboratory TRC at *Paderborn University*. For the first trial runs, standard laboratory TRC parameters were chosen according to Table 1.

Table 1: *Applied standard parameters for TRC*

Material	EN AW-1070 / P1020	
Mass	15	[kg]
Melting temperature	620 to 645	[°C]
Casting temperature	685	[°C]
Casting velocity	8	[m per minute]
Casting gap	1.45	[mm]
Release agent	Pyrotek Nekote 35XL	
Roller cooling	2.6 (11 °C)	[m³/h]
Roller diameter	370	[mm]
Roller width	200	[mm]

While conducting the first trials with the TRC, the sprayed coating separated from the substrate roller during the process (Figure 4 and Fig. 5). The residing parts of the sprayed coating showed a nominal thickness of 2 mm, whereas the substrate surface showed a peened surface, leading to the conclusion that the surface preconditioning is the culprit as well as the waiting time between HVOF and tempering. Hence, the HVOF surface was reapplied, with peening according to the changes mentioned above. Also, the spray surface thickness was reduced to 0.6 mm to eliminate any problems with residual stresses and tempering.

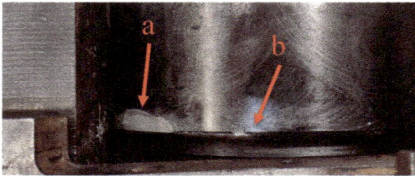

Fig. 4: *Image of broken-off parts of the sprayed surface of one roller in the TRC machine; a) break-out, b) crack*

Fig. 5: *Image of a broken-off part of the HVOF coating during the first trial*

In addition, tempering was performed immediately after the application of the HVOF surface and grinding to appropriate thickness and roughness at *UPB*. The oven used was a POV 125/600 convection oven from *Fresenberger Industrieofenbau* capable of heating up to 600 °C with inert gas application. Table 2 summarizes the tempering parameters.

Utilizing the modified parameters, the application of the sprayed surface was successful. 12 TRC trials with the parameters in Table 1 were conducted and a total of 135 m aluminum strips were cast.

Sheet Metal 2025
Materials Research Proceedings 52 (2025) 35-42

Materials Research Forum LLC
https://doi.org/10.21741/9781644903551-5

Table 2: Tempering parameters

Temperature	530	[°C]
Inert gas	Nitrogen	
Inert gas rate	5	[l per minute]
Duration of operation	5 hours at 530 °C	
Cooling	Slow (in oven)	

Results and Discussion

Table 3 shows the results of the different heat treatment temperatures. Only the heat treatment at 530 °C could ensure the desired hardness of at least 51 HRC regarded in the research project. The measurements were conducted in HV10 using a *KB 30 FA* automated hardness tester and converted to HRC.

Table 3: *Hardness measurements after tempering*

Tempering temperature [°C]	450	500	530	580
Mean hardness [HV10]	508	519.4	541.4	525.6
Mean hardness [HRC]	49.6	50.4	51.8	50.7

Fig. 6 and Fig. 7 show two exemplarily chosen coatings to emphasize the possibility of setting different layer thicknesses and porosities leading to different heat transfer conditions to tailor the resulting microstructure of the cast strip and therefore, the mechanical properties as well.

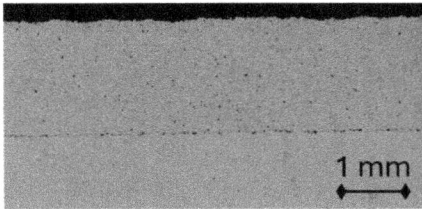

Fig. 6: *Cross section of a sample with a high layer thickness and low porosity*

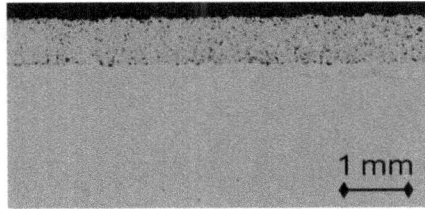

Fig. 7: *Cross section of a sample with a low layer thickness and high porosity*

As the rollers cannot be dismantled from the Twin-Roll-Caster after each trial and therefore, no data could be gained utilizing SEM, other measuring techniques had to be employed. Pictures of the surface were taken after each trial at the same spot using a *Keyence VHX-1000* light microscope. Furthermore, the roughness of the roller as well as the cast strip was measured with a *Mitutoyo SJ-210* digital tactile measuring apparatus. The initial roughness of the ground surface was tested at *Benninghoff Oberflächentechnik*. The measurements were performed at the same place utilizing an indentation on the drive shaft.

Fig. 8 and Fig. 9 depict the surface of the rollers after 6 and 12 trials, respectively. No degradation is visible.

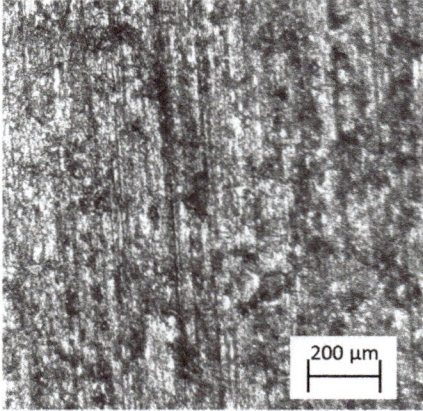

Fig. 8: *Microscopic image of the surface of the renewed roller after 6 trials*

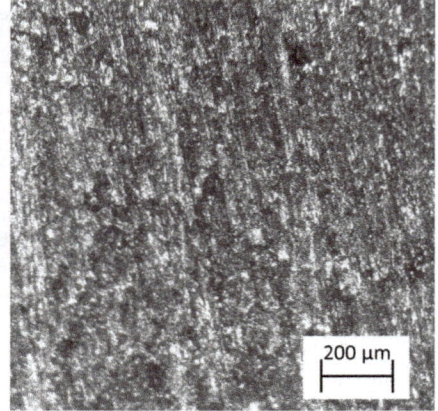

Fig. 9: *Microscopic image of the surface of the renewed roller after 12 trials*

Fig. 10 shows the evolution of the surface roughness of the cast strip throughout the 12 trials, measured evenly at the middle and the end of the cast strip. Besides trial 11, which shows a slightly higher roughness, a significant deviation is not detectable. Degradation of the roller's surface is therefore not traceable. The cast strip was also hardness tested, showing no deviation in hardness which was always almost 25 HV1.

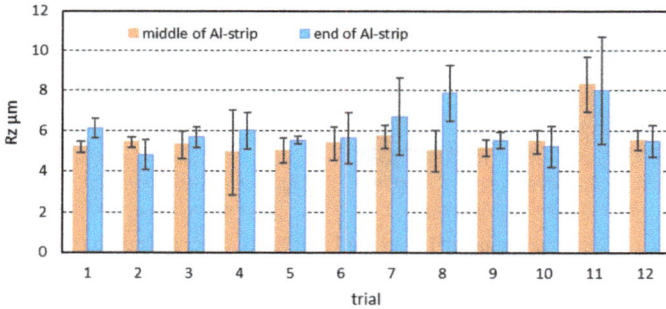

Fig. 10: *Mean surface roughness of cast strips, measured in the middle and at the end of each aluminum strip, middle of Al-strip (orange), end of Al-strip (blue)*

Fig. 11 shows a comparison between the roughness of the roller's surface and the cast strip. The strip has always a higher roughness than the roller. This can be derived from the fact that the separation force leads to a rougher surface. No significant degradation could be found throughout the entirety of the trials. Hence, the roughness stays constant, which in turn is a hint of neglectable changes in the surface conditions of the rollers as well as the cast strip.

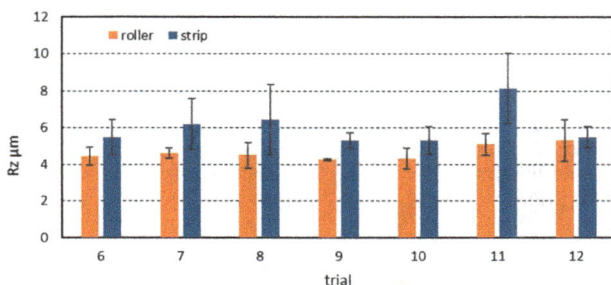

*Fig. 11: Comparison of surface roughness throughout the 12 trials,
roughness of roller (orange), roughness of cast strip (blue)*

The alteration of the surface preconditioning enables the coating material to adhere to the peened surface. A further consideration, if thicker HVOF-sprayed surfaces should not adhere, is to alter the time between HVOF spraying and grinding to tempering to a minimal amount, to diminish the chance of crack dissipation over time. A rougher surface due to a bigger peening material would lead to a higher surface roughness enabling mechanical clamping, but also introduces a higher porosity at the connection layer, thus hindering the heat transfer.

A significant depreciation of the surface roughness of the roller is a sign of aluminum pressed into the grinding groves, this would ultimately lead to casting failures, as the molten aluminum will have an even higher affinity to aluminum than to the steel rollers.

A reduced usage of release agent in comparison to a standard roller was not expected. A higher surface roughness leads to an initial memory effect, in which the release agent tends to build up in the groves. However, the amount washed out to the cast strip will remain constant. Furthermore, a higher amount of release agent on the roller's surface also hinders heat transfer which in turn affects the resulting microstructure and mechanical properties of the cast strip.

Conclusion

This research work aimed to identify whether an HVOF-sprayed surface can withstand the environmental conditions of the TRC process to reenable a worn-down roller for further rolling campaigns. To address this, a 1.2344 HVOF layer on a worn-down TRC roller consisting of 1.2367 at a layer height of 0.6 mm thickness was applied. This layer could withstand the circumstances of the TRC process as can be seen in the production of thin strip material in a length of about 135 m during 12 trials in total. The use of a release agent, *Pyrotek Nekote 35 XL* was necessary to prevent sticking of the strip-cast material to the steel rollers.

According to the results, repairing Twin-Roll-Caster rollers with HVOF spray is possible, and is also economically viable. Therefore, it is essential to spray a thicker HVOF layer onto the surface to elongate the time between reapplication. Further investigation on an industry-sized caster needs to be performed to acknowledge the length of a rolling campaign of three weeks. The main results can be summarized as follows:

- Repair is, as of now, possible with 0.6 mm new material
- Renewed surface can withstand the circumstances of twin-roll-casting
- Release agent is still necessary

Following test results, the application of 1.2344 via HVOF onto a preconditioned surface is purposeful to extend the usability and therefore, save money, material and energy. A transfer of the results to standard roller mills is possible, but the circumstances of those processes must be considered, e.g. the discontinuity of most rolling processes in comparison to TRC.

Acknowledgement
The authors like to thank Arne Tim Göbel and Dr.-Ing. Dietrich Voswinkel for their help in conducting the TRC trials and microstructural analyses. In addition, the authors would like to thank the Federal Ministry for Economic Affairs and Climate Action for financial support within the scope of the "Central Innovation Programme for Small and Medium-sized Enterprises (SMEs)" project KK5011519SH2.

References

[1] A. Kundu, P. Biswas, A. Mallik and S. Das, Electromagnetic twin roll casting alloy sheets: an overview, JOM, Vol. 74, No. 12, 2022 https://doi.org/10.1007/s11837-022-05490-y

[2] Haraldsson, J; Johansson, M. (2018): Review of measures for improved energy efficiency in production-related processes in the aluminium industry - From electrolysis to recycling; Renewable & Sustainable Energy Reviews 93, S. 525-548 https://doi.org/10.1016/j.rser.2018.05.043

[3] M. Stolbchenko, Zwei-Rollen-Gießwalzen und thermomechanische Behandlung von dünnen Bändern aus der Aluminiumlegierung EN AW-6082, Dissertation 2021

[4] W. Püttgen, M. Pant, W. Bleck, I. Seidl, R. Rabitsch, C. Testani, Selection of Suitable Tool Materials and Development of Tool Concepts for the Thixoforging of Steels, steel research international: Volume 77, Issue 5, May 2006, pp. 342-348 https://doi.org/10.1002/srin.200606396

[5] T.S. Sidhu, S. Prakash, R.D.Agrawal, State of the Art of HVOF Coating Investigations-A Review; Marine Technology Society Journal, Volume 39, Number 2, Summer 2005, pp. 53-64(12) https://doi.org/10.4031/002533205787443908

[6] Sino Steel Engineering & Equipment Co. Ltd: Walzenachsen. http://www.sinosteelrolls.com/product/277291192

[7] M. Yun, S. Lokyer, J.D. Hunt, Twin roll casting of aluminium alloys. Mater. Sci. Eng. A 280, 116-123 (2000) [4] R.J. Ong, J.T. Dawley and P.G. Clem: submitted to Journal of Materials Research (2003) https://doi.org/10.1016/S0921-5093(99)00676-0

[7] DIN EN 602, 2004/07

[8] DIN EN ISO 11607, 2024/02

[9] M. Dündar, B. Beyhan, O. Birbaşar, H. M. Altuner & C. Işıksaçan, Surface Crack Characterization of Twin Roll Caster Shells and Its Influence on As-Cast Strip Surface Quality, Light Metals 2013, pp. 491- 495 https://doi.org/10.1007/978-3-319-65136-1_84

[10] https://novelispae.com/twin-roll-casting-machine/

[11] American Society of Materials. "Introduction to Thermal Spray Processing," Handbook of Thermal Spray Technology, ASM International, Cleveland, Ohio, 2004

[12] S. Amin, H. Panchal, A Review on Thermal Spray Coating Processes, International Journal of Current Trends in Engineering & Research (IJCTER), Volume 2 Issue 4, April 2016 pp. 556 - 563

Sheet Metal 2025
Materials Research Proceedings 52 (2025) 43-50

Materials Research Forum LLC
https://doi.org/10.21741/9781644903551-6

Experimental investigations on a process adapted material testing method for hydroforming of tubular components

Jonas Reblitz[1,a] * and Marion Merklein[1,b]

[1]Institute of Manufacturing Technology, Friedrich-Alexander-Universität Erlangen-Nürnberg, Egerlandstraße 13, 91058 Erlangen, Germany

[a]jonas.reblitz@fau.de, [b]marion.merklein@fau.de

Keywords: Hydroforming, Aluminum, Tube Bulge Test

Abstract. In the automotive sector, an important strategy for reducing CO_2 emissions is lightweight construction. In this regard, body-in-white parts offer a high potential for weight reductions by substituting conventional steel parts with tubular, aluminum-based components. Due to their high strength-to-weight-ratio and high crashworthiness, tube profiles are often used as safety-relevant car body components. As lightweight design becomes increasingly important, complex part geometries are required, which can be realized by hydroforming. Especially for structural components, an exact determination of the material properties is necessary, in order to achieve a high prediction accuracy of FE-simulations. Thus, a failure-free component production has to be enabled. However, due to a varying stress and strain state, the tensile test according to DIN EN ISO 6892 as a standard characterization method is only insufficiently suitable for the characterization of tube profiles for a hydroforming process. For this purpose, a new testing method has been developed and investigated within this paper, which provides new insights regarding the material behavior of tubular components. The testing-rig is based on a bulge test, which reproduces the stress and strain state of a hydroforming process. By comparing the results of the material characterization with those of a tensile test, the differences in the mechanical properties depending on the testing method are evaluated. During the test, the surface strains and the internal pressure are varied and recorded. Based on these findings, a more accurate prediction of FE-simulations is aspired in future research work.

Introduction

One current challenge is the global reduction of CO_2 emissions. To meet this challenge, the EU plans to reduce the emissions of carbon dioxide up to 90% in 2040 compared to 1990 [1]. The aim is to become climate-neutral until 2050 [2]. In this context, the transporting sector offers a high potential for CO_2 savings since it is responsible for 26% of total emissions within the EU [3]. For an effective reduction of emissions, electrification can be a suitable option. However, electrically powered vehicles have a high weight due to the battery, which moreover has to be protected against mechanical loads and crash impacts. Nevertheless, a reduction of car weight leads to further savings in energy consumption. 100 kg additional weight leads to a reduction of range of 3.5% and an increase of the required engine power of 6% [4]. Nevertheless, a high percentage of aluminum is already used for the battery enclosure in current vehicles [5]. Due to the high strength-to-weight ratio, a further weight reduction can be achieved by the application of profiles made out of 7xxx aluminum alloys. The increasingly complex profile geometries can be realized by hydroforming. Their use for structural applications requires precise numerical design. The material data is particularly important for a good prediction accuracy of the FEM simulation. A common method in material testing is the tensile test. However, this is only suitable to a limited extent for modeling the tube properties for hydroforming due to the deviating testing direction and strain paths. [6] Therefore, a tube bulge test tool is set up for a process-related characterization method and the derived material properties are compared with those from the tensile test. Tubular semi-

Content from this work may be used under the terms of the Creative Commons Attribution 3.0 license. Any further distribution of this work must maintain attribution to the author(s) and the title of the work, journal citation and DOI. Published under license by Materials Research Forum LLC.

Sheet Metal 2025
Materials Research Proceedings 52 (2025) 43-50

Materials Research Forum LLC
https://doi.org/10.21741/9781644903551-6

finished parts of the alloys AA7020 and AA7075 are used for this purpose, which are first analyzed by means of direction-dependent grain sizes and the wall thickness distribution. Afterwards, the setup of the tube bulge tool for the characterization of tube hydroforming is then presented. Furthermore, the optical strain measurement and the associated possibility of local strain evaluation are discussed. Finally, the differences in the strain paths and flow curves between the tensile test and the tube bulge test are evaluated and the necessity of a process-related characterization method is verified.

Properties of extruded tubular semi-finished parts

Semi-finished parts. Semi-finished tubular products made from the alloys AA7020 and AA7075 in the T6 condition were examined. The semi-finished tubes have a diameter of 60 mm, a wall thickness of 5 mm and a length of 300 mm. Since extruded tubes are characterized by direction-dependent material properties and wall thickness variations, the average grain sizes were analyzed with the aid of linear intercept method according to ASTM E 112, as shown in Fig. 1. Both alloys show a fine-grained microstructure with grain orientations in both the circumferential and axial directions. The wall thickness distribution of the semi-finished tube products is measured using the ATOS optical measuring system from Carl Zeiss GOM Metrology GmbH. The measurements show a sheet thickness gradient in the circumferential direction. There is a continuous wall thickness gradient, with the lowest value being 4.93 mm, whereas the maximum value is 5.18 mm. Consequently, the extrusion process causes a variation in wall thickness of 0.25 mm. The hydroforming process results in increased thinning, particularly in the area of the lowest wall thickness [7].

a)

$n = 3$		AA7020	AA7075
Average grain size in [µm]	Axial	7.00 ± 0.94	4.17 ± 1.25
	Radial	4.31 ± 0.29	2.29 ± 0.26
	Circum-ferential	9.39 ± 2.46	3.20 ± 0.12

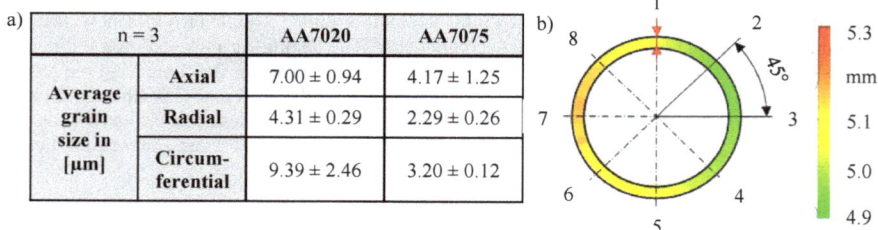

Fig. 1. *a) Average grain size and b) wall thickness distribution of the tubular semi-finished parts*

To compare the material properties determined from the tube bulge test with those from the tensile test, the semi-finished products are put into the W-temper condition. For this purpose, they are solution annealed in a furnace at 460 °C for AA7020 and 480 °C for AA7075 and then quenched in water. By dissolving the alloying elements in the supersaturated solid solution, a significantly improved formability is achieved compared to the T6 state. A negative strain rate sensitivity is present for AA7020 [8] and AA7075 [9] in the W-temper condition. This phenomenon is caused by the PLC (Portevin-Le Chatelier) effect [10].

Tube bulge test as a process-related material testing method for hydroforming

Experimental setup. A tube bulge tool is developed and designed for material characterization of tubes via hydroforming. The tool can be inserted into a hydraulic press type Hydrap HPDZb 630 so that the clamping force is applied hydraulically. This setup is intended to better approximate the deformation behavior during hydroforming. In contrast to material testing using a tensile test, the main deformation during hydroforming usually occurs in a circumferential direction. The tube bulge tool is designed for maximum inner pressures up to 100 MPa with exchangeable tool inserts so that the test conditions can be adapted. In this study, a bulging length of 60 mm was used; a maximum of 120 mm is possible. In addition, test temperatures of the oil and the tools up to 200 °C

Sheet Metal 2025   Materials Research Forum LLC
Materials Research Proceedings 52 (2025) 43-50   https://doi.org/10.21741/9781644903551-6

can be covered and the strain measurement is performed with the aid of an optical strain measurement system. The corresponding setup is shown in Fig. 2.

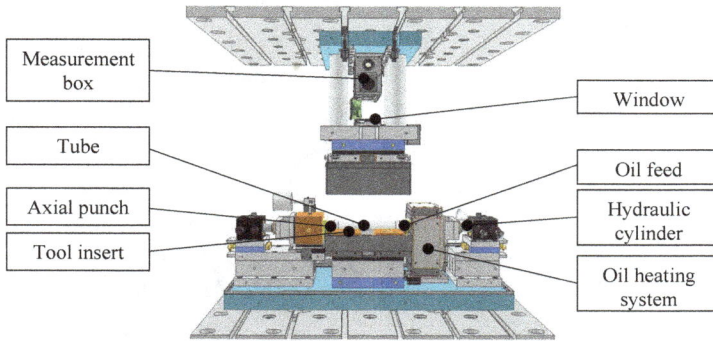

Fig. 2. Experimental setup of the tube bulge tool

Usually, the strain is determined analytically via the bulging height [11]. Those experimental setups do not enable the evaluation of local effects in deformation since the derived strain is averaged over the tube diameter [12]. The optical strain measurement also enables strain rate control during forming, which could lead to more precise material data for FE simulations. But first the tubes are placed in the tool inserts. The axial punches then move into the tube ends via hydraulic cylinders so that it can be filled with the thermal oil. Afterwards the tool is closed and the tube ends are expanded and sealed by the axial punches. Thus, the tube ends are clamped, which does not allow any slipping or additional pressure to support the bulging in the forming zone. The semi-finished tube products are then bulged until they burst. In addition, evacuation devices are installed to extract the oil vapors produced during warm tests. The optical strain measurement system is located in the upper half of the tool. The measurement setup is shown in Fig. 3 a). The measuring box contains two cameras for optical measurement of the strain during the tube bulge test. There is also a pyrometer for recording the temperature on the semi-finished tube product during the test. The measuring setup is protected against damage and oil leakage by a protection glass. A lamp is located under the measuring box, which shines directly into the measuring channel and thus illuminates the tube in the free bulging zone. The light source is also protected by a protection glass. It is located directly below the lamp to prevent any reflection visible to the cameras. Overheating of the measuring setup during a warm test is prevented by fans in the measuring box. The optical strain measurement system is used to calculate the strain field for the measuring area. In addition to the strain measurement, the internal tube pressure is measured using a pressure sensor, as well as the temperature of the tube in the expansion zone and the oil.

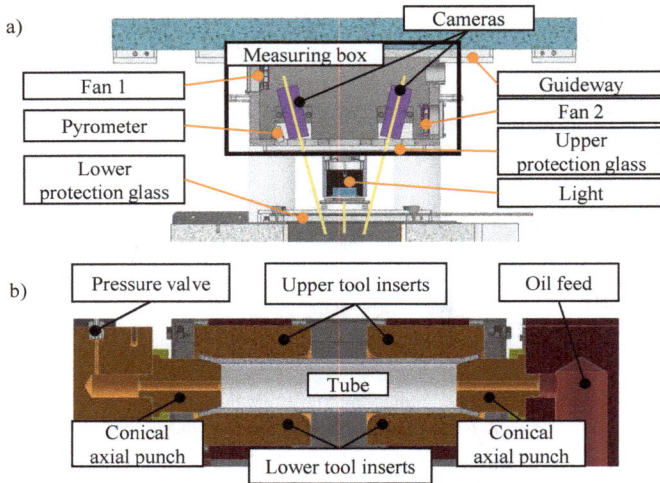

Fig. 3. *a) Setup of the measurement box and b) of the tool inserts*

Analytical model. An analytical model is used to derive flow curves from the experimentally determined measured values. The measured input variables are the internal pressure, the plastic strain in both the axial and circumferential directions and the radii of curvature. These are approximated by the strain measurements, which leads to more accurate values than the analytical determination that is usually used for setups without optical strain measurement. In the setup from Liu et al. [13] also an optical strain measurement is used. However, the radii of curvature are determined via FE simulations, which might lead to deviations. In [6, 14] the bulging geometry is measured after the forming process only for the final state. Formulas based on membrane theory are applied to calculate the effective strains and stresses [13], which are shown with the corresponding input and output variables in Fig. 4.

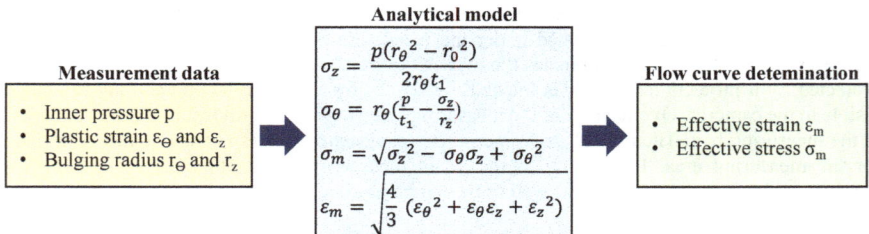

Analytical model

Measurement data		Flow curve detemination
• Inner pressure p • Plastic strain ε_θ and ε_z • Bulging radius r_θ and r_z	$\sigma_z = \dfrac{p(r_\theta{}^2 - r_0{}^2)}{2r_\theta t_1}$ $\sigma_\theta = r_\theta(\dfrac{p}{t_1} - \dfrac{\sigma_z}{r_z})$ $\sigma_m = \sqrt{\sigma_z{}^2 - \sigma_\theta \sigma_z + \sigma_\theta{}^2}$ $\varepsilon_m = \sqrt{\dfrac{4}{3}\left(\varepsilon_\theta{}^2 + \varepsilon_\theta \varepsilon_z + \varepsilon_z{}^2\right)}$	• Effective strain ε_m • Effective stress σ_m

Fig. 4. *Input and output of the analytical model for flow curve derivation*

σ_z represents the stress in axial tube direction, whereas σ_θ is the circumferential stress component. r_θ and r_0 describe the current and the initial tube radius. The bulging radius in axial direction is named r_z. ε_t in wall thickness direction is determined by assuming volume constancy. Thus, the current wall thickness t_1 can be calculated. The last step of the flow curve derivation is the compensation of elastic deformation.

Local strain analysis. Optical strain measurement offers the possibility of local strain evaluation. This is illustrated in Fig. 5 using the example of the aluminum alloy AA7020 in W-

Sheet Metal 2025
Materials Research Proceedings 52 (2025) 43-50

Materials Research Forum LLC
https://doi.org/10.21741/9781644903551-6

temper condition for an oil volume flow of 24.5 cm³/s. The area of the local strain evaluation can be flexibly adapted. Additionally, it is necessary to measure the local bulging radii. Depending on the selected evaluation area, different initial wall thicknesses are assumed, based on the values measured for the wall thickness as shown in Fig. 1 b). In the regions between two measuring points, the wall thickness is determined by interpolation using a fourth-degree polynomial. The high strain values, reaching up to 0.28, for the main deformation show that evaluation 1 is in the area of the necking that led to the crack, while evaluation 2 with strain values of approximately 0.2 is positioned next to that area. According to Fig. 5 b), the flow curve in the area of the necking shows lower values from a true strain of 0.1 than in the adjacent area. There is an average strain rate of 0.02 s⁻¹ and an increased strain rate of 0.05 s⁻¹ in the area of the necking. These differences show that a negative strain rate sensitivity in combination with higher strain rates in the necking could lead to varying flow stresses. The local analysis of the strain hardening behavior can be used for a more accurate numerical modeling of the forming behavior. Furthermore, this methodology enables the investigation of the material properties of welded tubes, both for the weld seam and for the surrounding areas as well as the base material.

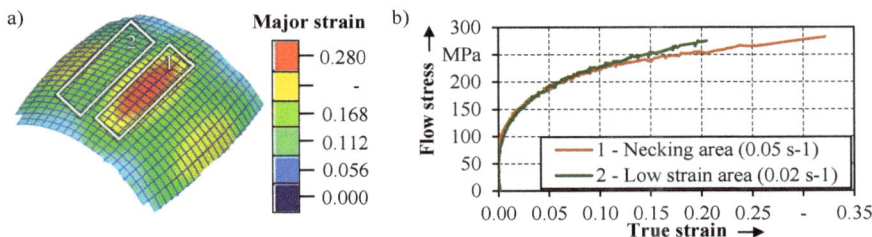

Fig. 5. a) Strain measurement with local evaluation and b) the resulting flow curves

Comparison of flow curves derived from tube bulge test and tensile test

In order to illustrate the necessity of the tube bulge test in comparison to the tensile test (DIN EN ISO 6892), the flow curves for two aluminum alloys are compared as a function of the respective test methods. The tube bulge tests were carried out at a constant oil flow rate of 24.5 cm³/s, the tensile tests were performed with a comparable resulting strain rate. Depending on the alloy, different mean strain rates result. For the alloy AA7020 the strain rate is 0.061 s⁻¹ while for AA7075 it is 0.015 s⁻¹. These differences could possibly be due to the lower formability of AA7075. According to Fig. 6, the results from the tube bulge test are compared with the tensile test. The A50 tensile specimens used for this were cut from the semi-finished tube products by laser in the same way as in [8] and tested in curved clamping jaws in a thermomechanical simulator type Gleeble 3500 according to DIN EN ISO 6892. The mean strain rates from the tensile tests were 0.08 s⁻¹ for AA7020 and 0.069 s⁻¹ for AA7075. For both materials, the flow stress for the tube bulge test is lower than for the tensile tests. However, the differences for AA7075 are smaller than for AA7020. Since both materials show a negative strain rate sensitivity in W-temper condition [8, 9], the strain rate curves and flow curves are analyzed in detail, as shown in Fig. 6. The differences in the hardening behavior could be attributed to the varying strain rate curves. For AA7020 there is a continuously increasing strain rate for a true strain bigger than 0.06. For these values, also a higher scattering in strain rate is visible. In this section also the flow stress for the tube bulge test decreases, which might be an indication for a correlation between increasing strain rate and decreasing flow stress. Even though there is a lower mean strain rate for the tube bulge test compared to the tensile test, the maximum strain rate with values up to 1.2 s-1 is significantly higher than for the tensile test with maximum 0.15 s-1. The strain rate and thus the flow stress during hydroforming of tubes can not be properly represented using conventional testing methods.

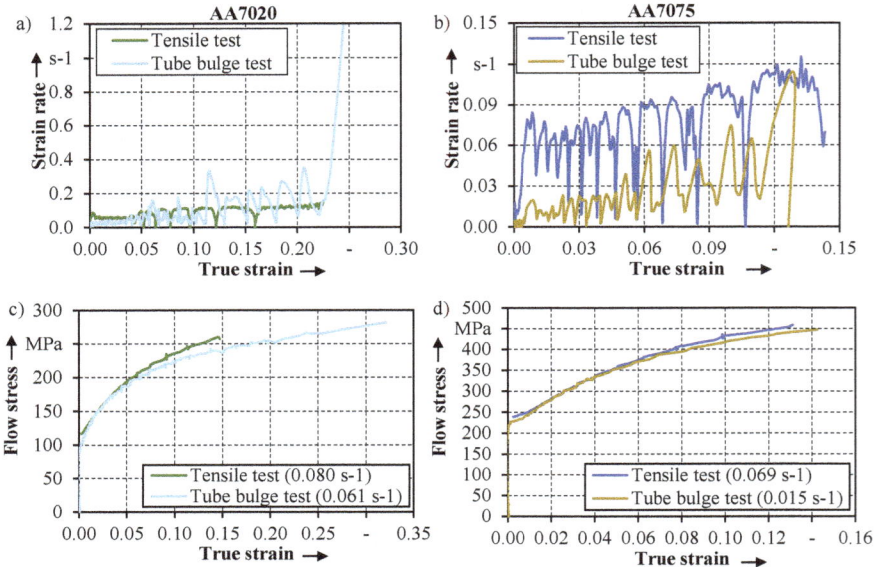

Fig. 6. *Comparison of tube bulge test and tensile test regarding strain rate for a) AA7020 and b) AA7075 as well as flow curves for c) AA7020 and d) AA7075*

In contrast, the tube bulge test of AA7075 shows a lower strain rate than for the tensile test with a stronger increase at a true strain of 0.06, whereas the maximum values of approximately 0.12 s^{-1} are comparable. As a consequence, the flow curves exhibit only minor differences. For both alloys there was an increase of strain rate visible for true strains of 0.06. At this strain also the strain paths change, as shown in Fig. 7. While the tensile test shows a linear curve with positive major strain and negative minor strain, the tube bulge tests show only very small negative minor strain in the axial direction of the tube at the beginning of the test. From a major strain of 0.1 in the circumferential direction, only slight positive minor strain is observed. This results in a plane strain condition. In general, varying tube bulge lengths as well as the subsequent feed of the punches can shift the strain path in the direction of the tensile test [15]. The decrease of slope resulting in a strain path in direction of positive biaxial strain states occurs similar to the increase of strain rate at a major strain of 0.06. Reason for that might be the increasing contact area between the tube and tool inserts, which leads to a decreasing forming zone. This could generate higher strain rates and higher axial strains due to frictional compensation of compression stresses from the axial punches and a strong tube expansion in the bulging zone. The deviations in the flow curves between the two testing methods could be due to the different strain rate curves in conjunction with the negative strain rate sensitivity. Additionally, there are changes in strain path, which cannot be represented by tensile test and differences in testing direction. Due to the semi-finished product properties and the boundary conditions given for tube hydroforming, material testing in the tube bulge test is a promising method for precise material modeling.

Sheet Metal 2025
Materials Research Proceedings 52 (2025) 43-50

Materials Research Forum LLC
https://doi.org/10.21741/9781644903551-6

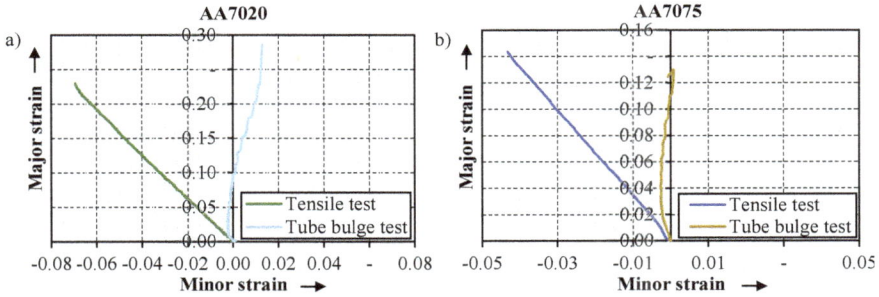

Fig. 7. Comparison of tube bulge test and tensile test regarding strain paths for a) AA7020 and b) AA7075

Summary

The use of high-strength aluminum profiles for structural applications in the transportation sector has a high potential for lightweight design. This results in savings in energy consumption and CO_2 emissions during the usage phase. In combination with hydroforming, complicated component geometries can be generated, which allows component weights to be reduced while at the same time achieving functionalization. In this context, precise component and process design using FE-simulation is essential. So far, however, tensile tests have mainly been used to characterize the material behavior for tube hydroforming. Due to deviating boundary conditions between the test method and the forming process, a process-related characterization method was investigated.

In this context, a tube bulge tool was constructed. Due to its modular design, it allows different expansion lengths and can be operated at temperatures of up to 200 °C by using thermal oil. In particular, the integrated optical strain measurement enables an extended modeling accuracy. On the one hand, the radii of curvature can be approximated in the axial and circumferential directions using optical strain measurement, so that there is less deviation compared to analytical modeling. This results in a more precise flow curve determination. On the other hand, the setup enables local strain and geometry evaluation. This means that local effects can also be recorded. These can be caused by the sheet thickness gradient in the circumferential direction. Due to the resulting differences in the strain rate distribution, a locally varying strain hardening behavior can be determined as a result of the negative strain rate sensitivity. Finally, the results for a flow curve determination were compared using a tube bulge test and a tensile test for AA7020 and AA7075. The tube bulge test provided lower flow stresses for both alloys than the tensile test. This could be due to the varying strain rate curves and the change in strain paths due to altering contact conditions between the tube and the tool inserts. While the tensile test is characterized by a two-dimensional strain state, the investigated tube expansion is characterized by a plane strain state with non-linear strain paths. In addition, the test direction varies so there are different grain orientations in the direction of the major strain. Therefore, there is a need for further research into the investigation of different tube bulge lengths and their influence on the determined material properties. In addition, the influence of the differently derived material data on the prediction quality of the FE-simulation for the process design must be evaluated.

References

[1] Information on https://commission.europa.eu/strategy-and-policy/priorities-2019-2024/european-green-deal/delivering-european-green-deal_en#making-transport-sustainable-for-all, (October 17, 2024).

Sheet Metal 2025   Materials Research Forum LLC
Materials Research Proceedings 52 (2025) 43-50   https://doi.org/10.21741/9781644903551-6

[2] Information on https://commission.europa.eu/strategy-and-policy/priorities-2019-2024/european-green-deal_en, (October 17, 2024).

[3] Information on http: https://www.destatis.de/Europa/EN/Topic/Environment-energy/CarbonDioxideRoadTransport.html, (October 17, 2024).

[4] Information on https://www.cargroup.org/wp-content/uploads/2021/02/Aluminum-Battery-Enclosures-Constellium-February-2021-FINAL.pdf, (October 17, 2024).

[5] Information on https://www.audi-technology-portal.de/en/download?file=1946, (October 17, 2024).

[6] P. Bortot, E. Ceretti, C. Giardini, The determination of flow stress of tubular material for hydroforming applications, Journal of Materials Processing Technology 203 (2008) 381-388. https://doi.org/10.1016/j.jmatprotec.2007.10.047

[7] J. Reblitz, R. Trân, V. Kräusel, M. Merklein, Evaluation of the properties of AA7020 tubes generated by a heat treatment based hydroforming process, Materials Research Proceedings 25 (2023) 221-228. https://doi.org/10.21741/9781644902417-28

[8] J. Reblitz, F. Reuther, R. Trân, V. Kräusel, M. Merklein, Numerical and Experimental Investigations on the Mechanical Properties of Milled Specimens from an AA7020 Tube, Key Engineering Materials 926 (2022) 1949-1958. https://doi.org/10.4028/p-oxs247

[9] F. Reuther, T. Lieber, J. Heidrich, V. Kräusel, Numerical Investigations on Thermal Forming Limit Testing with Local Inductive Heating for Hot Forming of AA7075, Materials 14 (2021) 1882. https://doi.org/10.3390/ma14081882

[10] F. Ostermann, Anwendungstechnologie Aluminium, third ed., Springer Vieweg, Berlin, 2014. https://doi.org/10.1007/978-3-662-43807-7

[11] A. Khalfallah, M. C. Oliveira, J. L. Alves, T. Zribi, Mechanical Characterization and Constitutive Parameter Identification of Anisotropic Tubular Materials for Hydroforming Applications, International Journal of Mechanical Sciences 104 (2015) 91-103. https://doi.org/10.1016/j.ijmecsci.2015.09.017

[12] Y. M. Hwang, C. W. Wang, Flow stress evaluation of zinc copper and carbon steel tubes by hydraulic bulge tests considering their anisotropy, Journal of Materials Processing Technology 209 (2009) 4423-4428. https://doi.org/10.1016/j.jmatprotec.2008.10.033

[13] J. Liu, X. Liu, L. Yang, H. Liang, Determination of flow stress of thin-walled tube based on digital speckle correlation method for hydroforming applications, International Journal of Advanced Manufacturing Technologies 69 (2013) 439-450. https://doi.org/10.1007/s00170-013-5039-1

[14] Z. He, S. Yuan, Y. Lin, X. Wang, W. Hu, Analytical model for tube hydro-bulging test, part I: models for stress components and bulging zone profile, International Journal of Mechanical Sciences 87 (2014) 297-306. https://doi.org/10.1016/j.ijmecsci.2014.05.009

[15] M. Mohammadi, J. S. Karami, S. J. Hashemi, Forming limit diagram of aluminum/copper bi-layered tubes by bulge test, Int J Adv Manuf Technol 82 (2017) 1539-1549. https://doi.org/10.1007/s00170-017-0225-1

Incremental forming

Sheet Metal 2025
Materials Research Proceedings 52 (2025) 52-58

Materials Research Forum LLC
https://doi.org/10.21741/9781644903551-7

Supporting toolpath generation for double sided incremental forming of polyhedron parts

Hans Vanhove[1,a] *, Arnoud Van Hees[1,b] and Joost R Duflou[1,c]

[1]Department of Mechanical Engineering, Katholieke Universiteit Leuven / Member of Flanders Make, Celestijnenlaan 300B, B-3001 Leuven, Belgium

[a]hans.vanhove@kuleuven.be, [b]arnoud.vanhees@student.kuleuven.be, [c]joost.duflou@kuleuven.be

Keywords: Incremental Sheet Forming, Tool Path, Double Sided

Abstract. Double Sided Incremental Forming (DSIF) enhances formability and accuracy by using a dynamic supporting tool synchronized with the forming tool, eliminating the need for custom dies. Generating a toolpath for the supporting tool aimed at consistent support is challenging, particularly around sharp corners of a contour. This paper presents an algorithm for generating such supporting toolpath for DSIF of polyhedron parts. It calculates the toolpath points based on the forming toolpath, adding, moving and reordering path points to manage internal and external corners. This approach enables synchronized tool movement, improving surface quality and preventing tool collisions.

Introduction

Incremental sheet forming, as a flexible sheet metal forming process, has seen much research over the last two decades towards improving accuracy and formability. Strategies such as compensation of unwanted plastic or elastic deformation, multistage forming or physically altering the material locally for better control of the deformation zone (eg. heat assisted forming) have proven to improve ISF, but are complex to master and are currently still struggling to lift accuracy and formability of ISF up to required industrial standards. Using a static supporting die provides a solution for many problematic forming features at the cost of fabricating dedicated dies. The challenging die design, compensating for spring back, often results in a number of iterations.

Double Sided Incremental Forming (DSIF) with a dynamic supporting tool has gained interest by different research groups [1, 2]. DSIF comes in two main variants as described by Meier. et al.[3]. DSIF with peripheral supporting tool, where specific areas of the part known for over forming are supported, such as transition zones between two planes with different wall angles. DSIF with a local supporting tool, on the other hand, is characterized by the use of a secondary tool in a synchronized motion immediately at the opposite side of the sheet with a controlled squeezing factor. Using this type of tooling setup allows to increase both accuracy and formability of incremental sheet forming [4] while remaining a flexible process relying on programmable toolpath control rather than dedicated tooling. Even though the control of the deformation zone is improved in DSIF, the process still requires several variables to be selected. In addition to the process parameters for single point ISF, the location, squeezing area and force applied by the supporting tool are influencing the forming result [5].

The most common variant of DSIF uses a hemispherical tipped tool for both the forming and supporting tool. Industrial robots or dedicated platforms consisting of two multi axis positioning devices are typically used. These systems require a toolpath for both tools as a list of tool center points (TCP) specified in cartesian coordinates between which the tools move in a linear fashion. For each forming toolpath point a corresponding point is assigned in the supporting toolpath. This is especially difficult in areas with great angle changes of the toolpath, as the forming path has multiple corresponding supporting positions in such 'corner points'. These challenges become

Content from this work may be used under the terms of the Creative Commons Attribution 3.0 license. Any further distribution of this work must maintain attribution to the author(s) and the title of the work, journal citation and DOI. Published under license by Materials Research Forum LLC.

Sheet Metal 2025   Materials Research Forum LLC
Materials Research Proceedings 52 (2025) 52-58   https://doi.org/10.21741/9781644903551-7

apparent when trying to maintain a constant supporting force in corners of workpieces as described by Möllensieb et. al.[1]

Literature lacks a clear description on the algorithms generating the supporting toolpath points and challenges involved in synchronizing both toolpaths for complex parts and for workpieces requiring the tool to pass distinct corners. An explicit understanding of the strategies used for the synchronization of both tools is crucial for a good interpretation and reusability of the final forming results and lessons learned.

This paper aims to describe the supporting toolpath generating algorithms used by KU Leuven.

Corresponding points for forming and supporting tool
Moving two tools synchronously requires both an offline programming of concurrent toolpaths consisting of corresponding points that have to be reached simultaneously and an online control system to ensure in-sync timing of the equipment. This paper will focus on the offline programming. More specific on generating the coordinates of the supporting toolpath (P_S) from an existing forming toolpath in the form of X,Y and Z coordinates of the forming tool center point (P_F) and the wall angle (γ) at the contact point between the forming tool and the wall of the part (Fig. 1a).

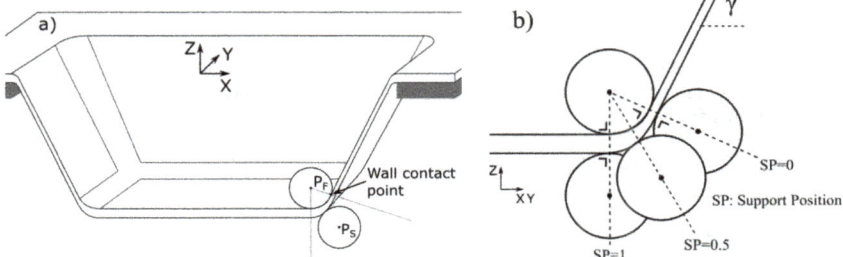

Figure 1. a) Cross section view of double sided incremental forming of a pyramidal part, b) close-up of different possible support positions along the contact zone of the forming tool.

The support position finally determines at what point the counter pressure is applied along the contact zone of the forming tool and the sheet (Fig. 1). SP is a numerical value between 0 and 1, where 0 corresponds to the position at the border of the contact zone and the already formed wall of the part, and 1 corresponds to the border between contact zone and the bottom of the part.

The algorithms described in this paper suppose that both tools operate in the same coordinate system where the toolpath conours lay in the XY plane and the Z axis is oriented towards the forming side. The toolpath contours consist of the tool center points between which the tool moves in straight lines.

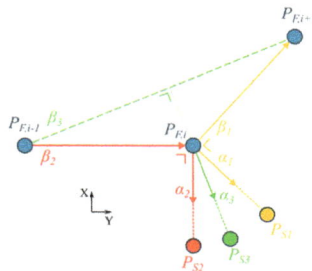

Figure 2 Different offset directions for the supporting tool.

Sheet Metal 2025   Materials Research Forum LLC
Materials Research Proceedings 52 (2025) 52-58   https://doi.org/10.21741/9781644903551-7

The offset direction in the XY plane of the supporting tool with respect to the forming tool and its direction of movement can be determined in different ways. Three such methods are depicted in Fig. 2, where the offset direction α of the supporting tool relative to point $P_{f,i}$ of the forming tool is determined. Point $P_{F,i-1}$ represents the previous point of the forming tool path, and point $P_{F,i+1}$ the subsequent point. In the developed algorithm, all directions are expressed as angles relative to a fixed coordinate system in the XY plane. In the first method, the direction α_1 is determined based on the direction β_1, which is the angle between point $P_{F,i}$ itself and the next point of the tool path $P_{F,i+1}$. The calculation of β_1 and α_1 is given in equations Eq. 1 to Eq. 4.

$$\Delta[x] = P_{F,i+1,x} - P_{F,i,x}. \tag{1}$$

$$\Delta[Y] = P_{F,i+1,Y} - P_{F,i,Y}. \tag{2}$$

$$\beta 1 = \arctan\left(\frac{\Delta[y]}{\Delta[x]}\right). \tag{3}$$

$$\alpha_1 = \beta_1 + \frac{1}{2} * \pi. \tag{4}$$

In the second case the offset direction is not perpendicular to the direction of the current and succeeding point, but between the current and preceding point ($P_{F,i}$ and $P_{F,i-1}$). The third option uses the preceding and succeeding point to calculate the offset direction ($P_{F,i-1}$ and $P_{F,i+1}$). In the remainer of this study, the first option will be used.

The offset distance is determined separately for the offset in XY plane and the Z axis (Fig. 3). Bothare dependent on the effective wall angle γ_{eff} which is defined by the support position (Fig. 1).

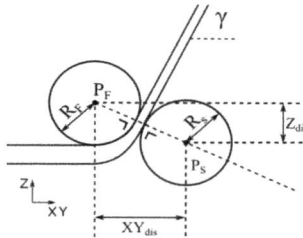

Figure 3. Section view depicting the XY and Z offset of the supporting tool with respect to the forming tool.

$$\gamma_{eff} = \gamma * SP. \tag{5}$$

$$XY_{dis} = (R_F + R_S + t) * \sin(\gamma_{eff}). \tag{6}$$

$$Z_{dis} = (R_F + R_S + t) * \cos(\gamma_{eff}). \tag{7}$$

This results in following transformation matrix for calculating the toolpath list for supporting points.

$$P_{s,i} = \begin{bmatrix} P_{s,i,x} \\ P_{s,i,Y} \\ P_{s,i,z} \end{bmatrix} = \begin{bmatrix} P_{F,i,X} \\ P_{F,i,Y} \\ P_{F,i,Z} \end{bmatrix} + \begin{bmatrix} XY_{dis} * \cos\alpha \\ XY_{dis} * \sin\alpha \\ -Z_{dis} \end{bmatrix}. \tag{8}$$

Internal and external corners

Equation (Eq. 8) only provides correct support for tool path segments with limited angles. When the angle between consecutive tool path segments of a contour in the X Y plane increases, proper support is not maintained. Fig. 4 illustrates that, in extreme cases, the supporting robot can even penetrate the metal sheet or cross the path of the forming robot. Fig. 4a shows a shape with internal and external corners of 90 degrees, respectively marked by red and purple arrows. Fig. 4b shows the tool path computed by the standard algorithm described above at an external angle. This figure demonstrates that, between the first and second points, the supporting tool must travel a significantly greater distance than the forming tool. Moreover, the support distance is not correct and does not respect the minimal offset required between the two tools. The challenges with internal corners are represented in Fig. 4c. The second point of the supporting path even lies on the path of the forming robot for this case.

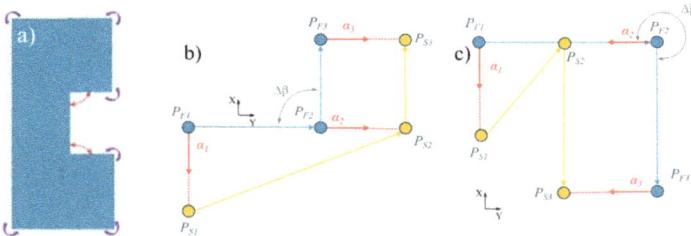

Figure 4 a) Example part showing internal corners in red and external corners in purple, b) issues occuring in external corners and c) internal corners.

To resolve the issues occurring with internal and external corners a $\Delta\beta_{max}$ variable is introduced. When the angle between two consecutive forming toolpath segments is greater than $\Delta\beta_{max}$ specific alterations to the path are required to assure sufficient accuracy. These alterations will be described later in the paper. First, internal and external corners need to be detected. For each toolpath point $\Delta\beta$ is calculated as the angle difference between the preceding and succeeding path segments and transformed to $0 \le \Delta\beta < 2\pi$. Equations Eq. 9 and Eq. 10 compose the algorithm that decides whether the corners are external or internal. If neither statement fits $\Delta\beta$, the standard equation is used to generate the supporting path.

$$External\ if \begin{cases} \Delta\beta > \Delta\beta max \\ \quad and \\ \Delta\beta < \pi \end{cases}. \tag{9}$$

$$Internal\ if \begin{cases} \Delta\beta > \pi \\ \quad and \\ \Delta\beta > 2 * \pi - \Delta\beta max \end{cases}. \tag{10}$$

External corners. In order to resolve issues with external corners extra points need to be added to the supporting toolpath, as depicted in Fig. 5a. First, the number of additional points is calculated by dividing $\Delta\beta_{max}$ by a selected angular change between the consecutive supporting path segments $\Delta\beta_{inc}$. Then the directions of the additional points are computed following equation Eq. 12 and

Sheet Metal 2025   Materials Research Forum LLC
Materials Research Proceedings 52 (2025) 52-58   https://doi.org/10.21741/9781644903551-7

their cartesian coordinates following equation Eq. 13. This algorithm results in a tool path as shown in Fig. 5a. The first additional point P_{S2} is positioned along direction α_1, making the paths of both robots equal in length up to the corner section, which enables synchronized and proper support. Following this point, additional points are added along a circular trajectory around the corner point, while the extra points on the forming path remain in the same position (P_{F2}-P_{F6} in the example shown) Beyond the corner, the tool path continues along the route defined by the standard algorithm.

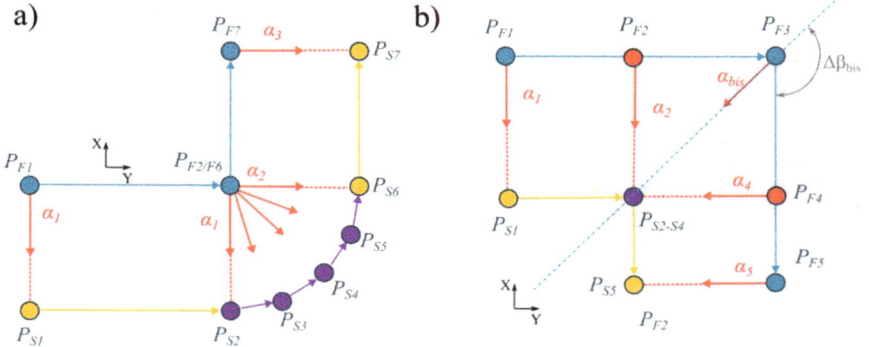

Figure 5 Solutions for generating toolpaths for double sided incremental forming for a) external corners and b) internal corners

$$n = \left\lfloor \frac{\Delta\beta}{\Delta\beta_{inc}} \right\rfloor. \tag{11}$$

$$For\ i\ in\ (0:n):\ Angle[i] = \alpha_{i-1} + \frac{\Delta\beta}{n} * i. \tag{12}$$

$$\forall\ Angle\ [i]:\ P_{Sext,i} = \begin{bmatrix} P_{Sext,i,x} \\ P_{Sext,i,Y} \\ P_{Sext,Z} \end{bmatrix} = \begin{bmatrix} P_{F,i,X} \\ P_{F,i,Y} \\ P_{F,i,Z} \end{bmatrix} + \begin{bmatrix} XY_{dis} * \cos(Angle[i]) \\ XY_{dis} * \sin(Angle[i]) \\ -Z_{dis} \end{bmatrix}. \tag{13}$$

Internal corners. Internal corners form a more complex problem as the forming tool would have to curve around the supporting tool. Altering the forming path would however change the final geometry of the part in an undesired way. The proposed solution can be seen in Fig. 5b. By adding a point in the bisector of segments [P_{F1}-P_{F2}] and [P_{F2}-P_{F3}] of the original trajectory seen in Fig. 4. First half of the angular change is determined as $\Delta\beta_{bis}$. Then the absolute angle of the bisector (α_{bis}) is calculated as $\alpha_{i-1} + \Delta\beta_{bis}$. In order to accurately position point P_{S2} to P_{S4}, the distance to the corner point needs to be computed (Eq. 14). The coordinates of the supporting toolpath are finally processed following equation (Eq.15), where $P_{F,angle}$ is the corner point of the forming path.

$$XY_{dis,bis} = \frac{XY_{dis,i}}{\sin\Delta\beta_{biss}}. \tag{14}$$

$$P_{S,bis} = \begin{bmatrix} P_{S,bis,X} \\ P_{S,bis,Y} \\ P_{S,bis,Z} \end{bmatrix} = \begin{bmatrix} P_{F,angle,X} \\ P_{F,angle,Y} \\ P_{F,angle,Z} \end{bmatrix} + \begin{bmatrix} XY_{dis,bis} * \cos(\alpha_{biss}) \\ XY_{dis,bis} * \sin(\alpha_{biss}) \\ -Z_{dis} \end{bmatrix}. \tag{15}$$

Sheet Metal 2025
Materials Research Proceedings 52 (2025) 52-58

Materials Research Forum LLC
https://doi.org/10.21741/9781644903551-7

To ensure that both robots remain synchronized for as long as possible, points P_{F2} and P_{F4} (Fig. 5b) are added to the forming path. The addition of these points keeps the robots synchronized along the path before and after the corner. The position of these points is determined by drawing a perpendicular line from the corner point of the supporting tool to the forming tool path. These additional forming tool points are determined by Equations (Eq. 16) to (Eq. 19).

$$\Delta\beta_{perp} = \frac{\Delta\beta-\pi}{2}. \tag{16}$$

$$XY_{dis,perp} = XY_{dis,bis} * \cos(\Delta\beta_{perp}). \tag{17}$$

$$P_{F,perp1} = \begin{bmatrix} P_{F,perp1,X} \\ P_{F,perp1,Y} \\ P_{F,perp1,Z} \end{bmatrix} = \begin{bmatrix} P_{F,angle-1,X} \\ P_{F,angle-1,Y} \\ P_{F,angle-1,Z} \end{bmatrix} + \begin{bmatrix} XY_{dis,perp} * \cos(\alpha_{angle-1}) \\ XY_{dis,perp} * \sin(\alpha_{angle-1}) \\ 0 \end{bmatrix}. \tag{18}$$

$$P_{F,perp2} = \begin{bmatrix} P_{F,perp2,X} \\ P_{F,perp2,Y} \\ P_{F,perp2,Z} \end{bmatrix} = \begin{bmatrix} P_{F,angle,X} \\ P_{F,angle,Y} \\ P_{F,angle,Z} \end{bmatrix} + \begin{bmatrix} XY_{dis,perp} * \cos(\alpha_{angle}) \\ XY_{dis,perp} * \sin(\alpha_{angle}) \\ 0 \end{bmatrix}. \tag{19}$$

In cases of sharp internal angles, this algorithm may encounter an issue. If the distance between the original points immediately before or after the corner point and the corner point itself is insufficient relative to the thickness of the metal sheet and the diameter of the tools, the standard offsetting algorithm will place points beyond the bisector point (P_{S2} and P_{S6} in Fig. 6). In this case the tool would initially pass the bisector point and then return to it. A similar problem occurs at the point after the corner. This would result in poor support. To avoid this problem, a final check is implemented, in the algorithm for internal angles, for points that are located on the segments between the added points on the forming tool path and the corner point of the forming tool path. In case points are detected that meet these conditions, the corresponding points on the supporting tool path are shifted to the bisector point. Finally, the order of points on the forming path is adjusted to ensure a smooth movement. The result of the modified algorithm is shown in Fig. 6b.

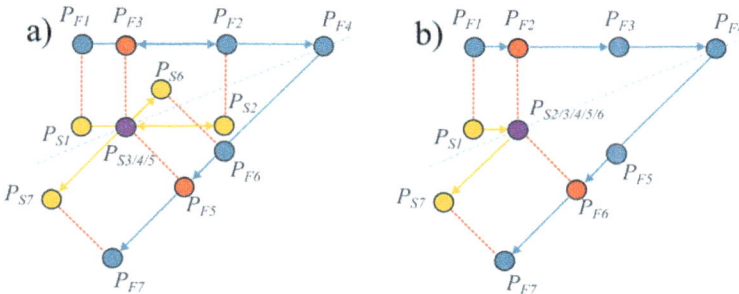

Figure 6 a) Challenge occurring in sharp internal corners with high point resolution and b) solution through displacing and reordering the tool path points.

Conclusions
This paper described the procedure used by KU Leuven for generating a supporting toolpath for DSIF of polyhedron parts. First a standard offsetting algorithm is introduced which can be used on a wide variety of parts. The different offset angles for a point are explained with respect to the movement direction of the forming tool. It is however illustrated that this algorithm lacks

Materials Research Forum LLC
https://doi.org/10.21741/9781644903551-7

effectiveness in the corners of contouring toolpaths. In the paper, the challenges for internal and external corners are described, followed by required alterations to the toolpath generation specific for these two cases.

Acknowledgement

The study performed in the framework of the STIFF project, facilitated by the KU Leuven C2 research fund.

References

[1] D. Möllensiep, J. Schäfer, F. Pasch, et al., Cluster analysis for systematic database extension to improve machine learning performance in double-sided incremental sheet forming, Int. J. Adv. Manuf. Technol. 133 (2024) 4301-4315. https://doi.org/10.1007/s00170-024-14014-8. https://doi.org/10.1007/s00170-024-14014-8

[2] K. Praveen, CH S., N. Venkata Reddy, Effect of support force on quality during double-sided incremental forming: an experimental and numerical study, Int. J. Adv. Manuf. Technol. 122 (2022) 4275-4292. . https://doi.org/10.1007/s00170-022-09871-0

[3] H. Meier, V. Smukala, O. Dewald, J. Zhang, Two-point incremental forming with two moving forming tools, in: Sheet Metal 2007, vol. 344 of Key Engineering Materials, Trans Tech Publications Ltd., 2007, pp. 599-605. https://doi.org/10.4028/www.scientific.net/KEM.344.599

[4] R. Malhotra, J. Cao, F. Ren, V. Kiridena, Z.C. Xia, N.V. Reddy, Improvement of geometric accuracy in incremental forming by using a squeezing toolpath strategy with two forming tools, J. Manuf. Sci. Eng. 133 (2011). https://doi.org/10.1115/1.4005179

[5] D. Möllensiep, T. Gorlas, P. Kulessa, B. Kuhlenkötter, Real-time stiffness compensation and force control of cooperating robots in robot-based double-sided incremental sheet forming, Prod. Eng. 15 (2021) 683-699. https://doi.org/10.1007/s11740-021-01052-4

Sheet Metal 2025
Materials Research Proceedings 52 (2025) 59-66

Materials Research Forum LLC
https://doi.org/10.21741/9781644903551-8

Revisiting formability limits in incremental sheet forming

Margarida GRALHA[1,a], Bernardo COLAÇO[2,b], João P. MAGRINHO[1,c], Énio CHAMBEL[2,d] and M. Beatriz SILVA[1,e*]

[1]IDMEC, Instituto Superior Técnico, Av Rovisco Pais 1, 1049-001 Lisboa, Portugal

[2]Academia Militar, R. Gomes Freire 203, 1169-203 Lisboa, Portugal

[a]margarida.gralha@tecnico.ulisboa.pt, [b]colaco.ba@academiamilitar.pt, [c]joao.magrinho@tecnico.ulisboa.pt, [d]enio.chamber@academiamilitar.pt, [e]beatriz.silva@tecnico.ulisboa.pt

Keywords: Aluminium, Incremental Sheet Forming, Formability

Abstract. The digital transition in manufacturing is driving significant advancements toward mass customization, enabling the production of tailored products at lower costs and shorter lead times. Single Point Incremental Forming (SPIF) is a promising technology in this context, offering the flexibility needed for producing complex, customized geometries without the need for dedicated tooling. Unlike conventional forming processes, SPIF relies on unique deformation mechanics that allow for higher formability, with the fracture forming line (FFL) as the process limit. This work revisits the analysis of formability limits in SPIF for aluminium sheets, revealing that fracture strains can occur beyond the FFL strain levels.

Introduction

Single point incremental forming (SPIF) is an innovative and flexible manufacturing process that allows the production of complex sheet geometries without the need for dedicated tooling. Unlike conventional deep-drawing, which requires expensive and geometry-specific tooling, SPIF has a more flexible approach where a universal tool follows a programmed path to progressively deform a desired shape [1, 2]. This eliminates the need for dedicated dies, significantly reducing both setup costs and lead times while allowing for greater adaptability in design changes. These characteristics make SPIF suitable for small-batch or prototype production. However, the industrial adoption of SPIF remains limited, primarily due to the need for a deeper understanding of the deformation mechanics to achieve final parts with accurate tolerances.

One significant characteristic of the SPIF process is its ability to increase formability beyond the forming limit curve (FLC) [3]. This phenomenon is directly related to the unique mechanics of deformation in SPIF. Several authors have explored this topic, providing alternative explanations [4, 5]. Silva et al. [6] developed an analytical model showing the suppression of necking in the process, which explains the enhanced formability. Thus, the process limit is governed by the fracture forming line (FFL) [7], rather than by necking as in conventional forming processes. This approach was validated for aluminium sheet AA1050-H111 in a previous study by the authors [6].

In the present work, this approach is analysed for the AA6082-O sheet, and the results revealed significant differences compared to previous studies. A literature review on the application of this methodology to a wide range of materials highlights key trends and offers guidance for future research.

Material and Methods

The experimental work was conducted on aluminium AA6082-O sheets with a thickness of 2 mm. Mechanical characterization was performed using tensile tests on water jet cut specimens aligned at 0°, 45° and 90° relative to the rolling direction (RD). The tensile tests were performed on an

Content from this work may be used under the terms of the Creative Commons Attribution 3.0 license. Any further distribution of this work must maintain attribution to the author(s) and the title of the work, journal citation and DOI. Published under license by Materials Research Forum LLC.

Sheet Metal 2025   Materials Research Forum LLC
Materials Research Proceedings 52 (2025) 59-66   https://doi.org/10.21741/9781644903551-8

INSTRON 5900 universal testing machine according to the ASTM Standard E8/E8M-16a [8] considering the standard geometry with a reference length of 50 mm.

Table 1 presents the mechanical properties of the material, including the modulus of elasticity E, the yield strength σ_Y, the ultimate tensile strength σ_{UTS}, the elongation at break, and the anisotropy coefficient r at 0°, 45° and 90° to the RD. The stress-strain curves resulting from the tensile tests were approximated by Ludwik-Hollomon's equation. The average property x values were calculated from,

$$\bar{x} = \frac{x_0 - 2x_{45} + x_{90}}{4}.$$ (1)

Table 1 – Mechanical properties of the aluminium AA6082-O sheets.

	E [GPa]	σ_Y [MPa]	σ_{UTS} [MPa]	A [%]	r	K [MPa]	n
0° RD	78.91	43.61	115.24	30.34	0.67	215.1	0.254
45° RD	73.30	44.47	117.67	32.91	0.61	221.9	0.260
90° RD	78.25	44.72	116.67	30.75	0.73	221.7	0.260
Average	**77.94**	**44.32**	**116.81**	**31.73**	**0.66**	**220.2**	**0.259**

The formability tests used to determine the formability limits by necking (FLC) and fracture (FFL) included tensile tests, hydraulic bulge tests with circular and elliptical dies, Nakajima tests, and double-notch tension tests (DNTT). The characteristics of the formability tests conducted are presented in Table 2, and allow the determination of the formability limits considering strain loading paths from uniaxial tensile ($\beta = d\varepsilon_2/d\varepsilon_1 = -0.5$) to equibiaxial strain paths ($\beta = 1$).

The double-notched tensile tests were performed on an INSTRON 5900 universal testing machine, while the Nakajima and bulge tests were conducted on an ERICHSEN 145/60 universal testing machine. The Nakajima tests followed the procedures and recommendations outlined in ISO 12004-2 [9].

Strain measurements for the tensile, DNTT, and Nakajima tests were conducted using a digital image correlation (DIC) system, specifically the Q-400 3D model from Dantec Dynamics, equipped with two 6-megapixel resolution cameras featuring 50.2 mm focal lenses and f/8 apertures. To enable surface strain measurements, the specimens were painted with a stochastic black speckle pattern on a uniform white background. The correlation algorithm used was the INSTRA 4D software, operating at an image acquisition frequency of 10 frames per second. A facet size of 13 pixels with a grid spacing of 7 pixels was applied.

Strain measurements for the bulge tests were conducted using circle grid analysis (CGA), where the specimens were electrochemically etched with a grid of 2.5 mm diameter circles prior to deformation. The major and minor in-plane strains were measured using a digital camera measuring system, specifically the GPA-100 model from ASAME.

The FLC was determined using the methodologies outlined in ISO 12004-2 [9] and Martínez-Donaire et al. [10]. The methodology to determine the FFL involved measuring the specimens' thickness before and after deformation using a Mitutoyo TM-505B optical microscope to obtain the 'gauge length' strains ε_3^f. The minor in-plane strains ε_2^f were assumed to remain constant after necking, under plane strain deformation, and thus obtained from the last measurement of the DIC system or the minor in-plane strain of an ellipse near fracture of the CGA. Considering incompressibility the major strains ε_1^f are obtained from

Sheet Metal 2025   Materials Research Forum LLC
Materials Research Proceedings 52 (2025) 59-66   https://doi.org/10.21741/9781644903551-8

$$\varepsilon_1^f = -\varepsilon_2^f - \varepsilon_3^f \tag{2}$$

this methodology is detailed in Centeno et al. [11].

Table 2 – Schematic representation of the experimental sheet formability tests and their parameters performed in the aluminium AA6082-O sheets.

Test		Dimensions [mm]	State of Stress	State of Strain	Formability Limit
Tensile		$l_c = 80$ $l_0 = 50$ $w_0 = 12.5$	$\sigma_1 > 0$ $\sigma_2 = \sigma_3 = 0$	$\varepsilon_1 > 0$ $\varepsilon_2 = \varepsilon_3 < 0$	FLC FFL
Bulge		$d_0 = 175$ $l_0 = 50$ $w_0 = 12.5$ Circular: Ø 100 Elliptical: 100:90, 100:80	$\sigma_1 \geq \sigma_2 > 0$ $\sigma_3 = 0$	$\varepsilon_1 \geq \varepsilon_2 > 0$ $\varepsilon_3 < 0$	FLC FFL
Nakajima		$d_0 = 210$ $r_0 = 50, 57.5,$ $72.5, 80$	$\sigma_1 > \sigma_2 \geq 0$ $\sigma_3 = 0$	$\varepsilon_1 \geq 0$ $\varepsilon_1 > \varepsilon_2$ $\varepsilon_2 > -\varepsilon_1/2$ $\varepsilon_3 < 0$	FLC FFL
Double Notched Tension		$w = 50$ $L = 125$ $l_0 = 10$	$\sigma_1 > 0$ $\sigma_2 < 0$ $\sigma_3 = 0$	$\varepsilon_1 > 0$ $\varepsilon_2 = 0$ $\varepsilon_3 < 0$	FFL

The SPIF tests were conducted on blanks with dimensions of 250 mm x 250 mm, using a Leadwell V-40 vertical machining center equipped with a rig, backing plate, clamping system and a hemispherical tip forming tool with a diameter of 12 mm. The experimental plan included two typical SPIF geometries: a truncated cone (Fig. 1a) and a truncated pyramid (Fig. 1b), designed to obtain strain loading paths in plane strain ($\beta = 0$) and biaxial strain ($\beta > 0$). Both geometries featured a variable wall angle that progressively increased with depth (Fig. 1c).

The process parameters considered were a constant feed of 1000 mm/min, and a free spindle. The process variables considered were a step down of 0.2 and 0.5 mm and two different tool trajectories, an incremental step down and a helicoidal with a scallop of 0.2 mm. Table 3 outlines the experimental plan, with two repetitions performed for each condition to ensure repeatability.

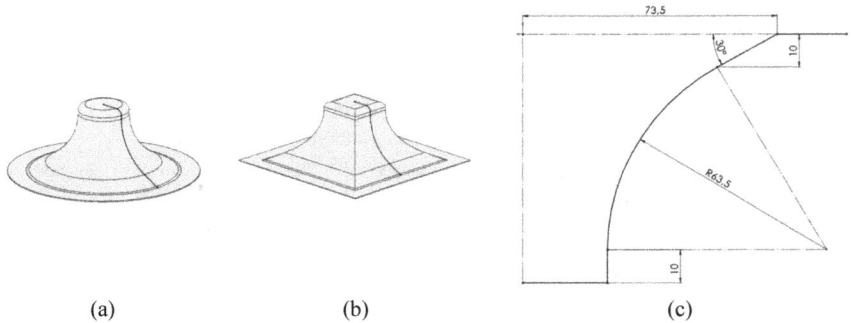

(a)   (b)   (c)

Figure 1 – Geometry of the SPIF parts included in the experimental plan: (a) truncated cone, and (b) truncated pyramid; (c) detail of the cross-section of the geometries showing the increase in the wall angle with depth.

Measurement of the SPIF parts deformation was conducted by CGA, with a grid of 2.5 mm of diameter circles electrochemically etched onto the surface. The tests were performed using Castrol Iloform TDN 81 forming fluid as lubrication between the tool and the sheet.

Table 3 – Experimental plan for the SPIF tests.

Reference	Geometry	Step down [mm]	Tool trajectory	Spindle [rpm]	Feed rate [mm/min]	Tool Diameter [mm]
Cone A	truncated cone	0.2	constant step down	free	1000	12
Cone B		0.5				
Cone C		0.2	helicoidal			
Pyramid A	truncated pyramid	0.2	constant step down			
Pyramid B		0.5				
Pyramid C		0.2	helicoidal			

Results and Discussion

Formability limits by necking (FLC) and by fracture (FFL) determined considering the tests and methodologies previously presented are presented in Fig. 2.

The FFL can be approximated by a straight line defined by $\varepsilon_1 + 0.41\varepsilon_2 = 1.2$. To account for possible errors associated with the experimental measurements in determining the FFL, an uncertainty margin of 10% was considered and is represented in Fig. 2.

The set of experiments was conducted on the truncated conical and pyramidal geometries following the experimental plan presented in Table 3. Open markers in Fig. 3a and 4a refer to the experimental in-plane strains measured by CGA along meridional cross-sections of the final part (Fig. 3b and 4b), while solid markers refer to the experimental fracture strain measurements obtained from thickness measurements in the crack (Fig. 3c and 4c).

Sheet Metal 2025

Materials Research Proceedings 52 (2025) 59-66

Materials Research Forum LLC

https://doi.org/10.21741/9781644903551-8

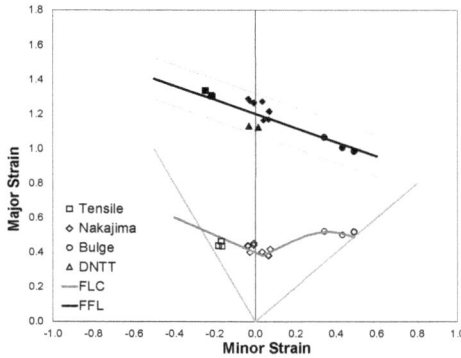

Figure 2 – Formability limits of aluminium AA6082-O sheets with 2 mm thickness. Open markers correspond to failure by necking, while solid markers to failure by fracture.

Results from the first set of experiments on the conical geometry, presented in Fig. 3, confirm the characteristic plain strain deformation mode typical of conical geometries formed by SPIF. The three different test conditions (Table 3) show no significant variation in the fracture strain pairs; however, these values are notably higher than the FFL obtained from conventional formability tests. To account for this discrepancy, a parallel line to the FFL was drawn through the fracture points, and the difference was quantified as a 61% increase.

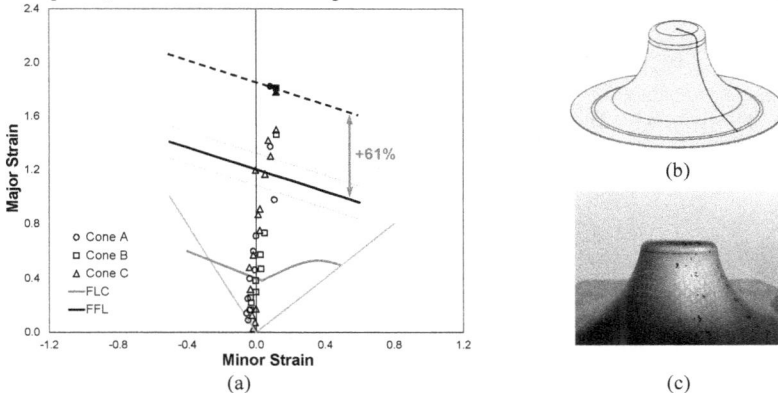

(a)

(b)

(c)

Figure 3 – (a) Experimental strain measurements associated to meridional cross-sections of the truncated conical parts presented in Table 3, (b) identification of the cross-section, and (c) detail of the fracture.

The results for the pyramidal geometry set are presented in Fig. 4. The experimental in-plane strains were measured along two different meridional cross-sections: one on the part wall (indicated by the black line in Fig. 4b) and the other at the corner where fracture occurred (indicated by the red line in Fig. 4b), as shown in Fig. 4c.

The three different test conditions (Table 3) show that the deformation mode in the wall cross-section is plain strain, while the deformation mode in the corner cross-section is biaxial strain, as expected [6]. The fracture occurred at the corner in all three test conditions, with the fracture strain pair of condition 3 showing a maximum increase of 35% compared to the FFL. These results reveal that, for the truncated pyramidal geometry, the tool trajectory significantly influences the strains at the corner of the part.

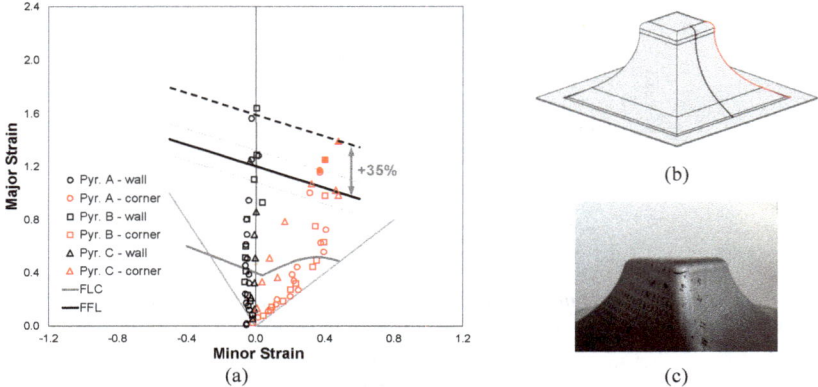

Figure 4 – (a) Experimental strain measurements associated with meridional cross-sections of the truncated pyramidal parts presented in Table 3, (b) identification of the cross-sections, and (c) detail of the fracture.

To better understand the significant increase in the process formability observed beyond the FFL in this work, the authors reviewed relevant literature that considers the SPIF process limit to be governed by the onset of fracture. For the FFL determination, these studies typically relied on thickness measurements at the fracture. Table 4 summarizes the key information collected, identifying the main materials and process parameters.

Previous work from McAnulty et al. [12] reviewed the influence of various SPIF parameters on process formability and concluded that these parameters are highly interdependent. To quantify the strength of these relationships, the authors utilized the Pearson correlation coefficient r_{xy}, defined as:

$$r_{xy} = \frac{\sum_{i=1}^{n}(x_i - \bar{x})(y_i - \bar{y})}{\sqrt{\sum_{i=1}^{n}(x_i - \bar{x})^2}\sqrt{\sum_{i=1}^{n}(y_i - \bar{y})^2}} \tag{3}$$

where n is the sample size, x_i and y_i are the individual sample points indexed by i, and $\bar{x}$ and $\bar{y}$ the samples means given by $\bar{x} = \frac{1}{n}\sum_{i=1}^{n} x_i$.

The parameter with the highest correlation was the thickness-to-tool radius ratio, t/R_{tool} showing a positive and moderate value of 0.76. This result is promising, considering that the data collected includes various materials, lots and thicknesses, and the work of different researchers using distinct equipment and techniques.

In the future, expanding this analysis will be essential to gaining a deeper understanding of its viability and determining the range of applications. This could involve exploring the influence of various materials, tool geometries, and process parameters under a broader set of conditions to identify the optimal operating window for specific industries. Additionally, further experimental studies and computational models could help validate the correlation between key SPIF parameters. Such research would contribute to improving the predictability and reliability of SPIF in industrial applications, making it a more robust process for small-batch and custom manufacturing.

Table 4 – Summary of the main parameters retrieved from the literature in experimental SPIF research that considered the FFL as the process limit.

Material	Thickness [mm]	Work hardening n	Geometry	R_{part}/R_{tool}	t/R_{tool}	Deviation from FFL
AA1050-H111 [6]	1.0	0.04	cone and pyramid	20.60	0.25	+12.6%
AA2024-T3 [13]	1.2	0.13-0.20*	cone	8.88	0.24	+15.8%
AA6082-O	2.0	0.26	cone and pyramid	13.75	0.33	+61.0%
AISI 304 [14]	0.8	0.59	cone	8.88	0.16	+26.8%
DP590 [15]	1.0	0.36	cone and pyramid	8.75	0.10	+7.1%
Titanium grade 1 [16]	0.6	0.05-0.08*	cone and pyramid	20.60	0.15	+7.6%
Titanium grade 2 [17]	0.7	0.07-0.10*	cone	20.60	0.18	+3.4%
Copper [18]	0.8	0.17	cone and pyramid	13.75	0.13	+3.9%

*Values retrieved from [19]

Conclusion

The conventional formability tests established the formability limits by necking (FLC) and by fracture (FFL) for aluminium AA6082-O with a thickness of 2 mm. Experimental SPIF tests performed on truncated conical and pyramidal geometries confirmed the typical deformation modes of plane strain and biaxial strain. The process strategy had no significant influence on the process limit for conical geometries, while a notable influence was observed for pyramidal geometries. Experimental SPIF tests revealed a process window extending beyond the determined FFL, and a positive and moderate correlation was identified between the thickness-to-tool radius ratio and formability increase. Expanding this analysis in the future will be essential to better understand SPIF's viability, validate parameter correlations, and enhance its reliability for industrial, small-batch, and custom manufacturing applications.

References

[1] J. Jeswiet, F. Micari, G. Hirt, A. Bramley, J. Duflou, J. Alwood, Asymmetric single point incremental forming of sheet metal, CIRP Annals 54-2 (2005) 88-114. https://doi.org/10.1016/S0007-8506(07)60021-3

[2] J.R. Duflou, A.M. Habraken, AM., J. Cao, et al. Single point incremental forming: state-of-the-art and prospects. Int J Mater Form 11 (2018) 743-773. https://doi.org/10.1007/s12289-017-1387-y

[3] G. Hirt, S. Junk, N. Witulski, Incremental Sheet Forming: Quality Evaluation and Process Simulation, 7th ICTP International Conference on Technology of Plasticity, October 27-November 1, 2002, Yokohama, Japan, paper no. 343.

[4] W.C. Emmens, A.H. van den Boogaard, An overview of stabilizing deformation mechanisms in incremental sheet forming, J. Mater. Process. Technol. 209 (2009) 3688-3695. https://doi.org/10.1016/j.jmatprotec.2008.10.003

[5] R. Malhotra, L. Xue, T. Belytschko, J. Cao, Mechanics of fracture in single point incremental forming, J. Mater. Process. Technol. 212 (2012) 1573-1590.
https://doi.org/10.1016/j.jmatprotec.2012.02.021

[6] M.B. Silva, M. Skjoedt, A.G. Atkins, N. Bay, P.A.F. Martins, Single-point incremental forming and formability-failure diagrams, J. Strain Analysis 43 (2008) 15-35.
https://doi.org/10.1243/03093247JSA340

[7] J.D. Embury, J.L. Duncan, Formability maps, Annu. Rev. Mater. Res. 11 (1981) 505-521.
https://doi.org/10.1146/annurev.ms.11.080181.002445

[8] ASTM E8/E8M-22 (2022) Standard test methods for tension testing of metallic materials.

[9] ISO 12004-2 (2021) Metallic materials-Sheet and strip-Determination of forming-limit curves-Part 2: Determination of forming-limit curves in the laboratory.

[10] A. Martínez-Donaire, F. García-Lomas, C. Vallellano, New approaches to detect the onset of localised necking in sheets under through-thickness strain gradients, Mater. Des. 57 (2014) 135-145. https://doi.org/10.1016/j.matdes.2014.01.012

[11] G. Centeno, A.J. Martínez-Donaire, D. Morales-Palma, C.Vallellano, M.B. Silva, P.A.F. Martins, Novel experimental techniques for the determination of the forming limits at necking and fracture, in: J.P. Davim (Ed.), Materials Forming and Machining, Woodhead Publishing, Amsterdam, 2015, pp. 1-24. https://doi.org/10.1016/B978-0-85709-483-4.00001-6

[12] T. McAnulty, J. Jeswiet, M. Doolan, Formability in single point incremental forming: A comparative analysis of the state of the art, CIRP-JMST 16 (2017) 43-45.
https://doi.org/10.1016/j.cirpj.2016.07.003

[13] G. Centeno, A.J. Martínez-Donaire, C. Vallellano, L.H. Martínez-Palmeth, D. Morales, C. Suntaxi, F.J. García-Lomas, Experimental study on the evaluation of necking and fracture strains in sheet metal forming processes, Procedia Eng. 63 (2013) 650-658.
https://doi.org/10.1016/j.proeng.2013.08.204

[14] G. Centeno, I. Bagudanch, A.J. Martínez-Donaire, M.L. García-Romeu, C. Vallellano, Critical analysis of necking and fracture limit strains and forming forces in single-point incremental forming, Mater. Des. 63 (2014) 20-29. https://doi.org/10.1016/j.matdes.2014.05.066

[15] S. Pandre, A. Morchhale, G. Mahalle, N. Kotkunde, K. Suresh, S.K. Singh, Fracture limit analysis of DP590 steel using single point incremental forming: experimental approach, theoretical modelling and microstructural evolution, ACME 21:95 (2021) 1-20.
https://doi.org/10.1007/s43452-021-00243-1

[16] V.A.M. Cristino, M.B. Silva, P.K. Wong, P.A.F. Martins, Determining the fracture forming limits in sheet metal forming: A technical note, J. Strain Analysis 52-8 (2017) 467-471.
https://doi.org/10.1177/0309324717727443

[17] M.B. Silva, P. Teixeira, A. Reis, P.A.F. Martins, On the formability of hole-flanging by incremental sheet forming, Proc. Inst. Mech. Eng. Part L 277-2 (2013) 91-99.
https://doi.org/10.1177/1464420712474210

[18] K. Jawale, J.F. Duarte, A. Reis, M.B. Silva, Characterizing fracture forming limit and shear fracture forming limit for sheet metals, J. Mater. Process. Technol. 255 (2018) 886-897.
https://doi.org/10.1016/j.jmatprotec.2018.01.035

[19] ASM Handbook. Volume 2, Properties and selection: Nonferrous alloys and special-purpose materials, Metals Park, Ohio, ASM International, 1990.

Sheet Metal 2025
Materials Research Proceedings 52 (2025) 67-75

Materials Research Forum LLC
https://doi.org/10.21741/9781644903551-9

SPIF accuracy improvement by FEM analysis of multi-step tool trajectories with experimental validation

Cristian Cappellini[1,a] *, Claudio Giardini[1,b] and Sara Bocchi[1,c]

[1]University of Bergamo, Department of Management, Information and Production Engineering, Via Pasubio 7/b, 24044 Dalmine (BG), Italy

[a]cristian.cappellini@unibg.it, [b]claudio.giardini@unibg.it, [c]sara.bocchi@unibg.it

Keywords: Incremental Sheet Forming, Accuracy, Finite Element Method (FEM)

Abstract. This study focuses on the application of Single Point Incremental Forming (SPIF) in modern manufacturing, particularly in batch production where flexibility and cost-efficiency are crucial. SPIF enables forming various shapes with reduced tool-workpiece contact and minimal forming forces. A critical factor affecting product quality is the tool trajectory, which influences material springback and accuracy, leading to issues like geometric deviations and defects such as the pillow effect. To address these challenges, the research proposes Finite Element Method (FEM) simulations to analyze alternative SPIF multi-step tool paths. Focusing on frustum cones made of AA1050-H24 aluminum alloy, the study compares simulation results with experimental data, highlighting a roughing-finishing approach that improves geometric accuracy and reduces defects, making it viable for industrial use.

Introduction

Following the Industry 4.0 paradigms, enterprises shifted from mass to batch production, where higher flexibility and lower asset costs are mandatory [1]. Considering this, the capability of working several materials in die-less condition with a generic tool, individuates Single Point Incremental Forming (SPIF) as an appropriated forming alternative [2]. Despite dedicated machines has been developed to accomplish SPIF, its application on generic 3-axis CNC and robot-based machines, further enhancing process flexibility, was demonstrated [2]. SPIF is performed by a sequential movement of the forming tool on a blank, by following different paths at increasing penetration depths, as a function of desired shape and axial deepness. This results in a reduced tool-workpiece contact area, with low forming forces and increased material formability [3]. Numerous studies recognized tool trajectory as one of the most influencing parameters on the quality of the final product [4-6]. Each selected tool trajectory, in fact, causes a determined material displacement, influencing material work-hardening and related springback [7]. This leads to reduced accuracy, defined as the deviation of the part geometry from the nominal one. In this sense, multi-step tool path methodology demonstrated to beneficially affect material formability [8], while issues regarding accuracy remain still open. Despite in the production of cylindrical parts, multi-stage SPIF allows to obtain high precision vertical walls, it introduces rigid body motion (RBM) phenomena, resulting in stepped features that compromise the bottom quality [9]. Multi-pass associated stepped features were also observed in [10,11], hence, with the aim of reducing RBM, a multidirectional tool path was proposed in [12], but with poor results. Similarly, pillow defect, a bulging effect on the bottom, was detected in the manufacturing of conical squared cup exploiting sequential tool steps [13], revealing the achievement of a flat bottom part as one of the most difficult task. In order to reduce expensive and time-consuming experimental tests, Finite Element Method (FEM) simulation method was widely employed for analyzing SPIF process. In [14] simulations applying different element types were developed for predicting SPIF forces, with good results of both solid and shell elements. A reliable SPIF loads forecast was achieved in [15] as well, where a sub-modeling technique was implemented to refine the blank-tool contact area.

Content from this work may be used under the terms of the Creative Commons Attribution 3.0 license. Any further distribution of this work must maintain attribution to the author(s) and the title of the work, journal citation and DOI. Published under license by Materials Research Forum LLC.

Sheet Metal 2025   Materials Research Forum LLC
Materials Research Proceedings 52 (2025) 67-75   https://doi.org/10.21741/9781644903551-9

Alternative multi-stage tool paths were studied by FEM in [16] improving thickness distribution in vase shaped parts. For validating a deflection model, FEM was used in [17] allowing to estimate springback on walls and in the contact zone. FEM models studying the effect of process parameters in vacuum assisted SPIF, obtaining a close comparison with experimental results, were developed [18]. A good match was also achieved in [19] where the effects of varying dimensions shaped tools on the amount of pillow defect were evaluated.

Considering this, the present work proposes a FEM investigation of the influence of alternative SPIF multi-step tool trajectories on the geometric accuracy of frustum cones made of AA1050-H24 aluminum alloy. Aimed to minimize wall deviation and pillow effect, these strategies were analyzed by experimentally validated FEM simulations. The comparisons of nominal and simulated geometries, the resulting thickness distributions, and the estimated forming loads, allowed to individuate a roughing-finishing approach giving encouraging quality outcomes even exploitable under an industrial point of view.

The analyzed multi-step tool trajectories
Five different tool paths (Fig. 1), one concerning a single-step and four regarding multi-step strategies, developed for producing frustum cone geometry (Fig. 2a), were investigated by FEM simulations. Fig. 1 reports the values of the employed geometrical parameters as well.

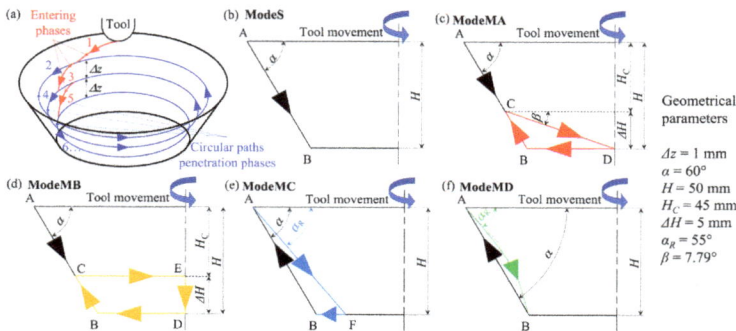

Fig. 1 Schematic of the analyzed tool path strategies with geometrical parameters' values.

Fig. 1b reports the employed single-step path ModeS (helical tool path described by Fig. 1a). In ModeS the tool follows a circular path at a determined axial depth. After this, the tool moves downward of a quantity Δz (step down) to cover a new circular path. This procedure is repeated until reaching the bottom of the frustum cone (depth H) with a circular path diameter in accordance with the desired wall angle α. ModeMA (Fig. 1c) concerns a multi-step strategy where the initial step is analogous to ModeS until a cone depth of H_C. Form this position, the tool moves following a lower wall angle β up to the bottom depth H at constant reduced Δz values according to the $\tan(\beta)/\tan(\alpha)$ ratio. Then, the tool describes a spiral path to enlarge the bottom at the desired radial position (point B). Subsequently, the tool displacement follows the α angle by the initial constant value of Δz up to depth H_C. ModeMB (Fig. 1d) represents a tiny modification of multi-step ModeMA in which, at depth H_C, the tool shifts to the cone center (point E) and then axially penetrates to bottom depth H. The movement of the tool over D-B and B-C lines is the same performed in ModeMA. As suggested in [16], the remaining penetration depth ΔH was chosen equal to the 10 % of the height H, for minimizing RBM, resulting in a β angle connecting C with the bottom center D. ModeMC (Fig. 1e) consist of a multi-step strategy where a first roughing step, concerning a wall angle α_R narrower than α, is achieved following ModeS to the bottom depth

Sheet Metal 2025   Materials Research Forum LLC
Materials Research Proceedings 52 (2025) 67-75   https://doi.org/10.21741/9781644903551-9

H (point F). Afterward, the bottom surface is enlarged to desired point B by a spiral tool movement where the finishing step, realized for achieving the desired wall angle α (B-A) at constant Δz, starts. ModeMD (Fig. 1f) denotes again a roughing-and-finishing multi-step strategy where, the former step realizes an increasing wall angle from the top (α_R in point A) to the desired angle α at the bottom (point B), while the latter is performed at constant α and Δz from bottom to top. The choice of forming the bottom was performed trying to reduce the amount of pillow defect. The α_R value in ModeMC was selected to achieve a reduced length of F-B on the bottom to contain sheet stretching and thinning.

FEM simulations setup and experimental validation
The FEM simulation setup was calibrated and then validated by comparing the experimental results (final geometry, forming forces, and material thickness) with the simulated ones, in the production of frustum cone (Fig. 2a). The calibration referred to ModeS strategy, while the final validation to ModeMD. The experiments were performed on a MC-60 Evolution 3-axis CNC machine having provide with a 18i-M GE Fanuc numerical control. The initial blank was a 1 mm in thickness AA1050-H24 aluminum sheet (Table 1), having a squared shape with a cantilevered edge length of 150 mm. Along the SPIF process, the blank was supported by a specifically designed equipment, allowing to acquire the forming forces, shown in Fig. 2b. For measuring the three components of the force, a Kistler 9257BA load cell, having a range of ±5 kN for *X-Y* components and from -5 kN to +10 kN for the *Z* one, was exploited. It was then connected to a control unit (Kistler Type 5223A) sending the forces' signals to a Hottinger Brüel & Kjær® Catman Data Acquisition Software. This latter was used for realizing a dedicated interface able to acquire the signals at a frequency of 200 Hz. The applied tool was a hemispherical one with a radius *r* = 10 mm, made of AISI 1045 steel. During the SPIF tests it moves with both constant feed rate of 600 mm/s and Δz of 1 mm. It was free to rotate and no spindle speed was assigned to it. The tests were performed in lubricated conditions using grease. In conformity to the cone geometry and tool dimensions, NC programs were prepared exploiting the Autodesk Fusion 360® CAD/CAM interface. The geometries of the produced frustum cones were acquired by the Hexagon® RA-8525-7 scanning system, consisting of a scanner (model AS1) coupled to a 7-axis arm. The achieved point cloud geometry was then fixed and exported as a STL file utilizing PolyWorks® software. Once the frustum cone geometry was attained, it was trimmed on the diametral section for measuring the thickness distribution.

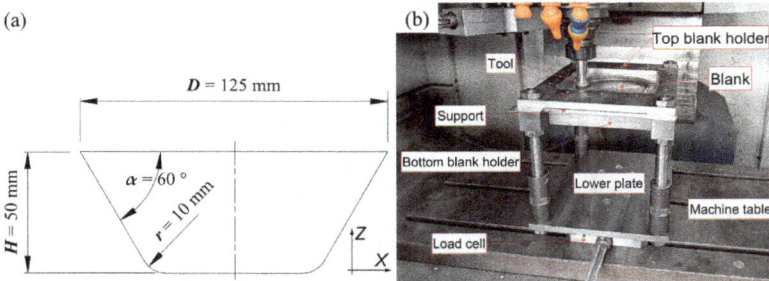

Fig. 2 (a) Frustum cone geometry and dimensions, and (b) SPIF equipment and setup.

Table 1 Chemical composition and mechanical properties of the blank material.

AA1050 H24 chemical composition						
Element %	Al	Si	Cu	Fe	Mg	Mn
in weight	99.50	<=0.25	<=0.05	<=0.40	<=0.05	<=0.05

Sheet Metal 2025

Materials Research Proceedings 52 (2025) 67-75

Materials Research Forum LLC

https://doi.org/10.21741/9781644903551-9

For analyzing the effect of the proposed tool path strategies on the accuracy of the final part, FEM simulations of the SPIF process were developed by means of ABAQUS dynamic explicit software (Fig. 3). The support and the tool were modeled as analytically rigid surfaces. The former was fixed in the space, while the latter was moved accordingly to the related simulated strategy CN program. Shape and dimensions of them were defined in line to the experimental tests. The blank was modeled as a square shaped elasto-plastic deformable surface having a section thickness property of 1 mm. Each edge had a length of 150 mm and was fixed in the three spatial directions. It was discretized with a mesh consisting of squared shape shell elements. The dimension of these latter was varied in a range of 0.75-2.00 mm, at intervals of 0.25 mm, to perform a mesh convergency analysis. Its results revealed a forming load settlement at 1.25 mm; thus, this value was selected, leading to a total number of 14400 elements. The material flow stress σ was represented by the Hollomon power law [20] of Eq. 1, where ε, K, and n were the plastic strain, the strength coefficient, and the strain hardening exponent, respectively.

$$\sigma = K \cdot \varepsilon^n = 157 \cdot \varepsilon^{0.04} [MPa] \tag{1}$$

As reported in [15], since Hill's anisotropic yield criteria does not give significant FEM results improvement, Von Mises yield criteria was applied. The material elastic modulus and Poisson coefficient were set equal to 68 [GPa] and 0.33. Penalty method with a friction coefficient of 0.05 was set for modeling the blank-tool and the blank-support contact interfaces. A time step dt of 1×10^{-4} [s] was selected and mass scaling throughout the steps was applied to the whole model each 10 increments. Considering the lubrication condition and the low deformation heating generated during SPIF, FEM were performed in isothermal conditions.

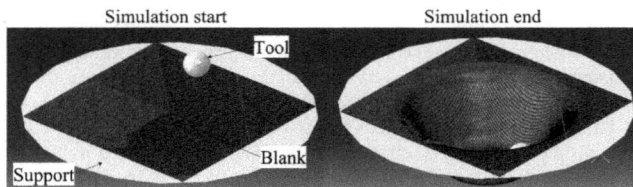

Fig. 3 FEM setup; start (on the left) and end (on the right) of the simulation for ModeS.

Fig. 4 shows the comparison of FEM and experimental (EXP) results for calibration (ModeS) and validation (ModeMD). The geometric comparison (Fig. 4a) was performed by importing the STL files of FEM and EXP achieved shapes in the GOM Inspect® software. After a best alignment phase, GOM depicts the deviations between geometries, permitting the point-to-point measurement of these latter and furnishing their distribution. The black line represents the experimental profile, while the colored one is referred to simulation. Negative or positive deviation labels' values correspond respectively to a geometry smaller or greater than the nominal one. The good superpositions of geometry, thickness distribution, and loads, are clearly visible, for both calibration and validation cases, enforcing FEM setup capability.

Sheet Metal 2025
Materials Research Proceedings 52 (2025) 67-75

Materials Research Forum LLC
https://doi.org/10.21741/9781644903551-9

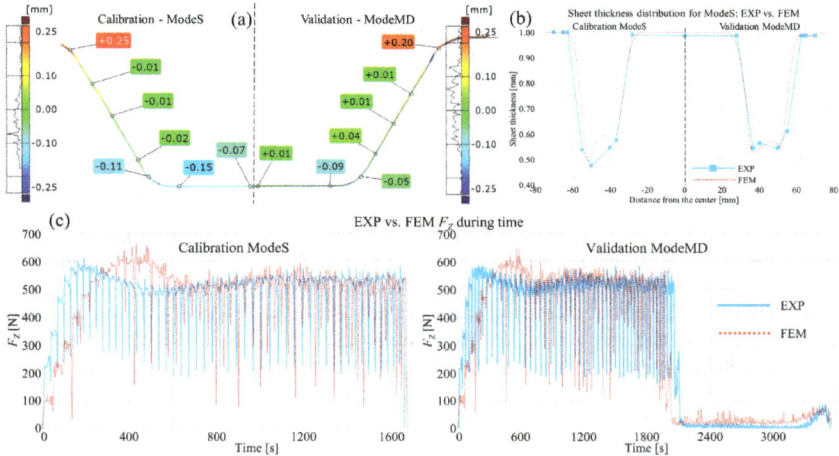

Fig. 4 EXP-FEM comparison of (a) geometry, (b) thickness and (c) force, for FEM validation

Results and discussion

All the strategies of Fig. 1 were simulated, exploiting the validated FEM setup, to individuate the one giving the best wall and bottom accuracy. The comparisons among FEM (black line) and nominal (colored) geometries, for the worst (ModeMA) and the best (ModeMD) strategy, are shown in Fig. 5. Table 2 summarizes the comparison results. $Wall_{Total}$ (Eq. 2) represents the maximum diametral deviation of the conical wall and is calculated by the summation of the correspondent two sides radial deviations ($Wall_1$ and $Wall_2$). *Pillow* is the bulge due to pillow effect, determined by Eq. 3 where, $Bottom_1$ and $Bottom_2$ are the deviations in the correspondence of the connecting radius amongst conical and bottom wall, while $Bottom_{Center}$ is the deviation on the revolution axis.

$$Wall_{Total} = Wall_1 + Wall_2 \tag{2}$$

$$Pillow = \frac{Bottom_1 + Bottom_2}{2} - Bottom_{Center} \tag{3}$$

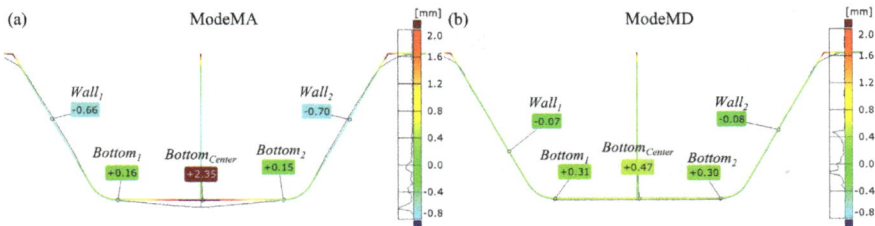

Fig. 5 GOM analysis of FEM resulting geometries for (a) the worst and (b) the best tool path strategy.

Table 2 Geometrical deviations of wall and bottom surface for the applied tool path strategies.

Mode	Wall₁	Wall₂	Wall_Total	Bottom₁	Bottom₂	Bottom_Center	Pillow
			Deviation value [mm]				
S	-0.57	-0.58	-1.15	+0.50	+0.47	+0.67	-0.19
MA	-0.66	-0.70	-1.36	+0.16	+0.15	+2.35	-2.19
MB	-0.65	-0.65	-1.30	+0.22	+0.23	+2.25	-2.02
MC	+0.08	+0.07	+0.15	+0.51	+0.52	+1.78	-1.26
MD	-0.07	-0.08	-0.15	+0.31	+0.30	+0.47	-0.16

Pillow results of ModeMA, ModeMB, and ModeMC shows that a direct contact of the tool in forming the bottom has an adverse effect on pillow accuracy. This is ascribable to the rigid body motion (Fig. 6) of the bottom material, as a consequence of the radial displacement of the tool after the axial penetration. In this case, the directions of the radial (F_r) and axial (F_z) forming force components are positioned outward and downward respectively. Therefore, the bottom is enlarged to the desired diameter but at excessive depth. Moreover, the rigid motion is increased by the material work-hardening at which it is subjected [12]. *Wall_Total* values are lower for ModeMC and ModeMD highlighting the good influence of roughing-finishing approach. Even if the similar values of *Pillow* for ModeS and ModeMD are the lowest, the analysis of *Wall_Total* suggests ModeMD as the most profitable strategy.

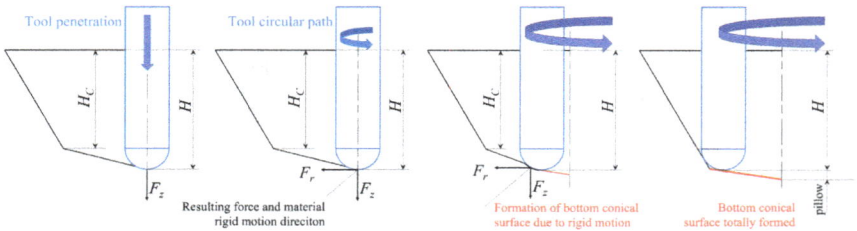

Fig. 6 Rigid motion of the material on the bottom part (red segment) due to the acting forces.

Fig. 7a graphs the z-components (F_Z) of the forming load for the totality of the analyzed strategies. In all the cases a force peak of 600 [N] is observable around a depth of 10 mm. During the remaining roughing phase, F_Z settles around 500 [N]. When working the bottom, after a first decreasing of F_Z, it reaches a second peak at 450 [N], while the axial finishing loads are lower than 100 [N] for all the strategies.

The sheet thickness distributions are represented in Fig. 7b, where the theoretical thickness *t*, calculated by the sine law of Eq. 4, in which α and t_i are the wall angle and the initial sheet thickness respectively, is reported as well.

$$t = t_i \cdot \sin\left(\frac{\pi}{2} - \alpha\right) \tag{4}$$

Concerning Fig. 7b, directly working on the bottom results to be deleterious even on thickness distribution, with a reduction up to 0.9 mm in the central area; not detectable for ModeS and ModeMD. Conical walls show low thickness peaks in line with Eq. 4 but, a lower peak and a more uniform distribution is observable for ModeMD. As observed in [16], due to the low wall angle of roughing step, ModeMC presents the highest wall thickness, while, the lowest one is visible in the bottom region as a consequence of the material stretching in the F-B movement (Fig. 1e). Considering the better outcomes in terms of wall and bottom accuracy, homogeneity of thickness

Sheet Metal 2025
Materials Research Proceedings 52 (2025) 67-75

Materials Research Forum LLC
https://doi.org/10.21741/9781644903551-9

distribution, and forming loads, ModeMD results the most suitable strategy, even if a higher process time is requested.

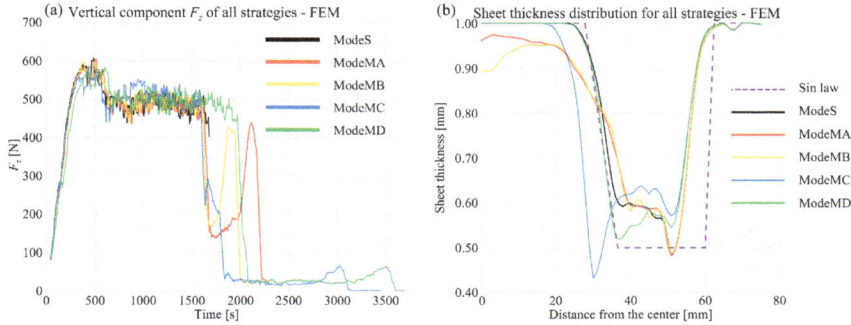

Fig. 7 Comparison of FEM resulting Fz (a) and thickness (b) for all the analyzed strategies.

Conclusions and future developments

In this paper, the influence of multi-step tool path strategies in SPIF manufacturing of AA1050-H24 frustum cones was analyzed by FEM simulations. In order to validate the simulation setup, the experimental results, in term of geometry, force and thickness, for a single-step methodology (ModeS), were compared with the numerical ones. Aimed to individuate the tool trajectory giving the best aforementioned results, four multi-step strategies, two concerning the complete forming of the cone bottom (ModeMA, ModeMB) and two consisting roughing-finishing passages (ModeMC, ModeMD), were then simulated. The FEM evidence pointed up the harmful influence on the pillow defect when, totally (ModeMA, ModeMB) or partially (ModeMC), forming the bottom, as a consequence of material rigid body motion. Moreover, a bottom reduction and an inhomogeneous distribution of thickness were detected. In contrast, a more uniform thickness and an enhanced accuracy resulted from the application of ModeMD, revealing it as the best strategy for the studied geometry. With the intention of reducing process times, additional works will be performed by increasing tool feed and varying process parameters. Likewise, supplementary investigations of the proposed strategies will be carried out on different materials.

References

[1] G. Mandaloi, A. Nagargoje, A.K. Gupta, G. Banerjee, H.Y. Shahare, P. Tandon, A Comprehensive Review on Experimental Conditions, Strategies, Performance, and Applications of Incremental Forming for Deformation Machining. Manuf. Sci. Eng. 144(11) (2022) 110802. https://doi.org/10.1115/1.4054683

[2] A.M. Rohit, C. Pandivelan, A critical review of incremental sheet forming in view of process parameters and process output, Adv. Mater. Process. Technol. 8(2) (2021) 2039-2068. https://doi.org/10.1080/2374068X.2021.1878730

[3] A. Gohil, B. Modi, Review of the effect of process parameters on performance measures in the incremental sheet forming process, Proc. Inst. Mech. Eng. B J. Eng. Manuf. 235(3) (2021) 303-332. https://doi.org/10.1177/0954405420961215

[4] S. Choudhary, A. Mulay, Influence of Tool Size and Step Depth on the Formability Behavior of AA1050, AA6061-T6, and AA7075-T6 by Single-Point Incremental Forming Process, J. Mater. Eng. Perform. 33 (2024) 3283-3298. https://doi.org/10.1007/s11665-023-08231-7

[5] S.M. Najm, I. Paniti, Study on Effecting Parameters of Flat and Hemispherical end Tools in SPIF of Aluminium Foils, Tehnički vjesnik, 27(6) (2020) 1844-1849. https://doi.org/10.17559/TV-20190513181910

[6] T. Trzepiecinski, S.M. Najm, V. Oleksik, D. Vasilca, I. Paniti, M. Szpunar, Recent Developments and Future Challenges in Incremental Sheet Forming of Aluminium and Aluminium Alloy Sheets, Metals 12 (2022) 124. https://doi.org/10.3390/met12010124

[7] H. Lu, H. Liu, C. Wang, Review on strategies for geometric accuracy improvement in incremental sheet forming, Int. J. Adv. Manuf. Technol. 102 (2019) 3381-3417. https://doi.org/10.1007/s00170-019-03348-3

[8] Z. Liu, Y. Li, P.A. Meehan, Tool path strategies and deformation analysis in multi-pass incremental sheet forming process, Int. J. Adv. Manuf. Technol. 75 (2014) 395-409. https://doi.org/10.1007/s00170-014-6143-6

[9] M. Skjoedt, M.B. Silva, P.A.F. Martins, N. Bay, Strategies and limits in multi-stage single-point incremental forming, J. Strain Anal. Eng. Des. 45(1) (2010) 33-44. https://doi.org/10.1243/03093247JSA574

[10] D. Xu, R. Malhotra, N.V. Reddy, J. Chen, J. Cao, Analytical prediction of stepped feature generation in multi-pass single point incremental forming, J. Manuf. Proc. 14 (2012) 487-494. https://doi.org/10.1016/j.jmapro.2012.08.003

[11] Z. Liu, Y. Li, P.A. Meehan, Vertical Wall Formation and Material Flow Control for Incremental Sheet Forming by Revisiting Multistage Deformation Path Strategies, Mater. Manuf. Proc. 28(5) (2013) 562-571. https://doi.org/10.1080/10426914.2013.763964

[12] G. Buffa, M. Gucciardi, L. Fratini, F. Micari, Multi-directional vs. mono-directional multi-step strategies for single point incremental forming of non-axisymmetric components, J. Manuf. Process. 55 (2020) 22-30. https://doi.org/10.1016/j.jmapro.2020.03.055

[13] S. Wu, L. Gao, Y. Matsuoka, S. Rashed, Y. Zhao, N. Ma, Multi-step toolpath approach of improving the dimensional accuracy of a nonaxisymmetric part in incremental sheet forming and its mechanism analysis, J. Mech. Sci. Technol. 36(4) (2022) 1975-1985. https://doi.org/10.1007/s12206-022-0333-1

[14] C. Henrard, C. Bouffioux, P. Eyckens, H. Sol, J.R. Duflou, P. Van Houtte, A. Van Bael, L. Duchene, A.M. Habraken, Forming forces in single point incremental forming: prediction by finite element simulations, validation and sensitivity, Comp. Mech. 47 (2011) 573-590. https://doi.org/10.1007/s00466-010-0563-4

[15] P. Eyckens, J.R. Duflou, A. Van Bael, P. Van Houtte, The significance of friction in the single point incremental forming process, Int. J. Mater. Form. 3(1) (2010) 947-950. https://doi.org/10.1007/s12289-010-0925-7

[16] S. Wu, Y. Ma, L. Gao, Y. Zhao, S. Rasheda, N. Ma, A novel multi-step strategy of single point incremental forming for high wall angle shape, J. Manuf. Proc. 56 (2020) 697-706. https://doi.org/10.1016/j.jmapro.2020.05.009

[17] Z. Chang, M. Yang, J. Chen, Geometric deviation during incremental sheet forming process: Analytical modelling and experiment, Int. J. Mach. Tools Manuf. 198 (2024) 104160. https://doi.org/10.1016/j.ijmachtools.2024.104160

[18] M. Murugesan, H.W. Youn, J.H. Yu, W. Chung, C.W. Lee, Investigation of forming parameters influence on pillow defect in a new vacuum-assisted incremental sheet forming

Materials Research Forum LLC
https://doi.org/10.21741/9781644903551-9

process, Int. J. Adv. Manuf. Technol. 127 (2023) 5531-5551. https://doi.org/10.1007/s00170-023-11854-8

[19] B.B.L. Isidore, G. Hussain, S.P. Shamchi, W.A. Khan, Prediction and control of pillow defect in single point incremental forming using numerical simulations, J. Mech. Sci. Technol. 30 (2016) 2151-2161. https://doi.org/10.1007/s12206-016-0422-0

[20] R. Perez-Santiago, N.J. Hendrichs, G. Capilla-González, E. Vázquez-Lepe, E. Cuan-Urquizo, The Influence of the Strain-Hardening Model in the Axial Force Prediction of Single Point Incremental Forming, Appl. Sci. 14 (2024) 5705. https://doi.org/10.3390/app14135705

Sheet Metal 2025
Materials Research Proceedings 52 (2025) 76-84

Materials Research Forum LLC
https://doi.org/10.21741/9781644903551-10

Investigating intermediate shapes for multi-stage forming of cranial implants: The influence of two intermediates stages

M. Vanhulst[1,a] * and J.R. Duflou[1,b]

[1]Department of Mechanical Engineering, KU Leuven / Flanders Make, Celestijnenlaan 300B, B-3001 Leuven, Belgium

[a]marthe.vanhulst@kuleuven.be, [b]joost.duflou@kuleuven.be

Keywords: Incremental Sheet Forming, Accuracy, Formability, Multi-Stage Forming, Medical Implants

Abstract. This study investigates the use of two intermediate shapes, constructed with an automatic design strategy, in Single Point Incremental Forming (SPIF) to improve thickness distribution and geometric accuracy for cranial implants. Previous work showed a trade-off between these two factors when using only one intermediate stage. In this research, five three-stage experiments were conducted, along with both a two-stage and a single-stage control. The intermediate shapes are constructed using z-scaling and z-translation techniques and the design is based on the outcome of experiments with only one intermediate shape. Building further upon these previous results, the last intermediate stage is selected for optimal geometric accuracy and kept constant for all experiments for comparison purposes. An earlier stage is added and optimized further to improve the thickness distributions. Results show that while multi-stage forming improved the thickness distributions in the last intermediate stage, this benefit vanished in the final forming stage. Geometric accuracy was slightly better in the two-stage process compared to the three-stage experiments.

Introduction

Incremental Sheet Forming (ISF) is a flexible sheet metal forming technique that shows significant potential in producing complex and customized parts or small batches [1]. One prominent ISF variant, Single Point Increment Forming (SPIF), utilizes a simple setup with a hemispherical tool following a pre-defined toolpath on the sheet surface. However, challenges related to formability and geometric accuracy have limited its broader adoption in industrial applications [2]. Unlike conventional methods like stamping and deep drawing, ISF offers enhanced flexibility and cost reduction by eliminating the need for expensive dies. Particularly applications where unique or patient-specific geometries are required, such as medical implants, would benefit from its flexibility, low cost and customization [3]. Research has been conducted on multiple implants, including ankle support [4], facial and cranial [5], [6], [7], clavicle [8] and knee [9] implants. Hieu et al. [6] researched facial implants, but posed that the main challenges include complex designs. In order to make more challenging designs with steep wall angles, multi-stage forming can be used. Duflou et al. [10] proposed a multi-stage strategy with two steps to increase the process limits for a cranial implant. They designed their intermediate shapes as two double curved surfaces and made sure the full initial blank was fully processed before forming the final step. This multi-stage strategy was necessary to increase the process limits when applying compensation to increase the geometric accuracy, as compensation drastically increased the wall angles to be formed. Later, Fahad et al. [11] also studied multi-stage forming of a cranial implant, again in three steps. The intermediate stages were constructed based on scaling and lofting techniques, and were designed by taking three rules into account: (1) each preform has a lower wall angle and total depth than the previous one, (2) the preforms were selected to drive the material along all three axes and (3) the gap between the toolpath shapes was kept relatively small, with a maximal difference in depth of

Content from this work may be used under the terms of the Creative Commons Attribution 3.0 license. Any further distribution of this work must maintain attribution to the author(s) and the title of the work, journal citation and DOI. Published under license by Materials Research Forum LLC.

Sheet Metal 2025

Materials Research Proceedings 52 (2025) 76-84

Materials Research Forum LLC

https://doi.org/10.21741/9781644903551-10

10 mm. In their work, they saw that the geometric accuracy was higher when applying multi-stage forming. They also observed a slightly lower minimal thickness of the final shape, as well as a shift in its location compared to the single-stage formed part.

Both of these studies assume three forming stages but do not explore intermediate shape design or its impact on the final outcome. To extend SPIF to more complex and steep geometries, multi-stage strategies are crucial, yet further research is needed to understand the influence of the intermediate shapes and optimize their construction. For that reason, a previous study [12] was conducted to investigate the role of intermediate shape design, focusing on only one intermediate shape constructed using generic strategies like z-scaling and z-translation. It revealed a trade-off in multi-stage forming using one intermediate shape: optimizing for either thickness distributions or geometric accuracy. This paper aims to address this trade-off by proposing two intermediate shapes instead of one: one for the optimization of the thickness, the other one for the optimization of the geometric accuracy.

Experiments and methods

This study focuses on a cranial implant, which is embedded towards the flat clamping zone. The goal of the study is to extend the process limits and allow production of implants with wall angles exceeding the critical wall angle of the sheet. Investigating optimal, generic strategies to extend the process limits are crucial for industrial adaptation, as it allows for production of more complex, steeper shapes. However, in order to avoid failed experiments, an implant shape that does not exceed the critical wall angle is selected, thereby allowing for comparison with conventional single-stage forming. The intermediate stages were constructed using z-scaling and z-translation, as proposed in the previous study [12]. That study demonstrated that intermediate shapes with better thickness distributions were not the best ones in terms of final geometric accuracy. The study showed that the intermediate shape closest to the final shape gave the best geometric accuracy of the final implant geometry. Only two of the proposed intermediate shape strategies, the ones that only formed a small part of the final shape in the intermediate stage, resulted in slightly better final thickness distributions, almost similar to the single-stage forming in terms of minimal thickness. This minimal thickness is used as an indication of where failure might occur.

Strategies. Building on the insights of the previous study, a new experimental campaign was developed to optimize the intermediate shape design further and combine the optimal intermediate shape strategy for geometric accuracy with possible intermediate shapes to increase the process window. The first intermediate shape aims to optimize for a lower minimal thickness and more uniform thickness distributions throughout the part, crucial for both structural integrity and mechanical performance. The second intermediate shape targets enhanced geometric accuracy, minimizing deviations from the desired final shape. For this last intermediate stage, the shape with the best geometric performance from the previous study [12] is utilized. It is constructed by z-scaling the part until a 10 mm height difference with the final shape is reached. This second stage is kept constant across all experiments for comparison purposes.

For the first intermediate shape, designed to improve the thickness distribution, further analysis was needed. The best intermediate shapes in terms of final thickness distributions were not necessarily the ones with the lowest wall angles. This is likely due to a high z-translation, after which the edges of the intermediate shape are cut-off closer to the areas with the highest wall angles, but further from the backing plate. Since the area between the edges of the intermediate shape and the backing is not directly formed by the tool and has a lack of support, this leads to deviations of the blank without forming this area. Hence, we aim to use an intermediate shape that only forms the middle area, to avoid multiple forming passes in the critical area. To reach this, the next two guidelines for the intermediate shape can be extracted from the previous study: to avoid forming the area with critical angles, (1) the difference in height between the final shape and the intermediate shape should at least be 30 mm and (2) the part should be translated at least 20 mm in the z-direction.

Based on these findings, combinations of z-scaling and z-translation are used to construct the first intermediate shape, aiming to further optimize it for uniform thickness distributions (see Table 1 and Fig. 1). This results in a total of five experiments consisting of two intermediate stages before forming the final stage. These are compared to a two-stage experiment, where the first intermediate shape is skipped, and a single-stage experiment.

Table 1. Construction parameters of the intermediate stages for all multi-stage experiments.

Exp number	First intermediate shape			Second intermediate shape		
	z-scaling [mm]	z-translation [mm]	Height difference [mm]	z-scaling [mm]	z-translation [mm]	Height difference [mm]
1	0	30	30	10	0	10
2	10	20	30	10	0	10
3	0	40	40	10	0	10
4	10	30	40	10	0	10
5	20	20	40	10	0	10
2 stages	10	0	10	-	-	-

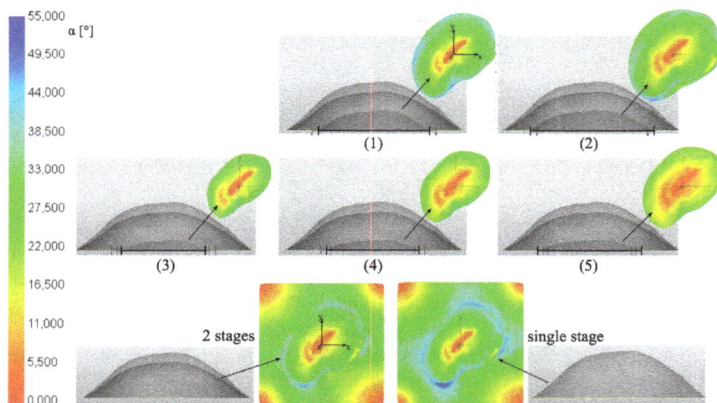

Fig. 1. Intermediate shapes for a cranial implant, constructed using z-scaling and z-translation as outlined in Table 1. The wall angles of the first stage are shown for all experiments. For the single-stage experiment, this first stage corresponds to the final shape.

Materials and set-up. The experiments were conducted using pure Zinc sheets, as a cost-effective alternative for Titanium. The 1 mm thick sheet of 225x225 mm is clamped in a 182x182 mm backing plate and Nuto H46 hydraulic oil is applied for lubrication. The parts are formed with a hemispherical tool of 12 mm, that moves at a feed rate of 2000 mm/min without spindle rotation on a vertical 3-axis CNC milling machine. After forming, the parts are sprayed with white developer spray for diffuse reflection and measured in a clamped state on a Coord MC16 CMM with LC60Dx Laser Line Scanner. These results are exported and processed in the GOM inspect software, where the clamping rig is used for alignment. The deviations are calculated as normal projections from the CAD model to the measured front mesh (as perceived by the tool, i.e., inside

Sheet Metal 2025

Materials Research Forum LLC

Materials Research Proceedings 52 (2025) 76-84

https://doi.org/10.21741/9781644903551-10

the formed part). The thickness distribution is determined using normal projections from the front to the back of the measured mesh.

Results and Discussion

Geometric accuracy. Fig. 2 visualizes the geometric accuracy of the final stage across all experiments. As can be seen, the maximal underforming (red) is higher for all multi-stage experiments compared to the single-stage experiment. However, the mean absolute deviation in Table 2 shows that the single-stage experiment performs worse.

This trend can also be seen in the violin plot of the deviation distributions in Fig. 3, where the single-stage experiment shows a broader spread of geometric deviations, as the sheet flattens more compared to multi-stage forming. Interesting to observe is that the two-stage experiment performed better in terms of mean absolute deviation than the three-stage experiments. On the other hand, the violin plot in Fig. 3 also shows that there is no big difference in the shape of the deviation distribution between the different multi-stage experiments.

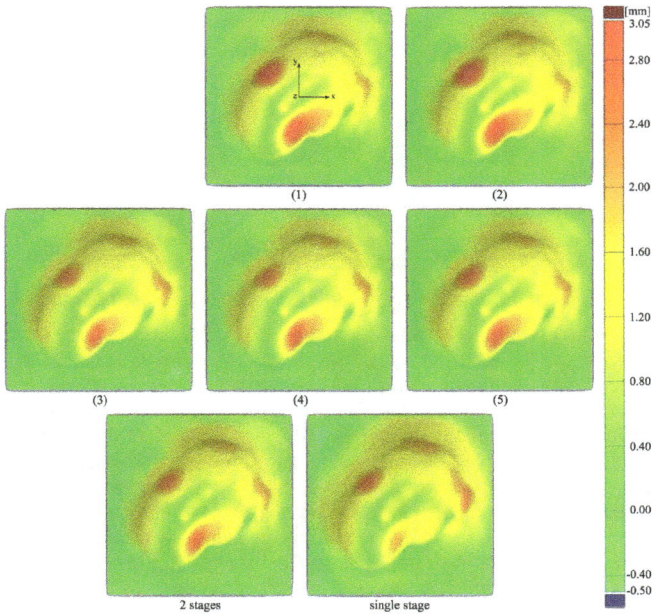

Fig. 2. Geometric deviations of the final shapes of all experiments. Positive values (red) indicate underforming, while negative values represent overforming.

Sheet Metal 2025

Materials Research Proceedings 52 (2025) 76-84

Materials Research Forum LLC

https://doi.org/10.21741/9781644903551-10

Table 2. Resulting geometric deviations and minimal thicknesses for all experiments. Positive deviations indicate underforming, negative indicate overforming.

Exp. number	z-scaling [mm]	z-translation [mm]	Height difference [mm]	Max wall angle in the first stage (CAD) [°]	Maximal underforming [mm]	Maximal overforming [mm]	Mean absolute deviation [mm]	Minimal thickness in the last intermediate stage [mm]	Minimal final thickness [mm]
1	0	30	30	49.02	3.01	-0.31	0.80	0.68	0.51
2	10	20	30	50.26	3.04	-0.28	0.80	0.68	0.50
3	0	40	40	37.52	2.60	-0.45	0.73	0.66	0.50
4	10	30	40	36.28	2.66	-0.39	0.74	0.67	0.50
5	20	20	40	32.42	2.74	-0.36	0.74	0.67	0.50
2 stages	10	0	10	50.39	2.58	-0.41	0.63	0.62	0.47
Single-stage	0	0	0	56.40	2.31	-0.43	0.82	-	0.52

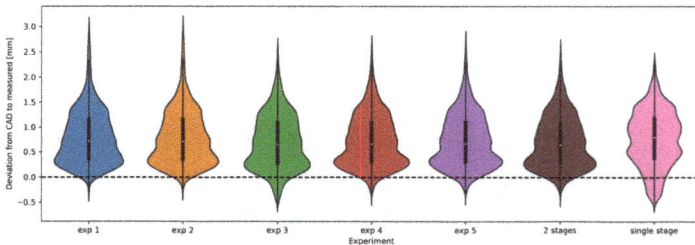

Fig. 3. Violin plot of the deviation distributions for the final shapes of all experiments.

Thickness distributions. Table 2 shows the minimal thickness for the last intermediate shape and the final full shape for all experiments. Since the last intermediate stage is kept constant amongst all experiments, the thickness distributions of the last (or second) intermediate stage in the three-stage experiments can be compared to the first (and only) intermediate stage of the two-stage experiment. The intermediate shape of the two-stage experiment is therefore a single-stage outcome. As can be seen in Table 2, the minimal thickness for the multi-stage experiments is 0.66-0.68 mm, which differs from the lower minimal thickness in the single-stage intermediate shape of 0.62 mm. Fig. 4 shows the thickness distributions across the intermediate shapes. Here, it is clear that experiments 1 and 2 lead to a more uniform thickness distribution in the second stage, compared to the intermediate step of the two-stage experiment. On the other hand, even though the minimal thickness is higher in the second intermediate stage of experiments 3, 4 and 5, compared to forming it in one step, it does not lead to a more uniform thickness distribution across the final part.

When studying the thickness distributions and minimal thickness of the final stage for all experiments in Fig. 5 and Table 2, the more uniform thickness distributions of the previous intermediate stage do not lead to an improved minimal thickness and more uniform distributions in the final stage. Even more, the minimal thickness of the final stage is almost the same for all experiments, including the single-stage experiment. However, the minimal thickness of the two-stage experiment is slightly lower than the others, and the single-stage experiment slightly higher. No harsh conclusions can be made here, since the differences are close to or within the measurement accuracy, which is around 15 micrometers for each side of the part. Together with this measurement accuracy – which is worse for the thickness measurements than for the geometric accuracy, since the thickness calculations are based on two measurements – the process

Sheet Metal 2025   Materials Research Forum LLC
Materials Research Proceedings 52 (2025) 76-84   https://doi.org/10.21741/9781644903551-10

repeatability should also be considered. A small repeatability test is performed by forming the single-stage experiment two times, and measuring the two resulting geometries immediately after each other while keeping the measuring conditions constant (same calibration, temperature, scanner path, filtering, etc.). In this small process repeatability study the minimal thickness was 0.52 mm for both, but showed a difference of 20 micrometers between the mean thicknesses (with 0.85 versus 0.87 mm). Considering this, the minimal thickness after the final stage of all experiments do not deviate significantly from each other.

Hence, the study shows that the initial advantage of more uniform thickness distributions and a higher minimal thickness, resulting from the optimized first shape, vanishes after forming a second intermediate shape that focuses on a better geometric accuracy. The strategy of combining the optimization for geometric accuracy with the optimization for more uniform thickness distributions did still result in a trade-off between the two, with no significant thickness improvement in the final stage for any of the optimized strategies. Hence, the additional first stage for better thickness distributions is unnecessary in this case.

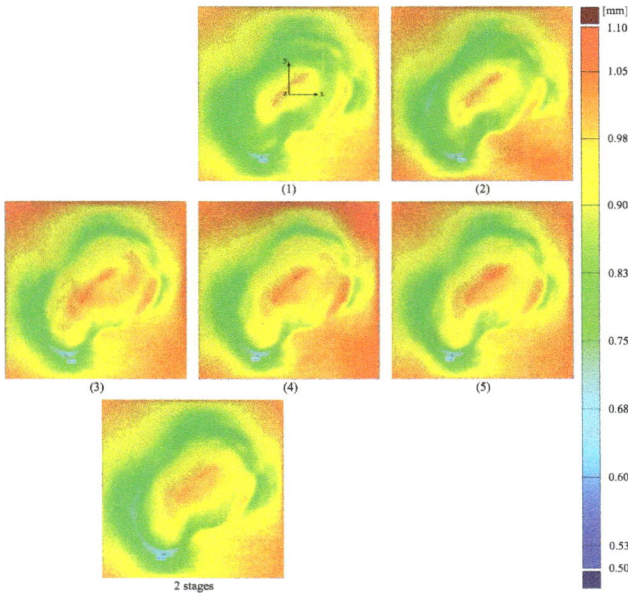

Fig. 4. Measured thickness distributions of the second stage for all three-stage experiments, compared to the first stage of the two-stage experiment. In all of these experiments, the same final intermediate shape is formed (see Fig. 1 and Table 1).

Sheet Metal 2025

Materials Research Proceedings 52 (2025) 76-84

Materials Research Forum LLC

https://doi.org/10.21741/9781644903551-10

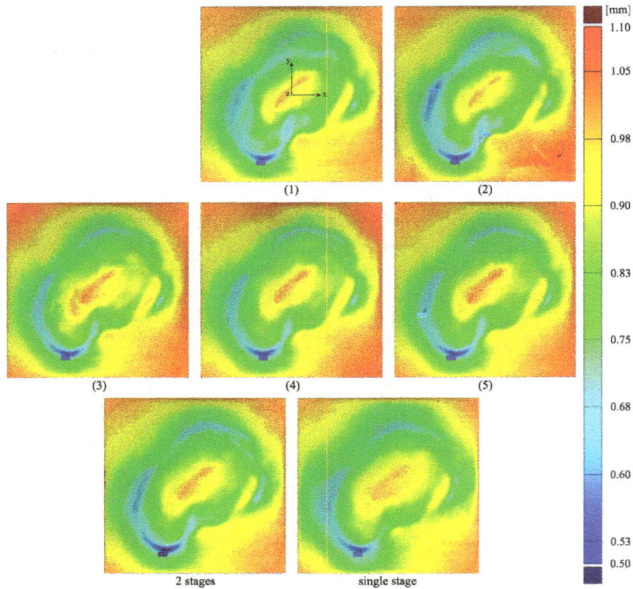

Fig. 5. Thickness distributions of the final shape for all experiments.

Summary

Multi-stage forming has been widely accepted as a way to enhance the process limits of SPIF. This study builds further on a previous work that showed the influence of the intermediate shape design when forming a cranial implant in two stages compared to single-stage forming, in terms of geometric accuracy and thickness distributions. However, the previous study showed that a trade-off should be made to achieve optimal thickness distributions or geometric accuracy when designing the intermediate shape. This follow-up study was conducted to tackle the trade-off by using a first, further optimized stage to achieve optimal thickness distributions, followed by a second stage optimized for better geometric accuracy.

The key findings are:
- In terms of geometric accuracy, the sheet has less outspoken features after forming it in one stage compared to any of the proposed multi-stage forming strategies. This shows in the deviation distributions and in the mean absolute deviation of the final shapes. On the other hand, the underforming of the part is substantially more in the multi-stage forming experiments.
- When comparing the last intermediate step of the three-stage experiments, the minimal thickness is higher in these multi-stage experiments than when this intermediate stage is produced in one step. Even though more uniform thickness distributions can be observed for some experiments, this is not reflected anymore after forming the final shape. Hence, the last step, which was optimized to improve the geometric accuracy, resulted in similar thickness distributions of the final shape for all experiments. Unfortunately, the proposed multi-stage strategies with one or two intermediate stages did not significantly improve the process window.

- A trade-off between optimization for geometric accuracy and thickness distributions remains. Out of the five three-stage experiments, the two shapes with the biggest height difference performed best in terms of thickness distributions in the last intermediate stage, although this effect vanished after forming the last step. The geometric accuracy, on the other hand, was for these experiments less good compared to the other three-stage experiments and the two-stage experiment.

In conclusion, adding a third stage optimized for thickness distributions is in this case not deemed necessary, as it did not significantly improve the final results and even resulted in a lower geometric accuracy. Future work will explore cases exceeding the critical forming angle and will focus on the combination of compensation and multi-stage forming to address the trade-offs and further enhance process performance.

Acknowledgment

Marthe Vanhulst was supported by a Predoctoral Strategic Basic Research Fellowship of the Research Foundation—Flanders (FWO) with project 1S47622N.

References

[1] J. R. Duflou *et al.*, "Single point incremental forming: state-of-the-art and prospects," Nov. 01, 2018, *Springer-Verlag France*. https://doi.org/10.1007/s12289-017-1387-y.

[2] P. Eyckens, S. He, A. Van Bael, P. Van Houtte, and J. Duflou, "Forming Limit Predictions for the Serrated Strain Paths in Single Point Incremental Sheet Forming," pp. 141–147, 2007.

[3] Z. Cheng, Y. Li, C. Xu, Y. Liu, S. Ghafoor, and F. Li, "Incremental sheet forming towards biomedical implants: a review," *Journal of Materials Research and Technology*, vol. 9, no. 4, pp. 7225–7251, Jul. 2020. https://doi.org/10.1016/J.JMRT.2020.04.096.

[4] G. Ambrogio, L. De Napoli, L. Filice, F. Gagliardi, and M. Muzzupappa, "Application of Incremental Forming process for high customised medical product manufacturing," *J Mater Process Technol*, vol. 162–163, no. SPEC. ISS., pp. 156–162, May 2005. https://doi.org/10.1016/J.JMATPROTEC.2005.02.148.

[5] R. Araújo, P. Teixeira, L. Montanari, A. Reis, M. B. Silva, and P. A. Martins, "Single point incremental forming of a facial implant," *Prosthet Orthot Int*, vol. 38, no. 5, pp. 369–378, 2014. https://doi.org/10.1177/0309364613502071.

[6] L. C. Hieu *et al.*, "Integrated Approaches for Personalised Cranio-Maxillofacial Implant Design and Manufacturing," vol. 27, Springer, Berlin, Heidelberg, 2010, pp. 119–122. https://doi.org/10.1007/978-3-642-12020-6_29.

[7] G. Ambrogio, R. Conte, L. De Napoli, G. Fragomeni, and F. Gagliardi, "Forming approaches comparison for high customised skull manufacturing," *Key Eng Mater*, vol. 651–653, pp. 925–931, 2015. https://doi.org/10.4028/www.scientific.net/KEM.651-653.925.

[8] H. Vanhove, Y. Carette, S. Vancleef, and J. R. Duflou, "Production of thin Shell Clavicle Implants through Single Point Incremental Forming," *Procedia Eng*, vol. 183, pp. 174–179, 2017. https://doi.org/10.1016/J.PROENG.2017.04.058.

[9] P. K. Bhoyar and A. B. Borade, "The use of single point incremental forming for customized implants of unicondylar knee arthroplasty: a review," *Research on Biomedical Engineering*, vol. 31, no. 4, pp. 352–357, Oct. 2015. https://doi.org/10.1590/2446-4740.0705.

[10] J. R. Duflou, A. K. Behera, H. Vanhove, and L. S. Bertol, "Manufacture of Accurate Titanium Cranio-Facial Implants with High Forming Angle Using Single Point Incremental Forming," *Key Eng Mater*, 2013. https://doi.org/10.4028/www.scientific.net/KEM.549.223.

Sheet Metal 2025   Materials Research Forum LLC
Materials Research Proceedings 52 (2025) 76-84   https://doi.org/10.21741/9781644903551-10

[11] M. I. Fahad, A. Z. M. Shammari, and R. binti M. Nasir, "A Study on the Effect of Multistage Toolpath in Fabricating a Customized Cranial Implant in Incremental Sheet Metal Forming," *Al-Khwarizmi Engineering Journal*, vol. 19, no. 3, pp. 72–87, 2023. https://doi.org/10.22153/kej.2023.03.003.

[12] M. Vanhulst, S. Waumans, H. Vanhove, and J. R. Duflou, "Investigating intermediate shapes for multi-stage forming of cranial implants," *J Manuf Process*, vol. 127, pp. 1–8, Oct. 2024. https://doi.org/10.1016/J.JMAPRO.2024.07.126.

Joining

Sheet Metal 2025
Materials Research Proceedings 52 (2025) 86-92

Materials Research Forum LLC
https://doi.org/10.21741/9781644903551-11

Investigation failure behavior in the shear tensile test with respect to the arrangements of clinched joints

Eugen Wolf[1,a*] and Alexander Brosius[1,b]

[1]TUD Dresden University of Technology, Chair of Forming and Machining Technology, George-Bähr-Straße 3c, 01069 Dresden, Germany

[a]eugen.wolf1@tu-dresden.de, [b]alexander.brosius@tu-dresden.de

Keywords: Joining, Sheet Metal, Clinching

Abstract. This paper focuses on the failure behavior of specimens with various configurations of clinched joints under shear tensile loading. The primary objective is to assess the influence of the joining direction and the spatial arrangement of clinched joints on their mechanical performance. A number of experiments was conducted, focusing on three clinched joints arranged in different configurations, each varying in terms of joining direction and spacing. These configurations were subjected to shear tensile tests, with force-displacement curves recorded for each sample to provide a detailed characterization of their structural response. The experimental findings indicate that the specific arrangement of the clinched joints, in terms of joining direction, has a marginal impact on the overall failure behavior. This suggests that intricate modifications to the joining direction are unnecessary to achieve improved mechanical performance in such applications. These results offer valuable insights for the design of clinched joint assemblies, indicating that simplified joining strategies may suffice without compromising structural integrity under shear loading.

Introduction

Collaborative Research Center (CRC285), entitled "Method Development for Mechanical Joinability in Versatile Process Chains", the German Research Foundation provides the opportunity to examine clinching as a subject of investigation within the framework of basic research. As defined in DIN 8593 of the German Institute for Standardization, clinching is categorized within the fifth group of joining processes. This mechanical joining process creates a permanent connection through plastic deformation without additional components such as screws or rivets, thereby allowing for the joining of similar or dissimilar sheets without damaging their surfaces. Based on [1] CRC285 explores scientific methods for versatility in three areas of mechanical joinability: joining suitability, joining safety and joining possibility. The research outlined in [2] on joining safety focuses on the optimal development of materials, structures and processes to ensure reliable connections, considering efficiency and load-bearing capacity to develop robust and dependable solutions. According to [3] most joints in vehicle bodies are subjected to mainly static shear loads. This raises the question of how the overall stiffness of the body is influenced by varying the joining direction and the arrangement of clinched joints. The shear tensile test is used to investigate the failure behavior of joined specimens, considering various types of failure. Based on reference [4] a clinched joint may fail in a number of ways. This can occur through a pure neck fracture, through pure unbuttoning or through mixed failure behavior. The latter can be further divided into two subcategories: firstly, a neck fracture with plastic deformation and secondly, unbuttoning with a crack just on one side of the neck. Unlike the findings in [5], the concept of failure behavior can be defined in a more expansive manner. Prior to the occurrence of cracks or fractures, plastic deformation results in a loss of geometric integrity. Even after the load has been removed, the deformation remains undesirable. The post-failure behavior of joined specimens can be delineated into three stages: beginning of plastic deformation, crack formation without fracture and complete separation of the joined specimens.

Content from this work may be used under the terms of the Creative Commons Attribution 3.0 license. Any further distribution of this work must maintain attribution to the author(s) and the title of the work, journal citation and DOI. Published under license by Materials Research Forum LLC.

Sheet Metal 2025
Materials Research Proceedings 52 (2025) 86-92

Materials Research Forum LLC
https://doi.org/10.21741/9781644903551-11

Experimental Setup

The reproducibility of the experimental investigation is ensured through the utilization of the TruMatic 1000 fiber punch laser machine from TRUMPF for the preparation of specimens for the shear tensile test. In accordance with the findings reported in reference [6], the implementation of automated specimen preparation and subsequent clinched joints necessitates an adjustment to the existing machine utilized for punching and forming operations. In order to achieve this, the typical clinching tools were combined with those provided by the machine manufacturer. Similar to a conventional clinching tool, it comprises a die with a diameter of 8 mm and a depth of 1.6 mm (TOX BD8016) combined with a punch of 5 mm in diameter (TOX A50100). An inverted arrangement of punch and die is also possible, which offers advantages for alternative specimen configurations. This indicates that the punch can move in the negative z-direction from above at the upper sheet and in the positive z-direction from below at the lower sheet. The use of these two combinations of punch and die allows for the change of joining direction without the need to turn the sheets over. However, the sequence of the clinched joints is not part of this investigation.

Specimen Preparation. The shear tensile test specimens are composed of two sheets of the same material, HCT590X+Z, with a thickness of $t = 1.5$ mm and a zinc coating. This steel has a two-phase structure consisting of a ferritic base structure with an embedded martensitic second phase. This dual-phase steel exhibits a markedly low yield strength ratio in conjunction with a remarkably high tensile strength and pronounced work hardening. This material is distinguished by its favorable cold formability and is commonly employed in the automotive industry, where clinching is also utilized. The specimens are designed with specific geometric dimensions to arrange clinched joints in a 3 x 5 matrix, while ensuring the necessary minimum distance of 20 mm between them. Notably, there are several differences compared to specimens defined by DVS/EFB 3480-1 of the German Association for Welding and Allied Processes. To accommodate the placement of three clinched joints in close proximity and at a perpendicular angle to the load direction, the width of the specimen has been increased by 15 mm. The specimen exhibits an overlap length of 84 mm, which allows for the arrangement of five clinched joints in a linear configuration, parallel to the load direction and positioned one behind the other. In order to carry out further investigations, the arrangement of the clinched joints on the specimens will be reorganized by extension or modification. These specimens are designed to ensure that the tests conducted are comparable to those planned. Fig. 1 shows the variation 5 of the specimen.

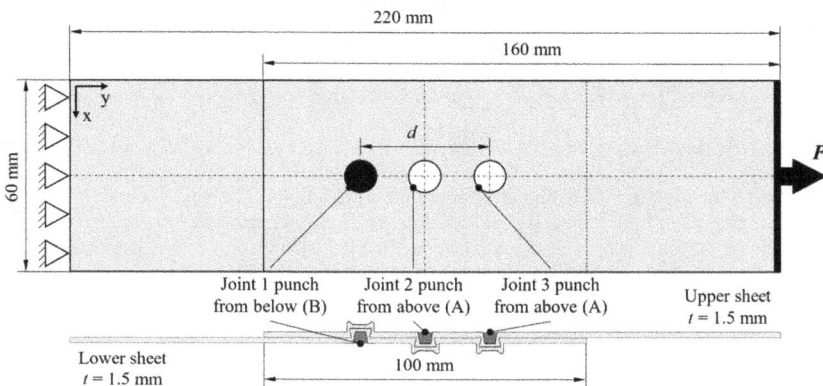

Fig. 1: Example of a specimen under shear loading

Sheet Metal 2025

Materials Research Forum LLC

Materials Research Proceedings 52 (2025) 86-92

https://doi.org/10.21741/9781644903551-11

A specimen is joined with three clinched joints, each of which can be varied in two joining directions. Depending on the combinatorics, eight variations are possible. However, these eight variations can be reduced to three through the utilization of the symmetry inherent to the specimen's geometry. The symmetry of the specimen clamping in the shear tensile test results in an equally symmetric load. Tab. 1 shows all eight variations, including corresponding illustrations and substitutions. The substituted variations are shaded in gray.

Tab. 1: Variation of joining direction from above (A) or from below (B)

Variation	Joint 1	Joint 2	Joint 3	Illustration	Substitution
1	A	A	A		Variant 1
2	A	A	B		Variant 2
3	A	B	A		Variant 3
4	A	B	B		Variant 2
5	B	A	A		Variant 2
6	B	A	B		Variant 3
7	B	B	A		Variant 2
8	B	B	B		Variant 1

The eight variations can be replaced by three variants. Variant 1 substitutes variation 1 and variation 8 (with a 180° rotation around the x-axis). Variant 2 substitutes variation 2 and variation 4 (with a 180° rotation around the x-axis), as well as variation 5 (with a 180° rotation around the x-axis and a 180° rotation around the y-axis) and variation 7 (with a 180° rotation around the y-axis). Variant 3 substitutes variation 3 and variation 6 (with a 180° rotation around the x-axis). In the process of reducing the number of variations, the focus is on the arrangement of the clinched joints rather than the specimens as a whole, as the overlap length is sufficiently large. This reduction in the number of variants not only minimizes the effort required for specimen production and testing but also conserves resources and reduces waste. In consideration of the three resulting variants, the spacing parameter d is introduced, which enables the distance between clinched joints to be modified. This parameter pertains to the distance between the two outer clinched joints. Two discrete values, $d = 40$ mm for variants 1-3 and $d = 80$ mm for variants 4-6, yield six distinct variants.

Specimen Testing. The purpose of this study is to investigate the failure behavior of the joined specimens with the shear tensile test. Prior to conducting the test, it is imperative to ascertain that each clinched joint adheres to the stipulated requirements based in [7] on three geometric features: bottom thickness, neck thickness and interlock. The bottom thickness can be evaluated without causing damage to the material. In accordance with [8], a stable joining process may also be indicative of compliance with the tolerance range of the other two geometric features, provided that the bottom thickness is within the specified tolerance range. Accordingly, the specimens must be subjected to a shear tensile test in order to determine force-displacement curves. This quasistatic and destructive test, as defined by the German Association for Welding and Allied Processes, is conducted with a position-controlled load at a velocity of 5 mm/min. The shear tensile test was performed using the Inspekt 250 tensile testing machine from HEGEWALD & PESCHKE and the test was considered complete once the specimen had undergone complete separation. In order to

Sheet Metal 2025   Materials Research Forum LLC
Materials Research Proceedings 52 (2025) 86-92   https://doi.org/10.21741/9781644903551-11

ensure a static evaluation of the investigation, seven specimens were tested for each of the six variants.

Results and Discussion

The force-displacement curves of the six different variants display a remarkable degree of similarity. The curves of one variant are nearly indistinguishable from those of another. To demonstrate this point, all 42 curves are illustrated in a single diagram. Fig. 2 shows the diagram with curves for all six variants, with the locations of maximum force indicated. The maximum forces exhibited by the system are found to be in relatively close proximity to one another, with a mean value of approximately $F = 12.5$ kN.

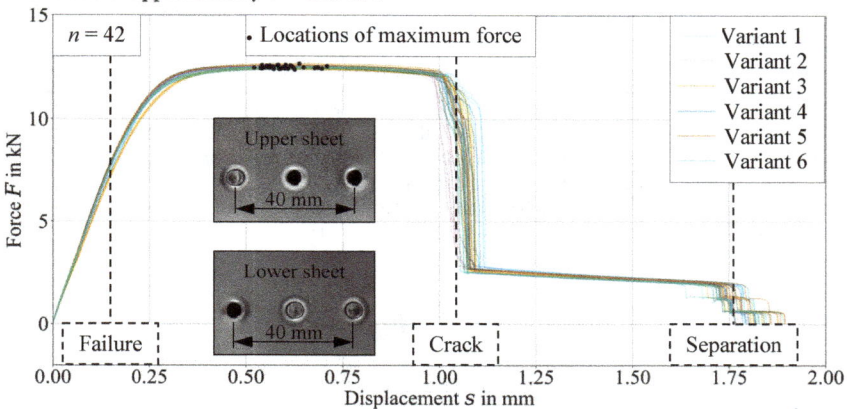

Fig. 2: Force-displacement curves and images of neck fracture for variant 2

In the initial phase, the specimens display linear behavior within the elastic deformation region before failure, leading to the onset of the first stage of post-failure behavior. The plastic deformation occurs at this transition point, as an irreversible alteration in the specimen's geometry has already taken place. Subsequently, the phenomenon of plastic deformation persists and the force attains its maximum value on the plateau that has been reached. The abrupt decline in force marks the onset of the second stage of post-failure behavior. At all three clinched joints, a unilateral neck crack occurs almost simultaneously, without the specimen undergoing complete separation. In accordance with [9] this indicates that the load-bearing capacity is essentially lost. Subsequent loading results in additional plastic deformation, with the buttons deforming to a similar extent. The conclusion of the test is marked by three successive force drops, which result the onset of the third stage of post-failure behavior in the complete separation of the specimens due to the formation of neck fracture at the clinched joints. In reference to [10] the mark shows the crack initiation in a region of the highest global maximum principal stress values. The two parts of the separated specimens display no evidence of bonding, indicating that the rigid geometry of the specimens did not permit any bending during the shear tensile test. Based on [11] the surrounding geometry has an influence on the stress distribution in the clinched joint. In an arrangement of clinched joints arises a mutual influence. This influence depends on distance between the clinched joints. This distance is an indefeasible control variable when designing components because an increase leads to an increase in stiffness of the joined specimen. The linear increase in the quotient of force and displacement in the area of elastic deformation can be employed to ascertain the stiffness of the specimens. Tab. 2 shows the statistical metrics of the stiffness for the variant 1. Based on the slope of the linearized line in the elastic range of the force-displacement curve the

Sheet Metal 2025

Materials Research Proceedings 52 (2025) 86-92

Materials Research Forum LLC

https://doi.org/10.21741/9781644903551-11

stiffness was calculated under application of three different coefficient of determination. An elevated coefficient of determination will result in an increased stiffness, given that the linearized curve will become steeper. This also leads to an increase in the values for deviation and variance.

Tab. 2: Statistical metrics of the calculated stiffness for variant 1

	Coefficient of determination $R^2 = 0.990$	Coefficient of determination $R^2 = 0.995$	Coefficient of determination $R^2 = 0.999$
Average $\bar{c}$	43.47 kN/mm	45.78 kN/mm	48.74 kN/mm
Deviation σ	1.57 kN/mm	1.74 kN/mm	1.99 kN/mm
Variance σ^2	2.48 kN^2/mm^2	3.02 kN^2/mm^2	3.96 kN^2/mm^2

Fig. 3 shows the calculated stiffness of the six variants in comparison with $R^2 = 0.999$, which represents the coefficient of determination. The illustration depicts two key elements. Firstly, it was observed that a change in the joining direction had no significant influence on the stiffness of the specimens. This finding is consistent across specimens with $d = 40$ mm and $d = 80$ mm. Secondly, an increase in the distance between clinched joints resulted in an elevated stiffness of the specimens. All results demonstrate only minor fluctuations.

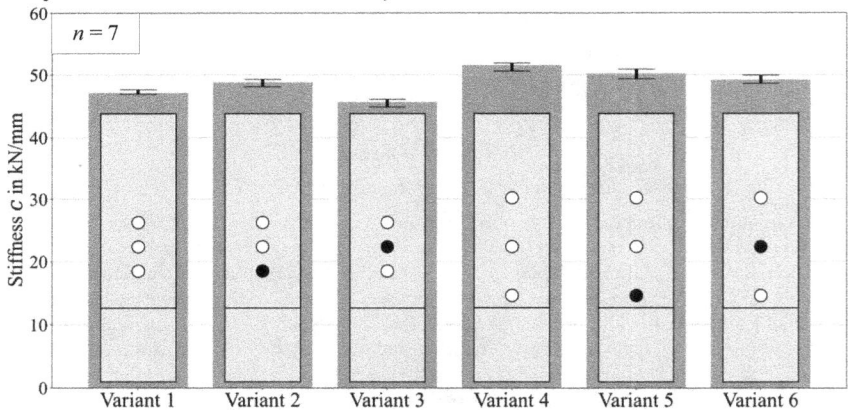

Fig. 3: Calculated stiffness of six variants in comparison

Conclusion and Outlook

The findings of this investigation indicate that the variation of the joining direction within an arrangement of clinched joints has no notable influence on the failure behavior of the specimens that have been joined. The conclusion of this investigation refers exclusively to the specimens that were subjected to investigation, and therefore, it is suggested that the joining direction should remain consistent when designing components, particularly in the event of a shear tensile load. It is evident that a noticeable effect results from changing the clinched joint distance. An increase in clinched joint distance has been observed to result in an increase in the stiffness of the specimens that have been joined, irrespective of the direction in which the clinched joints have been formed. Therefore, it is imperative to consider the impact of the clinched joint distance. These findings highlight the need for further research, including a more detailed examination of the influence of clinched joint distance. An additional avenue for further investigation into the post-failure behavior

Materials Research Forum LLC
https://doi.org/10.21741/9781644903551-11

is to reduce the stiffness of the joined specimens by modifying the specimen geometry. The implementation of the experiment in a finite element simulation can facilitate the application of the novel approach, as generally outlined in [12] for load path analysis. In addition, [13] shows the special application for the load path analysis. This approach enables the redistribution of loads, as described in reference [14], with the objective of averting overloading or oversizing.

Acknowledgement
Funding by the German Research Foundation (DFG) – Collaborative Research Center 285/2 – Project-ID 418701707, subproject B01 is thankfully acknowledged.

References

[1] G. Meschut, M. Merkein, A. Brosius, M. Bobbert, Mechanical joining in versatile process chains, Production Engineering 16 (2022) 187-191. https://doi.org/10.1007/s11740-022-01125-y

[2] S. Martin, A. A. Camberg, T. Tröster, Probability Distribution of Joint Point Loadings in Car Body Structures under Global Bending and Torsion, Procedia Manufacturing 47 (2020) 419-424. https://doi.org/10.1016/j.promfg.2020.04.324

[3] S. Martin, T. Tröster, Joint point loadings in car bodies – the influence of manufacturing tolerances and scatter in material properties, 24th International Conference on Material Forming (2021). https://doi.org/10.25518/esaform21.3801

[4] B. Schramm, S. Martin, C. Steinfelder, C. R. Bielak, A. Brosius, G. Meschut, T. Tröster, T. Wallmersperger, J. Mergheim, A Review on the Modeling of the Clinching Process Chain - Part I: Design Phase, Journal of Advanced Joining Processes 6 (2022) 100133. https://doi.org/10.1016/j.jajp.2022.100133

[5] X. He, Clinching for sheet materials, Science and Technology of Advanced Materials 18 (2017) 381-405. https://doi.org/10.1080/14686996.2017.1320930

[6] C. Steinfelder, A. Brosius, Experimental investigation of the cause and effect relationships between the joint and the component during clinching, Materials Research Proceedings 25 (2023) 147-154. https://doi.org/10.21741/9781644902417-19

[7] C. Steinfelder, J. Kalich, A. Brosius, U. Füssel, Numerical and experimental investigation of the transmission moment of clinching points, IOP Conference Series: Materials Science and Engineering 1157 (2021) 012003. https://doi.org/10.1088/1757-899X/1157/1/012003

[8] H. Peng, C. Chen, X. Ren, J. Wu, Development of clinching process for various materials, International Journal of Advanced Manufacturing Technology 119 (2022) 99-117. https://doi.org/10.1007/s00170-021-08284-9

[9] M. Carboni, S. Beretta, M. Monno, Fatigue behaviour of tensile-shear loaded clinched joints, Engineering Fracture Mechanics 73 (2006) 178-190. https://doi.org/10.1016/j.engfracmech.2005.04.004

[10] L. Ewenz, C. R. Bielak, M. Otroshi, M. Bobbert, G. Meschut, M. Zimmermann, Numerical and experimental identification of fatigue crack initiation sites in clinched joints, Production Engineering 16 (2022) 305-313. https://doi.org/10.1007/s11740-022-01124-z

[11] S. Martin, K. Kurtusic, T. Tröster, Influence of the Surrounding Sheet Geometry on a Clinched Joint, Key Engineering Materials 926 (2022) 1505-1515. https://doi.org/10.4028/p-09md1c

Sheet Metal 2025
Materials Research Proceedings 52 (2025) 86-92

Materials Research Forum LLC
https://doi.org/10.21741/9781644903551-11

[12] C. Steinfelder, A. Brosius, A New Approach for the Evaluation of Component and Joint Loads Based on Load Path Analysis, Production at the leading edge of technology (2020). https://doi.org/10.1007/978-3-662-62138-7_14

[13] C. Zirngibl, S. Martin, C. Steinfelder, S. Wartzack, Methodical approach for the design and dimensioning of mechanical clinched assemblies, Materials Research Proceedings 25 (2023) 179-186. https://doi.org/10.21741/9781644902417-23

[14] S. Martin, C. R. Bielak, M. Bobbert, T. Tröster, G. Meschut, Numerical investigation of the clinched joint loadings considering the initial pre-strain in the joining area, Production Engineering 16 (2022) 261-273. https://doi.org/10.1007/s11740-021-01103-w

Sheet Metal 2025
Materials Research Proceedings 52 (2025) 93-100

Materials Research Forum LLC
https://doi.org/10.21741/9781644903551-12

Non-destructive testing in versatile joining processes

Michael Lechner[1*], Thomas Borgert[2], Matthias Busch[3], Arnold Harms[1],
Pia Holtkamp[4], Fabian Kappe[4], David Römisch[1] and Simon Wituschek[1]

[1]Institute of Manufacturing Technology, Friedrich-Alexander-University Erlangen-Nürnberg,
Egerlandstr. 13, 91058, Erlangen, Germany

[2]Forming and Machining Technology, Paderborn University, Warburger Straße 100, 33098
Paderborn, Germany

[3]Institute of Manufacturing Metrology, Friedrich-Alexander-University Erlangen-Nürnberg,
Nägelsbachstr. 25, 91052 Erlangen, Germany

[4]Laboratory for Materials and Joining Technology (LWF), University of Paderborn, Pohlweg 47-
49, 33098 Paderborn, Germany

*Michael.Lechner@fau.de

Keywords: Computed Tomography, Metrology, Mechanical Joining

Abstract. Mechanical joints are traditionally analyzed through destructive micrograph analysis, which may compromise internal geometry and morphology, as evidenced by radial cracks in semi-tubular self-pierce riveting. In contrast, industrial X-ray computed tomography (XCT) offers a non-destructive method for component diagnosis, providing volumetric insights without damaging the sample and enabling dimensional measurement. The DFG-funded Collaborative Research Center TRR 285 is exploring XCT's application in assessing mechanical joinability across various joining processes and materials, particularly in multi-material systems like steel-aluminum joints. XCT faces challenges in accurately capturing multi-material compositions, leading to artifacts that complicate interface detection. This research aims to validate XCT for joint investigations, yielding quantitative characteristics that surpass those from traditional micrograph analysis.

Introduction

Mechanical joints are usually analyzed for a robust process design by means of a destructive micrographic analysis in a single 2D plane using a microscope [1]. However, the loss of applied stresses and an influence on the internal geometry and morphology cannot be excluded during the sample preparation of the force- and form-fit connection, which has been proven, for example, in the case of semi-tubular self-pierce riveting in the form of preparation-related radial cracks [2].

Industrial X-ray computed tomography (XCT) is still a relatively new technology in component diagnosis [3]. As with the generally known medical diagnostic application, the greatest advantage lies in the volumetric view into the interior without obvious destructive effects and also has the potential to enable dimensional measurement. The DFG-funded Collaborative Research Center TRR 285 "Method development for mechanical joinability in versatile process chains" is investigating this type of non-destructive testing (NDT) in the context of "Metrology for joining processes and joints".

XCT is based on linear attenuation by X-ray absorption in the component, depending on the component thickness and the material. The challenge in computed tomographic investigations of joints therefore lies, in particular, in their possible composition, as different materials are predominantly less accurately captured in a CT measurement than mono-material components [4]. This leads to artifact formation , impairing interface detection and thus geometric evaluation [5]. The joints investigated are multi-material systems with steel-aluminum sheet metal joints that are manufactured using different versatile joining processes. The processes used are the two adapted

Content from this work may be used under the terms of the Creative Commons Attribution 3.0 license. Any further distribution of this work must maintain attribution to the author(s) and the title of the work, journal citation and DOI. Published under license by Materials Research Forum LLC.

Sheet Metal 2025   Materials Research Forum LLC
Materials Research Proceedings 52 (2025) 93-100   https://doi.org/10.21741/9781644903551-12

semi-tubular self-pierce riveting processes with multi-range [6] and tumbling kinematics [7], pin joining [8] and joining with adaptive friction elements [9]. The joining processes are characterized by different properties, both of the joining process and of the joint.

The aim of this work is to demonstrate the possibility of joint investigations based on CT technology and to determine quantitative characteristic values based on volumetric data, which are more accurate or comparable to the characteristic values determined by means of two-dimensional micrograph analysis.

Measuring principle of the XCT

In XCT, a specimen is positioned on a manipulatorsystem in the measuring device between an X-ray tube and a detector (see Fig. 1). During the measuring process, the test specimen is rotated by the manipulator. By connecting a voltage between the glowing cathode and the metallic anode of the X-ray tube, electrons are speeded up and abruptly braked, thus producing X-rays. Projections of the test specimen are generated in each angular position. According to Lambert-Beer's law (Equ. 1), X-ray radiation is absorbed during the penetration of material depending on the transmission length x and the material-specific absorption coefficient µ. An integration image is created at the detector by converting the residual energy I(x) into electrical signals.

$$I(x) = I_0 * e^{-\mu x} \tag{1}$$

A volumetric model of volume pixels (voxels) is then created from the projections in the reconstruction. Usually, reconstruction algorithms based on the filtered back projection are used. Depending on the resolution of the detector in the greyscale range (e.g. 16 bit), the generated model can reveal local information non-destructively through dynamic contrast adjustments. The generation of sectional images also enables measurement in the image – according to [10] due to the information combination regarding in the reconstruction process correctly termed as measurement "at the image" [10].

For XCT, the extraction of a surface is an essential step to perform dimensional measurements on features of the part surface, as with tactile coordinate measuring devices [5]. Therefore, it is necessary to find a suitable filter for the reconstructed volume that converts the grey values of the reconstructed volume into a binary data set that correctly defines the component surface according to the geometric feature size on the component surface. The process described is shown again graphically in Fig. 1.

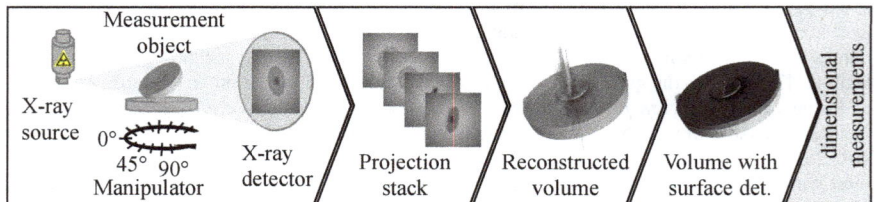

Figure 1: Measuring principle of the XCT using the example of a riveted joint

The common method is using a threshold. The grey value information of each voxel is converted into a histogram depending on the occurrence on the ordinate with increasing grey values on the abscissa. This results in characteristic accumulations which represent the background and the component, with exactly the grey value between the peaks is used as the threshold value. While the described method for the determination of the surface corresponds to the generally accepted standard, the determination of the surface for multi-material joints is not resolved in a

Sheet Metal 2025
Materials Research Proceedings 52 (2025) 93-100

Materials Research Forum LLC
https://doi.org/10.21741/9781644903551-12

method that can be applied in all cases. One option is to extend the local adaptive method by adding another threshold value between a possible additional peak in the grey scale histogram.

Another option for surface detection is the use of the 3DPM method, which detects component edges on the basis of sectional images in the reconstructed volume using prewitt and median filtering, which is then used as a starting value in the iterative surface detection process in VGStudio [5]. This method is also intended to extract surfaces for demanding multi-material components. In contrast to tactile coordinate measuring technology, the replacement surface can then be scanned and dimensionally measured using a large number of antaspots.

Experimental setup of the non-destructive testing
Using each of the versatile joining processes specified, three multi-material joints are manufactured from the steel HCT590X+Z with a density of 7.85 kg/dm^3 and 1.5 mm sheet thickness and an aluminum alloy EN AW-6014 with a density of 2.80 kg/dm^3 and 2.0 mm sheet thickness. The joints are measured using CT and examined using metallographic analysis. To prepare for the measurement of the joints in the CT, the test specimens were cut to a diameter of 25 mm. The measurement itself was carried out in the Metrotom 1500 computer tomography system (from Carl Zeiss AG, Oberkochen, Germany) using a detector with a resolution of 2048 pixels x 2048 pixels, with one pixel measuring 200 μm x 200 μm. For fixation in the CT, the test specimens were positioned in a weakly X-ray absorbing plastic foam under an angle of about 0° to reduce the transmission lengths in the joint.

The CT system parameters used for joint inspection are 225 kV acceleration voltage, 118 μA (set at 100 μA) tube current, 2050 projections with 2000 ms integration time using a frame averaging of 3 and signal amplification of 16. A pre-filter of 0.25 mm copper is used to reduce beam hardening artifacts. The resulting voxel size is 14 μm with a spotsize of 22 μm and the scattered radiation is corrected by the system. Afterwards the reconstructed volumes generated from measurements are processed in VGStudio MAX Version 2022.2 (VGS).

Joining processes, measurement strategies and versatility
In the following chapter, the joining technologies used as well as the geometric parameters and, for selected quality parameters, their significance with regard to the joint strength of the respective joint are presented. The measurement strategies used are subsequently presented, the visualization of the individual joints made possible by CT is demonstrated and the dimensionally determined measured values of the micrograph analysis are compared and discussed. A versatile joining process is the tumbling superimposed semi-tubular self-pierce riveting, shown in Fig. 2.

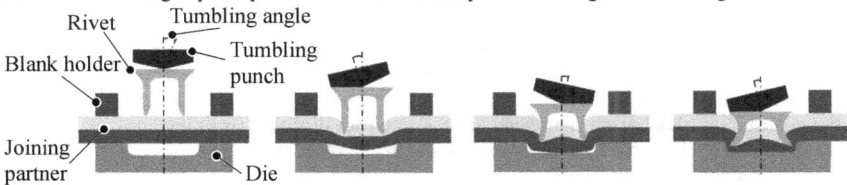

Figure 2: Tumbling superimposed self-piercing riveting process according to [7]

In this process, the punch with linear movement is replaced by a tumbling punch derived from the conventional riveting process. This punch can be adjusted by a tumbling angle and moved in freely configurable tumbling kinematic models. This enables precise control over the material flow, allowing the geometric joint properties to be specifically tailored. First, the two joining partners are clamped and the joining partner on the punch side is cut by driving in the rivet. Once the joining partner is completely cut through and the rivet begins to spread in the die-side sheet, the tumbling actuators are activated. As soon as the rivet is fully set, the tumbling actuator is reset and the punch is released. By being able to control the tumbling strategy, the geometric joint properties can be specifically influenced and the number of rivet and die changes can be reduced with varying boundary conditions. As with all mechanical joining processes, samples of joints are taken during process design using micrographs. This involves creating joints with different parameters and adjusting the geometric characteristics. The superimposition of the tumbling process induces stresses in the joint, which are released during the separating cut for the macrograph and cause a change in the geometric properties of the investigated joint. This process of releasing stresses in the joint can lead to incorrect depictions of the geometric joint formation. CT measurements provide an opportunity to eliminate this influence, as fully intact joints can be analyzed. Fig. 3 shows the CT images of a tumbling-superimposed joint.

Figure 3: CT-measurements a)/b) and macrograph c) of a tumbling superimposed self-piercing riveting joint

Another versatile joining process is the versatile self-piercing riveting. This enables the joining of different sheet thickness combinations without adjusting the die and rivet combination to increase tool actuation in combination with a multi-range rivet. The process sequence is shown in Fig. 4.

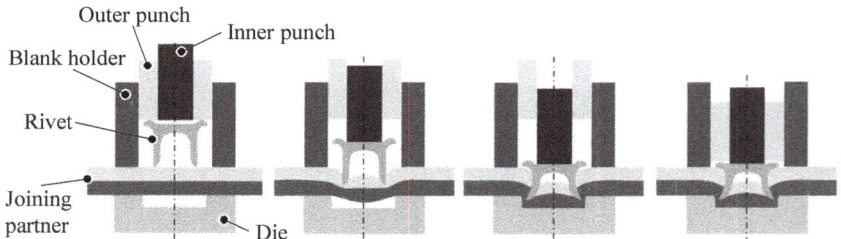

Figure 4: Versatile self-piercing riveting process

Firstly, the rivet and the sheets are positioned, and the sheets are fixed by the blank holder. In the second process step, the rivet penetrates the sheet metal on the punch side through the feed of the inner punch, punches through it and forms the interlock by spreading. After the setting process, the outer punch enables the rivet head to be formed for different sheet thicknesses. The last step describes the backstroke of the two punches and the blank holder. The surface of the riveted joints was determined using the 3DPM method. Fig. 5 shows the result after surface determination. To determine the minimum die-side material thickness, 1000 probing points were generated on the die side of the joint. The rivet foot end position was generated using probing points on a circular free-form surface using small expand (VGS). The shortest distance between the probing points of the respective data sets of the two surfaces was determined with Matlab and used as the minimum die-side material thickness t_r. The undercut was determined in the 3D volume using Gaussian circle fitting. The undercut u_G (Equation 2) is half the difference between the fitted Gaussian circle diameter at the material transition of the rivet between the sheets d_t and the fitted Gaussian circle root diameter d_f.

Sheet Metal 2025   Materials Research Forum LLC
Materials Research Proceedings 52 (2025) 93-100   https://doi.org/10.21741/9781644903551-12

$$u_G = \frac{(d_t - d_f)}{2} \qquad (2)$$

An exemplary comparison of a sectional view from the macrograph and that of a CT measurement with surface detection can be seen in Fig. 5. The figure shows the differences significantly that can be taken from a 2D and a 3D measurement.

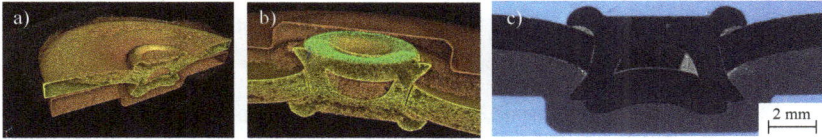

Figure 5: CT-measurements a)/b) and macrograph c) of a joint manufactured with the versatile self-piercing riveting process

The dimensional measurement values determined for the adaptively manufactured rivets and for the tumbled rivets are shown in Fig. 6. Comparing the undercut values of the manufactured riveted joints determined using macrograph with the undercut values determined using CT reveals only limited agreement. The CT values differ of up to 167 µm. One reason attributed to the CT technology and the joint design is that the transition area of the two joining partners at the rivet penetration at which the Gaussian circle is fitted may be too small. At the same time, the Gaussian circle fit is a type of fitting averaging. For the evaluation of the joint, however, it must be noted that the undercut is predominantly underestimated, which provides additional reliability for the load-bearing capacity. The missing trend of overestimating the undercut is non-existent in the tumbling superimposed process, as there is a comparatively high scatter in the process.

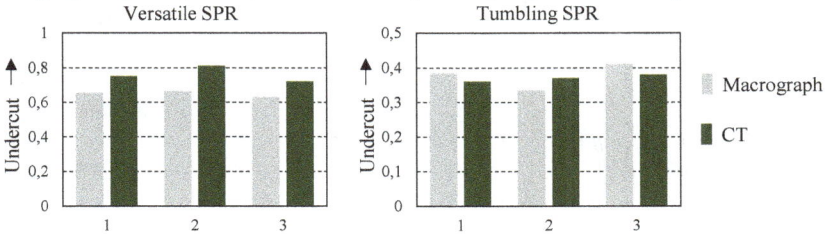

Figure 6: Comparison of macrograph and CT-measurement of three identical SPR joints on the undercut

Another versatile joining process is the Friction-Spun Joint Connectors process (FSJC), shown in Fig. 7. The production of joints using FSJC with different material and thickness combinations has proven to be effective, resulting in strong joints. The production of joints using FSJC involves a two-step process that begins with thermo-mechanical forming and is followed by joint production. In the initial step, a round bar made of aluminum or steel is clamped in a high-speed spindle, which can reach speeds of up to 15,000 rpm. During this phase, friction generated by the spinning bar creates heat, which reduces the material's strength, allowing it to be shaped into the desired FSJC geometry using a sintered carbide tool known for its durability and heat resistance. Once the FSJC is formed, it is utilized in the next stage to create the joint. The FSJC is driven into the joint while rotating and creates an undercut. The process is characterized by reaching high load-bearing capacities in shear and cross tensile test.

Sheet Metal 2025
Materials Research Proceedings 52 (2025) 93-100

Materials Research Forum LLC
https://doi.org/10.21741/9781644903551-12

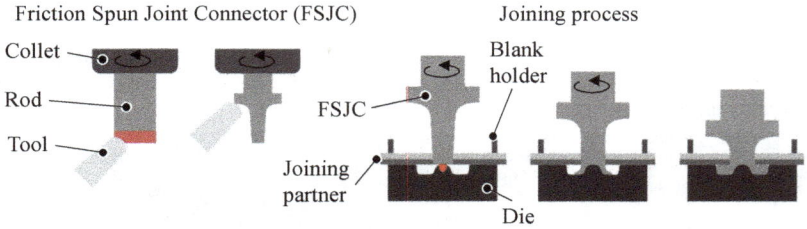

Figure 7: Friction-spun Joint Connectors joining process

In terms of steel content, the FSJC joints have a high proportion compared to the other joints. As a result, there is a reduction in contrast between the weaker X-ray absorbing aluminum and the background, causing a unclear detection of multi-material surface determination using CT with the given device settings. However, only the steel content is relevant for determining the inner and outer closing head diameter, using a higher threshold value at the valley before the steel peak was used for localization-adaptive surface detection.

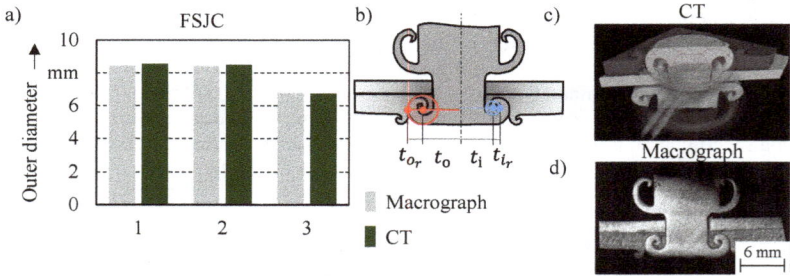

Figure 8: Geometric properties of three identical FSJC joints a), measuring points b), a CT-measurement of an FSCJ connection c) and a section view of a macrograph d)

The diameters were evaluated on the basis of the CT data by adapting two gates to the outermost and inner curves of the closing head. Fig. 8 shows the geometric joint properties of the outer radius, a section view of the macrograph and the CT measurement. The measurement results outline a similar picture of the joint quality, whereby the measured value deviation of the different measurement methods was 52 µm on average and 93 µm at most. A variation in measured values caused by stress relief can therefore be virtually ruled out. Pin joining is a new type of versatile joining process that can join dissimilar materials such as steel and aluminum, as well as metal and fiber-reinforced plastics, without the use of auxiliary joining parts [8]. Here, mostly cylindrical pin structures are extruded from the sheet metal plane using a forward extrusion process and subsequently used for joining. Various joining strategies can be used depending on the joining task, enabling a wide range of joining applications. The pin extrusion and joining process is shown in Fig. 9. One joining strategy is direct pin pressing, in which the extruded pin is pressed directly into the joining partner. This results in an upsetting of the pin, which creates an undercut inside joining partner. In this case, it is necessary for the pin to have a higher strength than the joining partner to be able to penetrate the sheet metal.

Sheet Metal 2025   Materials Research Forum LLC
Materials Research Proceedings 52 (2025) 93-100   https://doi.org/10.21741/9781644903551-12

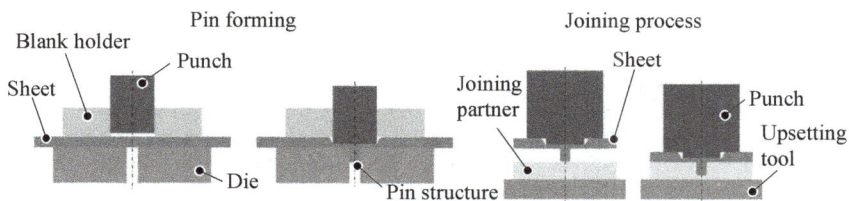

Figure 9: Pin joining process

Caulking can also be used as a joining strategy, where the pin is inserted through a pre-punched joining partner and upset on the head side. Compared to direct pin pressing, such a pronounced strength gradient between the joining partners is not necessary, as the pin does not need to penetrate the sheet metal. Thus, joining of materials of the same strength is possible. However, caulking requires further process steps for producing the pre-punched holes in the joining partner. Nevertheless, pin joining has a number of advantages that underline its versatility. For example, the pin height can be used not only to react to variations in the thickness of the joining partners, but also to adjust the load-bearing capacity of the joint to the requirements, as a higher pin height can achieve greater strength through increased work hardening in the pin. In addition, the number of pins can also be used to respond to the load-bearing capacity requirements, as the strength scales with the number of pins or the cross-section area.

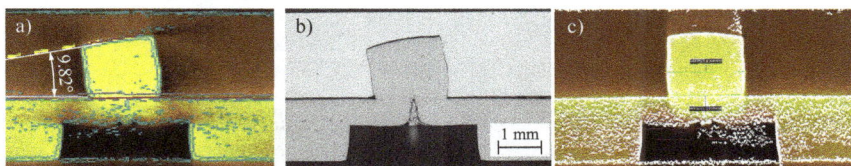

Figure 10. Macrograph b) and CT-measurements a)/b) of a pin joint with 2.0 mm initial pin height

Since pin buckling can occur in pin joints depending on the pin height and material combination used, it is important that these joints can be analyzed Conventional macrographs, however, suffer from limitations since they are only able to show one specific cutting plane. This limitation poses the risk that defects or faults in the joints, which would only be visible in other planes, remain unidentified. As shown in Fig. 10, the pin with a hight of 2.0 mm buckled during the joining process. Although this buckling is visible in this specific micrograph, there is a risk of incorrect assessment if the cut had been made in a different plane. A macrograph created at a 90° offset, could lead to an erroneous positive result, as the bend would not be visible. In contrast, the CT measurement enables a comprehensive three-dimensional view of the joint, which allows a more precise characterization. The deformation of the pin can be clearly recognized in the CT image, and the tilting and bending are quickly and reliably visible. This represents a significant advantage of CT measurement over the conventional method.

Summary

In summary, it can be stated for the examined joints that the CT measurement of geometric joint properties of mechanical joints, manufactured using versatile processes, provides a great advantage for the determination. In particular, the non-destructive character of the measurement method enables the detection of the present properties without affecting the joint. The investigations have shown that the deviations from the macrographs and the CT measurements vary considerably for different joining processes. This can be explained by the fact that individual processes introduce higher residual stresses into the joint, which cause a change in the geometric properties when cutting to produce the macrograph. This cannot occur when measuring using CT. However, CT has the disadvantage that it is strongly influenced by artifacts and the surrounding material. Thus, the contours must be determined using algorithms, which also introduce uncertainty into the measurement method. This means, that the possibility of measuring the geometric properties is significantly improved by CT measurements, but these are not useful or even necessary in all cases. This depends on what can be determined with the examination. In particular, the more elaborate description of the geometric properties of versatile joints, the increased number of properties, makes it possible to achieve a more detailed description of the joint using CT. These include, for example, asymmetries in semi-hollow self-piercing riveted joints, the orientation of the pins in the joint or the distribution of the steel content in an FSJC joint.

Funding
This work was funded by the German Research Foundation (DFG) within the scope of the transregional Collaborative Research Centre 285 on Method development for mechanical joinability in versatile process chains Project-ID TRR 285/2 – 418701707.

References
[1] Li, D., Chrysanthou, A., Patel, I., Williams, G., 2017. Self-piercing riveting-a review. Int J Adv Manuf Technol 92 (5-8), 1777-1824. https://doi.org/10.1007/s00170-017-0156-x

[2] Johannes Eckstein, 2009. Numerische und experimentelle Erweiterung der Verfahrensgrenzen beim Halbhohlstanznieten hochfester Bleche.

[3] Chiffre, L. de, Carmignato, S., Kruth, J.-P., Schmitt, R., Weckenmann, A., 2014. Industrial applications of computed tomography. CIRP Annals 63 (2), 655-677. https://doi.org/10.1016/j.cirp.2014.05.011

[4] Withers, P.J., Bouman, C., Carmignato, S., Cnudde, V., Grimaldi, D., Hagen, C.K., Maire, E., Manley, M., Du Plessis, A., Stock, S.R., 2021. X-ray computed tomography. Nat Rev Methods Primers 1 (1). https://doi.org/10.1038/s43586-021-00015-4

[5] Busch, M., Hausotte, T., 2022. Application of an edge detection algorithm for surface determination in industrial X-ray computed tomography. Prod. Eng. Res. Devel. 16 (2-3), 411-422. https://doi.org/10.1007/s11740-021-01100-z

[6] Kappe, F., Bobbert, M., Meschut, G., 2021. New Approach for Versatile Self Piercing Riveting: Joining System and Auxiliary Part. KEM 883, 3-10. https://doi.org/10.4028/www.scientific.net/KEM.883.3

[7] Wituschek, S., Lechner, M., 2022. Investigation of the influence of the tumbling angle on a tumbling self-piercing riveting process. Proceedings of the Institution of Mechanical Engineers, Part L: Journal of Materials: Design and Applications 236 (6), 1302-1309. https://doi.org/10.1177/14644207221080068

[8] Kraus, M., Frey, P., Kleffel, T., Drummer, D., Merklein, M., 2019. Mechanical joining without auxiliary element by cold formed pins for multi-material-systems, Proceedings of the 22nd international esaform conference on material forming, Vitoria-Gasteiz, Spain. 8-10 May 2019. https://doi.org/10.1063/1.5112570

[9] Wischer, C., Wiens, E., Homberg, W., 2021. Joining with versatile joining elements formed by friction spinning. Journal of Advanced Joining Processes 3, 100060. https://doi.org/10.1016/j.jajp.2021.100060

[10] Verein Deutscher Ingenieure, 2019. Accuracy of coordinate measuring machines - Characteristics and their testing: Form measurement with coordinate measuring machines.

Sheet Metal 2025
Materials Research Proceedings 52 (2025) 101-108

Materials Research Forum LLC
https://doi.org/10.21741/9781644903551-13

Analysis of the binding mechanisms depending on versatile process variants of self-piercing riveting

Stephan Lüder[1,a*], Pia K. Holtkamp[2,b], Simon Wituschek[3,c], Mathias Bobbert[2,d], Gerson Meschut[2,e], Michael Lechner[3,f] and Hans C. Schmale[1,g]

[1]Institute of Manufacturing, TUD Dresden University of Technology, Helmholtzstraße 10, 01069 Dresden, Germany

[2]Laboratory for material and joining technology, Paderborn University, Pohlweg 47-49, 33098 Paderborn, Germany

[3]Institute of Manufacturing Technology, Friedrich-Alexander-Universität Erlangen-Nürnberg, Egerlandstraße 13, 91058 Erlangen, Germany

[a]stephan.lueder@tu-dresden.de, [b]holtkamp@lwf.upb.de, [c]simon.wituschek@fau.de, [d]bobbert@lwf.upb.de, [e]meschut@lwf.upb.de, [f]michael.lechner@fau.de, [g]hans_christian.schmale@tu-dresden.de

Keywords: Joining, Sheet Metal, Self-Piercing Riveting

Abstract. The constantly increasing demand for climate protection and resource conservation requires innovative and versatile joining processes that improve adaptability to the joining task and robustness to enable flexible manufacturing on a production line. Therefore, the versatile SPR (V-SPR) and tumbling SPR (T-SPR) were developed. Using the example of a mixed material combination HCT590X+Z ($t_0 = 1.0$ mm) / EN AW-6014 T4 ($t_0 = 2.0$ mm), these processes were examined and compared with regard to the binding mechanisms form closure and force closure using micrographs, non-destructive resistance measurements and destructive torsion tests. For this purpose, a new sample geometry was defined, and the methods were adapted to the SPR process variants.

Introduction

The European Green Deal, which was introduced by the European Union in 2019 with the objective of achieving climate neutrality by 2050, requires the implementation of a multitude of initiatives across nearly all sectors of society and the economy [1]. The transportation sector, and in particular individual transportation, plays a vital role in contributing to greenhouse gas emissions. A strategy for reducing emissions within the mobility sector that is both feasible and effective involves the reduction of moving masses [2]. In order to achieve this, lightweight construction techniques are employed, which enable the fulfillment of mechanical requirements while maintaining a lower overall weight. One such technique is the use of lightweight materials, wherein specific materials are selected to meet performance criteria optimally. The combination of different materials results in the creation of hybrid structures, commonly referred to as multi-material systems. These systems are characterized not only by their different materials, each with different mechanical properties, but also by their geometric diversity. This approach allows the reduction of weight and the fabrication of assemblies that are tailored to specific requirements. Nevertheless, the fabrication of multi-material systems necessitates the implementation of robust joining processes, and the currently available methods frequently encounter limitations due to the inherent complexities involved [3]. Mechanical joining processes are frequently employed for these multi-material structures; however, they tend to be constrained in their versatility to varying process conditions and external disturbances. There is also a need for flexible production for smaller, individual batch sizes of one product (e.g., car) on a large-scale production line. Among

Content from this work may be used under the terms of the Creative Commons Attribution 3.0 license. Any further distribution of this work must maintain attribution to the author(s) and the title of the work, journal citation and DOI. Published under license by Materials Research Forum LLC.

Sheet Metal 2025   Materials Research Forum LLC
Materials Research Proceedings 52 (2025) 101-108   https://doi.org/10.21741/9781644903551-13

the most commonly used joining techniques is semi-tubular self-piercing riveting, which is distinguished by its relatively inflexible nature, as it demonstrates limited responsiveness to variations in process and disturbance parameters.

These novel process variants of self-piercing riveting are investigated with the aid of experimental methods, non-destructively by measuring the electrical resistance and destructively by means of torsion testing with regard to the force closure.

Self-piercing Riveting Processes. Self-piercing riveting (SPR) is a mechanical joining process with an auxiliary joining part that enables the joining of multi-layer joints of different types under the condition of two-sided accessibility [4]. The process sequence is divided into four continuous steps (Fig. 1). The first step describes the positioning of the sheets between the blank holder and the die. The blank holder force-fixes the sheets in place. In the second process step, the rivet penetrates and pierces the punch-sided sheet due to the continuous punch movement. The rivet shank absorbs the resulting slug of the punch-sided sheet. In the third process step, the die-sided sheet is displaced into the die, and the rivet begins to spread and upset, forming the interlock. The result is a form- and force-fit joint.

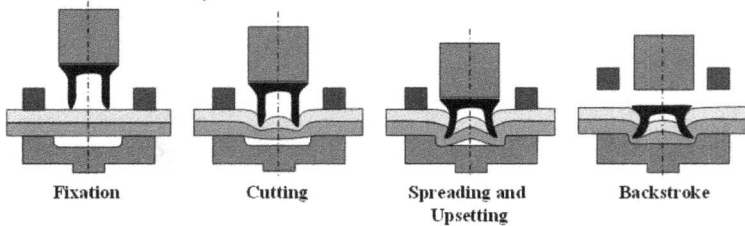

| Fixation | Cutting | Spreading and Upsetting | Backstroke |

Fig. 1: Process sequence of the SPR process according to DVS/EFB Merkblatt 3410 [4].

A significant disadvantage of the SPR is the rigid tool systems, which prevents it from reacting to changing boundary conditions, such as changes in sheet thickness. For this, it is necessary to adapt the rivet-die combination. To expand the use of self-piercing riveting across a broader range of applications, research is being conducted on more versatile and adaptable process variations.

One of these process variants is the versatile self-piercing riveting (V-SPR) [5]. This enables the connection of joints with different boundary conditions by integrating advanced tool actuator technology with a multi-range capable rivet. The process sequence is illustrated in Fig. 2. First, the sheets are fixed by the blank holder. The difference to the SPR is the split punch actuator. The rivet penetrates the sheet metal through the feed of the inner punch, pierces the punch-sided sheet and spreads. The outer punch enables the subsequent head forming to the respective punch-side sheet thickness.

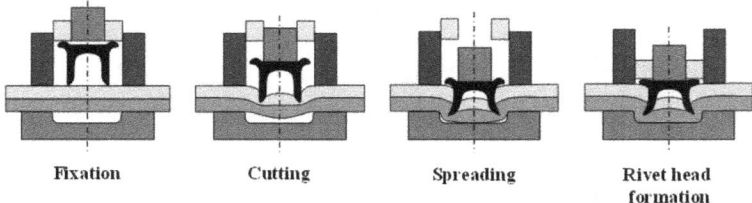

| Fixation | Cutting | Spreading | Rivet head formation |

Fig. 2: Process sequence of the V-SPR [5].

A highly adaptable joining technique is the tumbling superimposed semi-tubular self-pierce riveting process. In this method, the traditional linear punch utilized in semi-tubular self-piercing

Sheet Metal 2025
Materials Research Proceedings 52 (2025) 101-108

Materials Research Forum LLC
https://doi.org/10.21741/9781644903551-13

riveting is substituted with a tumbling punch. This tumbling punch can be modified through an adjustable tumbling angle α and can operate within various configurable kinematic models. Such versatility enables a considerable impact on the control of material flow, allowing for precise adjustments to the geometric properties of the joint. The sequence of the process is illustrated in Fig. 3. Initially, the two sheets to be joined are secured in place, and the rivet penetrates the punch-sided joining partner. Once the rivet completely pierces the joining partner and begins to expand within the die-sided sheet, the tumbling actuators are activated. Upon achieving the whole setting of the rivet, the tumbling actuator is deactivated, and the punch is released. By manipulating the tumbling strategy, it is possible to precisely control the geometric characteristics of the joint while also increasing the possible joining tasks with one rivet-die combination.

Fig. 3: Process sequence of the T-SPR.

Experimental Setup

The investigation of the various self-piercing riveting processes with regard to the binding mechanism of the force closure is carried out using two different test methods. As part of the non-destructive testing, the electrical resistance of the joint is first measured and then tested destructively in a torsion test. Both test methods are carried out on a newly defined specimen geometry, which enables the resistance measurement and the torsion test to be carried out on the same specimen. The specimen geometry is shown in Fig. 4.

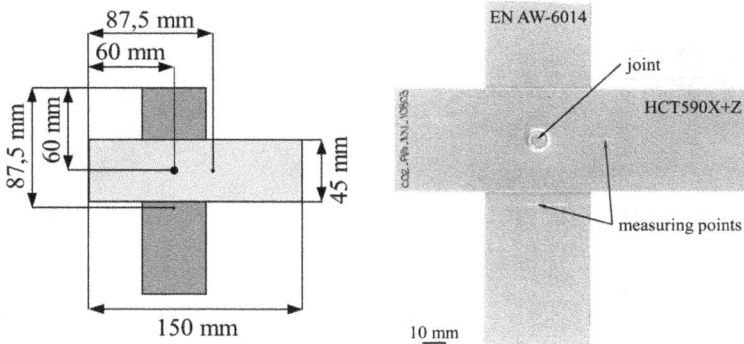

Fig. 4: Specimen geometry for resistance measurement and torsion test.

Fig. 5 shows the experimental setup for measuring the electrical resistance. A microohmmeter is used to determine the resistance of the joint using the four-wire measurement method [6,7]. For the clinching process, it has already been proven that this method is suitable for evaluating the force closure of the joint, as a high contact force leads to a low electrical resistance R of the joint

[8]. For the evaluation of the measurements on the SPR joints, the electrical conductance G of the joint is specified and calculated as follows.

$$G = \frac{1}{R}. \tag{1}$$

Fig. 5 also shows the experimental setup for the torsion test. The specimen is twisted using a tightening spindle, and the required torque is recorded. The evaluation of the breakaway torque enables the qualitative assessment of the force closure of the joints [9].

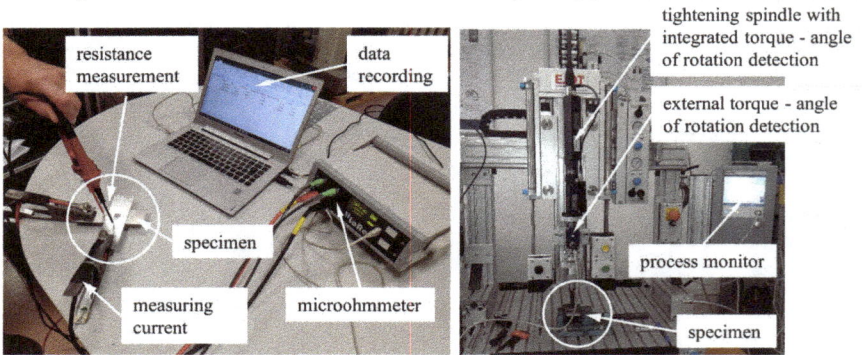

Fig. 5: Experimental setup of electrical resistance measurement (left) and torsion test (right).

Results

For this investigation, the material combination HCT590X+Z ($t_0 = 1.0$ mm) / EN AW-6014 T4 ($t_0 = 2.0$ mm) was joined with four different SPR process variants in delivery condition. Fig. 6 shows the micrographs of the process variants a) SPR, b) V-SPR, c) T-SPR $\alpha = 1°$, d) T-SPR $\alpha = 5°$.

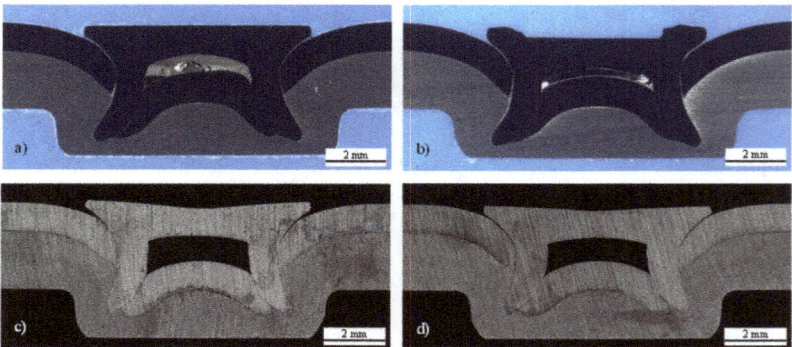

Fig. 6: Micrographs of the process variants a) SPR, b) V-SPR, c) T-SPR $\alpha = 1°$, d) T-SPR $\alpha = 5°$.

The geometric parameters which characterize the form closure of the SPR joints, such as interlock f, minimum die-sided material thickness t_r and rivet head position p_h were measured on 5 samples each and are shown in Fig. 7.

Sheet Metal 2025
Materials Research Proceedings 52 (2025) 101-108

Materials Research Forum LLC
https://doi.org/10.21741/9781644903551-13

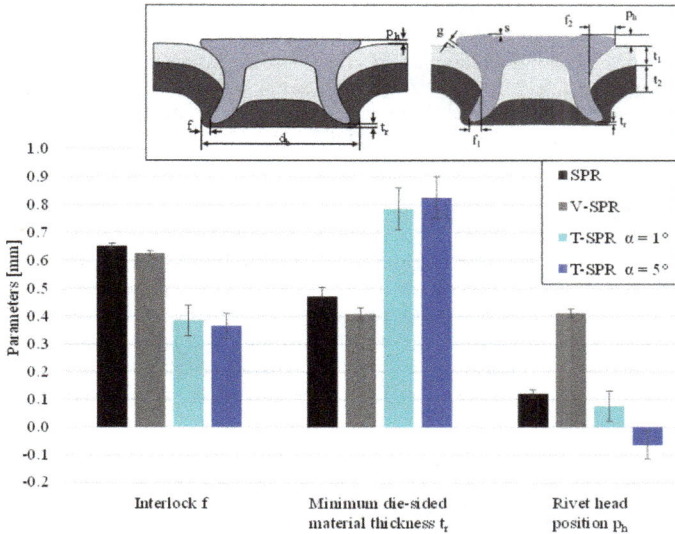

Fig. 7: Geometric parameters of SPR joints (n = 5).

SPR and V-SPR have a significantly higher interlock than the T-SPR process variants. This is due to the fact that the joining speed (SPR: 20 mm/s; V-SPR: 5 mm/s) is significantly higher with these processes, and the rivet penetrates deeper as a result, which leads to a higher rivet foot spread. With the tumbling process variants, the joining process is incremental, with a joining speed of just 5 mm/min. As a result, the rivet does not penetrate as deeply into the material, which is visible in the higher minimum die-sided material thickness. The influence of the tumbling angle α is particularly evident in the rivet head position. The larger tumbling angle leads to better contact of the rivet head.

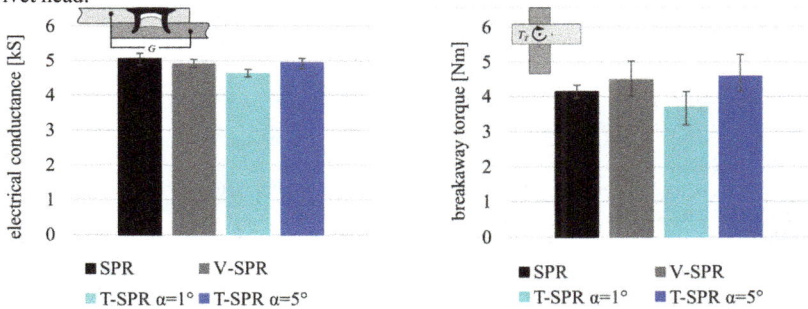

Fig. 8: Electrical conductance (left) and breakaway torque (right) of the SPR joints (n = 5).

The Results for the electrical conductance of the SPR joints are shown in Fig. 8. Five samples were tested for each SPR process. The highest conductance was determined in conventional SPR.

Sheet Metal 2025

Materials Research Proceedings 52 (2025) 101-108

Materials Research Forum LLC

https://doi.org/10.21741/9781644903551-13

However, all process variants are at a similar level and exhibit only slight variations. This proves a successful process design and no serious deviations with regard to force closure or gap formation within the test series. The most striking feature here is also the influence of the tumbling angle. In addition to the improved contact of the rivet head, the larger tumbling angle also results in an increased radial material flow, which leads to an improvement in the force closure, which is expressed in the improved conductance of the joint.

Fig. 8 also shows the breakaway torque, which is the maximum testing torque required to twist the sheets. The conventional SPR is a good reference and shows the lowest scatter. The versatile SPR processes show a slightly higher scatter, which results from the fact that these processes are more variable and, therefore, more complex, and the rivet has to be positioned manually. The highest breakaway torque is achieved with the V-SPR and T-SPR processes with a tumbling angle of 5°. The influence of the tumbling angle on the force closure of the joint can also be clearly identified in the torsion test, as the lowest breakaway torque was measured at a tumbling angle of 1°. In Fig. 9, the testing torque curve in the torsion test of the SPR joints for each sample is shown.

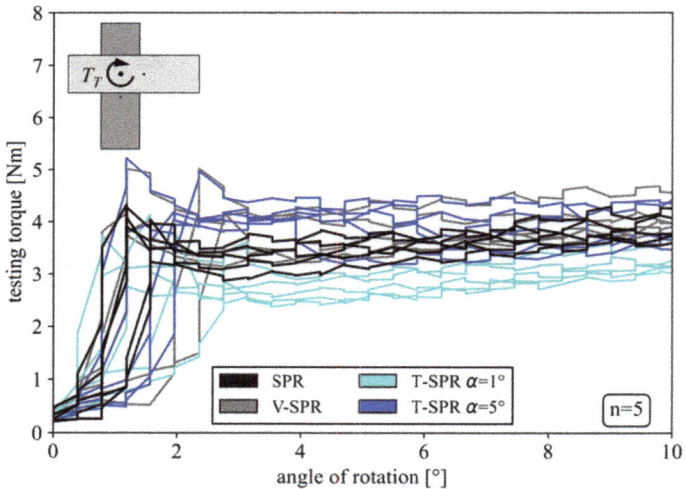

Fig. 9: Testing torque - angle of rotation curve in torsion test of the SPR joints.

The process-dependent scattering is visible here, which can be caused, for example, by a slightly off-center positioning of the rivet. The deviations in the angle of rotation, on the other hand, can be attributed to tolerances or clearance fit in the experimental setup. However, no significant outliers were detected in the test series, which proves the robustness of the processes.

In Fig. 10, the curves of the 5 samples were averaged to illustrate the qualitative comparison of the curves of the process variants. Due to the angular offset of the individual curves, however, the breakaway torque is somewhat distorted because maxima are partially reduced by averaging.

Sheet Metal 2025   Materials Research Forum LLC
Materials Research Proceedings 52 (2025) 101-108   https://doi.org/10.21741/9781644903551-13

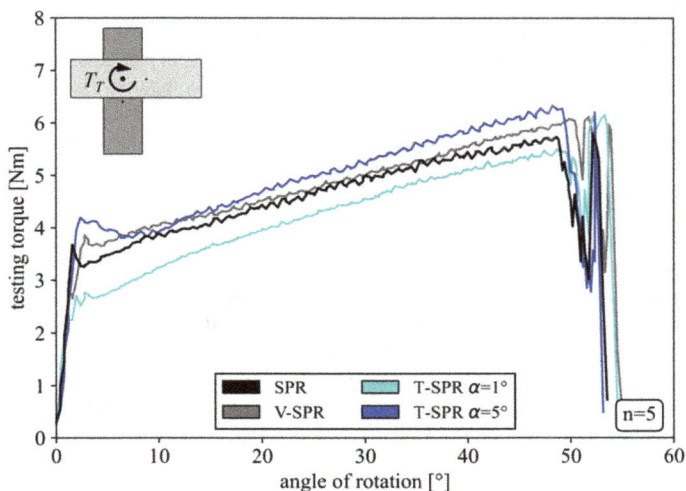

Fig. 10: Averaged testing torque - angle of rotation curve in torsion test of the SPR joints.

All process variants show a typical progression in the torsion test. After reaching the breakaway torque, the torque initially decreases and gradually increases as rotation continues. The SPR joint remains intact for all samples. The increase in torque can be explained by the grooving of particles on the surfaces and damage to the surfaces.

Regarding the force closure in the torsion test, the T-SPR $\alpha = 5°$ process variant shows the highest results, followed by V-SPR and outperforming the SPR reference along the entire curve. The lowest result in the torsion test is shown by the T-SPR $\alpha = 1°$ variant, which also underlines the thesis of the lower force closure due to the lower radial material flow.

Summary

In this investigation, it was proven that the methods for analyzing the binding mechanisms form closure and force closure, which have already been successfully tested for clinching, could also be adapted to self-piercing riveting. For this purpose, conventional SPR was compared as a reference with the innovative, versatile process variants V-SPR and T-SPR. The mixed material combination HCT590X+Z ($t_0° = 1.0$ mm) / EN AW-6014 T4 ($t_0 = 2.0$ mm) was selected as the joining task. The form closure was analyzed using micrographs. A new sample geometry was defined for the investigation of the influence of the process variant on the force closure, which makes it possible to first determine the electrical conductance of the joint on a sample by means of non-destructive resistance measurement and then to determine the breakaway torque destructively in the torsion test.

All the process variants examined were able to produce joints that met the requirements. This was demonstrated by measuring the geometric parameters to evaluate the form closure. The form closure is primarily characterized by the interlock. However, this cannot be characterized by either the resistance measurement or the torsion test. These methods only allow an assessment of the contact force in the joint and, thus, the force closure. Based on the electrical conductance of the joints, it has already been proven that the tumbling angle α in T-SPR has a significant influence on the force closure. This was also verified by the breakaway torque in the torsion test, which can be attributed to the increased radial material flow that leads to a higher contact force between the rivet and the parts to be joined. The tumbling angle and the adaptation of the rivet geometry have a significant influence on the binding mechanisms, as shown by the versatility of the variants T-SPR $\alpha = 5°$ and V-SPR with their higher averaged testing torques. In particular, the tumbling angle decisively influences the binding mechanisms, as shown in the comparison with $\alpha = 1°$.

Acknowledgment

The funding by the Deutsche Forschungsgemeinschaft (DFG, German Research Foundation) – Project-ID 418701707 – TRR 285/2, subprojects A04 and C02 is gratefully acknowledged. Data regarding the contents of the publication can be requested at www.trr285.de.

References

[1] European Commission, European Green Deal, Brussels, 2021.

[2] B. Bader, E. Türck, T. Vietor, Multi Material Design. A current overview of the used potential in automotive industries, in: K. Dröder, T. Vietor (Eds.), Technologies for economical and functional lightweight design, Springer Heidelberg, 2019, 3-13.

[3] G. Meschut, M. Merklein; A. Brosius et al., Review on mechanical joining by plastic deformation, J. Adv. Join. Process. 5 (2022) 100113. https://doi.org/10.1016/j.jajp.2022.100113

[4] DVS/EFB-Merkblatt 3410: Self-pierce riveting - Overview, Europäische Forschungs-gesellschaft für Blechverarbeitung e.V. (EFB), Düsseldorf, 2020

[5] F. Kappe, S. Wituschek, M. Bobbert, and G. Meschut, Determining the properties of multi-range semi-tubular self-piercing riveted joints, Prod. Eng. Res. Devel. 16 (2022) 363-378. https://doi.org/10.1007/s11740-022-01105-2

[6] S. Nikzad Khangholi, M. Javidani, A. Maltais, X.-G. Chen, Investigation on electrical conductivity and hardness of 6xxx aluminum conductor alloys with different Si levels, Proceedings of the 17th International Conference on Aluminium Alloys 2020 (ICAA17), Grenoble, France, 26-29 October 2020, MATEC Web of Conferences 326 (2020) 08002. https://doi.org/10.1051/matecconf/202032608002

[7] J. Zhang, M. Ma, F. Shen, D. Yi, B. Wang, Influence of deformation and annealing on electrical conductivity, mechanical properties and texture of Al-Mg-Si alloy cables, Mater. Sci. Eng., A 710, (2018) 27-37. https://doi.org/10.1016/j.msea.2017.10.065

[8] J. Kalich, U. Füssel, Influence of the Production Process on the Binding Mechanism of Clinched Aluminum Steel Mixed Compounds, J. Manuf. Mater. Process. 5 (2021) 105. https://doi.org/10.3390/jmmp5040105

[9] J. Kalich, U. Füssel, Design of clinched joints on the basis of binding mechanisms, Prod. Eng. Res. Devel. 16 (2022) 213-222. https://doi.org/10.1007/s11740-022-01108-z

Sheet Metal 2025
Materials Research Proceedings 52 (2025) 109-116

Materials Research Forum LLC
https://doi.org/10.21741/9781644903551-14

SE analysis as a tool for forming and medical technology

Sinan Yarcu[1,a *], Bernd A. Behrens[1,b], Sven Hübner[1,c] and Serdar Yalcin[1,d]

[1]Institute for Forming Technology and Forming Machines, Leibniz University Hannover, An der Universität 2, 30823 Garbsen, Germany

[a]s.yarcu@ifum.uni-hannover.de, [b]behrens@ifum.uni-hannover.de,
[c]huebner@ifum.uni-hannover.de, [d]yalcin@ifum.uni-hannover.de

Keywords: Acoustic Emission, Forming, Process Control

Abstract. Acoustic Emission Analysis (AE Analysis) represents an advanced method of non-destructive testing, enabling real-time detection of damage and defects during material loading. This paper provides a comprehensive overview of the application of AE Analysis in forming technology and its potential for process monitoring and quality assurance. Various research findings are presented, demonstrating the efficiency and accuracy of this method across different industrial and medical sectors.

Introduction

Acoustic emission (AE) analysis is an advanced and unique method of nondestructive testing (NDT). Unlike many other NDT techniques that are performed before or after a material is stressed, AE analysis is typically performed during stress. This makes it possible to detect damage and defects in materials in real time, often long before complete failure of the structure occurs. [1, 2]

The application of AE analysis spans several industries. In aerospace, it is used to monitor the structural integrity of aircraft and spacecraft. In the energy industry, it is used to inspect pressure vessels and pipelines. In the construction industry, it helps to ensure the safety of bridges and buildings. AE analysis is also used in manufacturing and materials research to ensure the quality of materials and to analyze their properties. [1, 2, 3]

A major advantage of AE analysis is the ability to observe damage processes throughout the loading process without interrupting or affecting the loading process. Compared to methods such as ultrasonic testing, which often require interruption of the load and access to all sides of the specimen, AE analysis requires only fixed sensors that do not need to be moved during the test. These characteristics make AE analysis particularly valuable for monitoring processes under real-world operating conditions.

Nevertheless, the application of AE analysis also presents certain challenges. The reproducibility of the tests can be limited by the random nature of the signal sources. Furthermore, the detection and analysis of weak signals requires the utilization of highly sensitive sensors and advanced data processing techniques to ensure the reliability of the resulting data. [1, 2] The operating principle of AE analysis is based on the detection of sound waves generated by the rapid release of energy within a material. The waves are detected by sensors attached to the surface of the object being tested and transmitted to the processing unit. AE analysis is considered 'passive' because the source of the sound waves is within the material and not generated externally. This characteristic distinguishes it from active methods such as ultrasonic testing, where an external signal source is used to identify defects.[1]

Acoustic emission signals are analysed to identify and describe the causes of their occurrence in order to draw conclusions about product or process faults. Various factors such as the characteristics of the measuring system and the choice of acquisition parameters (threshold value, sampling rate, filter settings, etc.) can influence the AE analysis.

Content from this work may be used under the terms of the Creative Commons Attribution 3.0 license. Any further distribution of this work must maintain attribution to the author(s) and the title of the work, journal citation and DOI. Published under license by Materials Research Forum LLC.

AE analysis in forming technology

The potential of acoustic emission analysis has been explored in a number of areas within the field of forming technology. The goal of these investigations was to examine the potential benefits and applications of AE analysis in various forming processes and to improve their efficiency. The following section provides an overview of the research projects that have been conducted in relation to AE analysis.

AE analysis for investigating tribological conditions in forming technology

In forming technology, acoustic emission analysis (AEA) provides a precise way of monitoring tribological processes in real time and identifying wear processes. A study by Kim et. al. on the cold forming process of zinc phosphate/stearate-coated steel shows that AE signals provide valuable information on the condition of the coating and the development of wear. [4]

Acoustic emission analysis has also been shown to be a promising method for real-time monitoring of wear processes in sheet metal forming. In another study by Sindi et al. AE signals during the forming process of AISI 316 stainless steel were analyzed using wavelet packet transform to accurately identify the different wear phases. [5]

Furthermore, the study by Buse et al. showed that a lack of lubrication in the deep drawing process leads to a significant increase in friction and AE activity, indicating an increased risk of surface damage. The results emphasize that reduced lubrication conditions can be identified by AE analysis. [6]

These studies underline the importance of AE analysis as an efficient tool for optimizing forming processes by enabling early detection of wear phases and thus extending the service life of tools and components. In addition to identify wear phases, all studies demonstrate how AE analysis can be used for continuous process monitoring in industrial forming technology. By tracking AE signals in real time, critical conditions such as severe scoring, fractures of friction contact points or insufficient lubrication can be detected early and corrective action can be taken before serious material damage or tool failures occurs. This not only allows precise adjustment of the process parameters, but also improves the efficiency and reliability of the forming process as a whole. The ability to directly correlate AE signals with specific wear mechanisms opens up new avenues for predictive maintenance and process optimization, which can be of great benefit in many industrial applications.

AE analysis in forging technology

Acoustic emission analysis (AEA) has been established in a number of studies as a promising tool for monitoring tool wear and early detection of material defects in metal forming. The study by Hawryluk et al. emphasizes the application of AE in hot die forging technology, with a particular focus on the early detection of material defects and the extension of tool life. The detection of acoustic waves generated by deformations or cracks enables continuous real-time monitoring of the process without the need to interrupt it. [7]

The work of Ha et al. extends the application of AE technology to the monitoring of cracking in aluminium alloys during the upsetting process. [8]

Another important study by Behrens et al. investigates the potential of AE to monitor process deviations during precision forging of aluminium alloys. Their research showed that AE can be successfully used to detect cracking and lubrication problems during hot and cold forging. [9]

Acoustic emission analysis as a method for process monitoring in sheet metal forming

The continuous demand for quality improvements and the use of high-strength and ultra-high-strength steels in sheet metal forming require monitoring methods that can provide information on the damage status of the component during the process. In this case, acoustic emission analysis is a particularly suitable online monitoring method.

Liang et al. have already investigated the application of AEA in sheet metal forming. In the study, two specific forming processes were analyzed more in detail: stretch forming and deep drawing. The study demonstrated that acoustic emissions can be effectively used for process monitoring in sheet metal forming by showing strong correlations with different process states such as yielding, deformation, wrinkling, necking and fracture. [10]

The investigations by Behrens et. al. showed an approach to the analysis of acoustic emissions in sheet metal forming as part of the deep drawing process. AE signals were recorded during the deep drawing process to analyze the influence of materials, geometries and process parameters. [6]

The results of the study show that the material properties have a significant influence on AE activity. The plate geometry also had an influence on the AE activity. Larger rectangular blanks produced higher AE activity than optimized geometries, indicating altered frictional conditions.

The study by Lai et. al. focused on the application of AE technology to monitor the hot forming process of press hardening, a process used to manufacture high-strength parts in the automotive industry. The tools used in this process are exposed to extreme thermal and mechanical stresses, resulting in wear, cracking and surface damage. [11]

AE monitoring made it possible to detect this damage at an early stage before it led to major problems. Of particular note is the ability of AE technology to detect the onset of cracking and adhesion by analysing high-frequency AE signals based on characteristic features. The results of this study showed that AE technology enables a significant improvement in mould monitoring, especially when combined with Industry 4.0 technologies for real-time monitoring and control.

The Sindi et. al. study investigated the application of Wavelet Packet Transform (WPT) to analyze AE signals in detail during the sheet metal forming process. By decomposing the AE signals into different frequency components, the energy distribution in these components could be analyzed and assigned to the corresponding damage mechanisms. [5]

In Baral et al. the AE technology was used to monitor plastic deformation and the occurrence of necking during sheet metal forming. Necking is a critical point in the forming process that often leads to material failure. Using the AE signals, the researchers were able to recognize the onset of necking before fractures occur. [12]

The AE signals showed that the plastic deformation of the material could already be detected by an increase in signal energy in the mid-frequency range. This made it possible to detect material failure at an early stage, which makes it possible to control and optimize the forming process in real time.

Acoustic emission analysis as an inline investigation for clinching

The IFUM also carried out preliminary investigations into crack detection during the clinching of high-strength steels and aluminium die castings and recorded the resulting acoustic emissions. In the course of the preliminary work, crack-free and cracked clinch joints were produced using the materials HX340LAD and AlSi10MnMg as the die-side sheet and HCT780 and CR240BH as the punch-side sheet. A TOX round point and an Eckold H-DF were used to create the clinching points. The sound emissions created during the clinching process were recorded using the AMSY-6 acoustic emission measurement system from Vallen Systeme GmbH and analyzed afterwards. Based on the investigations, it was established that the acoustic emissions occurring during the process can generally be classified into crack-free and cracked clinch joints using AEA, whereby different crack types could always be detected. The representative preliminary test results and the test setup are shown in Fig. 1.

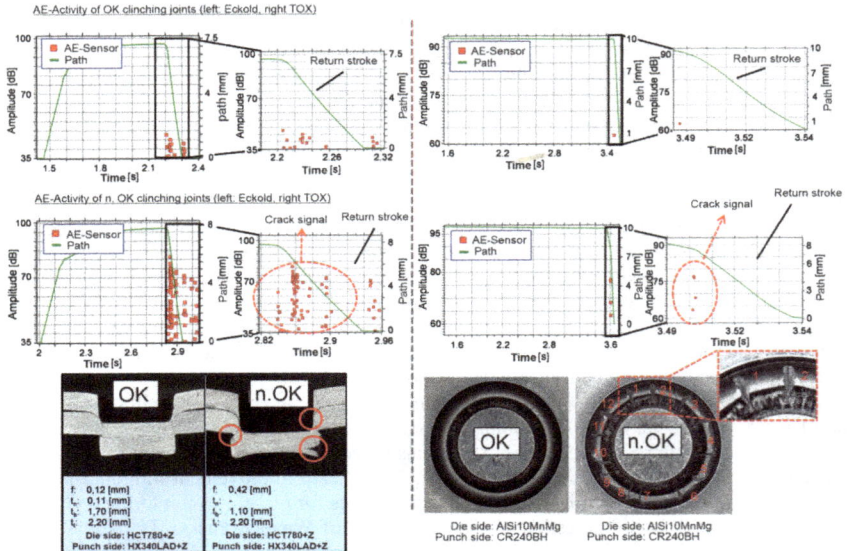

Figure 1: Results of preliminary investigations into crack detection during clinching using acoustic emission analysis

Fig. 1 shows the sound emissions occurring during clinching as amplitude in relation to the time past. The y-axis represents the maximum amplitude in dB of the respective acoustic emissions for a certain time window, the x-axis represents the time required to produce a clinch joint and the right vertical axis represents the path of the clinching press in mm. The sound emissions were not recorded continuously, but in an event-orientated manner. This means that acoustic emissions are only recorded when certain recording parameters are met. As can be seen from the diagram, a significantly increased AE activity manifests itself when the punch is moved out in the case of a cracked clinch joint, which is most likely due to spontaneous pressure reduction. This activity leads to the conclusion that there is one or several cracks. As part of the preliminary work, 3 OK. and 3 n.OK clinch connections were produced. The increased AE activity described above occurred reproducibly in each of the cracked clinch joints. The preliminary investigations carried out show the potential of AEA as a possibility for process monitoring during clinching and provide a good basis for the successful completion of the research project. Process monitoring of the clinching process using acoustic emission technology could therefore not only ensure the quality of the joints, but also protect the tools (punch, die) from damage.

This publication presents two important studies on the optimization of the clinching process, dealing with the monitoring and improvement of material changes and the service life of tools.

The study by Köhler et al. focuses on Transient Dynamic Analysis (TDA), an advanced method for monitoring material changes during the clinching process. TDA is based on the introduction of mechanical waves into the material to analyze its dynamic properties. It detects structural changes, such as cracking, by measuring the changes in frequency, amplitude and mechanical impedance. Unlike acoustic emission analysis (AEA), which passively records sound waves caused by spontaneous cracking or plastic deformation, TDA works actively by specifically introducing vibrations and measuring the reactions of the material. [13]

The study used TDA was used in combination with computed tomography (CT) to monitor the clinching process of aluminium sheets. Mechanical waves were introduced into the material by piezoelectric actuators, while sensors captured the structural changes during clinching. The results showed that TDA is capable of detecting subtle changes in the material at an early stage, such as the formation of cracks. As long as the material remained intact, the amplitudes of the measured vibrations steadily increased; however, when a crack began to form, the vibrations decreased significantly. These results confirm that TDA is an effective method for real-time monitoring of quality deviations and enables the detection of defects during the clinching process.

The study by Džupon et al. focused on the use of PVD coatings to improve the tool life in the clinching process. Various PVD coatings, including ZrN (zirconium nitride), CrN (chromium nitride), and TiCN (titanium carbonitride), were applied to the tools, and their wear behaviour during clinching of galvanized steel sheets was investigated. Finite Element Analysis (FEA) was used to identify the areas with the highest mechanical stresses during clinching and to simulate tool wear. The experimental tests confirmed the FEA predictions: the TiCN coating demonstrated the highest wear resistance, while CrN and ZrN exhibited defects in the highly stressed areas of the tools. These findings are crucial for optimizing tool life and ensuring quality control in the mechanical joining of lightweight materials. [14]

Together, these two studies provide valuable insights into advanced monitoring and protection methods that can make the clinching process more efficient and sustainable.

AE-Technologies: Application from forming technology to medical technology
After extensive research on the application of AE in metal forming, where AE is used to monitor and analyze material changes during the forming process, a promising approach has now emerged in a completely different field: orthopedics. The experience and knowledge gained in forming technology provide the scientific basis for utilizing AE as a monitoring tool for implants in the human body. Specifically, in total hip arthroplasty (THA), AE can be used for the early detection of implant loosening. This technology transfer between different disciplines opens up new possibilities and highlights the versatility and potential of AE as a diagnostic tool. This publication investigates whether the core principles of AE methods that have proven effective in metal forming can also be applied to medicine to improve the safety and longevity of hip implants. The following section provides a concise overview of previous studies on the application of AE technology in the medical field.

In biosensor AE systems, suitable coupling materials optimize signal transmission by reducing the acoustic impedance difference between the sensor and the target medium. In a study, different coupling media such as liquid, gel-based, dry and semi-dry materials were analyzed to determine the most efficient material for improved transmission in specific applications. [16]

In another study, AEA was used to enable precise monitoring of prosthesis wear by synchronizing AE signals with patient movement data. A validated AE device shows the highest signal activity during the support phase of the gait cycle and provides valuable biomechanical insight for early wear diagnosis of hip prostheses. [17]

Further investigations to optimize the AE technology, alternative coupling materials such as plasticine, wax and hot glue were tested, which are particularly suitable for temporary sensor installations. Plasticine proved to be the best material for reliable signal quality and easy handling in non-damaging applications on the skin. [18]

AEA also shows great potential in medical diagnostics for real-time monitoring of cartilage defects in knee joint arthrosis. This method enables reliable, cost-effective diagnosis and improves treatment options without any health risks for the patient. [15]

AE technology also enables the early detection of microcracks during the preparation of the femoral canal for cementless hip implants in orthopedic surgery. In a cadaver study, it showed

Sheet Metal 2025   Materials Research Forum LLC
Materials Research Proceedings 52 (2025) 109-116   https://doi.org/10.21741/9781644903551-14

potential in terms of real-time monitoring to prevent iatrogenic femoral fractures during surgery and to optimize surgical results. [19]

Based on previous studies and the current problem of a reliable loosening diagnostic gap for hip prostheses, new investigations were carried out within the SFB/TRR-298. When implants are used for joint replacement, loosening of the prosthesis over time is the main cause of component failure. Using conventional diagnostic tools such as X-ray and CT, this problem is often only recognized when the patient suffers of pain. As a result, preventive measures are no longer possible, which means that only a costly revision surgery needs to be performed. In this sense, this study deals with the correlation between axial torques introduced into implants and the resulting relative movements between implant and bone as well as the development of acoustic emission signals.

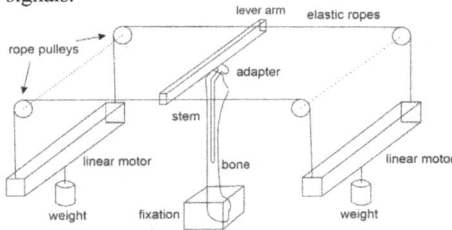

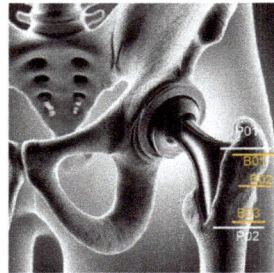

Figure 2: Schematic representation of the rotary measuring machine according to GÖRTZ et. al. [20]

Figure 3: Positions of the measuring pin in the bone (B01-B03) and in the prosthesis (P01, P02)

A test rig was constructed to simulate the physiological load situation at the implant-bone interface and to induce loosening-induced movements. A rotational measuring machine (see Fig. 2) was used to simulate micromovements between the implant and bone. The acquired movement data was precisely recorded using optimized control software based on LabVIEW. Inductive probes were used to measure the relative movements (Fig. 3), and different coupling media such as gel plates, pork and tofu were tested to optimize sound transmission. In a pilot study, a Sawbones femoral artificial bone with an implanted Metha-short-stem prosthesis was used, and the prosthesis was hydraulically inserted into the artificial bone. Relative movements and sound signals were recorded and analyzed at several measuring points. Measurable relative movements between implant and bone could be generated, which led to sound signals. The coupling of the media showed that ultrasound gel and FlexUS gel pads enabled good sound transmission, while pork and tofu exhibited greater signal scattering. FlexUS gel pads proved to be particularly suitable due to their pressure resistance and practicability. In the pilot study, measurable relative movements were detected in the area of primary stability, but the small number of sound signals detected and the small relative movements posed challenges for further research. The correlation between relative movements and acoustic signals was insufficient. Acoustic emission analysis is a promising method for the early detection of prosthesis loosening. Optimization of the coupling media and sensors could further improve the accuracy and reliability of the diagnosis.

Conclusion

This work explores the application of Acoustic Emission (AE) analysis across various industries, especially in forming and medical technologies. AE analysis, a non-destructive testing method, detects defects and damage in real time via acoustic signals during material loading, avoiding process interruption or the need for complete access to the object. With permanently installed sensors, AE analysis is ideal for industrial use, particularly in forming processes such as cold forming, sheet metal forming, and forging, where it monitors tribological conditions and wear. Studies show that AE signals provide insights into surface and internal material conditions, helping to extend tool life, reduce downtime, and optimize lubrication in processes such as deep drawing and precision forging of aluminum alloys.

Another key application area is clinching, a joining process where AE signals are used to distinguish between crack-free and cracked joints. The ability to detect cracking in real time during the clinching process significantly contributes to improving process safety and extending tool life.

AE analysis opens new perspectives in implant monitoring, particularly in orthopedics, making it a promising tool in medical technology. One application of AE is the monitoring of hip prostheses, allowing for the early detection of implant loosening. The use of AE technology enables the timely initiation of preventive measures after identifying complications at the implant interface. Consequently, AE can help extend the lifespan of implants in the human body. Various studies have shown that AE signals, generated by micromovements or microcracks in implants, can be detected long before clinical symptoms appear. The ability for real-time monitoring could play a crucial role in the early detection of implant failure in the future, significantly improving patient care.

Furthermore, AE analysis is applied in medical technology for the early detection of fractures and for monitoring bone healing processes. Particularly in the detection of fractures during surgical procedures and in identifying defects and damage within osteoporotic bones, AE analysis shows great potential. AE analysis faces challenges in signal processing and measurement reproducibility. For example, diagnosing hip prosthesis loosening involves weak signals that require sensitive sensors and advanced data processing. Qualitative analysis provides fast data, while quantitative methods offer deeper insight into crack formation and material failure mechanics.

AE analysis is a valuable tool for monitoring forming and medical technologies, enabling early defect detection for safer, more efficient manufacturing and prolonged implant life. Despite ongoing signal processing challenges, AE analysis offers significant benefits for advancing science and industry in these areas.

References

[1] C.-U. Grosse, M. Ohtsu: Acoustic Emission Testing, Basics for Research - Application in Civil Engineering; Springer Verlag Berlin Heidelberg; 2008

[2] S. Gholizadeh, Z. Leman, B.T.H.T. Baharudin: A review of the application of acoustic emission technique in engineering; Structural Engineering and Mechanics Vol. 54, No. 6; 2015; S. 1075-1095 https://doi.org/10.12989/sem.2015.54.6.1075

[3] S. Gholizadeh, Z. Leman, B.T.H.T. Baharudin: A review of the application of acoustic emission technique in engineering; Structural Engineering and Mechanics; Vol. 54; No. 6; 2015; pp 1075-1095 https://doi.org/10.12989/sem.2015.54.6.1075

[4] J.H. Kim, S. Kim, Y.H. Seo, W. Song: Acoustic emission analysis for monitoring tribological behaviors of steel coated with zinc phosphate/stearate for cold forging; Tribology International 196; 2024; https://doi.org/10.1016/j.triboint.2024.109641

[5] C.T. Sindi, M.A. Najafabadi M. Salehi: Tribological Behavior of Sheet Metal Forming Process Using Acoustic Emission Characteristics; Tribol Lett Vol. 52; 2013; S. 67-79 https://doi.org/10.1007/s11249-013-0193-z

[6] Buse, C.; Huinink, T.; El-Galy, I.; Behrens, B.-A.: Online Monitoring of Deep Drawing Process by Application of Acoustic Emission; 10th International Conference on Technology of Plasticity; 2011; S. 385-389

[7] A. Hawryluk, J. Ziemba, P. Sadowski: A Review of Current and New Measurement Techniques Used in Hot Die Forging Processes; Measurement and Control; Vol 50(3); 2017; pp. 74-86 https://doi.org/10.1177/0020294017707161

[8] M. Ha, J.H. Kim, S. Kim: Crack detection in upsetting of aluminum alloy using acoustic emission monitoring technology; The International Journal of Advanced Manufacturing Technology 124; 2023; S. 2823-2834 https://doi.org/10.1007/s00170-022-10628-y

[9] A. Santangelo, K, Wölki, C. Buse, A. Bouguecha; B.A. Behrens: Potentials of in situ monitoring of aluminum alloy forging by acoustic emission. Archives of Civil and Mechanical Engineering 16. 2016. S. 724-733 https://doi.org/10.1016/j.acme.2016.04.012

[10] A.Liang, D. Dornfeld: Characterization of Sheet Metal Forming Using Acoustic Emission; Engineering, Materials Science; Journal of Engineering Materials and Technology-transactions of The Asme Vol. 112; 1990; pp. 44-51 https://doi.org/10.1115/1.2903185

[11] C. F. Lai, H. I. Wong, C. H. Ng, S. N. M. Yahaya, S. Shamsudi: Review on Acoustic Emission Monitoring System for Hot Stamping Process; Recent Trends in Manufacturing and Materials Towards Industry 4.0. Lecture Notes in Mechanical Engineering. Springer, Singapore; 2021 https://doi.org/10.1007/978-981-15-9505-9_35

[12] M. Baral, A. Al-Jewad, A. Breunig, P. Groche, J. Ha, Y.P. Korkolis, B.L. Kinsley: Acoustic emission monitoring for necking in sheet metal forming; Journal of Materials Processing Tech. 310; 2022, 117758 https://doi.org/10.1016/j.jmatprotec.2022.117758

[13] D. Köhler, R. Stephan, R. Kupfer, J. Troschitz, A. Brosius, M. Gude: Investigations on Combined In-Situ CT and Acoustic Analysis during Clinching; Key Engineering Materials; Vol. 926; 2022; pp 1489-1497 https://doi.org/10.4028/p-32330d

[14] M. Dzupon, L. Kascak, E. Spisak, R. Kubik, J. Majernikova: Wear of Shaped Surfaces of PVD Coated Dies for Clinching; Metals; Vol. 515; 2017; https://doi.org/10.3390/met7110515

[15] M.S. Rashid, R. Pullin: The Sound of Orthopaedic Surgery-The Application of Acoustic Emission Technology in Orthopaedic Surgery: A Review; European Journal of Orthopaedic Surgery and Traumatology; 2014; S. 1-6.

[16] R. Manwar, L. Saint-Martin, K. Avanaki: An introduction to Acoustic Emission; Chemosensors; 2022; S. 181. https://doi.org/10.3390/chemosensors10050181

[17] T. FitzPatrick, G.W. Rodgers, L.J. King, G.J. Hooper: Development and validation of an acoustic emission device to measure wear in total hip replacements in-vitro and in-vivo; Control Engineering Practice; 2017; S. 287-297. https://doi.org/10.1016/j.bspc.2016.12.011

[18] M. Colombo, G. Vezzoli, C. Cinquemani: Frequency response of different couplant materials for mounting transducers; NDT & E International; 2005; S. 451-460.

[19] T. Pechon, R. Pullin, M.J. Eaton, S.L. Evans: Acoustic emission technology can warn of impending iatrogenic femur fracture during femoral canal preparation for uncemented hip replacement. A cadaveric study; Journal of Orthopaedic Research; 2010; S. 287-297.

[20] Görtz, W.; Nägerl, U.; Nägerl, H.; Thomsen, M.: Spatial micromovements of uncemented femoral components after torsional loads, In: Journal of biomechanical engineering (Bd. 124), 2002; H. 6, S. 706-713 https://doi.org/10.1115/1.1517565

Sheet Metal 2025
Materials Research Proceedings 52 (2025) 117-124

Materials Research Forum LLC
https://doi.org/10.21741/9781644903551-15

In situ computed tomography – Analysis of settling effects during single-lap shear tests with clinch points

Daniel Köhler[1,a *], Juliane Troschitz[1,b], Robert Kupfer[1,c] and Maik Gude[1,d]

[1]Institute of Lightweight Engineering and Polymer Technology, TUD Dresden University of Technology

[a]daniel.koehler3@tu-dresden.de, [b]juliane.troschitz@tu-dresden.de, [c]robert.kupfer@tu-dresden.de, [d]maik.gude@tu-dresden.de

Keywords: Joining, Sheet Metal, Computed Tomography

Abstract. In the field of mechanical engineering, destructive tests such as shear tests of mechanical joints are usually followed by imaging methods such as microsectioning or computed tomography (CT). They can help to interpret the measured load-displacement curves, analyze the failure behavior and validate numerical models. However, due to unloading, springback effects and crack closures can occur, which influence the state of the investigated specimen. In this context, in situ CT is able to explore the testing process with a specimen under load avoiding these influences. For in situ CT investigations, the displacement increase is interrupted at certain stop points. While the displacement is kept constant, the CT scan is performed. However, it was observed that the reaction force reduces during CT scanning, e. g. due to settling effects in the test setup. Although in situ CT is established now in research, little attention is paid to the uncertainties which arise from the discontinuous testing procedure. This study systematically explores the impact of these interruptions on the load-displacement behavior and the geometry of clinch points during tensile shear testing. To quantify the influence of the interruptions, loads at defined displacement levels and the final geometry are evaluated statistically. We found, that the load-displacement behavior of both test groups is similar. Despite some small but significant statistical deviations of the loads and the final geometry, our results show that, discontinuous testing has a high level of significance for the phenomena overserved in shear tests with clinch points.

Introduction

Clinching is a mechanical joining method that connects parts by cold-forming, eliminating the need for auxiliary joining elements. During shear testing, the failure behavior of clinched joints is characterized by typical deformation phenomena, such as rotation of the punch sided joining part and neck thinning, visualized in Fig. 1a [1]. To investigate the deformation and failure behavior of such mechanical joints, destructive testing is often followed by imaging methods like microsectioning. The results are used to assess the mechanical performance of the joint and deduct potential improvements [2]. In [3] the cut cross sections of a shear tested clinched joint are used to track the failure mechanisms by investigating the failure modes and the crack propagation paths. This was used to validate numerical models. In [4], shear tests are conducted for validation, however only the measured load-displacement behavior is used exhibiting deviations to the numerical model. Furthermore, the final deformation state of a joint is compared to a numerical model in [5].

However, this approach exhibits crucial disadvantages as there is no insight into the actual deformation phenomena occurring during the test. Moreover, due to unloading, springback effects and crack closures can occur, which influence the investigated state. Consequently, deviations between experiment and numerical models can only be explained to a limited extend. Computed tomography (CT), as a non-destructive imaging method enables the three-dimensional analysis of objects. It can be applied for damage analysis of joints e.g. for self-piercing riveting of

Content from this work may be used under the terms of the Creative Commons Attribution 3.0 license. Any further distribution of this work must maintain attribution to the author(s) and the title of the work, journal citation and DOI. Published under license by Materials Research Forum LLC.

fibre-reinforced plastics (FRP) and aluminium (Al) [6] or the analysis of phenomena in clinch points using FRP joining partners [7]. In order to investigate the deformations of a process, the method can be updated to in situ measurements. Here, the CT system is combined with, for example, a mechanical testing setup. The test is carried out up to specified stop points (SP). At each SP, the displacement is kept constant and a CT scan is conducted. This enables the investigation of the failure behaviour for example of thermomechanically manufactured joints with auxiliary joining element [8] or from clinch points [9]. However, it was observed that the reaction force decreases during CT scanning, likely due to settling effects in the test setup. Although in situ CT is now well-established in research, little attention has been given to these phenomena and the resulting uncertainties arising from the discontinuous testing procedure.

This study systematically explores the impact of the test interruptions on the load-displacement behavior and the geometry of clinch points during tensile shear testing. Clinched single-lap tensile shear specimens of aluminum sheets are used. For reference, the tests are conducted continuously (Conti), followed by tests featuring interruptions (StopGo) at defined SP. To investigate the influence of the interruptions, loads at defined displacements (evaluation points (EP)), and final geometric properties are evaluated statistically.

Materials and Methods

Experimental setup. The tensile shear tests were conducted according to ISO 12996 [1] using specimens of two overlapping 2 mm thick sheets made of aluminum EN AW 6014-T4[1]. They are clinched with the punch A50100 and the die BE8012[2]. The specimen's dimensions generally follow the norm and are shown in Fig. 1c. To fit the clamps in the in situ CT, the specimen width is reduced to 37 mm instead of the standard 45 mm. The clinched aluminum specimen was heat-treated to T6 state at 185 °C for 20 min. Furthermore, two shim plates are bonded onto the ends of the sheets to prevent specimen bending due to clamping. The testing is conducted in a testing machine[3] which is combined with the CT system FCTS 160-IS[4] (Fig. 1b). In this CT system, both the X-ray source FORE 160.01C TT (160 kV, 80 W) and the flat panel detector (3200×2300 pixels, 405 mm×290 mm active area) are mounted on a granite block which rotates around the specimen in order to take radiographs from different angles. The reconstruction uses the Feldkamp algorithm in CERA 2.2.1. and the CT images are evaluated in VG Studio Max 2.2.

Testing procedure. For reference, eight specimens are tested continuously with a cross beam displacement rate of 2 mm/min to a total crossbeam displacement of 1 mm (Conti specimen set, preload 60 N). After unloading the specimen, the clinch point is CT scanned (ex situ) (Table 1). To mimic the in situ CT process, another eight specimens (StopGo specimen set) are tested with process interruptions which is illustrated in Fig. 1a. Here, the crossbeam movement is interrupted and held in position at a displacement of 0.25, 0.50, 0.75 and 1.00 mm (stop points) for 15 min. After specimen unloading, an ex situ CT scan is conducted.

---

[1] Advanz™ 6F-e170, Novelis Inc., Atlanta, USA
[2] TOX PRESSOTECHNIK GmbH & Co.KG, Weingarten, Germany
[3] Z250, ZwickRoell GmbH & Co. KG , Ulm, Germany
[4] FineTec FineFocus Technologies GmbH, Garbsen, Germany

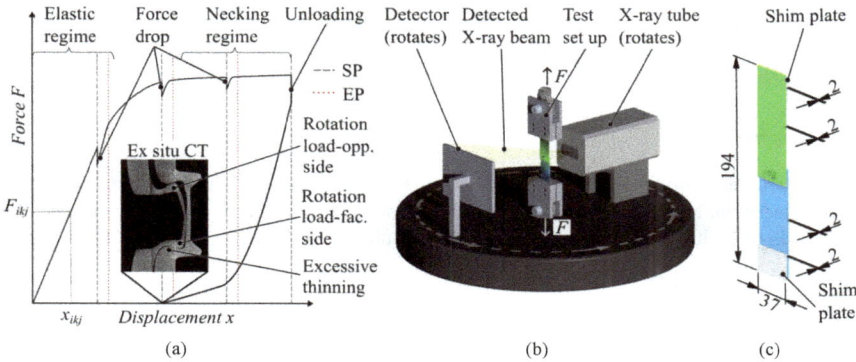

Fig. 1: Schematic process of the StopGo procedure and typical deformation phenomena during shear testing (a), CT set-up enabling ex situ CT scans (c) and specimen dimensions in mm (c).

Table 1: Applied parameters for the CT measurement and reconstruction

Parameter	Unit	Value
Acceleration voltage	kV	150
Tube current	μA	30
X-ray projections		1440
Exposure time	ms	625
Voxel size	μm	7.6
Physical filter	mm	0.3 (Cu)
Focal spot size	μm	4

Evaluation of the load-displacement behavior. First, the force displacement curves of both specimen sets are evaluated qualitatively and outliers are identified (1.5xIQR rule). Next, the coefficient of variation is computed to assess data consistency. The mean load values for all specimens in each set are then calculated at 400 evenly distributed displacement levels spanning the total displacement range. The differences between the mean force values of the two specimen sets are determined and normalized by the corresponding mean value of the Conti specimen set, resulting in the normalized differences of the means. This provides understanding of the relative deviations of the means and the variances of both procedures across the measurement.

The standard deviations s_{ik} of the loads of all samples at each displacement level are calculated for all tests, as shown in Eq. 1:

$$s_{ik}^2 = \frac{1}{n_k - 1} \sum_{j=1}^{n_k} F_{ijk} - \bar{F}_{ik}. \tag{1}$$

Where:
- i indexes the 400 displacement levels spanning the shear test displacement range ($[0, \dots, 1.00\,mm]$),
- j indexes the load values of the samples associated with the respective displacement level and the respective specimen set (Conti: $[F_{i1}, \dots, F_{i7}]$, StopGo: $[F_{i1}, \dots, F_{i8}]$),
- k indexes the two testing procedures (Conti and StopGo),
- n_k is the number of specimens for the respective specimen set (Conti: 7, StopGo: 8),
- F_{ijk} is the j-th load value at the i-th displacement level from specimen set k,

- $\bar{F}_{ik}$ is the mean load across all samples at i-th displacement level from specimen set k.

The mean values of the loads across all samples at each displacement level for each specimen set, $\bar{F}_{ik}$, are:

$$\bar{F}_{ik} = \frac{1}{n_k}\sum_{j=1}^{n_k} F_{ijk}. \tag{2}$$

The coefficients of variation $s_{ik,norm}^2$ of the loads are calculated for both specimen sets and for each displacement level as shown in Eq. 3. Furthermore, the normalized. differences between the mean loads of both procedures, $\bar{F}_{i,norm}$, for each displacement level are calculated as shown in Eq. 4:

$$s_{ik,norm}^2 = \frac{s_{ik}^2}{\bar{F}_{ik}} \tag{3}$$

and

$$\bar{F}_{i,norm} = \frac{\bar{F}_{i,Conti} - \bar{F}_{i,StopGo}}{\bar{F}_{i,Conti}}. \tag{4}$$

Due to the observed settling effects during the CT interruptions [9] larger load deviations at the stop points are expected. To investigate, how closely the load-displacement behavior after each interruption aligns with the continuous tests, statistical significance tests are performed at the EP, 0.05 mm after each stop point. Additionally, the maximum loads are assessed. The null hypothesis is postulated that the loads at the EP, as well as the maximum loads, from both procedures have equal variance and mean value. This hypothesis is tested using the following statistical methods:

1. Normal distribution: The loads at each EP and the maximum loads across all specimens of each specimen set are tested for normal distribution using the Shapiro Wilk [10] and the Anderson-Darling [11] test, as normality is a prerequisite for the subsequent analyses.
2. Variance equality: A two-sided F-test of equality of variances is used to assess whether the deviations in variance between both specimen sets are statistically significant.
3. Mean equality: For investigating the significance of deviations in mean values between the two procedures, a two-sided Student's t-test is applied when variances are equal, while Welch's test is used when variances are unequal.

These tests aim to verify whether there are significant deviations in the load-displacement behaviors between both procedures.

To determine possible geometrical differences, CT scans are performed for each specimen set after testing. For 3D surface detection, the surface of the specimen is generated based on the ISO 50 value, with erroneous voids and noise particles removed. The contours of the specimens in the sagittal plane are compared qualitatively. Furthermore, the final geometry is characterized by specific geometric properties that quantify phenomena in the punch-faced joining part. Particularly the geometric properties "gap on the load-opposing side", "gap on the load-facing side", and "excessive thinning" are measured according to Fig. 2b (the coordinate origin is defined according to Fig. 2a). These properties are evaluated using box plots and outliers are identified. To verify whether the final geometries of the two specimen sets differ significantly, statistical tests are conducted on these geometric properties, examining both variance equality and mean equality in the same way as in the evaluation of the load-displacement behavior.

Sheet Metal 2025
Materials Research Proceedings 52 (2025) 117-124

Materials Research Forum LLC
https://doi.org/10.21741/9781644903551-15

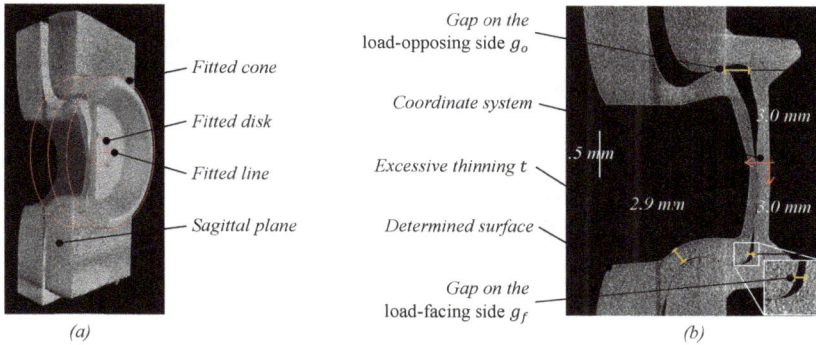

Fig. 2: Fitting of the geometries to the CT image (a), defining the coordinate system, visible in (b), and measurement of shear-test-specific geometric properties in the sagittal cross section (b).

Results and Discussion

Load-displacement evaluation. Overall, there is a good agreement between the curves of both specimen sets. However, the curves of the StopGo specimens scatter more strongly in the elastic regime compared to the continuously tested specimens (Fig. 3). Four specimens show a reduced gradient at the beginning of the test. In the necking regime, the deviations of the Stop-Go specimens at the SP, caused by the settling effects, are followed by a good realignment with the Conti specimens in further testing. The specimen C04_A_CV_373 is identified as an outlier and excluded from the statistical evaluation (1.5xIQR rule).

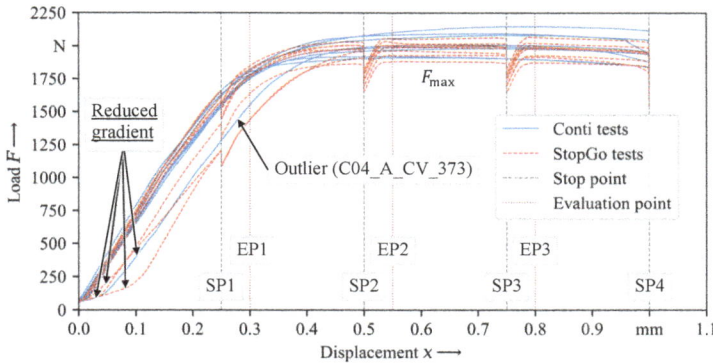

Fig. 3: Load-displacement diagram of Conti and StopGo tests

The coefficients of variation $s_{ik,norm}^2$ of the loads across the displacement indicate that the deviation between both specimen sets is highest in the elastic regime (Fig. 4a). However, the coefficients of both sets are similar and below 5 % in the necking regime. Likewise, the normalized differences of the means of the loads $\bar{F}_{i,norm}$ is highest in the elastic regime (Fig. 4b). Besides, the expected high differences at the SP, the normalized differences of the means are below 5 % in this phase indicating a similar load-displacement behavior.

Sheet Metal 2025   Materials Research Forum LLC
Materials Research Proceedings 52 (2025) 117-124   https://doi.org/10.21741/9781644903551-15

The loads at the evaluation points can be considered to be normally distributed (Shapiro-Wilk and Anderson-Darling test). The F-test of equality of variances revealed a significant deviation between both specimen sets at the EP1 supporting the previous evaluations. Even though the Welch test showed no significant deviation of the means (p-value: 0.18), a tendency in deviation can be noticed. Apart from EP1, no significant deviation in variance and mean value can be found at the other EP to the significance level of 5 %.

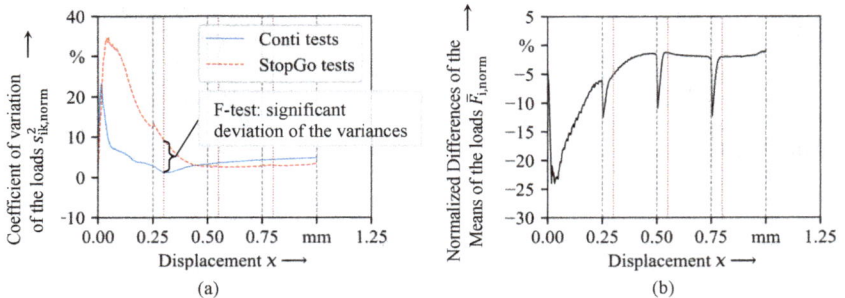

Fig. 4: The coefficients of variation of both specimen sets across the displacement (a). Normalized differences of the means of the loads of both procedures over the displacement (b).

Geometry evaluation. The specimen geometries of both specimen sets are evaluated using the surfaces extracted from the CT images. In Fig. 5, the contours of these surfaces in the sagittal cross section are compared. Overall, a good accordance of the contours can be seen. However, there are minor deviations on the punch-faced joining part on the load-opposing side. These deviations can also be seen in the box plots for the investigated geometric properties, shown in Fig. 6. The median values for g_o of both specimen sets clearly deviate between each other (Fig. y8a). Furthermore, the values for g_f of the StopGo specimen set exhibit a higher variance than the Conti set (Fig. 8b). Apart from that, two outliers can be identified in the data (1.5xIQR rule), which are excluded from the significance tests.

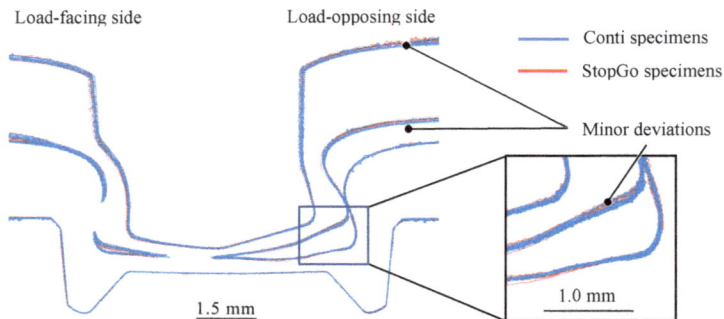

Fig. 5: Contours of Conti and StopGo specimens in the sagittal cross section of the CT images

The values of each geometric property of each specimen set are considered normally distributed (Shapiro-Wilk and Anderson-Darling test). The F-test revealed no significant deviation of the variances of both specimen sets apart from the excessive thinning (c). Here, the StopGo specimens exhibit a significantly higher variance (p-value: 0.036). The t-test showed a significant deviation

Sheet Metal 2025
Materials Research Proceedings 52 (2025) 117-124

Materials Research Forum LLC
https://doi.org/10.21741/9781644903551-15

(p-value: 0.022) for the gap on the load-opposing side (a) while no significant deviation of the means is found for the other geometric properties (b, c).

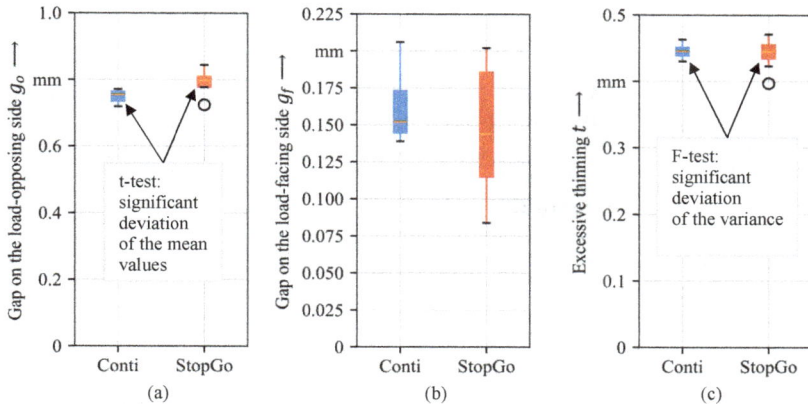

Fig. 6: Measured values for the gap on the load-opposing side (a), on the load-facing side (b) and excessive thinning (c).

Although this study evaluated eight specimens per testing procedure, three measured quantities exhibited outliers in either the continuous or the interrupted tests. Excluding these outliers influenced the results of the significance tests, thereby reducing the statistical robustness of this study. Besides that, four specimens displayed atypical load-displacement behavior at the beginning of the test, which cannot be caused by the investigated settling effect but by incorrect specimen mounting. This impact changed the load at the SP and probably the final geometry. Thus, further investigations are required to ensure validity of the results.

The deviations in the final geometry can be explained by relaxation mechanisms, as described in [12]. According to these findings, reducing the strain rate would decrease the stress relaxation. Thus, conducting discontinuous tests with lower strain rates on aluminum specimens could improve the equality with the respective continuous test.

Summary and Conclusions

In situ CT of clinch points under shear loading can visualize deformation phenomena in detail omitting the impact of unloading and specimen preparation. Process interruptions, which are inherent to in situ CT tests, cause settling effects. Their impact on the process is investigated using single-lap shear tests on clinched aluminum specimens. Continuous and interrupted tests were performed to evaluate whether these interruptions influence the load-displacement behavior and final geometry of the clinch point. The interrupted tests displayed slightly greater load variance in the elastic regime and relatively small geometric deviations at certain points. To enhance the validity of the results further experiments are necessary, which should also investigate the influence of strain rate. The overall results of the interrupted specimens, however, are comparable with the continuous specimens. Thus, these findings suggest that the discontinuous testing inherent to in situ CT is a valid and reliable approach for examining shear test behavior in clinched joints.

Acknowledgements

This research was funded by the German Research Foundation (DFG) within the project Transregional Collaborative Research Centre 285/2 (TRR 285/2, project number 418701707), sub-project C04 (project number 426959879).

References

[1] S.A. Nourani, G. Stilwell, D.J. Pons, Shear testing of clinch joints at different temperatures: Explanation of the failure sequence, Journal of Advanced Joining Processes 7 (2023) 100140. https://doi.org/10.1016/j.jajp.2023.100140

[2] Y. He, L. Yang, J. Dang, A. Gao, W. Zhang, Evaluation of Shear and Peel Strength of Al1060 Single-Lap and T-Lap Joints Produced by Rotated Clinching Process with Twin Rotating Punches, Materials (Basel) 15 (2022) 4237. https://doi.org/10.3390/ma15124237

[3] A. Barimani-Varandi, A.J. Aghchai, F. Lambiase, Failure behavior in electrically-assisted mechanical clinching joints, Journal of Manufacturing Processes 68 (2021) 1683–1693. https://doi.org/10.1016/j.jmapro.2021.06.072

[4] S. Coppieters, H. Zhang, F. Xu, N. Vandermeiren, A. Breda, D. Debruyne, Process-induced bottom defects in clinch forming: Simulation and effect on the structural integrity of single shear lap specimens, Materials & Design 130 (2017) 336–348. https://doi.org/10.1016/j.matdes.2017.05.077

[5] B. Da Alves Silva, Failure analysis of novel aluminum-copper joints for electric vehicle battery applications. Master Thesis, Porto, 2024.

[6] W.G. Drossel, R. Mauermann, R. Grützner, D. Mattheß, Numerical and Experimental Analysis of Self Piercing Riveting Process with Carbon Fiber-Reinforced Plastic and Aluminium Sheets, KEM 554-557 (2013) 1045–1054.

[7] B. Gröger, D. Köhler, J. Vorderbrüggen, J. Troschitz, R. Kupfer, G. Meschut, M. Gude, Computed tomography investigation of the material structure in clinch joints in aluminium fibre-reinforced thermoplastic sheets, Production Engineering (2021). https://doi.org/10.1007/s11740-021-01091-x

[8] T. Borgert, D. Köhler, E. Wiens, R. Kupfer, J. Troschitz, W. Homberg, M. Gude, 2024. In-situ computed tomography analysis of the failure mechanisms of thermomechanically manufactured joints with auxiliary joining element. Proceedings of the Institution of Mechanical Engineers, Part L: Journal of Materials: Design and Applications, 14644207241232233. https://doi.org/10.1177/14644207241232233

[9] D. Köhler, R. Kupfer, J. Troschitz, M. Gude, In Situ Computed Tomography-Analysis of a Single-Lap Shear Test with Clinch Points, Materials (Basel) 14 (2021). https://doi.org/10.3390/ma14081859

[10] S.S. SHAPIRO, M.B. WILK, An analysis of variance test for normality (complete samples), Biometrika 52 (1965) 591–611. https://doi.org/10.1093/biomet/52.3-4.591

[11] T.W. Anderson, D.A. Darling, Asymptotic Theory of Certain "Goodness of Fit" Criteria Based on Stochastic Processes, Ann. Math. Statist. 23 (1952) 193–212. https://doi.org/10.1214/aoms/1177729437

[12] H. Hamasaki, Y. Morimitsu, F. Yoshida, Stress relaxation of AA5182-O aluminum alloy sheet at warm temperature, Procedia Engineering 207 (2017) 2405–2410. https://doi.org/10.1016/j.proeng.2017.10.1016

Sheet Metal 2025
Materials Research Proceedings 52 (2025) 125-132

Materials Research Forum LLC
https://doi.org/10.21741/9781644903551-16

Investigation on manufacturing-induced pre-deformation on the fatigue behaviour of clinched joints

Malte Christian Schlichter[1,a] *, Özcan Harabati[1,b], Max Böhnke[1,c],
Christian R. Bielak[1,d], Mathias Bobbert[1,e] and Gerson Meschut[1,f]

[1]Laboratory for Material and Joining Technology (LWF), Paderborn University, Pohlweg 47-49, 33098 Paderborn, Germany

[a]malte.schlichter@lwf.upb.de, [b]harabati@lwf.uni-paderborn.de,
[c]max.boehnke@lwf.uni-paderborn.de, [d]christian.bielak@lwf.uni-paderborn.de,
[e]mathias.bobbert@lwf.uni-paderborn.de, [f]gerson.meschut@lwf.uni-paderborn.de

Keywords: Mechanical Joining, Pre-Deformation, Fatigue

Abstract. The present study is an experimental analysis of the influence of pre-forming on the failure behaviour of clinched specimens under quasi-static and cyclic loading conditions. In this context, the geometric formation of the clinched joints is taken into account, with regard to the loading behaviour. The study also includes a comparison of the failure behaviour of quasi-static and cyclic tested specimen. Testing is done on non-pre-deformed and pre-deformed specimens. For this purpose, experimental investigations are carried out on two material combinations consisting of HCT590X steel sheet and EN AW-6014 T4 aluminium sheet. The focus is on the fatigue analysis of the clinched joints. The aim is to identify the failure modes under cyclic loading and the crack formation with regard to forming operations prior to the joining process. The investigations show that the cyclic load-bearing behaviour of the HCT590X joints is reduced by introducing a plastic pre-deformation of the to be joined parts.

Introduction

Clinching. Clinching (DIN 8593-5) is a widely used mechanical joining technique for sheet metals that relies on cold forming without requiring auxiliary elements. It is especially useful for implementing multi-material designs, which are crucial for lightweight construction in the automotive industry. The clinching process creates a mechanical interlock between sheet metals using a punch-die tool combination, supported by a blank holder to prevent slippage. The formation of the clinched joint is influenced by the mechanical properties of the sheet materials, sheet thicknesses, friction on the contact surfaces and tool-combination used [1]. For evaluating clinch joints, geometrical characteristics such as mechanical interlock f, neck thickness t_n, and bottom thickness t_b are assessed Fig. 1.

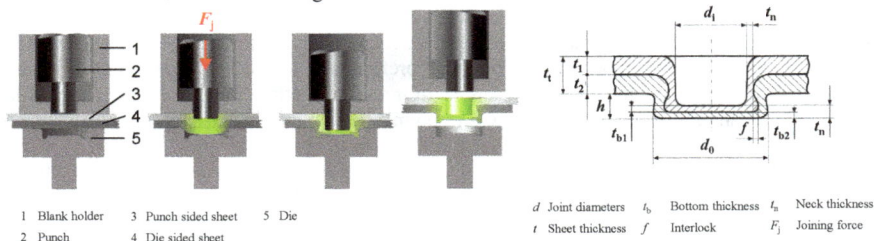

1 Blank holder 3 Punch sided sheet 5 Die
2 Punch   4 Die sided sheet

d Joint diameters t_b Bottom thickness t_n Neck thickness
t Sheet thickness f Interlock F_j Joining force

Figure 1: Clinching process and geometrical characteristics of a clinched joint in accordance with [1]

Content from this work may be used under the terms of the Creative Commons Attribution 3.0 license. Any further distribution of this work must maintain attribution to the author(s) and the title of the work, journal citation and DOI. Published under license by Materials Research Forum LLC.

Sheet Metal 2025   Materials Research Forum LLC
Materials Research Proceedings 52 (2025) 125-132   https://doi.org/10.21741/9781644903551-16

The failure mode of a clinched joint is influenced by joint design, forming history, and the types of loads endured. Four failure modes can occur and are distinguished: neck fracture, unbuttoning, and mixed modes as shown in Fig 2 a) - d) [2].

Lee et al. investigated, that the occurrence of neck fractures is highly depended on the punch diameter and the neck thickness. The neck thickness is the key factor for increasing the tensile strength under mixed loads when neck fracture occurred [3]. Lin et al. reported that evaluating the reliability of clinched structures primarily relies on phenomenological tests and assessing individual joint designs. To achieve reliable clinch results, the tool geometry must be optimised for each sheet metal combination due to the large number of influencing factors. [4].

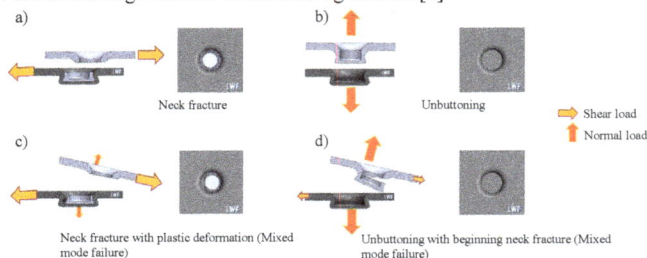

Figure 2: Description of the four failure modes of clinched joints, with indications of the type of loads applied [2].

Influence of plastic pre-deformation on clinched joints. The focus of this study is the influence of pre-deformed parts on the fatigue behaviour of clinched joints. Kurzok et al. shows that pre-deformation of the punch-sided material has a more pronounced impact on joining parameters than the die-sided material under static load [5]. A pre-deformation up to the uniform elongation of punch-sided materials resulted in a 70 % reduction in energy absorption for shear-load joints and a 15 % decrease in maximum loading capacity. He et al. [6] shows that the load bearing behaviour of clinched joints is significantly decreased (approx. 20 %) with a pre-deformation of 5 % uniaxial plastic elongation in tension direction. In [7] it was revealed that strain hardening and variations in sheet thickness significantly affect clinch joint properties. Specifically, a reduction in material thickness notably decreases the load-bearing capacity of the joint. Also, the neck thickness negatively correlates with interlock and the reduction in sheet thickness has a greater impact on both factors than strain hardening through pre-deformation. Correlations between the pre-deformation and quality-relevant parameters were compared with the maximum shear force using metamodels developed from the FEM calculations [8]. Hahn et al. observed that joints in body-in-white were often located in areas that had undergone previous deformation. Consequently, since only as-received material is used during optimization, tool optimization results differ from real outcomes on preformed material. To address this, new tool geometries must be adapted for each degree of pre-deformation. Frequent changes in tool combinations can reduce production efficiency and increase production costs [9].

Fatigue testing. An overview of relevant publications on cyclic investigations focussing on pre-deformation and clinched joints is provided. Studies have shown that DP600 and 6000x aluminium alloys exhibit strain hardening under pre-deformation [10]. As investigated by Le [11] and Al-Rubaie [12], pre-deformation up to 7% plastic strain leads to an increase in both quasi-static and cyclic loading behaviour due to strain hardening. However, a direct comparison of tensile specimens with joining specimens is not possible due to varying stress states while testing [13]. In view of HCXT590X, investigations have already been carried out on clinched joints that were not pre-deformed [14]. Ewenz et al. has already shown that due to the low adaptability of clinched joints in relation to the joining partners, optimisation of the process and geometry parameters as well as tool combinations is required even in the case of minor deviations [13]. Lin et al. established that pre-deformation of clinched materials leads to a change in the geometric characteristics and therefore requires a separate optimisation of tool combinations for pre-deformed clinched joints [4]. The failure behaviour of cyclically tested clinched joints made of HCT590X and aluminium alloys can be categorised into three types of failure for unprocessed material. These can be described as neck fractures, cracks in the die-side sheet and fretting [14]. Fretting is described as a frictional wear caused by multiple cycles of short amplitude reciprocating sliding motion at the interface of the material. Zhang et al. reported signs of fretting in the neck region and fretting wear around the joint, which appears to be influenced by the amount of applied load [15]. Ewenz was able to establish this dependence of the failure modes occurring on the force amplitude in his investigations for HCT590X. However, no conclusions were drawn as to the cause of the occurrence of these failure modes under cyclic loading [14]. The occurrence of failure modes, especially fretting, has not yet been comprehensively investigated for clinched joints. Furthermore, the influence of pre-deformed joining partners on non-optimised joints under cyclic load application has also not been investigated with regard to positive locking. However, this is essential for a better description of the frictional wear behaviour in clinched joints [16]. To illustrate this phenomenon in more detail, it is essential to conduct experimental studies on the fatigue behaviour of joints with pre-deformed materials to improve our understanding and enable the design of fatigue-resistant physical joints.

Experimental

For this investigation sheet metal HCT590X in sheet thickness of 1.5 mm and EN AW-6014 in temper T4 with a sheet thickness of 2.0 mm are used. The approximate ultimate tensile strength (UTS) R_m and elongation A_{80} of HCT590X are 618.5 Mpa and 33.2 %, for EN AW-6014 T4 244,4 Mpa and 27.9 %. Their chemical compositions and mechanical properties according to the manufacturers data sheets are shown in [17] and [18].

Single-lap shear specimens, with dimensions of 105 mm x 45 mm and an overlap length of 16 mm, were produced for each combination for static and cyclic testing. The joining device for producing the lap shear specimen is a servo-electrically operated unit TOX® PRESSOTECHNIK GmbH & Co. KG. The tool combinations used for production of each combination as well as the combination specifics are shown in *Table 1*. Exclusive deformation of the EN AW-6014 will not be demonstrated, due to insignificant influence of only die sided pre-deformed sheet in clinched joints, as shown in [7].

Table 1: Experimental design of the load bearing capacity test and fatigue tests based on the shear-lap specimen

Mat. Comb.	Material punch sided	Thickn. [mm]	Material die sided	Thickn. [mm]	Pre-deformation	Punch	Die
K1A	HCT590X	1.5	HCT590X	1.5	None	A50100	BD8016
K1B	HCT590X	1.5	HCT590X	1.5	Both parts		
K1C	HCT590X	1.5	HCT590X	1.5	Punch sided		
K2A	EN AW-6014	2.0	EN AW-6014	2.0	None	A50100	BE8012
K2B	EN AW-6014	2.0	EN AW-6014	2.0	Both parts		
K2C	EN AW-6014	2.0	EN AW-6014	2.0	Punch sided		

A defined homogeneous plastic deformation with a plastic strain of approximately $\varphi = 0.1$ was introduced in primary sheets out of HCT590X and EN AW-6014 T4 through uniaxial pre-deformation with a tensile testing machine according to [10]. Four shear load specimens and a joint sample were cut out of a 70 mm x 500 mm primary specimen post stretching in rolling direction. The stretching is based on the investigations described in [10]. All load-bearing capacity tests were performed on a Zwick Z100 tensile-compression testing machine with a test rate of 10

Sheet Metal 2025   Materials Research Forum LLC
Materials Research Proceedings 52 (2025) 125-132   https://doi.org/10.21741/9781644903551-16

mm/min. The fatigue tests were conducted on a resonance pulsation test system Rumul by Russenberger Prüfmaschine AG at a frequency of 70 Hz and under a force ratio of R = 0.1, with load cycles of 10^4 to $2*10^6$. To investigate the failure behaviour microscopic investigations were carried out on a Z16 APO Macroscope by Leica Microsystms GmbH.

Results

Geometrical formation. The results of the quasi-static tests in form of the ultimate tensile strength UTS as well as the geometric characteristics of the clinched joint combinations analysed can be found in detail in Fig. 3c)-h). Due to the induced pre-deformation the material thickness of the HCT590X reduced from 1.5 mm to 1.44 mm (-4.7 %). The EN AW-6014-sheet thickness reduced from 2.0 mm to 1.92 mm (-4.5 %). The results of the K1 combinations out of HCT590X-steel show an overall reduction in the geometrical characteristics, which were achieved by clinch joining the pre-deformed combinations K1B and K1C. In particular, the neck thicknesses and the mechanical interlocks of the combinations K1B and K1C were reduced in comparison to the reference combination K1A. The average reduction in parameters is approximately 0.03 mm. It should be noted that the interlock of combination K1C increased by 4.2 % compared to reference K1A. With regard to the neck thickness and bottom thickness of K2C, both parameters exhibited a reduction of 3.7 % in comparison to K2A. The parameters in the K2B combination demonstrate minimal variation in comparison to the reference, exhibiting a mere 2.3 % alteration in the bottom thickness.

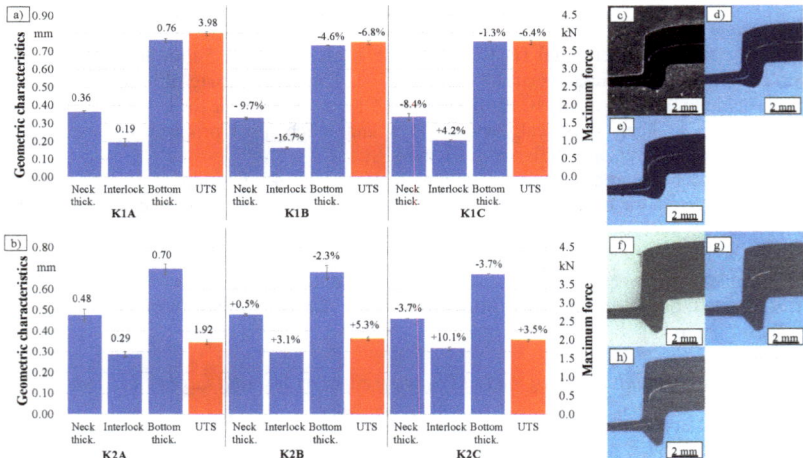

Figure 3: Geometrical characteristics and UTS of K1 (HCT590X/ HCT590X) a) and K2 b) (EN AW-6014 T4/ EN AW-6014 T4), as well as the cross sections of K1A (c), K1B (d), K1C (e), K2A (f), K2B (g) and K2C (h)

Quasistatic test and fatigue tests. The results of the loadbearing tests of both K1 A-C and K2 A-C are shown in Fig. 3 a) and b). For the K1A the UTS of 3988.2 N was achieved. The results for K1B and K1C show both as reduction in UTS of approximately 6.8 % and 6.4 % compared to K1A. This shows, that the tensile strength of the clinched joints reduces due to the pre-deformation of $\varphi = 0.1$ induced into the sheets beforehand. For the K2 combinations the opposite is seen. The UTS of the pre-deformed combinations K2B and K2C show a slight increase in their force-displacement-curves compared to ones of K2A. The UTS of all K2 variants slightly increase with 101.3 N and 68.6 N for K2B and K2C, compared to K2A.

In Fig. 4 the stress-cycle curves of all K1 and K2 combinations are plotted. Lifetimes are plotted logarithmically against the force amplitude F_A. A clear separation can be made between the curves of HCT590X and EN AW-6014 T4. The implemented force amplitudes of HCT590X are approximately twice as high as the ones achieved for the aluminium alloy specimens. The stress-cycle curve of K1A shows an inclination of $k_{K1A} = 19.22$. Compared to this the curves of K1B and K1C show a stepper incline of $k_{K1B} = 12.04$ and $k_{K1C} = 11.72$. Additionally, both curves show a shift to the left into lower life cycles for the same F_A values tested on K1A. For the K2 it can be seen, that the stress cycle curves of all three combinations lay close together. The curves o K2B shows a twice as high inclination ($k_{K2B} = 6.58$) as the one of K2A. For K2C the inclination is just a bit steeper ($k_{K2C} = 10.29$) compared to the reference. Both curves show also a slight shift into lower achieved live cycles for equivalent tested force amplitudes.

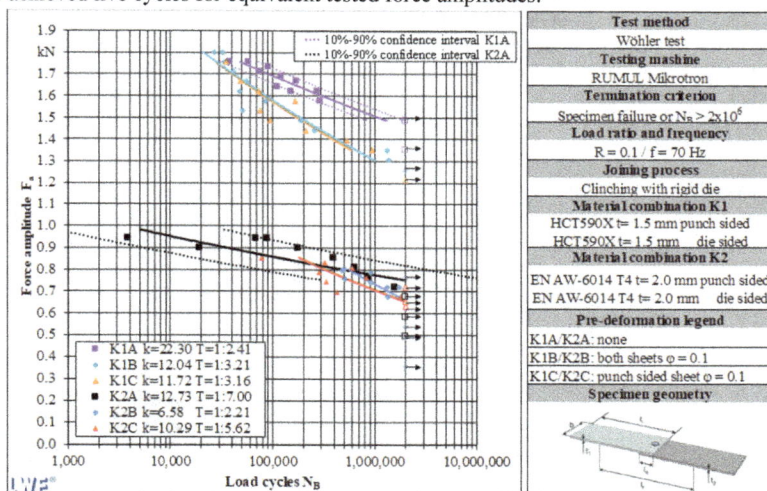

Figure 4: F-N curve of HCT590X clinched joint combinations (K1A, K1B and K1C) and EN AW-6014 T4 clinched joint combinations (K2A, K2B and K2C) with a schematic of the different failure modes, mode 1 (neck fracture), failure mode 2 ((neck fracture propagates to the base material) and fretting wear, the coloured areas represent the failure occurrence in the experiments

Failure behaviours and failure mode. In Fig. 5 the different failure modes for the load bearing test as well as for the fatigue testing of all six combination are shown. For the quasi static testing of the K1A and K1B specimens show all neck fracture as a primary failure mode, but three specimens show an additional unbuttoning. For K1C all five specimen show only neck fracture as the failure mode. In each of the three K2 combinations, only a neck fracture was identified as the dominant failure mode in the samples that had undergone static testing.

For the K1 and K2 samples that were tested under cyclic loading conditions primarily neck fracture occurred as a failure mode. Over 10^6 loading cycles for K1B and K2B specimen showed a failure behaviour of the die sided material similar to the failure mentioned in [14]. On all cyclic tested specimens of both material combinations different degrees of fretting wear could be observed (Fig. 5), but will be investigated in further studies.

Sheet Metal 2025
Materials Research Proceedings 52 (2025) 125-132

Materials Research Forum LLC
https://doi.org/10.21741/9781644903551-16

Discussion

Due to the pre-deformation ($\varphi = 0.1$) the sheet thickness of HCT590X reduced, therefore less material is in the forming zone of the clinched joints. This leads to a reduction of material that can be formed. The pre-deformation of the sheets leads additionally to a higher strain hardening effect of the steel, compared to reference combination K1A, therefore the geometrical characteristics of the HCT590X-HCT590X combinations that used pre-deformed material overall reduced. It is to mention, that the tool combinations used for the HCT590X combinations is only optimised for K1A. The alteration in material thickness results in the die utilized for the joint formation being incompletely filled, thereby preventing the punch-sided material from being formed in the same manner. Therefore, the interlock and neck thickness decrease in K1B and K1C but the bottom thickness is nearly unchanged because of the unchanged punch displacement [13]. With increasing plastic strain on pure HCT590X the maximum force increases due to strain hardening. Le and Kang showed that the same effect applies for pre-deformed HCT590X fatigue samples [11]. For material tensile specimen pure uniaxial stresses occur in comparison to clinched tensile specimen which show multiaxial stresses and are shear loaded. Therefore, a one to one comparison of the fatigue behaviours between both specimen types is not possible due to the loading conditions occurring in a clinched joint. Consequently, the reduction in UTS, as well as the shift of the K1B and K1C specimens to lower force amplitude levels and load cycles, can be partially attributed to the geometric characteristics of the formed clinched joints. It can be assumed that the initial deformation of $\varphi = 0.1$ together with the subsequent deformation within the joint formation results in a higher deformation of the materials, which has a significant negative influence on the static and cyclic loading behaviour of the K1B and K1C joints. In addition, the reduction in neck thickness of specimens K1B and K1C leads to a reduction in the effective neck ring area compared to K1A. This results in 14.2% and 21.8% higher local stress loads in the neck area of the deformed combinations. As a result of the increased local stress, the specimens fail at lower loading cycles than the K1A reference for the same force amplitude. The local stresses explicitly present in the neck area therefore need to be further investigated in the future, as they represent a significant influence. Further, the stress cycle curves of K1B and K1C are almost identical, with only a slight difference in incline of 0.32, suggesting that the pre-deformation of the punch sided sheet has the greatest influence on the results. This is supported by the results from [10] and [19] were it was shown that the pre-deformation of the die-sided sheet has no significant impact on the tensile strength of the clinched joint.

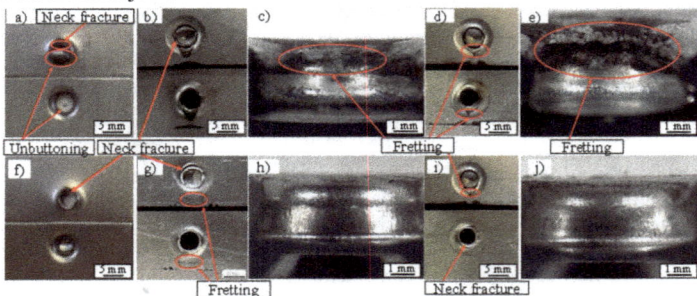

Figure 5: Description of the different failure behaviours of the HCT590X connections (a-e) and the EN AW-6014 T4 connections (f-g). The pictures show: a) K1A under static load, b) and c) K1B fatigue test 3300 N, d) and e) K1C fatigue test 3400 N, f) K2A static load, g) and h) K2B fatigue test 2600N, i) and j) K2C fatigue test 1850 N

For the EN AW-6014 T4 aluminium the UTS results shown in Fig. 1 b) indicate that the induced pre-deformation leads to a slightly increased (4 % - 5 %) tensile behaviour. This behaviour is based on the occurrence of strain hardening. Geometrical features only change significantly for the interlock of K2C in comparison to the reference. The fatigue behaviour of the EN AW-6014 do not change as much either. It is noteworthy that the curves of K2B and K2C fall within the 10 % to 90 % confidence interval of the K2A curve. However, the results demonstrated that a minimal pre-deformation of up to $\varphi = 0.1$ insignificantly impact the fatigue behaviour of the tensile specimens. As a result, the local stresses in the neck areas of the K2B and K2C specimens do not change as much, nor does the fatigue behaviour of those combinations. This findings were observed for a range of similar aluminium alloys [12].

Summary and Outlook

The study investigates the effects of pre-deformation ($\varphi = 0.1$) on the mechanical properties and fatigue behaviours of clinched joints made from HCT590X steel and EN AW-6014 T4 aluminium. In the case of HCT590X, the pre-deformation reduced sheet thickness, leading to less material in the forming zone and increased strain hardening. This altered the geometrical characteristics of the clinched joints, particularly decreasing interlock and neck thickness while leaving bottom thickness largely unchanged. This negatively impacted the static and cyclic loading behaviour of the K1B and K1C joints. Due to the increase in local stresses in the neck areas if the tested specimen. The stress cycle curves of these two joint combinations were nearly identical, indicating that pre-deformation of the punch-sided sheet has the most significant influence, while pre-deforming the die-sided sheet had minimal impact on tensile strength. For the EN AW-6014 T4 aluminium, pre-deformation slightly improved tensile behaviour due to strain hardening, but had little effect on fatigue behaviour, with results falling within the confidence interval of non-deformed samples. This aligns with findings from similar aluminium alloys, where minimal pre-deformation insignificantly affects fatigue performance.

In future investigations the focus will be set on the failure modes and failure criteria under cyclic loading conditions of pre-deformed sheet metals. Therefore, pre-deformed multi material joints as well as SPR joints need to be investigated. In addition, we intent to simulate the failure behaviour and failure modes of pre-deformed clinched joints under cyclic loading conditions to get a deeper understanding about the crack initiation point. The aim is to create a simulation that better describes the fatigue and wear behaviour of clinched joints.

Acknowledgements

The funding by the Deutsche Forschungsgemeinschaft (DFG, German Research Foundation) – Project-ID 418701707 – TRR 285/2, subproject A01 gratefully acknowledged. Moreover, we want to thank Janika Kroos and Erich Klassen for their support. Data regarding the contents of the publication can be requested at www.trr285.de.

Reference

[1] DIN Deutsches Institut für Normung e. V. DIN 8593-5: Manufacturing processes joining: Part 5: Joining by forming processing - Classification, subdivision, terms and definitions 2003. Berlin: Beuth Verlag GmbH.

[2] DVS - Deutscher Verband für Schweißen und verwandte Verfahren e.V. Merkblatt DVS/EFB 3420: Clinching - basics. 2021.

[3] Lee C-J, Lee S-K, Kim B-M, Ko D-C. Failure mode dependent load bearing characteristics of mechanical clinching under mixed mode loading condition. Procedia Engineering. 2017;207:938-43. https://doi.org/10.1016/j.proeng.2017.10.855

[4] Lin P-C, Lo S-M, Wu S-P. Fatigue life estimations of alclad AA2024-T3 friction stir clinch joints. International Journal of Fatigue. 2018;107:13-26.. https://doi.org/10.1016/j.ijfatigue.2017.10.011

[5] Hahn O, Kurzok JR. Forming technology joining of preformed sheet metals - Part 1: Steel (Umformtechnisches Fügen vorverformter Halbzeuge - Teil 1: Stahl). Aachen: Shaker; 1998.

[6] He X. Recent development in finite element analysis of clinched joints. Int J Adv Manuf Technol. 2010;48:607-12. https://doi.org/10.1007/s00170-009-2306-2

[7] Böhnke M, Bielak CR, Klassen E, Bobbert M, Meschut G. Experimental and numerical investigation of the influence of multiaxial loading conditions on the failure behavior of clinched joints. Proceedings of the Institution of Mechanical Engineers, Part L: Journal of Materials: Design and Applications. 2023;237:1444-57. https://doi.org/10.1177/14644207221145886

[8] Bielak CR, Böhnke M, Bobbert M, Meschut G. Further development of a numerical method for analyzing the load capacity of clinched joints in versatile process chains. ESAFORM 2021 2021. doi:10.25518/esaform21.4298. https://doi.org/10.25518/esaform21.4298

[9] Hahn, O., Kurzok, J.R., Dolle, N. DaimlerChrysler - Technologiekolloquium: Simulation der Fugetechniken Potentiale und Stuttgart, Germany, Januar). 2001:35-8.

[10] Bielak CR, Böhnke M, Bobbert M, Meschut G. Experimental and Numerical Investigation on Manufacturing-Induced Pre-Strain on the Load-Bearing Capacity of Clinched Joints. International Conference on Sheet Metal. 2022;926:1516-26. https://doi.org/10.4028/p-5d009y

[11] Le Q, Kang H, Kridli G, Khosrovaneh A, Yan B. Effect of prestrain paths on mechanical behavior of dual phase sheet steel. International Journal of Fatigue. 2009;31:607-15. https://doi.org/10.1016/j.ijfatigue.2008.03.028

[12] Al-Rubaie KS, Del Grande MA, Travessa DN, Cardoso KR. Effect of pre-strain on the fatigue life of 7050-T7451 aluminium alloy. Materials Science and Engineering: A. 2007;464:141-50. https://doi.org/10.1016/j.msea.2007.02.024

[13] Ewenz L, Kalich J, Zimmermann M, Füssel U. Effect of Different Tool Geometries on the Mechanical Properties of Al-Al Clinch Joints. International Conference on Sheet Metal. 2021;883:65-72. https://doi.org/10.4028/www.scientific.net/KEM.883.65

[14] Ewenz L, Bielak CR, Otroshi M, Bobbert M, Meschut G, Zimmermann M. Numerical and experimental identification of fatigue crack initiation sites in clinched joints. Prod. Eng. Res. Devel. 2022;16:305-13. https://doi.org/10.1007/s11740-022-01124-z

[15] Zhang Y, He X, Wang Y, Lu Y, Gu F, Ball A. Study on failure mechanism of mechanical clinching in aluminium sheet materials. Int J Adv Manuf Technol. 2018;96:3057-68. doi:10.1007/s00170-018-1734-2. https://doi.org/10.1007/s00170-018-1734-2

[16] Liu F, Chen W, Deng C, Guo J, Zhang X, Men Y, Dong L. Research advances in fatigue behaviour of clinched joints. Int J Adv Manuf Technol. 2023;127:1-21. doi:10.1007/s00170-023-11547-2. https://doi.org/10.1007/s00170-023-11547-2

[17] Salzgitter Flachstahl. HCT590X+Z: Mehrphasenstähle zum Kaltumformen - Dualphasenstähle. 2017. https://www.salzgitter-flachstahl.de/fileadmin/footage/MEDIA/gesellschaften/szfg/informationsmaterial/produktinformationen/feuerverzinkte_produkte/deu/hct590x.pdf.

[18] Novelis GLobal Automotive. EN AW-6014 T4 - Material Data Sheet: Novelis Advanz 6F - e170. 2019.

[19] Bielak CR, Böhnke M, Bobbert M, Meschut G, editor. Experimental und numerical Investigation on manufacturing-induced pre-strain on the load-bearing capacity of clinched joints; 2022. https://doi.org/10.4028/p-5d009y

Simulation

Sheet Metal 2025
Materials Research Proceedings 52 (2025) 134-141

Materials Research Forum LLC
https://doi.org/10.21741/9781644903551-17

A novel hybrid hot forming process concept for high strength aluminum alloys

Naveen K. Baru[1,a], Tobias Teeuwen[1,b], David Bailly[1,c] and Emad Scharifi[1,d]

[1]Institute of Metal Forming, RWTH Aachen University, Intzestraße 10, 52072 Aachen, Germany

[a]naveen.baru@ibf.rwth-aachen.de, [b]tobias.teeuwen@ibf.rwth-aachen.de,
[c]david.bailly@ibf.rwth-aachen.de, [d]emad.scharifi@ibf.rwth-aachen.de

Keywords: Metal Forming, Finite Element Method (FEM), Gas-Based Hot Forming

Abstract. The present study focuses on the numerical and experimental feasibility of a novel hybrid hot forming process involving solution heat treatment and a combined deep drawing and gas-based forming for high strength aluminum alloys. For this, FE simulations are first carried out by establishing a numerical model of the hybrid forming process in order to identify the process parameters to ensure macroscopic failure-free forming of a complex-shaped cross-die and counteract on localized thinning. In addition, an integrated hybrid forming test setup is realized to form the cross-die specimen for validating the numerical predictions. The numerical parameter studies suggest a suitable initial blank size for forming the cross-die, and the influence of the blank holder force and the friction coefficient on the maximum thinning at the cross-die radii. The following forming experiments show a high dimensional accuracy after deep drawing and gas-based forming steps. Comparing the individual forming steps moreover reveals that a higher local thinning is introduced during the gas-based forming. Finally, the comparison between the simulation and the experimental results demonstrates a good agreement for local thinning.

Introduction

High strength aluminum alloys are widely used as light-weight manufacturing material in automotive and aerospace industries to reduce the carbon dioxide emission and the fuel consumption by saving weight. The structural components for these transport sectors are traditionally realized via a large number of forming steps with complex strain path and strain hardening phenomena resulting in high spring back, localized thinning, thickness distribution or premature failure. This is due to the limited formability of high strength aluminum alloys at room temperature [1]. Moreover, different heat treatment steps are required to achieve the desired mechanical properties. For this reason, several forming processes at elevated temperatures are proposed for high strength aluminum alloys to overcome the undesirable phenomena and process limitations [2, 3]. These thermo-mechanical processes improve formability by using the thermally activated softening mechanisms during hot forming [4].

Superplastic forming (SPF) is proposed in this regard as an advanced forming process to produce large sized and complex-shaped as well as thin-walled components, which cannot be produced using traditional forming techniques at room temperature [5]. The process steps in SPF involve heating within the die, while gas pressure is applied to form the heated sheet into a single piece component according to the die cavity geometry [6]. In SPF, the stress concentration at the grain boundary between two grains or at triple junctions creates an accommodating slip in the neighboring grain, leading to the generation of dislocations at the neighboring grain boundary, where the dislocations can move freely across the grain without hindrance at the opposite grain boundary [7]. Hence, in addition to the reduced ultimate tensile strength (UTS) caused by softening mechanisms in high strength aluminum alloys at elevated forming temperatures (T_F), a significant rise in elongation to failure up to 200 % can be obtained [6]. SPF is carried out at elevated temperatures ($T_F > 0.5$ melting temperature, T_m) and low strain rates ranging from $0.0001\ \mathrm{s^{-1}}$ to

Content from this work may be used under the terms of the Creative Commons Attribution 3.0 license. Any further distribution of this work must maintain attribution to the author(s) and the title of the work, journal citation and DOI. Published under license by Materials Research Forum LLC.

0.001 s^{-1}. This however leads to a forming cycle of at least 30 min. Therefore, for increasing the production, quick plastic forming (QPF) is developed to mass-produce aluminum closure panels with a high dimension accuracy [3]. In contrast to SPF where a heated press with unheated forming tool system is used, QPF is carried out using a hydraulic press with a heated and insulated forming tool system at higher strain rates from 0.001 s^{-1} to 0.1 s^{-1}. This forming process is used in [3] for high strength aluminum alloys such as AA5083 in niche automotive applications.

Another promising gas-based hot forming process for high strength aluminum alloys is shown in [8], that could meet the requirement for large scale production. This novel forming process involves conductive heating and solution heat treatment (SHT) followed by rapid gas-based hot forming near to the SHT temperature. Hence, the required gas pressure during hot forming is reported to be very low and a nearly homogenous temperature distribution is guaranteed due to the heated tool system [8]. This process is applied to the precipitation hardenable aluminum alloy AA6010 and further evaluated concerning the microstructure and final mechanical properties [9]. After 15 min. SHT at 565 °C and hot forming, parts are quenched in water to obtain a high cooling rate and the supersaturated solid solution state. To obtain the full strength T6 condition as reported in [10] for AA6010, artificial aging is carried out at 205 °C for 45 min. An average UTS of 353 MPa and elongation until failure of 9.4 % is reported, which is comparable to other studies [9, 10]. However, this process already results in a local thinning over 60 % for the relatively smaller size of the parts, which makes it nearly impossible to realize deeper and complex geometries.

In order to exploit the potential of this hot forming process and increase its application spectrum, the localization of thinning within the part should be avoided. This can be done via a hybrid process concept involving pre-forming with punch followed by gas-based calibration of sharp geometrical features. Several hybrid hot forming processes based on SPF and QPF already exist, as mentioned in [11] along with the recent High Speed Blow Forming process. However, the duration of these processes is in the order of several seconds to minutes. In the current study, a novel hybrid hot forming process for rapid forming of high strength aluminum alloys is introduced and the process feasibly is demonstrated with help of simulations and laboratory experiments.

Process Setup and Concept
The numerical and experimental investigations are performed on high strength aluminum alloy AA6010 in the as-received T6 condition with a hardness of 120 HV1 and UTS of 330 MPa approximately. The current process setup (Fig. 1a) mainly consists of a die, a blank holder and a punch with integrated gas channels. This setup enables deep drawing and gas-based forming of sheet metal blank into a complex-shaped cross-die in Fig. 1b. This complex geometry with high depth and sharp radius cannot be achieved via a single deep drawing or gas forming step.

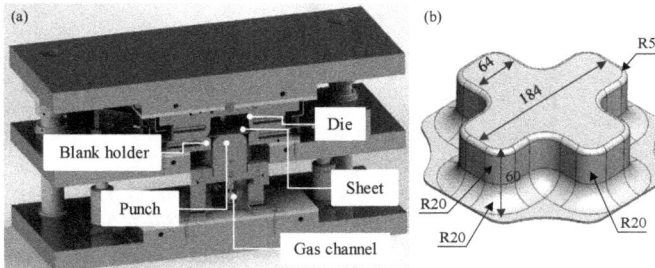

Fig. 1: (a) Process setup and (b) the complex cross-die specimen with dimensions in mm

This forming process set-up is moreover designed for elevated temperatures near to the SHT temperature of aluminum alloys. To achieve this, the forming tools are equipped with heating elements, cooling channels as well as the temperature measurement and control devices in order to control the temperature of forming tools and the consequent forming process precisely.

First, the sheet metal blank is positioned between the heated die and the heated blank holder, and conductively heated up to the SHT temperature, which is also the temperature of the forming tools. Then the blank is solution heat treated at 565 °C for 20 min in order to achieve a homogenous distribution of alloying elements. In the first forming step, the sheet metal blank is drawn into the die with the punch and in the next step, the bottom radius of the pre-formed cross-die is rapidly calibrated with help of pressurized gas entering through the gas-channels in the punch. During the deep drawing step, the blank holder holds down the blank in the flange region. At the end of the gas-forming step, the blank holder exerts a very high force on the flange to seal the forming chamber in order to retain the high pressure necessary for calibration of the bottom radius.

For this process, the blank holder force is an important parameter since it controls the amount of material drawn into the die, which further influences the thinning distribution. If the blank holder force is too low, too much material will be drawn-in, resulting in wrinkles as well as no sealing at the flange region during the subsequent gas forming step. This might also happen if the blank is too small. If the blank is too large, then the excess material in the flange region reduces the overall draw-in due to unfavorable drawing ratio, ultimately resulting in thinning and cracks at the part bottom radius. Therefore, it is important to select an appropriate blank size and blank holder force for an optimum result without part failure. Furthermore, the resulting part properties are also significantly influenced by the prevailing process conditions such as friction.

Therefore, in order to realize the current process, it is important to investigate the sensitivity to the prevailing process conditions and to select the controllable process parameters in an optimal manner so that the resulting part does not show signs of failure such as wrinkles, severe thinning or cracks. For this purpose, the corresponding FE model is setup and the simulations are performed.

Finite Element Model Setup and Simulations
The hybrid hot forming process is modelled and simulated using the FE software packages LS-PrePost and LS-Dyna R9.0. Due to symmetry of the cross-die, only a quarter of the original setup is modelled as shown in the Fig. 2 and constraints are defined accordingly at the symmetry planes. All of the four parts - die, blank holder, punch and the sheet metal blank are modelled with 3D thermal shell elements. For the tools, the outer surfaces are modelled with thermal shell elements with reduced integration formulation. For the sheet metal blank, the mid surface is modelled with help of thermal shell elements with full integration formulation with nine integration points through the thickness.

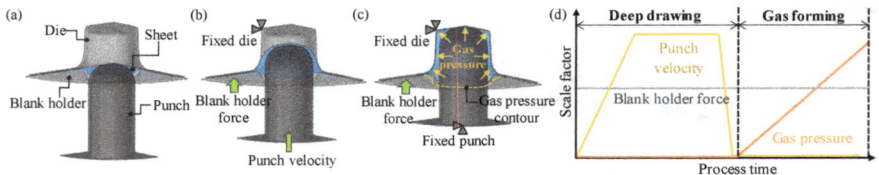

Fig. 2: 3D Model setup of the current hybrid hot forming process showing (a) the parts, the loads and boundary conditions during (b) deep drawing and (c) gas forming, and (d) schematic process parameter evolution over process time

Material Properties. The tools are modelled as rigid bodies using the Thermal_Isotropic material card with properties of tool steel whereas the blank is modelled as an elasto-plastic deformable body. For modelling the plasticity, the large strain flow curves of AA6010-T6 aluminum sheet near the SHT temperature of 565 °C determined via isothermal stacked layer compression   tests   is   taken   from   the   previous   work   [7]   and   assigned   via

Sheet Metal 2025
Materials Research Proceedings 52 (2025) 134-141
Materials Research Forum LLC
https://doi.org/10.21741/9781644903551-17

Thermo_Elasto_Viscoplastic_Creep card. Due to SHT and homogenization of the microstructure, the anisotropy of the sheet is neglected. For the time-being, no damage criterion is used and the failure is assumed to occur when the local part thinning is 60 %.

Process Duration, Loads and Boundary Conditions. The total process duration is two seconds, one second of deep drawing followed by one second of gas-forming. The die is fixed during the entire forming process, and it does not translate or rotate. The punch and the blank holder are allowed to move in vertical direction, but constrained from translating or rotating in other directions. During the deep drawing step, the punch movement is defined by a z-velocity of 60 mm/s. During the gas-forming step, the punch is fixed. The blank holder force, defined as a rigid body force in z-direction, is active during the entire forming process as shown in the Fig. 2d. During the gas-forming step, the gas pressure on the surface of the blank is modelled as a mask load on the sealed region of the blank within the blank holder contour. The load curve assigned for defining the gas pressure profile over the process time is shown in the Fig. 2d.

Interactions. Contact interactions are defined in the current model between the blank surfaces and the tools, i. e., die, punch and blank holder surfaces. These interactions are modelled using the Surface_to_Surface contact card. Moreover, Coulomb friction model is used to model both static and dynamic friction.

Thermal boundary conditions. Within the coupled thermo-mechanical model, two thermal boundary conditions are added. The first one is the constant temperature of the tools regulated by the heating cartridges and thermocouples during the entire process, modelled within FEM with help of the Temperature_Set card. The second one corresponds to the convection heat transfer from the workpiece surface to the pressurized gas during the gas-forming step, modelled with help of Convection_Set card. Within this, the environment (pressurized gas) temperature and the convection heat transfer coefficient are taken from the previous laboratory measurements [7].

Solver and simulation output. The coupled thermo-mechanical FE simulation is performed with help of an implicit solver. The resulting pre-formed cross-die after deep drawing and the fully calibrated cross-die after gas-forming for the converged FE model are shown in the Figure 3.

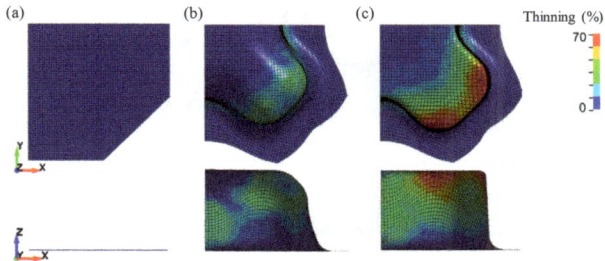

Fig. 3: Output from the FE simulation showing the part thinning distribution (a) initially, (b) after the deep drawing and (c) after the gas-forming

It can be seen that both the pre-formed and final states of the simulated cross-die specimen show max. thinning at the bottom radius. However, if the local thinning is too high, the part ruptures. Therefore, in the following, simulation-aided process sensitivity and parameter analyses are performed in order to optimize the process parameters, so that the max. thinning within the part could be minimized.

Simulation-Aided Process Parameter Study

Blank Size Selection. The first parameter study deals with finding a suitable initial size of the sheet metal blank. As shown in the Fig. 4, an octagonal blank shape according to the shape of cross-die is designed and its size is varied from 215 mm to 290 mm. For the smallest blank size of 215 mm,

Sheet Metal 2025   Materials Research Forum LLC
Materials Research Proceedings 52 (2025) 134-141   https://doi.org/10.21741/9781644903551-17

the blank completely draws into the die cavity, which would hinder the sealing of sheet during the gas-forming step. A sufficient amount of sheet material is required in the flange region, since the sealing in the real process is achieved by holding the sheet flange in between the die and blank holder surfaces. Also, a 240 mm blank is critical, due to very less remaining material in the flange region. On the other hand, the 290 mm blank still has too much leftover material in the flange region, and a high thinning (~ 66%) within the part, which might eventually result in part failure. Therefore, a blank size of 265 mm is chosen for the simulative and experimental investigations hereafter.

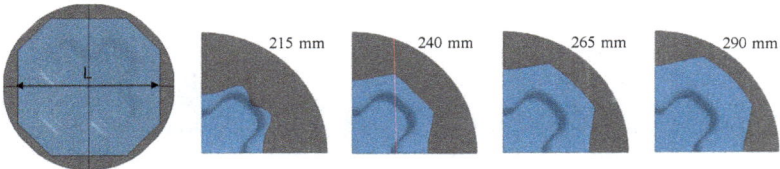

Fig. 4: Initial octagonal blank and top view of formed quarter part for different blank sizes.

Friction Sensitivity Analysis. A sensitivity analysis of friction between the blank and the tools is performed and the resulting thinning and draw-in are analyzed. For the current process and temperature conditions, graphite-based lubricants with a friction coefficient starting from 0.1 or Boron Nitride with a friction coefficient of up to 0.25 are available. Simulations with the previously determined blank size and a low blank holder force of 10 kN (Possible force range: 8 to 60 kN) are performed by varying the friction coefficients (μ = 0.10, 0.17 and 0.25) and the resulting maximum thinning and the sheet draw-in are analyzed as shown in Fig. 5.

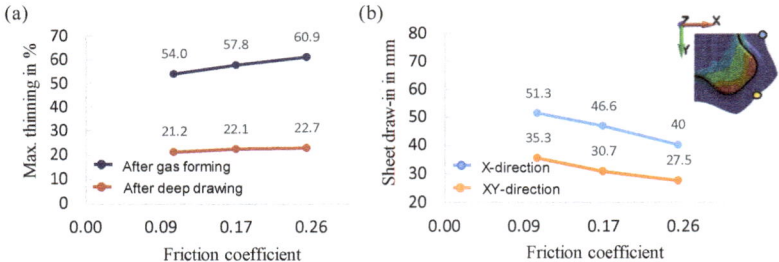

Fig. 5: (a) Maximum thinning and (b) sheet draw-in for different friction coefficient values

The variation of maximum thinning after deep drawing is small, but after gas forming a difference of 7 % is observed for the investigated cases. This is logical, since lower friction results in a more uniform material stretching. Due to the geometry of the blank and the cross-die, different types of material flow resulting in different strain states occur within the specimen, thereby resulting in different draw-in along X and XY directions. The sheet draw-in correlates with the thinning, where the draw-in decreases with increasing friction. This is due to higher frictional resistance in the flange region during deep drawing and the bottom and wall regions during gas forming. Assuming a safe cut-off thinning to be 60 % based on previous work, Boron Nitride lubricant could lead to part failure. Therefore, a graphite-based lubricant is chosen for the current process.

Sheet Metal 2025   Materials Research Forum LLC
Materials Research Proceedings 52 (2025) 134-141   https://doi.org/10.21741/9781644903551-17

Blank Holder Force Variation. For this study, the previously optimized blank size and a friction coefficient of 0.17 are taken and the blank holder force is varied from 10 to 50 kN, due to the pneumatic strut springs (Measured force range: 8 to 60 kN) built in the tool setup. The results from this parameter study are shown in Fig. 6.

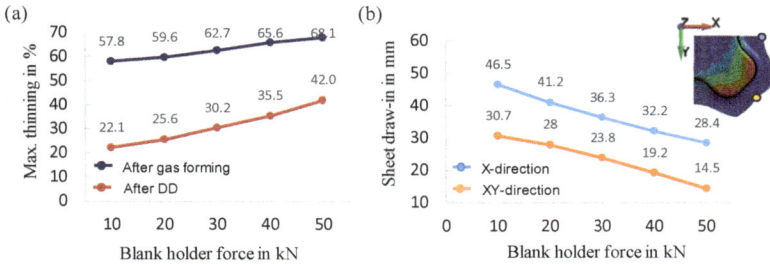

Fig. 6: (a) Maximum thinning and (b) sheet draw-in for different blank holder force values

As the blank holder force increases, the maximum increased from 22.1 % to 42 % after deep drawing. This is due to the increasing resistance to the material to flow into the die during deep drawing. This trend correlates with the draw-in, which decreases as the blank holder force increases. Again, considering a cut-off thinning of 60%, the blank holder force should be set to the lowest possible value in order to obtain a part without failure. This means that the blank holder force should be set to the least value possible to achieve good results. Therefore, 8 kN is applied within the experiments, even though the parameter study started from 10 kN.

Experimental Proof of Concept

Specimen and Tool Preparation. AA6010 aluminium sheet metal blanks with a thickness of 2 mm are prepared according to the previously optimized octagonal shape via waterjet cutting process. The edges of the blanks are then polished to remove burrs, and the surfaces are cleaned with ethyl alcohol to ensure optimal lubricant adhesion. A uniform coating of graphite-based lubricant is applied to both sides of each blank. All of the tool surfaces are also coated with the lubricant before they are heated up to the process temperature of 565 °C. During the entire experiment, the tool temperature and gas pressure are precisely controlled using a LabVIEW program developed for this purpose. Once the desired process temperature of 565 °C is achieved, the aluminum AA6010 blank is positioned between the tools and subjected to conductive heating followed by solution heat treatment for a minimum of 20 minutes.

Forming Process. During the forming process, the blank holder force of 8 kN and the punch velocity of 60 mm/s are set using the press panel control for the deep drawing stage. After the deep drawing stage, the gas forming starts, where the gas pressure profile is controlled manually to obtain a high pressure of 140 bars by fully opening the valve of the gas storage tank. After that, the gas valves are closed and the tools are driven apart to exert the cross-die specimen. The specimen is then quenched in water and cleaned from the graphite remains for further analysis. The following Fig. 7 (right) shows two specimens, the top one after deep drawing step, and the bottom one after the subsequent gas forming step.

Analysis of the Results. To analyze the thinning distribution within the specimen, three characteristic symmetry lines were drawn on the specimen as shown in Fig. 7. Afterwards, each specimen is cut approximately 1 cm close to the symmetry lines in order to measure the thickness along the three lines using a micrometer.

Sheet Metal 2025   Materials Research Forum LLC
Materials Research Proceedings 52 (2025) 134-141   https://doi.org/10.21741/9781644903551-17

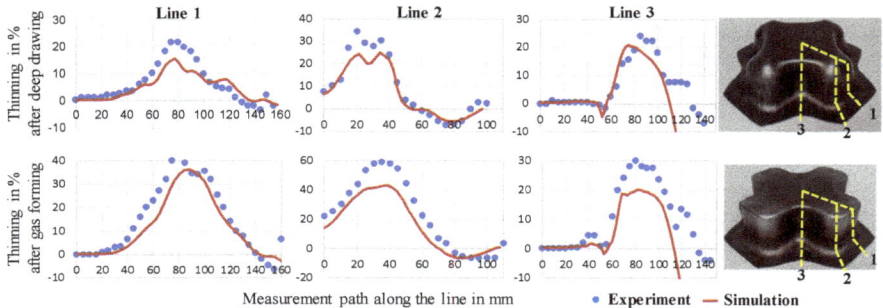

Fig. 7: Comparison of the thinning distribution along the three characteristic symmetry lines

Fig. 7 shows the comparison between the simulation and the experiment. It can be seen that the tendencies match quite well along all three symmetry lines, except for minor deviations. For all three lines, the thinning distribution at the cross-die specimen bottom (top surface in the above figure) is similar.

The difference in thinning increases and maximizes at the radius, and again decreases along the walls and minimizes at the flange. Along Line 3, the simulation shows high thickening at the flange region, which is not observed in experiments. This is due to the fact that during the experiment, there was a high blank holder force at the end of gas-forming process, locally compressing the flange area of the specimen resulting in thinning. Moreover, the high punch compression force acting on the bottom of the cross-die specimen in experiments resulted in inhomogeneous material stretching and increased localized thinning at the bottom radius area (Line-1 and Line-2) of the specimen, which is not considered within the relatively idealized simulation model. i. e., the simulation employs constant values that do not change throughout the process duration. Moreover, the Belytschko-Tsay shell elements cannot take the thickness stresses, thereby resulting in these differences in thinning.

Conclusion and Outlook
The current study provides insights into the feasibility of a novel rapid hybrid hot forming process for high strength aluminum alloys, especially for large-scale manufacturing of complex parts that could not be realized via individual deep drawing or gas-forming processes alone. The process combination enabled forming of a complex-shaped cross-die geometry by efficiently utilizing the thermally activated softening mechanisms during hot forming [12, 13]. The experimental and simulated local thinning distribution within the cross-die geometries resulted from the nature of stretching of the sheet material during the process agrees well except for minor deviations. Thus, the designed complex cross-die geometry offers a great potential for validation of the numerical model concerning the part properties such as part geometry, draw-in and local thinning. Hence, further parameter studies can be performed with the simulation model, which not only enable failure-free hot forming operations, but also enable further part thinning reduction via process optimization regarding the other process parameters, including temperature distribution, local surface enlargement dependent friction and local material flow.

Acknowledgements
This project (Lightness.NRW) was funded by "Europäischer Fonds für regionale Entwicklung" (EFRE). The authors of IBF further acknowledge excellent collaboration with the project partners

Sheet Metal 2025   Materials Research Forum LLC
Materials Research Proceedings 52 (2025) 134-141   https://doi.org/10.21741/9781644903551-17

SMS group GmbH, Ford-Werke GmbH, Hydro Extruded Solutions GmbH, Hydro Aluminum Rolled Products GmbH, HoD*forming* GmbH, CTES and Automotive Laboratory of FH Aachen.

References

[1] Y. Liu, L. Wang, B. Zhu, Y. Wang, Y. Zhang, Identification of two aluminum alloys and springback behaviors in cold bending, Procedia Manufacturing 15 (2018) 701-708. https://doi.org/10.1016/j.promfg.2018.07.303

[2] A. Foster, D. Trevor, J. Lin, WO 2010/032002.

[3] P.E. Krajewski, J.G. Schroth, Overview of Quick Plastic Forming Technology, MSF 551-552 (2007) 3-12. https://doi.org/10.4028/www.scientific.net/MSF.551-552.3

[4] K. Zheng, Y. Dong, D. Zheng, J. Lin, T.A. Dean, An experimental investigation on the deformation and post-formed strength of heat-treatable aluminium alloys using different elevated temperature forming processes, J. Mat. Pro. Tech. 268 (2019) 87-96. https://doi.org/10.1016/j.jmatprotec.2018.11.042

[5] F. Pitt, Developing a Superplastic Forming Application Using Aluminum Tube, J. Mat. Eng. and Perf. 13 (2004) 720-726. https://doi.org/10.1361/10599490421367

[6] A.P. Mouritz, Introduction to aerospace materials, Woodhead Publishing, Cambridge, 2012. https://doi.org/10.2514/4.869198

[7] J. Wongsa-Ngam, T.G. Langdon, Advances in Superplasticity from a Laboratory Curiosity to the Development of a Superplastic Forming Industry, Metals 12 (2022) 1921. https://doi.org/10.3390/met12111921

[8] N.K. Baru, T. Teeuwen, M. Teller, S. Hojda, A. Braun, G. Hirt, On appropriate Finite Element discretization in simulation of gas-based hot sheet metal forming processes, IOP Conf. Ser.: Mater. Sci. Eng. 1157 (2021) 12027. https://doi.org/10.1088/1757-899X/1157/1/012027

[9] T. Teeuwen, N.K. Baru, D. Bailly, G. Hirt, Investigation on evolution of microstructure and mechanical properties of heat-treatable 6010-S aluminium alloy during gas-based hot sheet metal forming process, IOP Conf. Ser.: Mater. Sci. Eng. 1284 (2023) 12005. https://doi.org/10.1088/1757-899X/1284/1/012005

[10] S. Taylor, S. Dhara, C. Slater, H. Kotadia, Identifying Optimal Hot Forming Conditions for AA6010 Alloy by Means of Elevated Temperature Tensile Testing, Metals 13 (2023) 76. https://doi.org/10.3390/met13010076

[11] O. Majidi, M. Jahazi, N. Bombardier, Finite Element Simulation of High-Speed Blow Forming of an Automotive Component, Metals 8 (2018) 901. https://doi.org/10.3390/met8110901

[12] E. Hornbogen, Hundred years of precipitation hardening, J. Light Metals 1 (2001) 127-132. https://doi.org/10.1016/S1471-5317(01)00006-2

[13] E. Scharifi, U. Savaci, Z.B. Kavaklioglu, U. Weidig, S. Turan, K. Steinhoff, Effect of thermo-mechanical processing on quench-induced precipitates morphology and mechanical properties in high strength AA7075 aluminum alloy, Materials Characterization 174 (2021) 111026. https://doi.org/10.1016/j.matchar.2021.111026

Sheet Metal 2025

Materials Research Proceedings 52 (2025) 142-149

Materials Research Forum LLC

https://doi.org/10.21741/9781644903551-18

Modeling of notch effects due to multi-material joints in automotive body components

Philipp Bähr[1,a*], Silke Sommer[1,b] and Gerson Meschut[2,c]

[1]Fraunhofer - Institute for Mechanics of Materials IWM, Freiburg, Germany

[2]Laboratory for Materials and Joining Technology (LWF), University of Paderborn, Paderborn, Germany

[a]philipp.baehr@iwm.fraunhofer.de, [b]silke.sommer@iwm.fraunhofer.de, [c]gerson.meschut@lwf.uni-paderborn.de

Keywords: Finite Element Method (FEM), Joining, Notch Effects

Abstract. Subject of the present paper is the investigation of notch effects due to mechanical joints in press-hardened steel components. It can be observed that joints within a component structure result in a stress concentration. This stress concentration in combination with the load history of the material due to the joining process may result in failure of the component under crash load. A self-piercing riveted (SPR) joint between a 22MnB5 steel sheet with a thickness of 1.2 mm and an EN-AW 6016 aluminum sheet with a thickness of 2.0 mm has been experimentally and numerically investigated. The main investigations are concerned with the press-hardened sheet metal, as this is used in safety-relevant areas of the car body. First, experimental and metallographic investigations of the punched hole in the 22MnB5 have been performed to assess the quality of the punched edge. Then an elastic-plastic material model with additional phenomenological damage model has been calibrated for the 22MnB5. With the calibrated material model, a simulation of the SPR joining process has been performed to determine the load history of the joint. Based on the results of this process simulation, simplified shell models have been created, which consider the preload caused by the joining process. It was shown, that the SPR joining process results in high plastic strain and pre-damage of the sheet metal. By taking this process-related pre-damage into account the accuracy of crash simulations can be significantly increased.

Introduction

The production of car bodies using multi-material design is a key technology for the realization of lightweight construction concepts. The load appropriate use of materials enables an optimized ratio of strength and weight. Therefore, the overall weight of the construction can be reduced while simultaneously increasing the safety. The reduction of weight is one of the most important drivers for an increased efficiency and therefore highly important for the development of sustainable mobility concepts. On the other hand, the use multi-material design challenges the joining technology since different material types must be joined. In particular, the traditional spot-welding process cannot be used to join different materials such as aluminum and steel. Thus, alternative joining technologies have been established to overcome these problems. Especially mechanical joining with self-piercing rivets (SPR) is a widely used technique, particularly used for joining dissimilar materials [1]. SPR joints have some advantages, which render them important for the mass production in the automotive sector [2]. Besides their positive characteristics SPR joints result in large deformation of the joined sheet metals. This in combination with the stress concentration at the edge of the punched hole (Fig. 1) can result in crack formation under crash load. Therefore, it is important to consider the load-bearing and failure behavior of the joint and

Content from this work may be used under the terms of the Creative Commons Attribution 3.0 license. Any further distribution of this work must maintain attribution to the author(s) and the title of the work, journal citation and DOI. Published under license by Materials Research Forum LLC.

Sheet Metal 2025
Materials Research Proceedings 52 (2025) 142-149

Materials Research Forum LLC
https://doi.org/10.21741/9781644903551-18

especially the notch effect due to the punched hole in the development process to ensure a reliable and safe product.

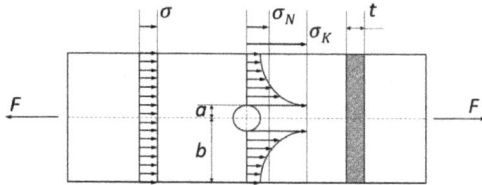

Fig. 1: Stress concentration due to a geometrical notch

Experimental Investigation

The investigations in this work are based on an exemplary connection between a press-hardened 22MnB5 steel sheet with a thickness of 1.5 mm and an EN-AW 6016 aluminum sheet with a thickness of 2.0 mm. To investigate the notch effect due to the SPR joint in the upper 22MnB5 sheet, two different specimen tests, namely tensile and punch tests, have been conducted. Both specimens were joined in the middle with an SPR, which has been removed from the joint prior to testing. The aim of these investigations was to determine the influence of the pre-damage due to the joining process on the load-bearing capacity of the punched edge. Therefore, additional specimens where investigated, where the hole was created via micro waterjet cutting (WJC). It is assumed that WJC will result in less damage to the material. Therefore, the parameters of the WJC process were selected so that the edge of the hole has the highest possible surface quality, nevertheless this process always results in surface imperfections that can contribute to crack initiation. All experimental results originate from the AiF/FOSTA research project (19751N/P1268) "Characterization and modeling of notch effects due to multi-material joints in car body components made of ultra-high strength steel" [3]. The used specimen geometry for the tensile tests is given on the left in Fig. 2. The specimens were tested with a tensile testing machine and the force-displacement behavior was evaluated. The displacement was measured locally on the specimen surface using two points at a distance of l_0. A total of five repetitions were carried out for the SPR specimens and three for the WJC specimens. The results can be seen on the right in Fig. 2. The elastic behavior of the specimens with the two different types of holes is nearly identical although the SPR hole shows a slightly higher stiffness. This could be due to the pre-strain at the hole edge caused by material hardening effects during the local cutting process. Furthermore, the specimens with a SPR hole show a lower maximum force as well as a lower displacement at fracture. This is a result of pre-damage due to micro cracks from the riveting process.

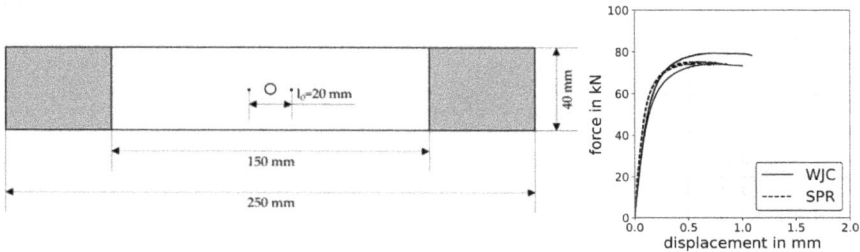

Fig. 2: Tensile specimen with central hole (left), experimental results of the tensile tests with central hole due to self-piercing rivet and micro water jet cutting (right)

The geometry of the punch test specimen can be seen on the left in Fig. 3. During the punch test, this specimen is held between two clamps and loaded by a punch with a spherical head coaxial to the center hole. A continuous measurement of the applied punch force is carried out during the test, and the punch displacement is also measured optically on the punch surface. Again, five specimens with holes manufactured via SPR and three specimens with WJC holes have been investigated. The measured force vs. displacement curves are given on the right in Fig. 3. A comparison of the curves shows a significant influence of the SPR process on the specimen stiffness. The SPR process leads to a bending off the specimen which increases the initial contact surface between the specimen and the punch, hence the increased stiffness because of higher friction. The plastic strain caused by the riveting process results in material hardening which has an influence on the deformation behavior of the SPR specimens as well. Finally, a major influence of the SPR process on the load bearing capacity can be observed. This is again a result of pre-damage at the hole edge due to the SPR process. The effects of pre-damage are greater for the punch test compared to the flat tensile test due to the radial loading of the hole edge. In the flat tensile test, only the areas of the hole tangential to the tensile direction are loaded, which results in a smaller influence on the specimen behavior.

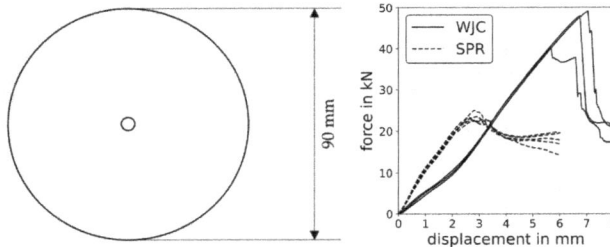

Fig. 3: Punch specimen with central hole (left), Experimental results of the punch tests with central hole due to self-piercing rivet and micro water jet cutting (right)

Material modeling

To describe the deformation behavior of the investigated materials in the numerical simulation an isotropic, elasto-plastic material model has been used. To describe the elastic deformation behavior this model allows the specification of a Young's modulus and a Poisson's ratio. The plastic deformation behavior is described with an arbitrary stress versus plastic strain curve which can be defined by the user. The basic shape of the stress versus plastic strain curve can be derived from the results of tensile tests, whereby the experimental results must be extrapolated with a suitable approach to describe the material behavior at high strains. For this reason, a modified Voce law according [4] has been used (Eq. 1):

$$\sigma_{Voce}(\varepsilon_{pl}) = \begin{cases} R_p + (A_1 + B_1\varepsilon_{pl})(1 - e^{k_1\varepsilon_{pl}}), & \varepsilon_{pl} < A_g \\ R_2 + (A_2 + B_2\varepsilon_{pl})(1 - e^{k_2\varepsilon_{pl}}), & \varepsilon_{pl} \geq A_g \end{cases} \tag{1}$$

The determined stress versus plastic strain curve for the investigated materials are shown in Fig. 4. Material failure for the press-hardened steel is considered with a phenomenological damage model [5]. In this model damage D is calculated incrementally from the current plastic strain increment $\Delta\varepsilon_{pl}$ and a defined fracture strain $\varepsilon_f(\eta, \xi)$ for the current stress state (Eq. 2):

$$D = \sum \frac{\Delta \varepsilon_{pl}}{\varepsilon_f(\eta, \xi)} \qquad (2)$$

To describe the stress state dependent fracture strain $\varepsilon_f(\eta, \xi)$ an approach according to Bai and Wierzbicki [6] is used (Eq. 2). In this approach the fracture strain is described with several exponential functions in dependence of the stress triaxiality η and the Lode angle parameter ξ. A symmetry of the fracture strain with respect to Lode angle parameter $\xi = 0$ is assumed. According to Eq. 3 the fracture strain is limited by two boundaries $\varepsilon_f^{(ax)}$ for $\xi = 1$ and $\varepsilon_f^{(0)}$ for $\xi = 0$:

$$\varepsilon_f(\eta, \xi) = \left(\varepsilon_f^{(ax)} - \varepsilon_f^{(0)}\right)\xi^2 + \varepsilon_f^{(0)} = (D_1 e^{-D_2 \eta} - D_3 e^{-D_4 \eta})\xi^2 + D_3 e^{-D_4 \eta} \qquad (3)$$

The Parameters D_1, D_2, D_3 and D_4 describe the shape of the failure surface and must be determined in accordance with the real material behavior. According to [6] there is a relationship between the stress triaxiality η and the Lode angle parameter ξ in case of the plane stress state (Eq. 4). The fracture strain for shell elements can therefore be described as a function of the stress triaxiality only:

$$\xi(\eta) = 1 - \frac{2}{\pi}\cos^{-1}\left[-\frac{27}{2}\eta(\eta^2 - \frac{1}{3})\right] \qquad (4)$$

The parameters D_1, D_2, D_3 and D_4 must be calibrated for the different element types based on experimental results of specimen tests with various geometries. Therefore, the stress-state dependent plastic strain history in the critical region of this specimen tests must be determined by simulation. With this information and the experimentally determined point of failure, the shape of the failure surface can be fitted. Fig. 4 shows the fitted failure surface with overlaid curve of fracture strain assuming a plane stress state according to Eq. 4. This assumption results in the characteristic shape with two local minima when the curve is plotted as a function of triaxiality only. The adjusted fracture curve for shell elements is shown in Figure 4 on the right.

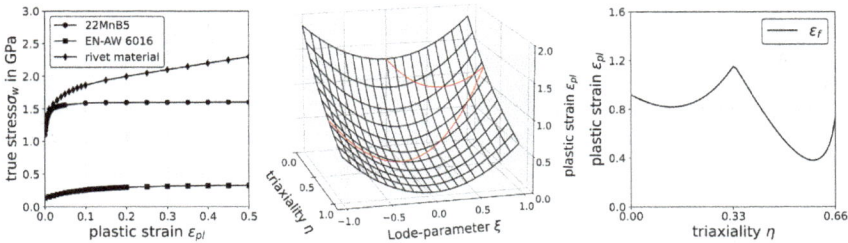

Fig. 4: Plasticity behavior of the investigated materials (left), fracture curve of 22MnB5 for shell elements (middle), fracture surface of 22MnB5 for solid elements (right)

Process Simulation

The calibrate material models from the previous chapter are used to simulate the SPR joining process. A two-dimensional, axisymmetric simulation model of the SPR joining process has been created (Fig. 5). The model is meshed with elements with an average size of 0.05 mm for the rivet and the two sheets. Punch, blank holder and die are meshed with coarser elements. Due to the large deformations of the sheets during the joining process, adaptive remeshing is used for these parts. The deformation behavior of the sheets and the rivet is modeled with the flow curves which can be found on the left in Fig. 4. For the upper steel sheet material failure is considered with the failure surface shown on the right in Fig. 4. The punch, blank holder and die are modeled as rigid bodies. The boundary conditions in the simulation are set in accordance with the real joining process. For the die the degrees of freedom of all nodes are fixed in all directions. The blank holder fixes the two sheets with a force of 15 kN applied in y-direction. The punch pushes on the rivet with a defined velocity of 2.0 m/s. The motion stops, when the contact force between the punch and the rivet reaches a value of 72 kN.

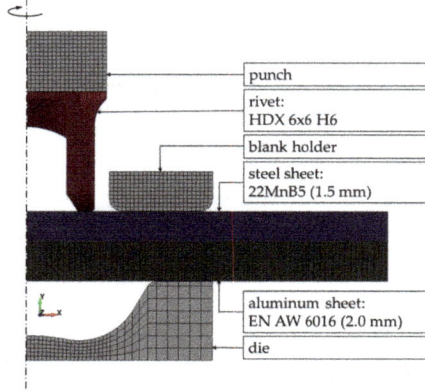

Fig. 5: Simulation model of the SPR joining process

Fig. 6 shows selected results from the simulation of the SPR joining process. An evaluation of the plastic strain distribution can be found on the left in Fig. 6. The highest strains are located in the lower aluminum sheet where the material is pressed into the die. The strain decreases with increasing distance from the middle of the joint. The rivet shows only minor deformations, mainly around the rivet foot. For the upper steel sheet, the highest deformations can be observed in the punched-out material underneath the rivet foot. Furthermore, higher plastic strains can be observed at the edge of the punched hole. Material damage is only calculated for the upper sheet. The damage distribution can be found on the right in Fig. 6. The highest damage values can be observed in the punched-out material between the rivet foot as well as at the edge of the punched hole in the upper sheet on the opposite side to the rivet head. The area in which the damage is localized is very narrow, which is why the elements next to the crack show relatively low damage values of 0.5-0.6.

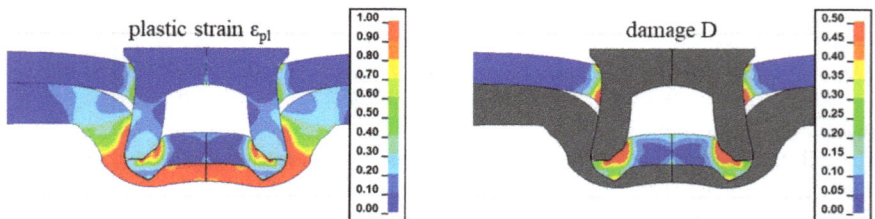

Fig. 6: Calculated distribution of plastic strain (left) and damage (right) of the SPR joint

Sheet Metal 2025
Materials Research Proceedings 52 (2025) 142-149

Materials Research Forum LLC
https://doi.org/10.21741/9781644903551-18

Specimen Simulation

Based on the results of the process simulations simplified FE shell models of the specimen tests shown in Fig. 2 and Fig. 3 have been created. The models are designed so that they can be used in a crash application. In order to reproduce the notch effect due to the SPR, a hole has been modeled within the specimen. Furthermore, a pre-strain and pre-damage of the shell elements surrounding the hole has been included. The distribution of these values is taken from the results of the process simulation and mapped on the individual integration points of the shell element. The damage distribution within these shell elements can be seen in Fig. 7.

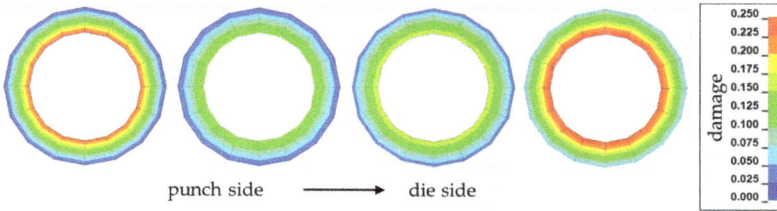

Fig. 7: Damage in the four integration point layers of the shell elements around the SPR

With the simplified model of the punched edge simulations of the specimen tests shown in Fig. 2 and Fig. 3 have been performed. The results of these simulation can be seen in Fig. 8. A comparison of the calculated force-displacement behavior for the tensile tests shows, that the initial damage due to the joining process must be considered to replicate the experimental fracture displacement (Fig. 8 right). The model without the initial damage shows an overestimation of the fracture displacement. Same is true for the punch test, whereby an overestimation of the maximum force can also be observed when no initial damage is considered.

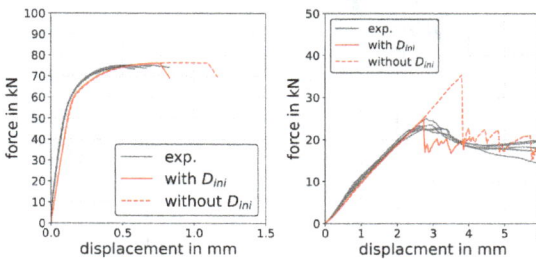

Fig. 8: Simulation results of the tensile (left) and punch tests (right) in comparison with the experimental results

Fig. 9 shows an evaluation of the damage evolution as well as the stress state within the critical element for the simulation of the tensile and punch test. In the damage evolution diagram the initialized damage and strain values can be clearly seen. After the onset of plastic deformation, a linear accumulation of damage can be observed until element failure occurs at a plastic strain of approximately 0.5. A comparison of the stress state shows that both tests result in similar triaxiality values of around 0.5. The plastic deformation takes place at an almost constant stress triaxiality until failure occurs at a plastic strain of approximately 0.5 (Fig. 9 right).

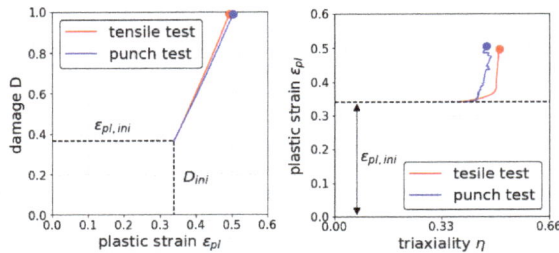

Fig. 9: Damage evolution (left) and stress state (right) of the critical integration point in the first failed element for the tensile and punch test simulation

Component Simulation

To validate the models presented above experimental component tests have been performed and simulated. The simulation model of these component tests can be seen in Fig. 10. The component itself consists of a press-hardened 22MnB5 profile with a SPR connection on top of the profile. The profile is loaded under 3-point bending load with the punch pushing on the opposite side to the SPR joint. This load-case results in a tensile stress at the edge of the SPR hole in the 22MnB5 profile and therefore failure due to the notch. The pre-damage of the hole edge is considered according to Fig. 7. The load-bearing behavior of the connection between the 22MnB5 profile and the attached EN-AW 6016 sheet is considered with an element free constrained based model [7].

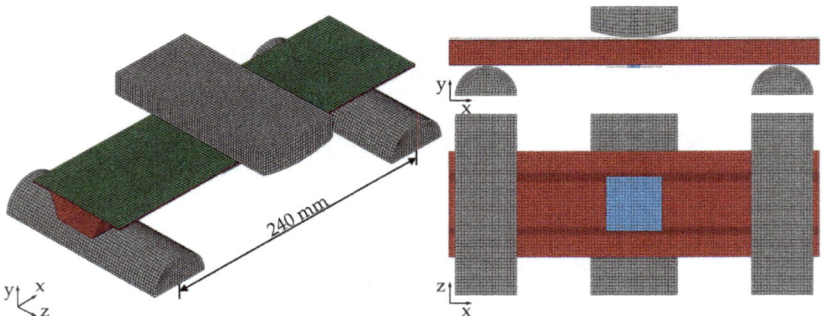

Fig. 10: Simulation model of the component tests

The results of the component simulations can be found in Fig. 11. A comparison of the force-displacement behavior with the experimental result, shows a good correlation for the model with included damage history (Fig. 11 left). The model without the damage history however shows an overestimation of the fracture displacement and therefore the energy absorption compared to the experimental result. An evaluation of the damage evolution within the critical element clearly shows the initialized damage and plastic strain values (Fig. 11 middle). From this starting point linear damage evolution can be observed until failure occurs at a plastic strain of approximately 0.5. The strain path shows an almost constant triaxiality of 0.5, which corresponds to the stress state of the specimen tests (Fig. 11 right).

Sheet Metal 2025
Materials Research Proceedings 52 (2025) 142-149

Materials Research Forum LLC
https://doi.org/10.21741/9781644903551-18

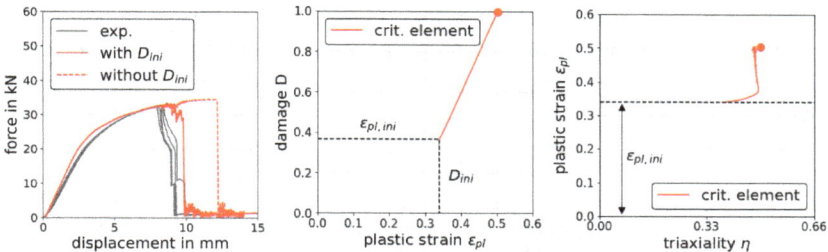

Fig. 11: Simulation results of the component tests in comparison with the experimental results (left), damage evolution (middle) and stress state (right) of the critical integration point in the first failed element

Conclusion

The results of this work showed, that SPR joints lead to a notch due to the punched hole and the process induced damage of the hole edge. Through simulation of the joining process this damage was numerically quantified. The determined damage was subsequently mapped on simplified specimen and component FE models. It was shown that this process induced damage influences the behavior of the models and must be considered to achieve predictable simulation results.

Acknowledgment

The IGF research project 19751 N (FOSTA P1268) by the Forschungsvereinigung Stahlanwendung e.V. (FOSTA) has been funded under the program for promotion of industrial research (IGF) by the Federal Ministry for Economics and Climate Action based on a decision of the German Bundestag.

References

[1] G. Meschut, M. Merklein, A. Brosius, D. Drummer, L. Fratini, U. Füssel, M. Gude, W. Homberg, P. A. F. Martins, M. Bobbert, M. Lechner, R. Kupfer, B. Gröger, D. Han, J. Kalich, F. Kappe, T. Kleffel, D. Köhler, C.-M. Kuball, J. Popp, D. Römisch, J. Troschitz, C. Wischer, S. Wituschek und M. Wolf, Review on mechanical joining by plastic deformation, J. Adv. Join. Process. 5 (2022). https://doi.org/10.1016/j.jajp.2022.100113

[2] A. Chrysanthou, X. Sun, Self-piercing riveting: Properties, Processing and Applications, Woodhead Publishing Limited, 2014.

[3] S. Sommer, P. Bähr, G. Meschut, E. Unruh, Characterization and modeling of notch effects due to multi-material joints in car body components made of ultra-high strength steel, final report for AiF-FOSTA-project 19751N/P1268 (2022).

[4] E. Voce, The relationship between stress and strain for homogeneous deformation, J. Inst. Met. 74 (1948), 537-562.

[5] F. Andrade, M. Feucht, A. Haufe, F. Neukamm, An incremental stress state dependent damage model for ductile failure prediction, Int. J. Fract. 200 (2016), 127-150. https://doi.org/10.1007/s10704-016-0081-2

[6] Y. Bai, T. Wierzbicki, A new model of metal plasticity and fracture with pressure and Lode dependence, Int. J. Plast. 24.6 (2008), 1071-1096. https://doi.org/10.1016/j.ijplas.2007.09.004

[7] M. Bier, S. Sommer, Simplified modeling of self-piercing riveted joints for crash simulation with a modified version of *CONSTRAINED_INTERPOLATION_SPOTWELD, 9th European LS-DYNA Conference (2013).

Sheet Metal 2025
Materials Research Proceedings 52 (2025) 150-158

Materials Research Forum LLC
https://doi.org/10.21741/9781644903551-19

Cross-process damage modeling: A process-chain case study of clinching and self-pierced riveting for aluminum connections

Özcan Harabati[1,a*], Christian Roman Bielak[1,b], Max Böhnke[1,c], Malte Christian Schlichter[1,d], Marc Brockmeier[1,e], Mathias Bobbert[1,f] and Gerson Meschut[1,g]

[1]Laboratory for material and joining technology (LWF), Paderborn University, Pohlweg 47-49, 33098 Paderborn, Germany

[a]harabati@lwf.uni-paderborn.de, [b]christian.bielak@lwf.upb.de, [c]max.boehnke@lwf.upb.de, [d]malte.schlichter@lwf.uni-paderborn.de, [e]mabro@mail.uni-paderborn.de, [f]mathias.bobbert@lwf.upb.de, [g]meschut@lwf.upb.de

Keywords: Mechanical Joining, Damage Model, FE-Simulation

Abstract. This study focuses on damage modeling across different mechanical joining processes within a process chain, specifically using clinching and self-pierce riveting (SPR). The aim is to apply a comprehensive model that captures the damage mechanisms and interactions in these technologies, optimizing them for enhanced performance and durability of aluminum joints. A GISSMO damage model was utilized, based on the stress states occurring during the joining process and a newly introduced damage testing method. This model was applied to both clinching and SPR processes. A detailed analysis of the stress states provided insights into their effect on the material. By incorporating these insights into the GISSMO model, improved accuracy in damage prediction was achieved. The model's application to clinching and SPR demonstrated its effectiveness in optimizing aluminum joint performance and durability, ensuring that the processes can be finely tuned to minimize damage and enhance joint quality.

Introduction

The application of mechanical joining techniques is becoming increasingly important in body-in-white manufacturing, as these technologies enable the realization of mixed-material joints. Clinching and self-pierce riveting (SPR) are two common mechanical joining processes. In clinching, no auxiliary joining element is used. In contrast, in SPR, the joining parts are joined with a semi-tubular rivet. In the joining technologies mentioned, high plastic strains occur in the joining parts. This limits the joinability of the parts to be joined. The SPR and clinching processes are shown in Fig. 1.

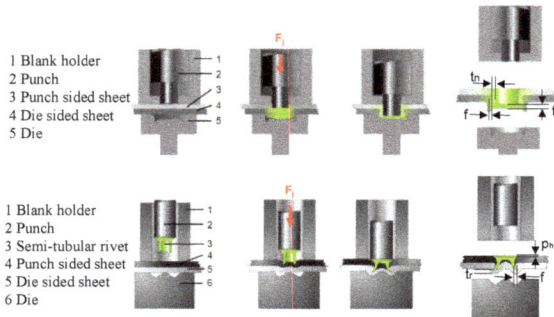

Fig. 1: Joining process flows of clinching and self-pierce riveting based on [1, 2]

Content from this work may be used under the terms of the Creative Commons Attribution 3.0 license. Any further distribution of this work must maintain attribution to the author(s) and the title of the work, journal citation and DOI. Published under license by Materials Research Forum LLC.

Sheet Metal 2025
Materials Research Proceedings 52 (2025) 150-158

Materials Research Forum LLC
https://doi.org/10.21741/9781644903551-19

In the SPR process, sheets are fixed between the die and blank holder, and a punch drives a rivet through the top sheet and expands it into the bottom sheet. In the clinching process, a punch and die are used to create a connection by forming of the materials. The quality of clinched joints is characterized by interlock (f), neck thickness (t_n), and bottom thickness (t_b). For SPR joints, quality is defined by interlock (f), minimum die-side material thickness (t_r), and rivet head position (p_h) [1, 2]. *Dario and Luca* [3] introduced a 3D FE Model illustrating the formation and failure mechanism of a clinch joint. It was noted that the model is able to predict the mechanical properties of the joint in terms of its tensile behavior by using the geometrical parameters of the tool and the friction coefficients. In [4] load capacity tests of a clinched joint under different load application angles were investigated by experiment und simulation. However, the fracture in the neck area at the clinch point could not be simulated because the damage model was not implemented. In [5] a numerical method was developed to investigate the influence of pre-deformation on the joinability of the examined material combinations. For this purpose, the process chain was modeled numerically. This method was extended to load-bearing capacity in [6]. A correlation between the pre-deformation influence and the shear tensile force was presented. A 3D FE Model for clinching process with a Hosfort-Coulomb failure model was presented in [7]. It was reported that the elements with highest numerically determined damage accumulation match with the failure location in the experiment. Numerical simulations were also performed for SPR joints. In [8], it was pointed out that the geometric separation criterion is well suited for SPR process simulations. However, this criterion reaches its limit for the sheet failure and also the separation direction. It has also been found that the transfer of residual stresses and damage from 2D process simulation to 3D load simulation increases the numerical prediction of SPR joint strength [9]. The influence of process parameters on the versatile self-pierce riveting process was also determined using FEM and published in [10]. Furthermore, a simulation process chain for SPR joints was modelled and the results published in [11–13]. The geometric separation criterion in SPR process and the damage model in load simulations were utilized as the failure mechanism for the sheet metal materials to be joined. *Otroshi et al.* [14] studied the stress state dependent damage model GISSMO to predict the damage development in the SPR process and to describe the material failure in the punch-side sheet. For this purpose, a strain rate-dependent 2D axisymmetric SPR FE model was considered. In [15], damage modeling in the 3D load simulation of mechanical joining processes was improved using the GISSMO model. The joint-induced pre-stressing was taken into account. The investigation of the joinability of additive manufactured components was demonstrated in [16]. The Cockroft-Latham criterion was applied for damage modeling and fracture prediction. None of the referenced studies used a unified damage model for the clinching and SPR joining technologies, nor for the shear load simulations of clinched and SPR joints.

Experimental
The TOX® PRESSOTECHNIK clinching system, model TZ-VSN, was used to produce the clinch joints. The joining speed was set to 2 mm/s, and the blank holder force was set to 785 N. A TOX® punch A50100 and a closed TOX® die BD8016 were used. The SPR joints performed on a TUCKER® TRT080 system with the servo-electric drive and C-frame. The joining speed was 20 mm/s with a blank holder force of 2,15 kN. The SPR joints were established with the die FM095 2115 and the rivet C 5.3x5.5 H4. The investigations in this study were performed with an aluminum alloy EN AW-6014 in condition T4 with a sheet thickness t = 2 mm. A standard C-rivet with H4 hardness and Almac coating was used for the SPR connection. Table 1 shows the chemical compositions of the aluminum and the rivet material. To determine the aluminum EN AW-6014 T4 flow curve for the numerical investigations, a layer compression test was carried out in accordance with DIN 50106 using two 3D DIC systems. The experimental test set-up was described in more details in [17].

Sheet Metal 2025
Materials Research Proceedings 52 (2025) 150-158

Materials Research Forum LLC
https://doi.org/10.21741/9781644903551-19

Table 1: Chemical compositions of investigated EN AW-6014 T4 and rivet [18]

Chemical composition EN AW-6014 T4 (wt.-%)								
Si	Fe max.	Cu max.	Mn	Mg	Cr max.	Zn max.	Ti max.	V max.
0.3-0.6	0.35	0.25	0.05-0.2	0.4-0.8	0.2	0.1	0.1	0.1
Chemical composition semi-tubular rivet (38B2) (wt.-%)								
Si max.	Fe	Cu max.	Mn	P max.	S max.	Cr max.	C	B
0.3	-	0.25	0.6-0.9	0.025	0.025	0.3	0.35-0.4	0.0005-0.0008

Fig. 2 shows the mechanical properties and extrapolated flow curves for the rivet material and the aluminum material to describe the material plasticity in process simulations.

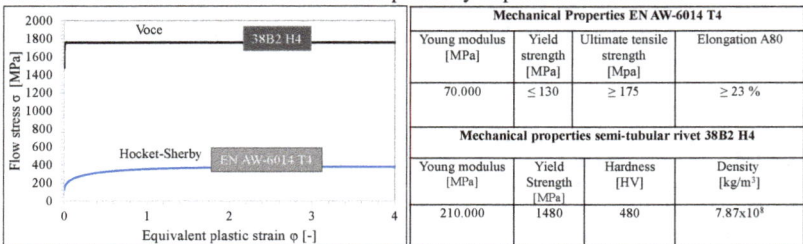

Fig. 2: Plasticity behavior and mechanical properties of EN AW-6014 T4 and 38B2 H4 [18]

The following figure shows the failure strain surface resulting from the calibration, depending on triaxiality and the lode angle [19]. The fracture strain depends on stress triaxiality within the Lode angle range of -1 to +1, flattening around a Lode angle of 0. It is observed that fracture strain decreases with increasing stress triaxiality, indicating that significantly higher fracture strains are achieved under compression-dominated stress conditions than under tension-dominated conditions.

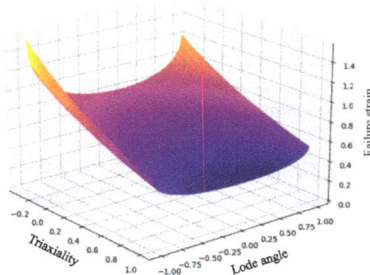

Fig. 3: Failure strain surface in stress triaxiality, Lode angle of EN AW-6014 T4 based on [19]

Numerical Simulation

The FE process simulation models are the main focus of this study. 2D rotational symmetry was used in the model setup. The rivet and tool geometries used in this study were created using an alicona Infinite Focus G5 measurement system. Fig. 4 provides information on the setup of the clinching and SPR FE models. The rigid part of the punch is used to move the punch and the rigid part of the hold-down clamp is used to move the hold-down clamp with a predefined blank holder

Sheet Metal 2025
Materials Research Proceedings 52 (2025) 150-158

Materials Research Forum LLC
https://doi.org/10.21741/9781644903551-19

preload force. To solve the problem with high element distortion, remeshing was activated via the CONTROL_ADAPTIVE [20] option card.

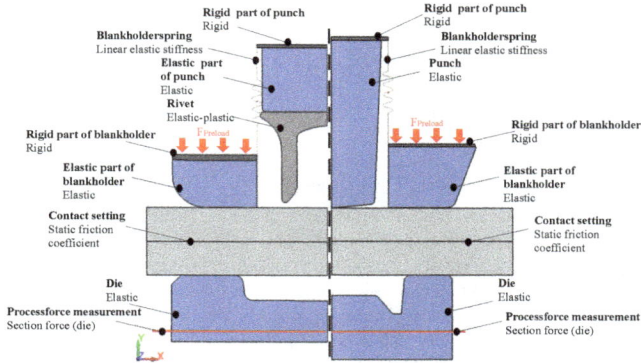

Fig. 4: Overview of the 2D FE-models

The joining process simulations were set up with displacement control. The spring back process was taken into account in order to better simulate the experimental conditions. Both process models were simulated with a damage accumulation and failure model, which is why the MAT_ADD_DAMAGE_GISSMO [21] material card was used.

Results and Discussion

In order to achieve the predictable shear tensile simulations, the FE models of both joining process must first be validated. The comparison of the simulatively determined contours (red) and simulative cross-sections of the clinching process and the SPR process with the experimental cross-section for the clinched and SPR joints is shown in Fig. 5.

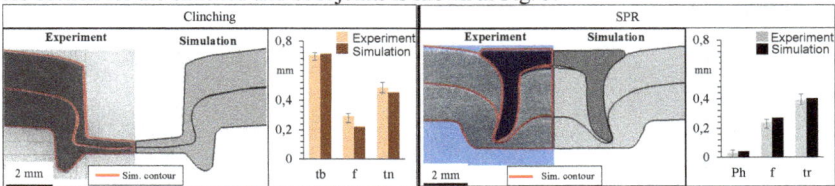

Fig. 5: Validation of the simulations of joining processes using geometric characteristics

It is shown that the clinching process simulation and the SPR process simulation, with the implemented and identical plasticity and damage modeling, predict the geometric properties with good accuracy. Furthermore, it can be observed that in both process simulations, the feed in behavior of the aluminum into the die and the flow behavior of the punch-sided sheet are reproducible. Another validation method for the FE models of both process simulations is the comparison of the process force-displacement diagrams. The curve comparisons of the two joining processes between experiment and simulation are shown in Fig. 6.

Sheet Metal 2025   Materials Research Forum LLC
Materials Research Proceedings 52 (2025) 150-158   https://doi.org/10.21741/9781644903551-19

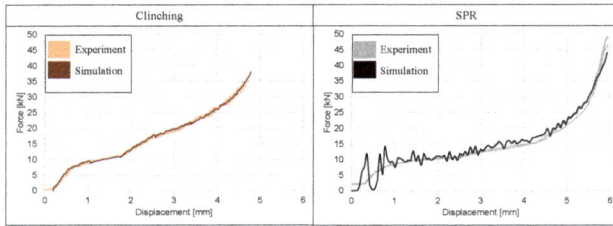

Fig. 6: Validation of simulation of joining processes using force-displacements diagrams

The results show a high level of agreement regarding the maximum force and the displacement at maximum force. For these reasons, the FE models of both joining processes can be considered valid and can therefore also be used for shear load simulations. After the 2D process simulations, the results (the formed mesh with plastic strain, pre-damage, residual stress etc.) were used in shear tensile load simulations to simulate the process chain. For this purpose, starting from the output file of the process simulations, a 3D model of the clinched and SPR joint point is created through mapping. In this process, the above-mentioned results are also transferred to the shear tensile model. In the next step, the 3D joint points are connected to the 3D shear tensile specimen via a TIED_CONTACT [20] to build 3D shear tensile models. Figure 7 shows the mapping method which was described in a similar way for the clinch connection in [6] and an overview of the procedure from 2D process simulation to 3D shear tensile model using the example of SPR.

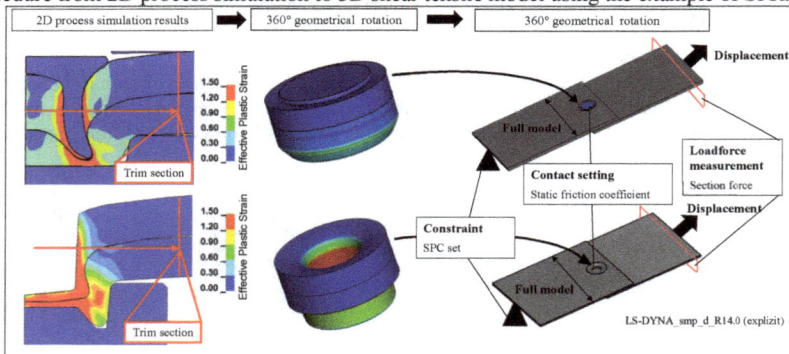

Fig. 7: Mapping from 2D to 3D to generate 3D shear tensile model based on [6]

The clinched and SPR shear tension models were fixed on one side and loaded on the other. The shear tension simulations were also simulated by displacement control. The force was measured using CROSS_SECTION. Remeshing was not necessarily due to the lower increase in plastic strain at the joints. The shear tensile simulations were also simulated with damage. Fig. 8 shows the results with regard to the force-displacement curve comparisons and the specimen deformations after experiment and simulation. By comparing determined shear load-displacement curves for both FE models, a good agreement is found with regard to the curve characteristics, the maximum force and the failure path. Therefore, it can be observed that the quasi-static load capacity of the clinched and SPR shear tensile specimens can be prognosed very well with the damage model implemented in this study. In addition, comparable sample deformations between experiment and simulation are also evident.

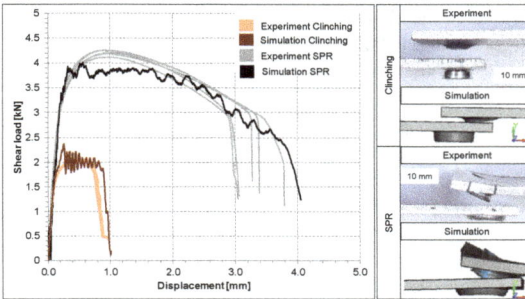

Fig. 8: Comprasion of the simulated and experimental shear load-displacement curves

The elements with the highest damage accumulation are in the neck area of the upper plate in the clinched shear tensile test and between the rivet foot and the upper plate in the SPR shear tensile test. This can be seen in Fig. 9. Therefore, in shear-tension models, the joints fail at the aforementioned elements.

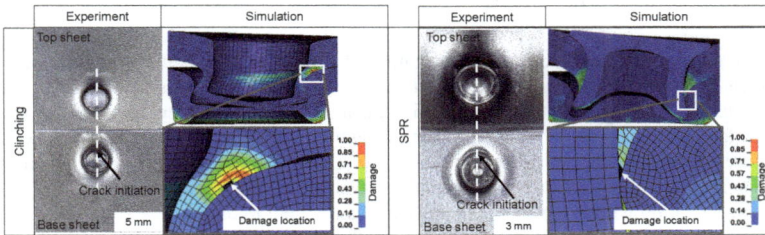

Fig. 9: Localization of the damage accumulation in the shear tensile tests

These elements were also selected in both joining processes to take into account the influence of pre-damage caused by the joining processes on the shear-tension test. The experimentally tested joints after shear tensile loading are shown on the left. It can be seen that the experimental failure location corresponds to the simulated failure location where the elements have the highest damage accumulation. To understand the local stress states occurring, the elements with the highest damage values were selected and evaluated during the both process chain (joining and shear tensile). Fig. 10 shows the plasticity curves of the selected elements during the complete process chain as a dependency of triaxiality and lode angle. By means of these parameters, the complex stress states that occur can be described. At the beginning, the selected elements of both joining process models show a tensile superimposed stress state. Compressive stress states occur during the subsequent process steps. In the SPR process in particular, the elements undergo significantly higher strain states (around 2) than in the clinching process. After reaching a degree of plastic strain of approx. 3, the elements return to a tensile superimposed and towards the end they experience a compressive stress state. In contrast, the elements in the clinching process remain in a tensile stress state at the end of the process. It can also be seen that at the beginning of the clinched shear tensile test, despite a change in the load direction, the selected elements continue to exhibit a tensile superimposed state of stress condition. As the elements are pulled over the sheet metal on the die side during the test, a shear stress is caused in the elements. Towards the end of the shear tensile test, the specimen tilts, which leads to unbuttoning at the failure point and in turn causes a stress state superimposed on the tensile stress. The stress triaxiality is approximately 0.12.

Sheet Metal 2025
Materials Research Proceedings 52 (2025) 150-158

Materials Research Forum LLC
https://doi.org/10.21741/9781644903551-19

The damage accumulation of the elements reaches the value 1, which leads to failure of the joint. In the FE-model, this means that the elements are deleted.

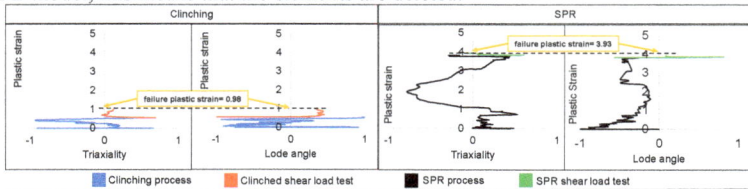

Fig. 10: Simulated stress states during the clinching and SPR processes and their shear tensile tests

At the beginning of the SPR shear tensile test, the change in load direction causes a transition of the selected elements form a compression-superimposed to a tension-superimposed stress state. In the further process of the test, the elements experience shear stresses due to the tilting of the shear tensile specimen, and at the point of failure of the SPR joint -similar to the clinch joint-unbuttoning occurs. The stress triaxiality is approximately -0.06. In the next time step in FE Model, the damage value reaches 1, which results that the elements are deleted and thus joint failure.

Summary

This study presents a comprehensive investigation of a clinching and self-pierce riveting process chain, including both the joining process and their corresponding shear tensile tests, using the finite element method. A significant aspect of this research is the implementation of the Hosford-Coulomb failure model, which was identified through a modified punch test as published in [7]. The failure model was applied in two process chains, allowing for a direct comparison of their performances under identical damage modeling conditions. The simulation results showed a high degree of correlation with experimental data. This was evident in the comparison of micro-section and process force-displacement diagrams, where the simulated outcomes closely matched the experimental results for both joining techniques. After validation of the process simulations, the clinched and SPR joints were mapped. This enabled the 3D shear tensile simulation models to be created. The results closely match the load-bearing capacity, failure points, and deformation behavior of the clinched and SPR specimens in shear tensile tests. In addition, differences in the stress states were found between the clinching process and the clinched shear tensile test compared to the SPR process and the SPR shear tensile tests. However, the fracture strain surface, created as a function of triaxiality and the lode angle, represents the fracture behavior of the aluminum material used to manufacture the clinched and SPR shear tensile samples precisely. Future research will expand on this study by integrating preforming steps. This will enhance the understanding of the influence of additional factors on the joining quality and structural integrity of the components involved. Investigations into the robustness of the numerical method are also required for predictive modelling of the process chain. No robustness analyses have yet been carried out in the current state of development. These are to be analysed in future studies, taking into account all process chain steps and varying process parameters.

Acknowledgement

The funding by the Deutsche Forschungsgemeinschaft (DFG, German Research Foundation) – Projekt-ID 418701707 – TRR 285/2, subproject A01 gratefully acknowledged. Data regarding the contents of the publication can be requested at www.trr285.de.

Sheet Metal 2025
Materials Research Proceedings 52 (2025) 150-158

Materials Research Forum LLC
https://doi.org/10.21741/9781644903551-19

References

[1] Deutscher Verband für Schweißen; Europäische Forschungsgesellschaft Blechverarbeitung: Taschenbuch DVS Merkblätter und-Richtlinien (2009)

[2] M. Neuser, F. Kappe, M Busch, O. Grydin, M. Bobbert, M. Schaper, G. Meschut, T. Hausotte, Joining suitability of cast aluminium for self-piercing riveting (2021),. IOP Conf Ser.: Mater Sci Eng 1157:12005. https://doi.org/10.1088/1757-899X/1157/1/012005

[3] D. Antonelli, L. Settineri, FEM simulation of clinched joints behavior (2001), Transactions of the North American Manufacturing Research Institute of SME

[4] S. Coppieters, P. Lava, R. van Hecke, S. Cooreman, H. Sol, P. van Houtte, D. Debruyne, Numerical and experimental study of the multi-axial quasi-static strength of clinched connections (2013). Int J Mater Form 6:437-451. https://doi.org/10.1007/s12289-012-1097-4

[5] C. R. Bielak, M. Böhnke, R. Beck, M. Bobbert, G. Meschut, Numerical analysis of the robustness of clinching process considering the pre-forming of the parts (2021). Journal of Advanced Joining Processes 3:100038. https://doi.org/10.1016/j.jajp.2020.100038

[6] C. R. Bielak, M. Böhnke, M. Bobbert, G. Meschut, Further development of a numerical method for analyzing the load capacity of clinched joints in versatile process chains (2021). 24th International Conference on Material Forming (ESAFORM 2021) https://doi.org/10.25518/esaform21.4298

[7] C. R. Bielak, M. Böhnke, J. Friedlein, M. Bobbert, J. Mergheim, P. Seinmann, G. Meschut, Numerical analysis of failure modeling in clinching process chain simulation (2023). Materials Research Proceedings Series, v. 25. Materials Research Forum LLC, Millersville https://doi.org/10.21741/9781644902417-33

[8] J. Eckstein, Numerische und experimentelle Erweiterung der Verfahrensgrenzen beim Halbhohlstanznieten hochfester Bleche (2009), Institut für Materialprüfung, Werkstoffkunde und Festigkeitslehre (IMWF) Universität Stuttgart und Materialprüfungsanstalt (MPA) Universität Stuttgart

[9] P. O. Bouchard, T. Laurent, L. Tollier, Numerical modeling of self-pierce riveting-From riveting process modeling down to structural analysis (2008). Journal of Materials Processing Technology 202:290-300. https://doi.org/10.1016/j.jmatprotec.2007.08.077

[10] F. Kappe, C. Zirngibl, B. Schleich, M. Bobbert, S. Wartzack, G. Meschut, Determining the influence of different process parameters on the versatile self-piercing riveting process using numerical methods (2022). Journal of Manufacturing Processes 84:1438-1448. https://doi.org/10.1016/j.jmapro.2022.11.019

[11] A. Rusia, M. Beck, S. Weihe, Simulation of self-piercing riveting process and joint failure with focus on material damage and failure modelling (2019). 12th European LS-DYNA Conference , Koblenz, Germany

[12] A. Rusia, S. Weihe, Development of an end-to-end simulation process chain for prediction of self-piercing riveting joint geometry and strength (2020). Journal of Manufacturing Processes 57:519-532. https://doi.org/10.1016/j.jmapro.2020.07.004

[13] Z. Du, B. Wei, Z. He, A. Cheng, L. Duan, G. Zhang, Experimental and numerical investigations of aluminium-steel self-piercing riveted joints under quasi-static and dynamic loadings (2021). Thin-Walled Structures 169:108277. https://doi.org/10.1016/j.tws.2021.108277

[14] M. Otroshi, M. Rossel, G. Meschut, Stress state dependent damage modeling of self-pierce riveting process simulation using GISSMO damage model (2020). Journal of Advanced Joining Processes 1:100015. https://doi.org/10.1016/j.jajp.2020.100015

[15] G. Meschut, M. Rossel, M. Otroshie, C. R. Bielak, Methodenentwicklung zur Verbesserung der Schädigungsmodellierung in der numerischen 3D-Belastungssimulation mechanischer Fügeverfahren unter Berücksichtigung der fügeinduzierten Vorbeanspruchung (2023), 1. Auflage. EFB-Forschungsbericht, vol 610. Europäische Forschungsgesellschaft für Blechverarbeitung e.V. (EFB), Hannover

[16] P. Heyser, R. Petker, G. Meschut, Development of a numerical simulation model for self-piercing riveting of additive manufactured AlSi10Mg (2023). Proceedings of the Institution of Mechanical Engineers, Part L: Journal of Materials: Design and Applications. https://doi.org/10.1177/14644207231158213

[17] M. Böhnke, F. Kappe, M. Bobbert, G. Meschut, Influence of various procedures for the determination of flow curves on the predictive accuracy of numerical simulations for mechanical joining processes (2021). Materials Testing 63:493-500 https://doi.org/10.1515/mt-2020-0082

[18] F. Kappe, C.R. Bielak, V. Sartison, M. Bobbert, G. Meschut F. Kappe, C. R. Bielak, V. Sartison, M. Bobbert, G. Meschut, Influence of rivet length on joint formation on self-piercing riveting process considering further process parameters (2021). ESAFORM 2021 https://doi.org/10.25518/esaform21.4277

[19] M. Böhnke, C. R. Bielak, J. Friedlein, M. Bobbert, J. Mergheim, G. Meschut, P. Steinmann, A calibration method for failure modeling in clinching process simulations (2023). Materials Research Proceedings Series, v. 25. Materials Research Forum LLC, Millersville https://doi.org/10.21741/9781644902417-34

[20] ANSYS (2024) LS-DYNA® Keyword User's Manual Volume I

[21] ANSYS (2024) LS-DYNA® Keyword User's Manual Volume II- Material Models

Sheet Metal 2025
Materials Research Proceedings 52 (2025) 159-167

Materials Research Forum LLC
https://doi.org/10.21741/9781644903551-20

Numerical and experimental investigation on full backward extrusion process in forming of pins from DC04 coil

LUO Keyu[1,a] *, VOGEL Marion[1,b], VOGEL-HEUSER Birgit[2,c] and
MERKLEIN Marion[1,d]

[1]Institute of Manufacturing Technology, Friedrich-Alexander-Universität Erlangen-Nürnberg, Egerlandstraße 13, 91058 Erlangen, Germany

[2]Institute of Automation and Information Systems (AIS), Technische Universität München, Boltzmannstraße 15, 85748 Garching bei München, Germany

[a]keyu.luo@fau.de, [b]marion.vogel@fau.de, [c]vogel-heuser@tum.de, [c]marion.merklein@fau.de

Keywords: Cold Forming, Sheet Metal, Simulation

Abstract. In pursuit of CO_2-neutral and resource-efficient production, the trend is increasingly moving towards near-net-shape geometries. In the field of cold forming, sheet-bulk metal forming is particularly suitable, as it combines the advantages of sheet metal forming with those of bulk forming. For preliminary investigations, strip material is used due to its flexibility, while coils are used from an industrial perspective. To gain a fundamental understanding of the process, previous analyses by Henneberg and Merklein [1] have explored the potentials and challenges of a sheet-bulk metal forming process. In order to minimize the losses caused by tool wear in this process, such as machine downtime, it is necessary to make an in-depth investigation of the influencing factors affecting tool quality. For this purpose, the forming process was numerically modeled using the simulation software Simufact.Forming. By determining various influencing factors, which can be inserted as simulation input parameters, the simulation should be generated as close to reality as possible. Upsetting tests are carried out using miniaturized specimens to determine the material properties of the steel DC04. The simulation model is validated via experimental tests conducted on a high-speed press. The quality of the workpiece and tool are studied through the examination of a continuous process. The trends identified in the experimental test are effectively reproduced in the numerical model despite minor deviations. It is shown that monitoring tool wear by analyzing the maximum force, the height of the formed pin and the diameter of the pin structure is relevant for the backward extrusion process.

Introduction

Based on its environmental balance, Germany ranks among the ten countries with the largest ecological footprint, underscoring the importance of environmentally and economically sustainable production nowadays [2]. This is particularly relevant in lightweight construction, which directly impacts CO_2 emissions [3]. To achieve higher functional integration, components with near-net-shape geometries need to be produced [4]. This is achievable through sheet-bulk metal forming (SBMF) [5]. While sheet metal forming offers high dimensional accuracy and material efficiency along with excellent surface quality, bulk forming focuses on high strength with significant shape changes [6]. The potentials arising from both forming processes are combined in SBMF [5]. SBMF includes several conventional forming operations. For cold forming processes, components are often formed at near room temperature. [7]. The workpiece is not actively heated during the forming process, thus preventing shrinkage and scaling in cold forming [8]. Sheet-bulk metal formed components are used in gearbox components for the automotive industry as well as in mechanical engineering for parts with complex geometries made of high-strength materials [8]. For preliminary tests and fundamental analyses, shorter sheet strips

Content from this work may be used under the terms of the Creative Commons Attribution 3.0 license. Any further distribution of this work must maintain attribution to the author(s) and the title of the work, journal citation and DOI. Published under license by Materials Research Forum LLC.

Sheet Metal 2025   Materials Research Forum LLC
Materials Research Proceedings 52 (2025) 159-167   https://doi.org/10.21741/9781644903551-20

are used, which are more flexible for investigating various parameters [9]. For the industrial production of sheet-bulk metal formed components, the material is provided in the form of a coil, which is used due to its high output and shorter processing times [10]. Both the potentials and challenges associated with a SBMF process using a coil are examined by Henneberg and Merklein in [1]. For this purpose, they use the typical SBMF processes of lateral and backward extrusion. Punches with and without circular cavities are used for the forming process. Both processes are numerically modeled and experimentally analyzed. The material flow has been identified as anisotropic due to two factors: the non-circular sheet geometry and the local pre-hardening of the strip resulting from the forming of pervious parts. In addition, it has been observed that anisotropic material flow can lead to uneven tool loading, which affects the dimensional accuracy of the components. In addition to the previously mentioned challenges, further considerations must be made with regard to tool wear when producing large quantities of parts. Since tool wear occurs particularly in forming processes with high effective surface loads, such as SBMF. As a consequence of the continuous operation of the process over an extended period, the dimensional accuracy of the parts is reduced, which also increases the scrap rate. Therefore, an early analysis of the causes and the onset of tool surface wear by means of real-time process monitoring of the part quality and the use of machine data is essential. The core objective of this research is to set up experimental and numerical model for the further development of an early warning system for tool wear and to conduct a preliminary assessment of the continuous forming process.

Experimental and Numerical Setup
This chapter gives an overview of the experimental and numerical setup for the backward extrusion process. First, the material characterization will be explained. The conventional low-alloy, cold-rolled deep-drawing steel DC04 (1.0338) with a thickness of 2 mm was utilized in this investigation. In order to determine the mechanical properties of material, an upsetting test with miniaturized cylindrical specimens was conducted in accordance with the methodology proposed by Hetz et al. [11]. In comparison to other conventional tests, it allows not only local component characterization but also material characterization under high strains [12]. The specimens with a height of 2 mm, which is the same as the sheet thickness in the real process and an upsetting ratio of h/d =1.5, resulting in a diameter of 1.33 mm, were extracted from the sheet plane on a microerosion machine (SX-200-HPM, Sarix SA) and subjected to upset testing at a strain rate of $\dot{\varphi}$ = 0.004 1/s using a universal testing machine (Z10, ZwickRoell AG). The experimental setup and the resulting flow curve are presented in Fig. 1.

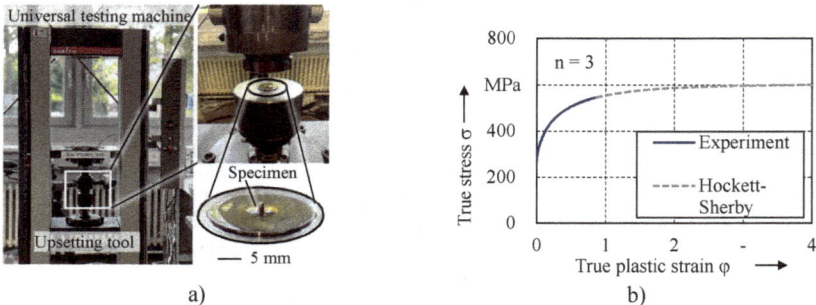

a)   b)

Fig. 1. a) Setup and b) flow curve of material characterization using the upsetting test with miniaturized cylindrical specimens

The initial true stress of the material is 262 ± 4 MPa, and the experimental determined maximum true plastic strain is $\varphi = 0.86$. To map the flow behaviors of SBMF with higher degrees

Sheet Metal 2025
Materials Research Proceedings 52 (2025) 159-167

Materials Research Forum LLC
https://doi.org/10.21741/9781644903551-20

of deformation, the Hockett-Sherby approach [13] was employed to extrapolate the flow curve up to a true strain of $\varphi = 4$. In this work, the process of forming pins from coil using full backward extrusion will be further investigated. The high-speed press (BSTA 50-95B2, Bruderer AG) with a nominal force of 500 kN has been used for the experimental research. It realizes requirements for high-volume production at a forming speed of 100 parts per minute. A force sensor (9106A, Kistler Group) and a displacement sensor (ST1278, Heidenhain AG) are integrated in the tooling system to monitor and record the force and displacement of the punch online. The blank is fed into the tooling system by a feed length of 20 mm per stroke through the feeding device on press. Before this step, the lubricant Beruforge 150 DL is applied on the blank in drops. Throughout the whole continuous forming process, the punch is in contact with the sheet blank with a stroke of 1 mm, resulting in the material flowing into the cavity of punch and forming workpieces. The experimental process setup is shown in Fig. 2.

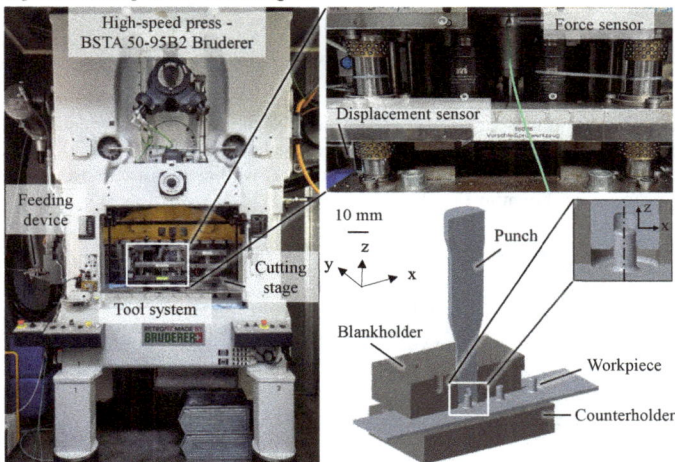

Fig. 2. Process setup of forming pins from coil using full backward extrusion

The simulation permits the formulation of detailed predictions regarding material behavior, stress distribution, and deformation patterns, thereby improving process optimization. In SBMF, particularly in the extrusion process, the Finite Element Method (FEM) has already been successfully applied by Johannes Henneberg [1]. In this work, the simulation software Simufact.Forming 2021.1, developed by Simufact Engineering GmbH, is applied. The sheet blank was meshed with hexahedral elements. To ensure a better trade-off between simulation accuracy and computational efficiency, the basic edge length was gradually adjusted, from 0.40, 0.45, 0.50, 0.55, to 0.60 mm. The actual edge lengths may deviate due to automatic refinements or internal mechanics during mesh generation. A mesh that is too coarse missed important details, while one that is too fine can consume excessive computational resources without significant gains in accuracy. This stepwise approach allowed for determining the appropriate complexity of the FE model, with 0.50 mm being chosen as the optimal value. To better capture the features of the forming zone, a hollow cylindrical refinement box with a diameter of 14 mm was created around the forming zone. The edge length in this area was reduced to 0.125 mm with two successive refinement levels. Investigations by Gröbel revealed that a friction factor $m = 0.1$ can be selected for the SMBF of DC04, which was used for present research [14]. The design of the simulation model and important set up parameters are shown in Fig. 3.

Sheet Metal 2025
Materials Research Proceedings 52 (2025) 159-167

Materials Research Forum LLC
https://doi.org/10.21741/9781644903551-20

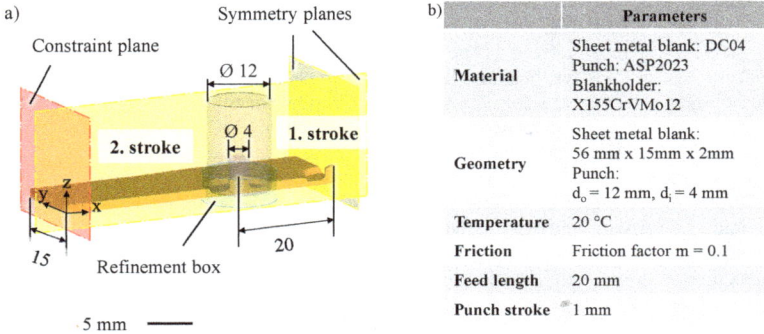

Fig. 3. a) Design of the simulation model and b) important set up parameters

Results

To ensure that the numerical model can reliably and accurately provide meaningful predictions which correspond to the real forming process, it was validated in terms of pin heights, diameters of the formed workpieces in the X- and Y-directions, as well as maximum process forces during the forming process, as shown in Fig. 4.

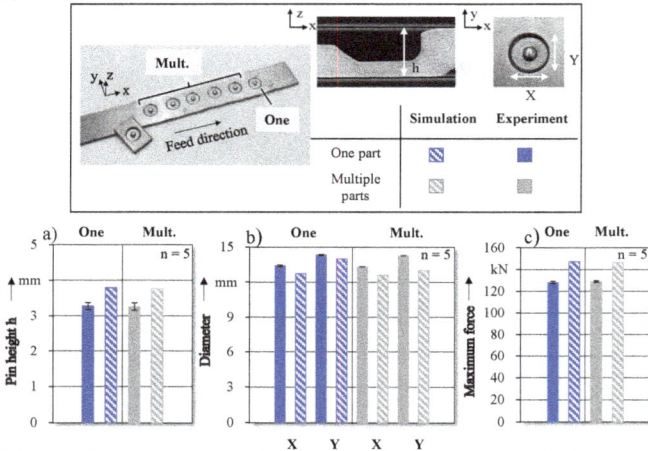

Fig. 4. Validation of the numerical model regarding to a) pin heights, b) diameters of the formed workpieces in X- and Y direction and c) maximum process forces

Furthermore, a comparison was conducted between the first workpiece (One) and the last five workpieces (Mult.) on one coil to account for the influence of the previous forming operation on the subsequent workpieces. A comparison between the experimental and numerical results indicated the presence of differences in the evaluated parameters. For example, the experimental pin height for one workpiece with 3.28 ± 0.10 mm is lower than the pin height 3.80 mm from the numerical model with a deviation of 0.52 mm. However, both do not reach the theoretical height, which indicates that some material flows out of the forming zone. Similar trends were observed in the maximum process forces, where the simulation consistently overestimated the experimental

Sheet Metal 2025
Materials Research Proceedings 52 (2025) 159-167

Materials Research Forum LLC
https://doi.org/10.21741/9781644903551-20

values with maximum deviations of 12.1 % - 13.1 %. Besides, the diameter in the Y direction is always higher than that in the X direction proving the anisotropy of the material. This was confirmed by both experimental and simulation results. All these differences can be attributed to the tribological conditions present in real SBMF processes, which are challenging to model with high accuracy in simulations [15]. Furthermore, the cumulative effects of the preceding forming operations on subsequent workpieces introduce additional complexity to the simulation, as the model may not fully account for the changes in material properties or wear of the tools throughout the process [16]. Despite these challenges, the numerical model captures the overall trends and provides valuable insights. For example, the pin height reduced from 3.80 mm to 3.77 mm when multiple workpieces were formed. The same results were obtained in the experiments. In conclusion, despite minor deviations, the numerical model effectively represents the experimental trends and provides valuable insights into the full backward extrusion process from coil.

In order to gain a deeper understanding of the forming process in continuous operation, a preliminary assessment was investigated in the following. Pre-experiments were conducted with 1000 strokes to simulate the effects of long-term continuous operation. These pre-experiments revealed significant findings into the behavior of the force, displacement curve, pin height and also diameters in X/Y direction during continuous forming. These parameters are also important indicators for the later assessment of wear of tools during the process.

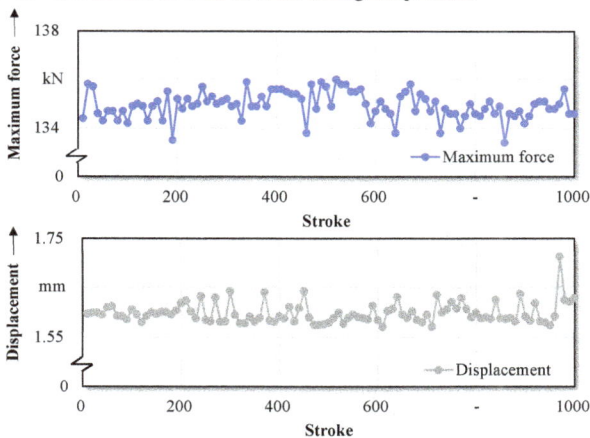

Fig. 5. Force / Displacement - stroke curves in continuous forming process

The important aspect of the continuous forming process is the behavior of the maximum force required during each stroke. The force applied to the workpiece and the corresponding displacement were measured for each stroke. As shown in Fig. 5, the maximum force exhibited significant fluctuations. These fluctuations were primarily attributed to the run-in phase of the system. At the beginning of the process, the irregularities of contact surfaces lead to unstable friction conditions, as well as unstable thermal effects, causing variations in force. The maximum force shows a slight decrease overall. Theoretically, as a result of working for long periods of time, the tool will wear and the surface roughness will increase. This leads to increased friction between the tool and the workpiece, which thus increases the force required to form the pins. However, a slight decrease occurs here. This reduction in force could be attributed to changes in the lubrication conditions throughout the operation. Initially, the tool and workpiece may have varying lubrication levels, which could create higher friction and require more force to form the workpiece. However, as the process continues, lubrication becomes more evenly distributed across the tool-workpiece interface, reducing the friction and subsequently lowering the force needed. The results of the slight decrease in maximum force also illustrate at the same time that forming only 1000 parts is not enough to describe a real change in the continuous forming process.

Furthermore, the displacement curve remains relatively stable with minimal variations, indicating consistent material flow during the forming process. The material properties of the workpiece can change as the process progresses. In SBMF, repeated strokes tend to alter the material's mechanical properties, such as its hardness and ductility. These changes occur due to strain hardening and localized plastic deformation, both of which can influence the force required for forming and the final shape of the workpiece [17]. The nearly consistent linear regression observed in Fig. 5 illustrates that, despite these material property changes, the process follows a predictable trend in terms of force and displacement.

Besides the analysis of the force and displacement curve, the geometrical component characteristics were also taken into account. The pin heights and diameter in X/Y direction were measured at regular intervals using Optical 3D measurement System (Infinitefocus XL200 G5, Bruker Alicona). As illustrated in Fig. 6 a), the height of pin exhibits considerable fluctuations. Potential explanations for this phenomenon include material flow and lubrication dynamics. In the early strokes, when lubrication is not yet uniformly applied across the workpiece surface, the material experiences higher friction. Consequently, the pin height is significantly higher than the average. As lubrication stabilizes throughout the operation, the pin height decreases a little bit and follows a more consistent increased trend line, as illustrated by the regression analysis in the graphs. The initial variation in pin height underscores the importance of continuous lubrication and material flow control in maintaining consistent quality in the formed workpieces. The increasing trend is also a good reflection of the increased friction between the tool and workpiece due to the gradual increase in tool wear, which leads to more material flowing into the cavity and an increase in the pin height.

In this continuous operation, besides pin heights, another notable observation is the difference in the diameters of the formed pins, measured in the X and Y directions. Fig. 6 a) also showed slight variations regarding to diameters across the 1000 strokes. Pre-hardening of the workpiece and tool wear both contribute to the slight differences. The continuous process tends to lead to higher values for the diameter in the Y-direction compared to the X-direction during the experiment. This anisotropic behavior suggests that there is a distinct material flow pattern that is influenced by the blank geometry and pre-hardening of the previous workpieces [1]. Furthermore, the diameters remain relatively stable after the initial strokes, with minimal changes observed in both directions. This can be caused by uneven tool wear. Optical roughness measurements of the punch surface before and after forming process which is shown in Fig. 6 b), indicated that after 1000 strokes, the punch surface, particularly in the feed direction, displays visible cracks in addition to some abrasive wear. This phenomenon can be attributed to the slight reduction in diameter observed in the X-direction. In contrast, the corresponding Y-direction shows an increase. This gradual wear, although minimal, emphasizes the significance of monitoring wear during the high-precision forming process, as it directly impacts the quality of the formed part.

Sheet Metal 2025   Materials Research Forum LLC
Materials Research Proceedings 52 (2025) 159-167   https://doi.org/10.21741/9781644903551-20

a) Geometry of workpieces

b) Punch surface quality

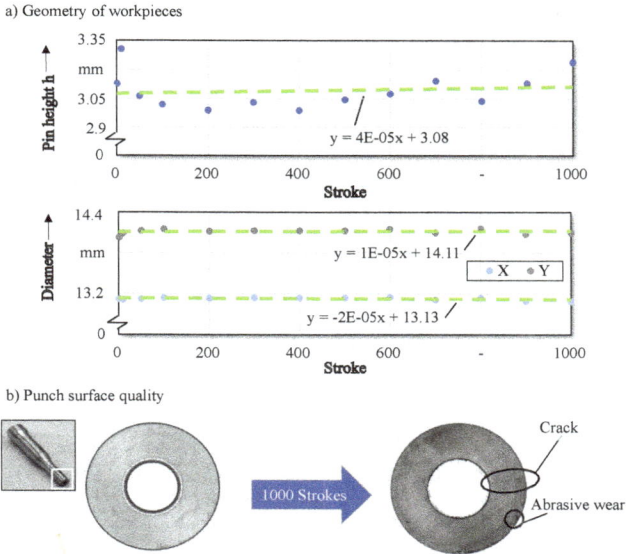

Fig. 6. a) Geometric variation of formed workpieces and b) punch surface quality in continuous forming process

Summary and Outlook

In order to provide an early analysis of the causes of tool surface wear during continuous operation forming processes, this paper presents a preliminary evaluation. Firstly, simulation model of a backward extrusion process was developed and validated by experimental testing. After validation, the simulation model provides a good overview of the process and effectively reproduces the experimental trends despite minor deviations. For a continuous process of 1000 strokes, it was shown that the maximum force required for forming decreases slightly, which was attributed to improved lubrication distribution over time, reducing friction. The displacement remained stable, indicating consistent material flow throughout. Additionally, tool wear and lubrication conditions played key roles in the variations of pin heights and diameters. Specifically, diameters in X/Y direction showed anisotropic behavior due to material flow influenced by geometry and pre-hardening. Furthermore, optical measurements revealed cracks and abrasive wear on the punch surface after 1000 strokes, underscoring the importance of wear monitoring in maintaining part quality. For further progress, a more detailed evaluation of material anisotropy is necessary to explain the results for diameters in X/Y direction. The numerical model will then be further refined. The pin extrusion test will be used to explore the effect of for example temperature or lubrication situation on wear. Further tests should be performed to increase the number of strokes and monitor more obvious tool wear effects. Both the evolution of the maximum force as well as pin height and diameter can be used as indicators to determine the wear behavior.

Acknowledgement

The project 'Data-based identification and prediction of the die surface condition and interactions in sheet bulk metal forming processes from coil' (Project nr.: 520194633) is part of the priority

Sheet Metal 2025   Materials Research Forum LLC

Materials Research Proceedings 52 (2025) 159-167   https://doi.org/10.21741/9781644903551-20

program SPP2422, funded by Deutsche Forschungsgemeinschaft (DFG, German Research Foundation). The authors would like to thank the DFG for their support.

References

[1] Henneberg, J.; Merklein, M.: Investigation on extrusion processes in sheet-bulk metal forming from coil. CIRP Journal of Manufacturing Science and Technology. 31 (2020) 561-574 https://doi.org/10.1016/j.cirpj.2020.08.007

[2] Sharif, A.; Bashir Uzmu; Mehmood, S.; Cheong, C. W.; Bashir, M. F.: Exploring the impact of green technology, renewable energy and globalization towards environmental sustainability in the top ecological impacted countries. Geoscience Frontiers. 15 (2024) 101895 https://doi.org/10.1016/j.gsf.2024.101895

[3] Ingarao, G.; Di Lorenzo, R.; Micari, F.: Sustainability issues in sheet metal forming processes: an overview. Journal of Cleaner Production. 19 (2011) 337-347 https://doi.org/10.1016/j.jclepro.2010.10.005

[4] Onodera, S.; Sawai, K.: Current cold-forging techniques for the manufacture of complex precision near-net-shapes. Journal of Materials Processing Technology. 35 (1992) 385-396 https://doi.org/10.1016/0924-0136(92)90329-Q

[5] Merklein, M.; Allwood, J. M.; Behrens, B.-A.; Brosius, A.; Hagenah, H.; Kuzman, K.; Mori, K.; Tekkaya, A. E.; Weckenmann, A.: Bulk forming of sheet metal. CIRP Annals. 61 (2012) 725-745 https://doi.org/10.1016/j.cirp.2012.05.007

[6] Altan, T.; Tekkaya, A. E.: Sheet Metal Forming - Fundamentals, ASM International: Materials Park, Ohio, 2012 https://doi.org/10.31399/asm.tb.smff.9781627083164

[7] Behrens, B.-A.; Bouguecha, A.; Lüken, I.; Klassen, A.; Odening, D.: Near-Net and Net Shape Forging. Comprehensive Materials Processing. (2014) 427-446 https://doi.org/10.1016/B978-0-08-096532-1.00323-X

[8] Doege, E.; Behrens, B.-A.: Handbuch Umformtechnik. Springer Berlin Heidelberg, 2016 https://doi.org/10.1007/978-3-662-43891-6

[9] Burggräf, P.; Bergweiler, G.; Kehrer, S.; Krawczyk, T.; Fiedler, F.: Mega-casting in the automotive production system: Expert interview-based impact analysis of large-format aluminium high-pressure die-casting (HPDC) on the vehicle production. Journal of Manufacturing Processes. 124 (2024) 918-935 https://doi.org/10.1016/j.jmapro.2024.06.028

[10] Birkert, A.; Haage, S.; Straub, M.: Umformtechnische Herstellung komplexer Karosserieteile. Springer Berlin Heidelberg, 2013 https://doi.org/10.1007/978-3-662-46038-2

[11] Hetz, P.; Kraus, M.; Merklein, M.: Characterization of sheet metal components by using an upsetting test with miniaturized cylindrical specimen. CIRP Annals. 71 (2022) 233-236 https://doi.org/10.1016/j.cirp.2022.03.010

[12] Peter Hetz, Marcel Rentz, Marion Merklein: Contact pressure-dependent friction compensation in upsetting tests with miniaturized specimens. Materials Research Proceedings. 25 (2023) 205-212 https://doi.org/10.21741/9781644902417-26

[13] J.E. Hockett; O.D. Sherby: Large strain deformation of polycrystalline metals at low homologous temperatures. Journal of the Mechanics and Physics of Solids. 23 (1975) 87-98 https://doi.org/10.1016/0022-5096(75)90018-6

[14] Gröbel, D.: Herstellung von Nebenformelementen unterschiedlicher Geometrie an Blechen mittels Fließpressverfahren der Blechmassivumformung. PhD Thesis FAU 317 (2019)

[15] Pilz, F.; Henneberg, J.; Merklein, M.: Extension of the forming limits of extrusion processes in sheet-bulk metal forming for production of minute functional elements. Manufactur. Rev. 7 (2020) 9 https://doi.org/10.1051/mfreview/2020003

[16] Behrens, B.A.; Wester, H.; Matthias, T.; Hübner, S.; Müller, P.; & Wälder, J.: Numerical Calculation of Tool Wear in Industrial Cold Forming Processes Using the Further Development of Wear Modelling, in: Merklein, M., et al.: Sheet Bulk Metal Forming: Research Results of the TCRC73, Springer Nature, Cham, 2020, pp: 535-552 https://doi.org/10.1007/978-3-030-61902-2_24

[17] Altan, T.; Ngaile, G.; Shen, G.: Cold and Hot Forging: Fundamentals and Applications, Vol. 1, ASM International: Materials Park, Ohio, 2005 https://doi.org/10.31399/asm.tb.chffa.9781627083003

Sheet Metal 2025

Materials Research Proceedings 52 (2025) 168-174

Materials Research Forum LLC

https://doi.org/10.21741/9781644903551-21

Modelling strategies for non-rotationally symmetric joints

Deekshith Reddy DEVULAPALLY[1,a] * and Thomas TRÖSTER[1,b]

[1]Paderborn University, Chair of Automotive Lightweight Design, Warburger Str. 100, 33098 Paderborn, Germany

[a]deekshith.reddy.devulapally@uni-paderborn.de, [b]thomas.troester@uni-paderborn.de

Keywords: Joining, Finite Element Method, Non-Rotationally Symmetric Joints

Abstract. Accurate Finite Element Modeling (FEM) of joints is essential in the design of complex mechanical systems such as automotive body-in-white (BIW) structures, as it plays a critical role in evaluating their performance. Although well-established techniques exist for modeling rotationally symmetric joints, there remains a significant gap in effectively modeling non-rotationally symmetric joints. These joints are particularly relevant in the automotive BIW, where they can better accommodate anisotropic loading conditions. In this study, strategies for modeling non-rotationally symmetric joints were explored using finite element simulations in LS-DYNA. The findings demonstrate that discrete beam elements can capture the anisotropic characteristics of such joints. Two models were tested: a single-beam model for stiffness periodicity every 90°, and a three-beam model for stiffness periodicity every 120°. Force responses, stress distribution, and sheet bending behaviors were analyzed, confirming that discrete beam elements can accurately represent direction-dependent stiffness. These results establish a foundation for developing advanced joint modeling strategies in complex mechanical systems.

Introduction

Joints are critical components in automotive BIW, directly influencing the structural integrity and overall performance of vehicles. They serve as primary pathways for transferring loads between structural elements, impacting safety, durability, and vehicle handling. The loading conditions that joints endure are complex, with many subjected to multi-directional forces. For example, a study on joint load distribution in car body structures revealed that the majority of joints are primarily subjected to shear forces under global bending and torsion conditions [1]. A subsequent study on steady-state vehicle simulations identified the need for direction-dependent joints in specific areas of the chassis [2]. Non-rotationally symmetric joints, exhibiting anisotropic behavior, offer a promising solution to these challenges by providing better resistance to torsional loads and allowing for targeted stiffness optimization in specific areas of the BIW. Further supporting this claim, another investigation explored the potential of non-rotationally symmetric joint geometries, specifically using elliptical shapes to implement anisotropic characteristics [3]. Their experiments demonstrated that varying the in-plane orientation angle of the elliptical joint altered its mechanical properties during shear testing. This research highlights the critical influence of joint geometry on performance and emphasizes the potential for optimizing joint designs to improve structural response in automotive applications.

In large assembly simulations, such as full vehicle analyses, accurately modeling joints is essential for predicting structural response. However, usage of detailed geometric models of joints in simulating large assemblies can be challenging. It requires highly refined meshes and complex material definitions to accurately capture the intricate behaviors of joints under various loading conditions. To address these challenges, simplified modeling approaches are used to achieve a balance between accuracy and computational efficiency [4]. These models effectively simulate various joint types commonly used in the automotive industry, including spot welds [5], bolted joints, adhesive joints [6], clinch joints, self-piercing rivets [7] and other joints. For instance, spot

Content from this work may be used under the terms of the Creative Commons Attribution 3.0 license. Any further distribution of this work must maintain attribution to the author(s) and the title of the work, journal citation and DOI. Published under license by Materials Research Forum LLC.

Sheet Metal 2025

Materials Research Forum LLC

Materials Research Proceedings 52 (2025) 168-174

https://doi.org/10.21741/9781644903551-21

welds can be represented using beam elements and solid elements [8], while adhesive joints are often modeled with cohesive zone models to capture the mechanical properties. Despite significant advancements in joint modeling techniques, the current methods including beam elements, cohesive zone models, and hybrid approaches are largely designed for conventional joints with rotationally symmetric geometries. These models are designed to efficiently simulate spot welds, bolted joints, riveted joints, and other standard connections, offering reasonable accuracy in predicting stiffness, failure, and damage under various loading conditions. However, they fall short when it comes to representing non-rotationally symmetric joints, which exhibit anisotropic behavior, particularly in the in-plane shear direction. The absence of specialized models for these types of joints limits the ability of engineers to capture the direction-dependent stiffness and complex load responses such geometries inherently possess.

In this paper, we employ discrete beam elements (beam formulation 6) in LS-DYNA R13 to model non-rotationally symmetric joints, focusing on their anisotropic properties. Stiffness values are assigned to represent the joints' direction-dependent behavior, and shear tests are conducted at various joint orientations to analyze the resulting forces. Additionally, we examine the effects of mesh size and joint geometry on the simulation results, aiming to understand how these factors influence the model's accuracy. This study aims to assess whether the discrete beam element is a viable approach for capturing direction-dependent stiffness in non-circular joints. While the discrete beam element is well-established for modeling joints with symmetrical behavior, its application to anisotropic joints remains unproven. Through these analyses, we evaluate the potential of discrete beams to effectively capture the complex behavior of non-rotationally symmetric joints.

Methodology

In this study, we utilize discrete beam elements within LS-DYNA to model non-rotationally symmetric joints. Discrete beams are particularly advantageous for representing complex joint behavior because they offer up to six degrees-of-freedom (DOF), unlike traditional spring elements which are limited to one DOF. The discrete beam formulation enables the capture of both translational and rotational effects, with resultant forces and moments output in the local (R, S, T) coordinate system [9]. This provides greater flexibility in simulating the directional stiffness and anisotropic properties of joints. However, it is important to note that this study considers only the translational stiffness of the discrete beam elements, while rotational stiffness is not accounted for.

A discrete beam element can be defined with zero or nonzero length, and the mass is determined by the product of the material density and volume (VOL in *SECTION_BEAM), independent of its length. The mass moment of inertia (INER) about the beam's axes must be provided when the rotational DOF are active, ensuring that rotational effects are appropriately captured during the simulation (see Fig. 1 for the section card of discrete beam).

Furthermore, discrete beams are compatible with a variety of material models in LS-DYNA, including nonlinear elastic, nonlinear plastic, and user-defined material models. This compatibility facilitates the simulation of both linear and nonlinear material responses, enabling a comprehensive representation of complex material behaviors in joint modeling. However, in the present study only the linear response is analyzed using *MAT_LINEAR_ELASTIC_DISCRETE_BEAM (*MAT_66). Orientation and shear force contributions are key aspects of discrete beam behavior. The beam's local system is defined by parameters such as SCOOR and CID, which control its initial orientation, and the development of shear forces and torques during simulation [10]. Depending on the application, torque contributions due to shear forces can be activated to simulate more realistic beam behavior under load. This flexibility is crucial for modeling non-rotationally symmetric joints, where direction-dependent stiffness must be accurately represented. However, in the present study, rotational

Sheet Metal 2025   Materials Research Forum LLC
Materials Research Proceedings 52 (2025) 168-174   https://doi.org/10.21741/9781644903551-21

stiffness is assumed to be negligible, as the beam is non-rotationally symmetric, and no rotation is expected in the joint.

Figure 1 Section card for discrete beam element

This study is entirely based on FEM simulations, focusing on the investigation of a shear lap specimen made of steel sheets joined by a discrete beam element. The dimensions of the specimen are 50 x 50 mm, with a sheet thickness of 1.5 mm and a joint diameter of 8 mm for the circular geometry. A 1 mm mesh size is applied for accurate representation of the joint and sheet behavior. In the following Fig. 2 the three nodes for the discrete beam and the local coordinate system can be seen.

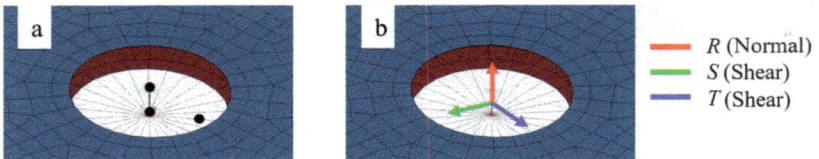

Figure 2 Discrete beam element (a) with three nodes and (b) local coordinate system

For all simulations, LS-DYNA R13 is used with the implicit solver, and LS-PrePost is employed for preprocessing tasks such as meshing and model setup. The default shell element type (Belytschko-Tsay formulation) is used with 5 integration points along the thickness to model the steel sheets. This shell element formulation is chosen due to its computational efficiency and ability to handle large deformation problems effectively. The units for all simulations are set to tons, millimeters, seconds, and newtons.

One edge of the steel sheet is fixed using an SPC constraint, restricting all degrees of freedom, while a displacement boundary condition is applied to the other sheet using the *PRESCRIBED_MOTION_SET keyword. A displacement of 0.1 mm in the X-direction is prescribed to simulate the applied load in 0.1 s Fig. 3 shows the boundary conditions for the setup along with the local coordinate system of the joint. The 0.1 mm displacement is chosen to represent a small deformation scenario, allowing for a focus on the stiffness and force response of the joint without introducing large nonlinear deformations.

Sheet Metal 2025   Materials Research Forum LLC
Materials Research Proceedings 52 (2025) 168-174   https://doi.org/10.21741/9781644903551-21

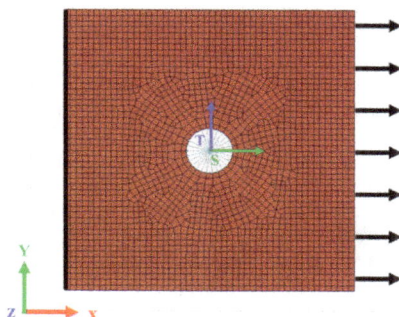

Figure 3 Lap joint specimen with boundary conditions

A linear elastic material card is used for the steel sheets, and *MAT_66 is used for the discrete beam element. The elastic properties of steel are provided as input, and for the discrete beam material, only translational stiffness values are defined, with no rotational stiffness, as no rotation is expected between the sheets. The rotational degrees of freedom of the discrete beam element are constrained to ensure no rotation occurs. Normal and shear stiffness values are assumed for the beam element, as this study aims to demonstrate the potential of the discrete beam for non-rotationally symmetric joints (Table 1).

The moving edge of the sheet is constrained so that displacement can occur only in the elongation direction. The end nodes of the discrete beam element are constrained to the nodes of the sheet using Constrained Nodal Rigid Body (CNRB), ensuring proper load transfer. Additionally, a third node is defined for the discrete beam element and constrained in all degrees of freedom to prevent rotational moments from appearing in the output. This third node also allows for adjusting the direction of stiffness by rotating the beam in LS-PrePost.

Table 1 Material properties for discrete beam

*MAT_LINEAR_ELASTIC_DISCRETE_BEAM_(TITLE) (066)			
1.	Density (RO)	R-Stiffness (TKR)	S-Stiffness (TKS)	T-Stiffness (TKT)
2.	7.85e-09 t/mm^3	1200 N/mm	1500 N/mm	1500 N/mm

The implicit solver is used in this study as it is well-suited for static and quasi-static problems, where loads are applied gradually and incrementally. It allows for larger time steps compared to explicit solvers, enabling efficient simulations while maintaining high accuracy. This makes the implicit method particularly useful in the present study.

Results and Discussions
To assess the effectiveness of discrete beam elements in representing non-rotationally symmetric joints, simulations were carried out using the model outlined in Section 2. Initially, a single discrete beam model was tested to evaluate the forces acting along the local S and T coordinates, which are orthogonal. In this model, identical translational stiffness values were assigned to both the S and T directions. The plots in following Fig. 4 illustrate the beam alignment and the resulting forces.

Sheet Metal 2025   Materials Research Forum LLC
Materials Research Proceedings 52 (2025) 168-174   https://doi.org/10.21741/9781644903551-21

In the first graph (a), the applied load was aligned with the S-coordinate, with the T-coordinate positioned 90° relative to S. As expected, no significant force was observed in the direction of T, as no load was applied along this component. However, the S component showed a force of 148 N, consistent with the applied load. Additionally, a small force in the R (normal) direction was recorded, even though no direct load was applied in that direction. This force can be attributed to the bending of the sheet edge, which induces a reaction in the R component. In the second graph (b), the beam was aligned at 45°, meaning the load was applied between the S and T components. As a result, both S and T components exhibited an equal force of 104 N due to the distribution of the load between the two directions. The R component continued to show a small force due to the bending of the sheet, similar to the first case. This demonstrates the discrete beam model's realistic representation of joint behavior under complex loading, accurately capturing both shear and normal forces.

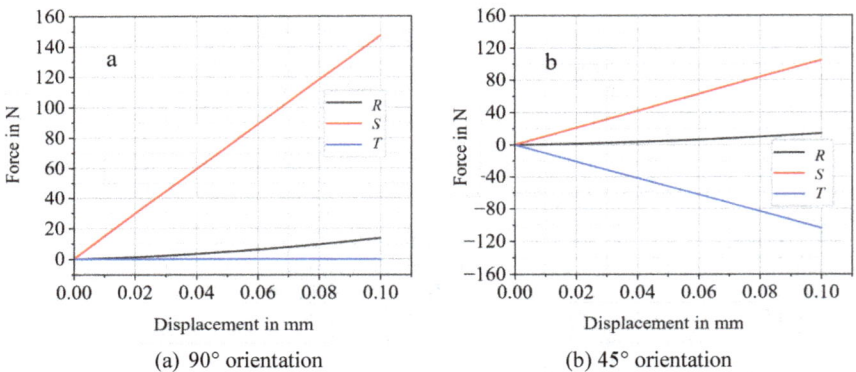

(a) 90° orientation   (b) 45° orientation

Figure 4 Force vs displacement plots comparing beam rotations

In the second model, three discrete beam elements were used instead of a single element. The relative alignment of the beams was investigated, with the S components of two beams arranged 120° apart. The alignment of the third beam's S component varied to observe its impact on the anisotropic behavior of the joint. The T components of all beams were assigned zero stiffness to avoid additional stiffness contributions from these directions. When the third beam's S component was also aligned 120° to the others, minimal change in behavior was observed, as it aligned with the applied load. However, when the third beam's alignment was varied, distinct anisotropic characteristics were exhibited.

To visualize the anisotropy, equivalent von Mises stress plots were analyzed. The stress distribution varied notably with changes in the third beam's alignment, highlighting directional dependency in the joint. Additionally, the Z-displacement of the sheets was examined to observe bending behavior (Fig. 5). The results indicated that the bending pattern of the sheet corresponded with the alignment of the third beam, further demonstrating the influence of anisotropy on the mechanical response.

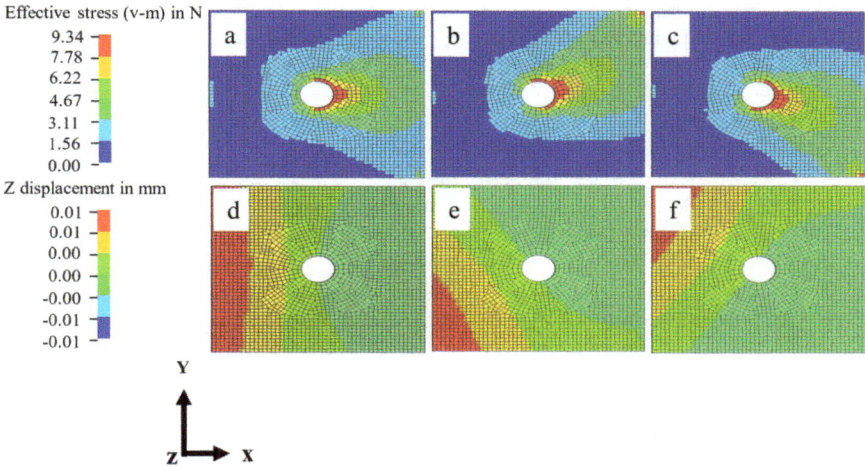

Effective stress (v-m) in N

9.34
7.78
6.22
4.67
3.11
1.56
0.00

Z displacement in mm

0.01
0.01
0.00
0.00
-0.00
-0.01
-0.01

Figure 5 Stress contour and Z displacement plots.
(a-c) effective stress in different orientations, (d-f) Z displacement in different orientations

Summary

This study investigated the potential of discrete beam elements to model non-rotationally symmetric joints with anisotropic characteristics. Using finite element simulations in LS-DYNA, a shear lap specimen was modeled with discrete beam elements to assess their ability to capture direction-dependent stiffness. Two models were tested: one using a single beam, suitable for joints with stiffness periodicity every 90°, and another employing three beams, designed for stiffness periodicity every 120°. The force responses in orthogonal local coordinates were analyzed, along with Z-displacement and von Mises stress contours of the steel sheets.

The results showed that discrete beam elements can effectively represent the anisotropic behavior of non-rotationally symmetric joints. The analysis of stress distribution and sheet bending confirmed that directional stiffness was accurately captured. Future work will focus on enhancing material modeling through user-defined subroutines, conducting comparative analyses with experimental data, and calibrating the model based on experimental results to ensure greater accuracy and reliability.

Acknowledgement

This research was funded by the Deutsche Forschungsgemeinschaft (DFG, German Research Foundation), project number 418701707/2 - TRR 285 Phase 2 (Sub-project B01). The authors gratefully acknowledge this financial support. They also extend their sincere thanks to the Paderborn Center for Parallel Computing (PC²) for providing the computational resources necessary to carry out this research.

References

[1] S. Martin, A. A. Camberg, and T. Tröster, "Probability Distribution of Joint Point Loadings in Car Body Structures under Global Bending and Torsion," Procedia Manuf, vol. 47, pp. 419–424, Jan. 2020. https://doi.org/10.1016/J.PROMFG.2020.04.324

Sheet Metal 2025   Materials Research Forum LLC
Materials Research Proceedings 52 (2025) 168-174   https://doi.org/10.21741/9781644903551-21

[2] S. Martin, J. Schütte, C. Bäumler, W. Sextro, and T. Tröster, "Identification of joints for a load-adapted shape in a body in white using steady state vehicle simulations," Forces in Mechanics, vol. 6, p. 100065, Feb. 2022. https://doi.org/10.1016/J.FINMEC.2021.100065

[3] D. R. Devulapally, S. Martin, and T. Tröster, "Non-rotationally symmetric joints – Mechanisms and load bearing capacity," Materials Research Proceedings, vol. 41, pp. 1650–1659, 2024. https://doi.org/10.21741/9781644903131-183

[4] D. Mundo, R. Hadjit, S. Donders, M. Brughmans, P. Mas, and W. Desmet, "Simplified modelling of joints and beam-like structures for BIW optimization in a concept phase of the vehicle design process," Finite Elements in Analysis and Design, vol. 45, no. 6–7, pp. 456–462, May 2009. https://doi.org/10.1016/J.FINEL.2008.12.003

[5] M. Palmonella, M. I. Friswell, J. E. Mottershead, and A. W. Lees, "Finite element models of spot welds in structural dynamics: review and updating," Comput Struct, vol. 83, no. 8–9, pp. 648–661, Mar. 2005. https://doi.org/10.1016/J.COMPSTRUC.2004.11.003

[6] M. Costa, G. Viana, R. Créac'hcadec, L. F. M. Da Silva, R. Créac'hcadec, and R. D. S. G. Campilho, "A cohesive zone element for mode I modelling of adhesives degraded by humidity and fatigu," Int J Fatigue, vol. 112, pp. 173–182, 2018. https://doi.org/10.1016/j.ijfatigue.2018.03.014

[7] L. Duan, Z. Du, H. Ma, W. Li, W. Xu, and X. Liu, "Simplified modelling of self-piercing riveted joints and application in crashworthiness analysis for steel-aluminium hybrid beams," J Manuf Process, vol. 85, pp. 948–962, Jan. 2023. https://doi.org/10.1016/J.JMAPRO.2022.11.068

[8] J. Z. Wu, "Beam-type versus solid-type spot weld in LS-DYNA," SAE Technical Papers, 2009. https://doi.org/10.4271/2009-01-0354

[9] "LS-DYNA ® KEYWORD USER'S MANUAL VOLUME I," 2023, Accessed: Oct. 10, 2024. [Online]. Available: www.lsdyna.ansys.com

[10]"LS-DYNA ® KEYWORD USER'S MANUAL VOLUME II Material Models," 2023, Accessed: Jan. 12, 2024. [Online]. Available: www.lsdyna.ansys.com

Sheet Metal 2025
Materials Research Proceedings 52 (2025) 175-183

Materials Research Forum LLC
https://doi.org/10.21741/9781644903551-22

Influence of thermal effects on clinch joining of sheet metal

Johannes Friedlein[1,a] *, Paul Steinmann[1,b] and Julia Mergheim[1,c]

[1]Institute of Applied Mechanics, Friedrich-Alexander-Universität Erlangen-Nürnberg, Egerlandstrasse 5, 91058, Erlangen, Germany

[a]Johannes.Friedlein@fau.de, [b]Paul.Steinmann@fau.de, [c]Julia.Mergheim@fau.de

Keywords: Thermal Effect, Modelling, Thermoplasticity

Abstract. Mechanical joining methods, such as clinching, are characterised by locally large plastic deformations of the sheet metal to be joined. The majority of the thereby inserted work is transformed into heat. The heat generation and temperature evolution are systematically studied herein by means of thermomechanical process simulations for joining the dual-phase steel HCT590X and the aluminium alloy EN-AW 6014. The thermal-induced softening of the material is incorporated by a suitable coupled thermoplastic constitutive model. It is observed how the tools significantly and importantly contribute to the heat exchange. They reduce peak temperature increases of 225 K (without heat transfer to tools) to less than 90 K for realistic behaviour of contact heat transfer. Overall, increases in temperature during clinch joining can be expected to remain below 90 K for steel-steel joints and around 50 K for aluminium-aluminium joints.

Introduction

Mechanical joining methods, such as clinching, assemble sheet metal by cold forming [1]. The sheets are mechanically interlocked by local controlled deformation. However, this involves high plastic strains of up to 300 % being focussed in a small volume of a clinch joint with diameters of 5–10 mm. While this plastic deformation is converted into heat, it is expected that the sheet temperature in the forming zone increases. Due to the comparatively slow heat conduction and the short process times of 1–2 s at travel speeds of ≈2 mm/s, transient thermal loading will occur.

Heat generation during forming of metals is a typical side effect often studied for more than a century [2,3]. The energy put into the material for forming is dissipated and primarily results in thermal energy. Typically a temperature-induced softening of the elastic and plastic material behaviour is observed, which is to be incorporated into the description of the constitutive behaviour if its effect is relevant.

The heat generation during flat clinching was studied in [4]. For this special mechanical joining method with flat die, a maximum temperature increase of 203 K was predicted for joining two DC04 steel sheets. Consequently, also the interlock formation was notably affected by the thermal effects. However, such a high temperature increase primarily results from the very short process time of about 0.1 s. Concerning the classical clinching process the accompanying heat generation has hardly been studied so far. However, the beneficial effect of warm forming conditions have already been investigated and utilised to improve the joint characteristics by external pre-heating [5], or to enable clinching of difficult-to-form sheets [6,7].

Herein, a thermomechanically coupled hyperelastic-based finite plasticity material model is utilised to study fundamental aspects of the heat generation and transfer during clinch joining. This is exemplarily conducted for steel-steel and aluminium-aluminium clinch joining processes.

Constitutive Modelling

The large irreversible strains during the clinch joining process are captured by a hyperelastic-based finite plasticity model. A multiplicative decomposition of the deformation gradient $F = F_e \cdot F_p \cdot F_\theta$ into elastic, plastic, and thermal parts is utilised based on [8]. Herein, isotropic behaviour is

Content from this work may be used under the terms of the Creative Commons Attribution 3.0 license. Any further distribution of this work must maintain attribution to the author(s) and the title of the work, journal citation and DOI. Published under license by Materials Research Forum LLC.

considered, such that $F_\theta = \exp(\alpha_\theta\,\Delta\theta)\,I$ with the relative temperature $\Delta\theta$ and the constant (temperature-independent) thermal expansion coefficient α_θ. Isotropic von Mises (J2) plasticity with nonlinear isotropic hardening is assumed

$$\Phi = \left\|\tau^{\mathrm{dev}}\right\| - \sqrt{2/3}\,\sigma_{\mathrm{flow}}(\alpha,\Delta\theta) \le 0 \tag{1}$$

$$\sigma_{\mathrm{flow}}(\alpha,\Delta\theta = 0) = \sigma_y - R(\alpha) = \sigma_y + K\,\alpha^{n_{p,K}} + \Delta\sigma_{y,\infty}[1 - \exp(-\omega\,\alpha^{n_{p,\omega}})] \tag{2}$$

$$\sigma_{\mathrm{flow}}(\alpha,\Delta\theta) = [1 - H_y\,\Delta\theta]\sigma_y - [1 - H_R\,\Delta\theta]R(\alpha) \tag{3}$$

with the initial yield stress σ_y and the accumulated plastic strain α. The six material parameters $(\sigma_y, K, n_{p,K}, \Delta\sigma_{y,\infty}, \omega, n_{p,\omega})$ describe a combination of power-law and exponential hardening. The used Kirchhoff stress τ is related to the Cauchy (true) stress $\sigma = 1/\det(F)\,\tau$. Details on the evolution equations, the spectral decomposition and the standard radial return mapping algorithm applied for the solution are omitted for brevity, but are detailed e. g. in [9].

In addition to the thermal expansion inherent to the thermal deformation gradient F_θ, further thermal effects are incorporated acting on elasticity and plasticity by the thermal softening parameters (H_E, H_y, H_R). The initial Young's modulus E_0 is decreased by $E = [1 - H_E\,\Delta\theta]E_0$ with increasing temperature to reproduce thermal elastic softening. Similarly, the initial yield stress and the hardening stress can be affected separately in Eq. (3).

The resulting thermoelastoplastic material model is implemented using the Mathematica-based AceGen package [10] as user-defined material model in LS-DYNA.

Heat equation and thermal solver

The partial differential equation of heat conduction $\rho\,c_p\,\partial\theta/\partial t = k\,\theta_{,ii} + Q$ is solved by the thermal solver in LS-DYNA with appropriate boundary and initial conditions [11]. Therein, ρ is the material density, c_p the specific heat capacity, k the thermal conductivity, and Q a heat source term. Q contains the internal heat generation due to plastic work and is automatically computed. Q can further include the friction energy. The amount of plastic work transformed into heat can be adjusted by the coefficient FWORK, acting as the Taylor-Quinney coefficient. To investigate limit values for heat generation and maximum temperatures FWORK=1 is assumed, compared to typical values of 0.8–0.9 [4, 12]. As the temperature affects the mechanical problem and the mechanical problem affects the heat source Q, the mechanical and thermal solvers are coupled and solved weakly staggered. This means that the thermal and mechanical problem are solved separately and no additional iterations are performed to ensure that these separate solutions are exact solutions to the actual coupled system. The fully implicit solver with identical mechanical and thermal time step sizes is used in LS-DYNA.

Material Parameters, Identification, and Reference Thermal Values

The material parameters for elastoplasticity at reference/room temperature (θ=20 °C) have been identified in [13] for the considered dual-phase steel HCT590X and the aluminium alloy EN AW-6014 T4, as reiterated in Table 1.

Table 1. Material parameters for elastoplasticity at reference/room temperature from [13].

	E_0 [GPa]	ν	ρ [kg/m³]	σ_y [MPa]	K [MPa]	$n_{p,K}$	$\Delta\sigma_{y,\infty}$ [MPa]	ω	$n_{p,\omega}$
HCT590X	205.8	0.29	7850	393.4	4.35	0.356	600.3	2.58	0.539
EN AW-6014 T4	70.4	0.34	2700	130.4	4.17	0.351	237.5	2.39	0.573

Sheet Metal 2025   Materials Research Forum LLC
Materials Research Proceedings 52 (2025) 175-183   https://doi.org/10.21741/9781644903551-22

Tensile tests at different, elevated, but constant temperatures up to θ=200 °C were conducted to investigate and calibrate the thermal influence. These experimental results have been utilised to identify the thermal softening parameters (H_E, H_y, H_R). No clear trend for the elastic softening H_E and the change of the initial yield stress H_y could be observed from the experimental results, such that both effects were deactivated (value 0). The remaining softening parameter H_R was inversely identified using 3D finite element simulations of the tensile tests and LS-Opt to also realistically incorporate the effect on necking. The experimental as well as fitted numerical results are summarised and compared in Fig. 1 for the dual-phase steel $(H_R = 2.12 \cdot 10^{-3}\ K^{-1})$ and the aluminium alloy $(H_R = 2.2 \cdot 10^{-3}\ K^{-1})$.

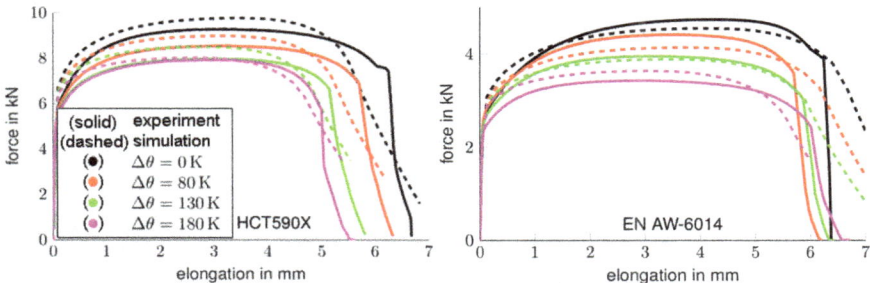

Fig. 1. Tensile tests – force-elongation responses for different temperatures (line colours). Experimental results (solid) vs numerical results (dashed) for dual-phase steel HCT590X (left) and aluminium alloy EN AW-6014 (right).

Additionally, thermal capacity, conductivity, and contact heat transfer need to be described. For this preliminary investigation of the underlying effects and the fundamental influence, we rely on values from literature to describe these processes and further study their sensitivities. The used thermal parameters are summarised in Table 2 and are based on different literature sources as stated in the table for a congeneric DP600 steel and EN AW-6016 T6 aluminium alloy. The contact heat transfer coefficients are based on general values for steel and aluminium. The clinching tools are made of tool steel with assumed values as given in Table 2. The contact heat transfer coefficient is assumed equal for the tool steel to steel/aluminium sheet combinations. Contact heat transfer coefficients strongly depend on the surfaces in contact and can deviate strongly [14]. Therefore, their influence is studied separately in the results section. Due to the high interface pressures of up to 2500 MPa during clinching [15] such high coefficients are accepted as reference. For future investigations, a more detailed thermal characterisation of the materials is recommended, for instance following the procedure in [4].

Table 2. Reference values for thermal capacity, thermal conductivity, and contact transfer.

	specific heat capacity c_p [J/(kg K)]	thermal conductivity k [W/(m K)]	heat expansion coefficient α [1/K]	Contact heat transfer coefficient h [W/(m² K)]
HCT590X	460 [16]	64 [16]	$10 \cdot 10^{-6}$ [16]	10000 [14]
EN AW-6014	1148 [17]	91.19 [17]	$25.4 \cdot 10^{-6}$ [17]	20000 [18]
tool steel	408	18	-	10000 [14]

Sheet Metal 2025   Materials Research Forum LLC
Materials Research Proceedings 52 (2025) 175-183   https://doi.org/10.21741/9781644903551-22

Clinch Joining Simulation Including Thermal Effects

Numerical setup. The process is modelled in LS-DYNA using an axisymmetric setup as shown in Fig. 2. The sheets are clamped between the blankholder and the die. The punch moves downwards in negative y-direction into the sheets. The tools (punch, die, blankholder) are modelled as elastic, and the sheets are equipped with the thermoelastoplastic material model described previously. For the sheets an element size of 40 μm is sustained by remeshing. The contact between the sheets and between the sheets and the tools uses *CONTACT_2D_AUTOMATIC_SURFACE_TO_SURFACE_MORTAR_THERMAL.

Thereby, it is possible to incorporate thermal contact, which transfers heat between the contact partners by heat conduction and radiation. Additionally, the energy dissipated by friction is included as additional source term for the thermal solver by setting FRCENG=1. The frictional energy is by default split up equally onto both contact partners. The friction coefficients have been inversely identified to fit to the experimental microsection and the process force as shown later. Heat radiation to the environment is omitted as its influence is expected to be low due to low overall temperatures and the setup enclosing the core zone by the die and blankholder.

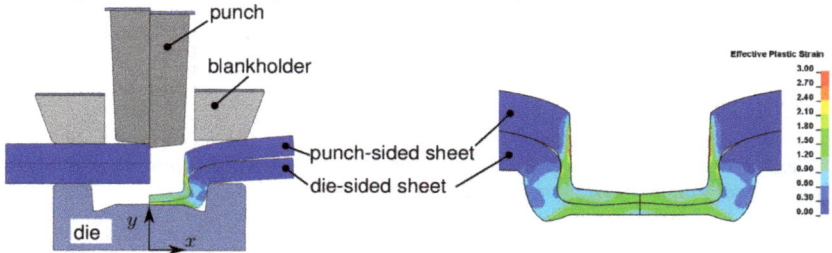

Fig. 2. Clinch joining – numerical setup. Axisymmetric model.

Steel-Steel. The heat generation and transfer is studied systematically using three setups of the joining process. The first setup (I) focusses on the understanding of heat generation and neglects all thermal contacts, see Fig. 3 (left). The results for setup I are shown in Fig. 3 for four snapshots of the joining process.

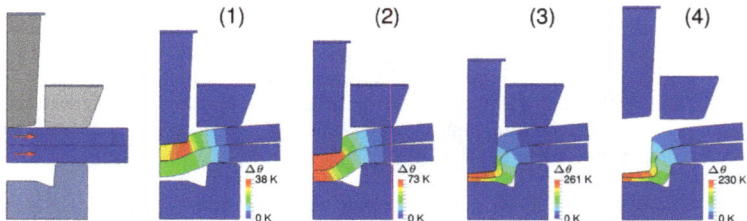

Fig. 3. Steel-steel clinching – without any contact heat transfer (setup I). Snapshots of the process simulation.

In the first phase (1) in Fig. 3, heat is primarily generated around the punch corner. Once the die-sided sheet is in contact with the bottom of the die in phase (2), the compression between punch and die generates a substantial amount of heat. Thereby, the temperature increase $\Delta\theta$ reaches up to 261 K in phase (3). Interestingly, even though the die-sided sheet is deformed stronger and also overall experiences a larger heat input, the maximum temperature is reached in the punch-sided sheet. It is surmised that the sheet thinning in the neck also acts as a bottleneck for the heat flux. Therefore, the heat is trapped in the centre of the punch-sided sheet, in contrast to the die-sided sheet, where the heat flows into the widening material whilst filling the ring channel. In the last phase (4), the tools are opened and the heat can conduct outwards to reach a steady state for each sheet. The steady state solution is approximately reached in the simulation by continuing the heat conduction for 10 s after unloading (not shown) to determine the final temperatures. Based on these steady state temperatures and the deactivation of most heat losses (tools, environment), it is possible to evaluate the actual heat input per sheet as $\Delta Q = m \cdot c_p \cdot \Delta\theta$. From the specific heat capacity, the sheet's volume (specimen radius 10.4 mm, thickness 1.5 mm) and the final temperature increase, the energy per sheet results as

$$m = \rho \cdot V = 7850 \, \frac{kg}{m^3} \cdot [10.4 \text{ mm}]^2 \cdot \pi \cdot 1.5 \text{ mm} \cdot \left[\frac{1 \text{ m}}{1000 \text{ mm}}\right]^3 = 0.00400 \text{ kg} \qquad (4)$$

$$\text{punch-sided sheet: } \Delta Q = 4.00 \cdot 10^{-3} \text{ kg} \cdot 460 \frac{J}{kg \, K} \cdot 20.7 \text{ K} = 38 \text{ J} \qquad (5)$$

$$\text{die-sided sheet: } \Delta Q = 4.00 \cdot 10^{-3} \text{ kg} \cdot 460 \frac{J}{kg \, K} \cdot 26.1 \text{ K} = 48 \text{ J} \qquad (6)$$

For comparison the overall energy input into the clinch joint is computed based on the integral over the process force vs displacement curve. The energy input results as 87 J. The small amount of energy missing (1.4 %) can be attributed to the friction energy partially assigned to the tools, residual stresses, and numerical discrepancies of contact and remeshing. For comparison, simulations with zero friction coefficients result in maximum temperature increases of 245 K, however the energy input also decreases to 81 J.

Setup II in Fig. 4 allows for heat exchange between both sheets, but neglects heat transfer to the tools. The phases are similar to setup I, but the maximum $\Delta\theta$ decreases to 225 K, the steady state is reached at equal sheet temperature $\Delta\theta$=23.4 K (coincides with average of setup I), due to their mutual heat exchange.

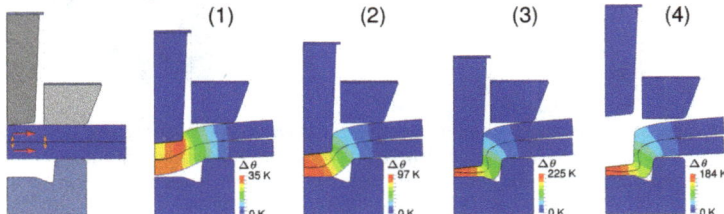

Fig. 4. Steel-steel clinching – with heat transfer between the sheets but not to the tools (setup II). Snapshots of the process simulation.

Setup III considers the full heat transfer, so also the heat exchange with the tools is incorporated in Fig. 5. In contrast to the previous setups, phase (2) shows how the temperature of the die-sided sheet notably decreases as the sheet comes into contact with the cold bottom of the die. The die receives a heat input of about 28 J and the blankholder just 2 J, where overall 91 J are input by mechanical work. Also the punch draws out a substantial amount of heat (about 11 J). The sheets receive an overall heat input of approx. 44 J, such that now roughly 7 % of energy is lost due to numerical discrepancies. The maximum $\Delta\theta$ in the sheets decreases to 88 K. The tool temperature reaches a maximum $\Delta\theta$ of 82 K on the bottom surface of the punch.

Sheet Metal 2025   Materials Research Forum LLC
Materials Research Proceedings 52 (2025) 175-183   https://doi.org/10.21741/9781644903551-22

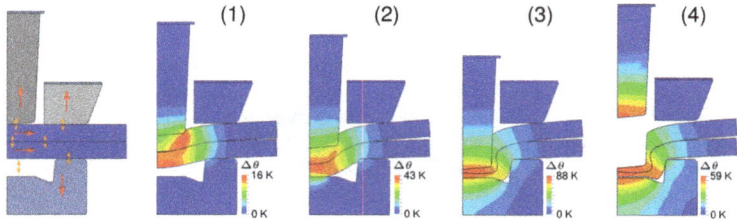

Fig. 5. Steel-steel clinching – with full contact heat transfer (setup III). Snapshots of the process simulation.

Due to the thermomechanically coupled constitutive model, the temperature also affects the mechanical behaviour of the sheet material. Therefore, as depicted in Fig. 6, the process force decreases due to the thermal-induced softening. However, for the realistic setup (III, with full contact heat transfer), the process force is only decreased by about 8 %. The joint contour in Fig. 6 (right) and its characteristic values (neck thickness, interlock) are hardly affected by the increased temperature. As the formation of the joint contour is primarily governed by the friction behaviour, it is expected that a more noticeable effect is visible once also a temperature-dependent friction behaviour is incorporated.

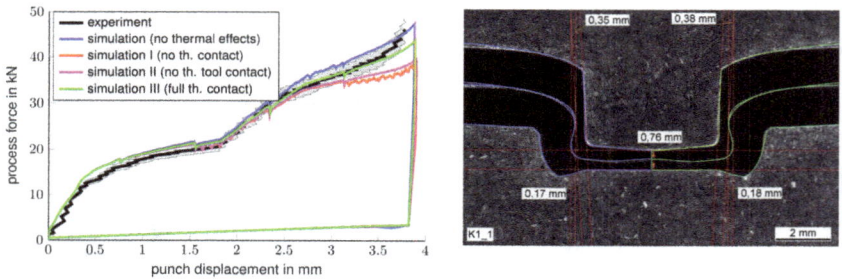

Fig. 6. Steel-steel clinching. Comparison to experiments. The process force (left) shows a stronger influence on thermal effects, where an increased temperature result in a decreased force. The joint contour (right) and its characteristic values are hardly affected by thermal effects (green) vs no thermal effects (blue).

The thermal parameters are, as mentioned previously, based on literature values. However, as for instance the contact heat transfer strongly depends on the contact conditions (friction, interface pressure, surface properties, …), parameter variations are performed next for setup III to ensure representative conclusions. First, the contact heat transfer coefficient, based on the reference value 10000 W/(m^2 K) with $\Delta\theta_{max}$=88 K, is varied for the sheet-sheet contact pair to {10, 100, 50000} W/(m^2 K) leading to $\Delta\theta_{max}$={94, 89, 85} K. Second, the contact heat transfer coefficient for the sheet-tool contact is varied based on 10000 W/(m^2 K) to {10, 50000} W/(m^2 K) resulting in $\Delta\theta_{max}$={126, 81} K. Consequently, even with minor contact heat transfer to the tools, the peak temperature values are greatly reduced compared to the adiabatic case with 225 K. Therefore, even though the results are sensitive to the sheet-tool contact heat transfer parameter, the qualitative conclusions and observed temperature ranges are reliable.

Aluminium-Aluminium. Lastly, clinching of two 2 mm thick aluminium alloy EN AW-6014 T4 sheets is considered. The temperature evolution during the process simulation is comparable to the steel-steel combination, although the distributions are a bit more diffuse. This can be attributed to the higher thermal conductivity of the aluminium alloy and is not shown for brevity. Fig. 7 depicts only phase (3) with the maximum temperatures for the different setups (I,II,II). The maximum increases in temperature are between $\Delta\theta$=208 K without any heat transfer (setup I) and 49 K with full heat transfer (setup III). The energy input is lower than for the steel-steel joint and about 64 J. The process force with full thermal contact is about 5 % lower than without considering thermal effects. The joint contour is only slightly affected in the bottom area of the characteristic "S"-shape, such that neither the neck thickness nor the interlock are notably affected.

Fig. 7. Aluminium-aluminium clinching – without any heat transfer (setup I), with sheet-sheet

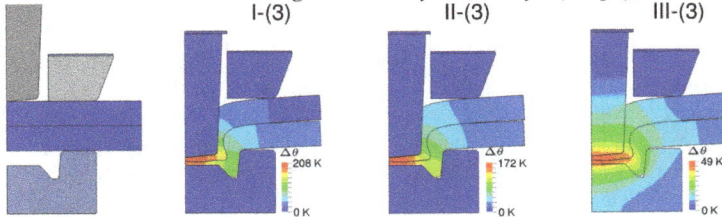

heat transfer (setup II), and with full contact heat transfer (setup III). Snapshots of the process simulation at the maximum temperature (phase 3).

Mixed material combinations of aluminium and steel are not considered here as it is assumed that the critical maximum temperatures are lower than for the initially investigated steel-steel clinch joint due to the higher thermal conductivity of the aluminium.

Summary

A coupled thermoelastoplastic material model was implemented and successfully applied in a 2D axisymmetric clinch joining simulation. This enabled to study the temperature evolution during clinch joining of steel and aluminium sheets. Clinch joining of two dual-phase HCT590X steel sheets and two EN AW-6014 aluminium alloy sheets was investigated separately. Heat conduction and heat transfer between the sheets and tools were considered. Attention was further devoted to the heat generation and the influence of contact heat transfer. For the steel-steel clinch joint, maximum temperature increases of 88 K were reached, when considering all thermal contact pairs. For the aluminium-aluminium joint, on the other hand, a lower peak increase of 49 K was achieved, due to its higher thermal conductivity.

Based on the results of this fundamental investigation, future work can incorporate the effect of temperature for instance on the friction behaviour, damage evolution, and residual stresses. It is also expected that thermal effects will gain importance for shorter process times than the here considered 1–2 s. However, then also viscous effects need to be taken into account as these oppose the thermal softening by a typical rate-dependent plastic hardening. To sum up, the effect of self-heating during classical clinch joining can be regarded as a secondary effect.

Acknowledgements

The funding by the Deutsche Forschungsgemeinschaft (DFG, German Research Foundation) – Project-ID 418701707 – TRR 285/2, subproject A05 is gratefully acknowledged. Moreover, we want to thank our colleague in the TRR285 Max Böhnke (A01) for conducting the tensile tests at elevated temperatures.

Sheet Metal 2025   Materials Research Forum LLC
Materials Research Proceedings 52 (2025) 175-183   https://doi.org/10.21741/9781644903551-22

References

[1] G. Meschut, M. Merklein, A. Brosius, M. Bobbert, Mechanical joining in versatile process chains, Production Engineering. 16 (2022) 187-191. https://doi.org/10.1007/s11740-022-01125-y

[2] W.S. Farren, G.I. Taylor, The heat developed during plastic extension of metals, Proc. R. Soc. (London) A, 107 (1925) 422-451. https://doi.org/10.1098/rspa.1925.0034

[3] J.M.P. Martins, D.M. Neto, J.L. Alves, M.C. Oliveira, L.F. Menezes, Numerical modeling of the thermal contact in metal forming processes, Int. J. Adv. Manuf. Technol. 87 (2016) 1797-1811. https://doi.org/10.1007/s00170-016-8571-y

[4] S. Härtel, M. Graf, T. Gerstmann, B. Awiszus, Heat generation during mechanical joining processes-by the example of flat-clinching, Procedia engineering, 184 (2017) 251-265. https://doi.org/10.1016/j.proeng.2017.04.093

[5] M. Džupon, Ľ. Kaščák, D. Cmorej, L. Čiripová, J. Mucha, E. Spišák, Clinching of high-strength steel sheets with local preheating, Applied Sciences, 13 (2023) 7790. https://doi.org/10.3390/app13137790

[6] J. Osten, P. Söllig, M. Reich, J. Kalich, U. Füssel, O. Kessler, Softening of high-strength steel for laser assisted clinching, Advanced Materials Research, 966 (2014) 617-627. https://doi.org/10.4028/www.scientific.net/AMR.966-967.617

[7] F. Xu, H. Wang, M. Gao, H. Liu, L. Xu, Connection of difficult-to-form sheets by clinching process: a review, Materials Science and Technology, 38 (2022) 622-644. https://doi.org/10.1080/02670836.2022.2062813

[8] S. Felder, N. Kopic-Osmanovic, H. Holthusen, T. Brepols, S. Reese, Thermo-mechanically coupled gradient-extended damage-plasticity modeling of metallic materials at finite strains, International Journal of Plasticity, 148 (2022) 103142. https://doi.org/10.1016/j.ijplas.2021.103142

[9] J. Bonet, R.D. Wood. Nonlinear continuum mechanics for finite element analysis. Cambridge university press (2016). https://doi.org/10.1017/CBO9781316336144

[10] J. Korelc, P. Wriggers, Automation of Finite Element Methods, Springer, Switzerland, 2016. https://doi.org/10.1007/978-3-319-39005-5

[11] Livermore Software Technology Corporation, LS-DYNA Theory Manual (2020).

[12] S. Schindler, P. Steinmann, J.C. Aurich, M. Zimmermann, A thermo-viscoplastic constitutive law for isotropic hardening of metals, Arch. Appl. Mech. 87 (2017) 129-157. https://doi.org/10.1007/s00419-016-1181-1

[13] J. Friedlein, M. Böhnke, M. Schlichter, M. Bobbert, G. Meschut, J. Mergheim, P. Steinmann, Material Parameter Identification for a Stress-State-Dependent Ductile Damage and Failure Model Applied to Clinch Joining, J. Manuf. Mater. Process. 8 (2024) 157. https://doi.org/10.3390/jmmp8040157

[14] V. A. Ustinov, Experimental investigation and modeling of contact heat transfer Dissertation, Rheinisch-Westfälische Technische Hochschule Aachen, 2020.

[15] M. Böhnke, M. Rossel, C.R. Bielak, M. Bobbert, G. Meschut, Concept development of a method for identifying friction coefficients for the numerical simulation of clinching processes, Int. J. Adv. Manuf. Technol. 118 (2022) 1627-1639. https://doi.org/10.1007/s00170-021-07986-4

[16] A. Chrysochoos, B. Berthel, F. Latourte, S. Pagano, B. Wattrisse, B. Weber, Local energy approach to steel fatigue. Strain, 44 (2008) 327-334. https://doi.org/10.1111/j.1475-1305.2007.00381.x

[17] L. Rose, A. Menzel, Optimisation based material parameter identification using full field displacement and temperature measurements, Mechanics of Materials, 145 (2020) 103292. https://doi.org/10.1016/j.mechmat.2019.103292

[18] F. Boeschoten, E.F.M. Van der Held, The thermal conductance of contacts between aluminium and other metals, Physica, 23 (1957) 37-44. https://doi.org/10.1016/S0031-8914(57)90236-7

Sheet Metal 2025
Materials Research Proceedings 52 (2025) 184-191

Materials Research Forum LLC
https://doi.org/10.21741/9781644903551-23

High-cycle fatigue testing and parameter identification for numerical simulation of aluminum alloy EN AW-6014

Chin Chen[1,a] *, Malte Christian Schlichter[2,b], Sven Harzheim[1,c], Martin Hofmann[1,d], Mathias Bobbert[2,e], Gerson Meschut[2,f] and Thomas Wallmersperger[1,g]

[1]Institute of Solid Mechanics, TU Dresden, George-Bähr-Straße 3c, 01069 Dresden, Germany

[2]Laboratory for Material and Joining Technology (LWF), Paderborn University, Pohlweg 47-49, 33098 Paderborn, Germany

[a]chin.chen@tu-dresden.de, [b]malte.schlichter@lwf.upb.de, [c]sven.harzheim@tu-dresden.de, [d]mathias.bobbert@lwf.upb.de, [e]martin.hofmann@tu-dresden.de, [f]gerson.meschut@lwf.uni-paderborn.de, [g]thomas.wallmersperger@tu-dresden.de

Keywords: Fatigue, Numerical Simulation, Clinched Joints

Abstract. As a widely used sheet metal in clinched joints within the automotive industry, the aluminum alloy EN AW-6014 has been the focus of numerous studies. High-cycle fatigue (HCF) is a critical aspect when assessing the durability of clinched joints. In the present work, the HCF behavior of EN AW-6014 T4 was explored both experimentally and numerically. To model the fatigue behavior, Lemaitre's two-scale damage model was used. Two key parameters, damage strength and damage exponent, are necessary for numerical investigations of HCF behavior. These parameters were determined through experiments with flat specimens and subsequently validated within a numerical model of clinched joints. The numerical results for fatigue match the experimental ones of the clinched joints quite well.

Introduction

The fatigue durability of mechanical components is of critical importance due to the growing demand for safety and cost-efficiency [1]. High-cycle fatigue (HCF) is a significant phenomenon that affects the long-term durability of materials subjected to cyclic loading, particularly in industries such as automotive [2] and aerospace [3], where components endure millions of stress cycles at relatively low amplitudes. As a result, investigating the HCF behavior and predicting the fatigue life are essential tasks for these industries. Fatigue experiments remain the most common method for accomplishing this. Wöhler [4] first introduced the concept of S-N curves, which are still a vital tool for assessing the fatigue behavior of new materials [5]. In addition to experimental approaches, numerical simulations are increasingly used to predict damage accumulation during cyclic loading. Lemaitre et al. [6] developed a two-scale damage model to simulate HCF behavior, while Roters et al. [7] offered insights into multi-scale modeling and the application of crystal plasticity to simulate fatigue at the microstructural level. These numerical methods offer time- and cost-efficient alternatives for studying HCF in materials.

The demand for lightweight materials such as EN AW-6014 has grown significantly, as vehicle manufacturers strive to reduce weight, improve fuel efficiency, and lower emissions, all while maintaining safety and performance standards [8]. Aluminum alloys from the 6xxx series, including EN AW-6014, are increasingly used in the automotive industry due to their excellent formability, strength, low cost, and corrosion resistance [9]. Thanks to its good formability, EN AW-6014 is frequently applied in clinched joints [10]. Numerous studies on EN AW-6014-based clinched joints have been conducted, including investigations into the effects of galvanic corrosion on fatigue behavior [11] and the performance of clinched joints throughout their operational

Content from this work may be used under the terms of the Creative Commons Attribution 3.0 license. Any further distribution of this work must maintain attribution to the author(s) and the title of the work, journal citation and DOI. Published under license by Materials Research Forum LLC.

lifecycle [12]. Therefore, identifying parameters for the HCF model is crucial for facilitating further research, especially for numerically calculating the fatigue life of clinched joints.

In the present work, Lemaitre's HCF model [6] is applied to simulate the HCF of flat EN AW-6014 samples in the heat treatment condition T4, using experimental data for identification. The obtained material parameters are then used for HCF calculations of clinched specimens. The present work is structured as follows. In the next section, the numerical model for high-cycle fatigue is summarized. Then, the experimental data is presented, and the material parameters for the HCF model are determined. Finally, results of the fatigue life prediction of clinched specimens are presented based on the identified parameters, and a brief summary is given.

High-Cycle Fatigue

In high-cycle fatigue, plasticity is confined to the microscale, meaning no plastic deformation occurs at the macroscale. However, as plastic strain accumulates at the microscale, damage gradually increases. Once the damage reaches the critical threshold after a certain number of loading cycles, crack initiation occurs ultimately leading to material failure. This section briefly introduces (i) S-N curves for estimating fatigue life and (ii) the two-scale damage model.

S-N Curves. Through the Stress Amplitude-Number of Cycles curves (S-N curves), fatigue life can be predicted. The concept of S-N curves was first introduced by Wöhler [4] and can be expressed as:

$$lgN = a - b \cdot S_a,\tag{1}$$

where a, b are undetermined coefficients and S_a is the stress amplitude. Basquin [13] proposed a new form of the equation as:

$$N = N_D \cdot \left(\frac{S_a}{S_D}\right)^{-k},\tag{2}$$

where S_D is the fatigue strength and N_D is the fatigue life. This curve can be derived through the statistical analysis of fatigue test results from specimens, such as the horizon method and the pearl string method [14]. Although the pearl string method cannot capture the scatter of the fatigue life, it will be used in the present work, as it requires fewer samples compared to the horizon method, thus saving time and costs.

Two-Scale Damage Model. The two-scale damage model proposed by Lemaitre [6] is a crucial tool for computing fatigue life. This model is based on the principle that mesoscopic stresses applied to a representative volume element (RVE) induce microscopic plastic strain within it. All calculations are conducted for the RVE. Notably, the microscopic damage occurring inside the RVE does not influence the macroscopic behavior. Once the microscopic damage reaches its critical value, the RVE is considered to have failed, leading to crack initiation. The relationship between mesoscopic and microscopic stress is described by the localization law of the self-consistent scheme [15]:

$$\sigma_\mu = \sigma - aE\varepsilon^{\mu p}.\tag{3}$$

In these equations, the superscript μ represents microscopic variables. $\varepsilon^{\mu p}$ denotes the microscopic plastic strain tensor, E is the Young's modulus and a is the Eshelby parameter for a spherical inclusion [16].

The von Mises yield criterion is expressed as:

$$f^\mu = (\tilde{\sigma}^\mu - X^{\mu D})_{eq} - \sigma_f = 0, \quad \text{with } dX^{\mu D} = \frac{2}{3}Cd\varepsilon^{\mu p}(1 - D),\tag{4}$$

Sheet Metal 2025
Materials Research Proceedings 52 (2025) 184-191

Materials Research Forum LLC
https://doi.org/10.21741/9781644903551-23

where $\tilde{\sigma}^{\mu} = \sigma^{\mu}/(1 - D)$ represents the effective stress, σ_f is the fatigue strength, C is the kinematic hardening parameter and D is the damage, where $0 \leq D \leq 1$. $X^{\mu D}$ describes the kinematic hardening, which is important to consider in high-cycle fatigue. σ_{eq} denotes the equivalent (eq) von Mises stress $\sigma_{eq} = (3/2\sigma^D : \sigma^D)^{1/2}$, where σ^D is the deviatoric stress.

The damage evolution law is formulated as:

$$dD = \left(\frac{Y^{\mu}}{S}\right)^{s} dp^{\mu}, \quad \text{if } p^{\mu} \geq p_{\mathrm{D}}, \tag{5}$$

where Y^{μ} is the energy density release rate, S the damage strength, s the damage exponent, p_{D} the damage threshold [17] and p^{μ} the accumulated plastic strain at the microscale.

In each time increment, the stresses at the microscale change in response to the external force. This can lead to an increase in microscopic plastic strains if the yield function f^{μ} becomes positive at the end of the time increment, indicating that damage has increased. Therefore, all variables at the microscale must be updated at t_{n+1}.

The system of equations, which must be satisfied for each time increment, and a simplified form are summarized in Lemaitre at al. [6]. Solving in closed form allows for stable and rapid computation. In this model, the damage strength S and the damage exponent s significantly influence the simulation results. Therefore, identifying these two parameters is essential for future investigations of high-cycle fatigue behavior of components. The present work focuses on the identification and validation of these parameters.

Fatigue Testing and Clinched Joints
In the present work, fatigue testing of flat specimens and clinched joints made from EN AW-6014 T4 was conducted. Fatigue experiments on the flat specimens were initially conducted to determine essential fatigue properties, such as fatigue strength and the Wöhler curve, which are crucial for investigating the material's fatigue behavior and for subsequent parameter identification.

Fig. 1 shows the testing environment and the test results, along with the geometry of the flat specimens. For the fatigue testing a resonance pulsation test system Rumul Mikrotron by Russenberger Prüfmaschine AG is used. A frequency of 75 Hz under a force ratio of R = 0.1, with load cycles between 10^4 to $2*10^6$ are used. For testing, the EN AW-6014 T4 specimens with a thickness of 2.0 mm according to SEP1240 [18] (Fig. 1) were manufactured. The force amplitudes can be converted into stress amplitudes. The Wöhler curve was computed through the pearl string method. From the Wöhler curve, it can be inferred that the fatigue strength of EN AW-6014 T4 with a stress ratio $R = 0.1$ is approximately 1260 N, or 63 MPa. For parameter identification, the log-log slope k of the Wöhler curves in the finite life fatigue region, given by $k = \log\left(\sigma_{a,max}/\sigma_{a,min}\right)/\log\left(N_{min}/N_{max}\right)$, was calculated and resulted in $k = -0.158$.

In addition to flat specimens, fatigue testing of clinched joints was conducted to describe their fatigue properties and to validate the results. The clinched specimens were joined using a servo-electric TZ-VSN clinch unit from TOX® PRESSOTECHNIK GmbH & Co. KG, operating at a joining speed of 2 mm/s. A A48100 punch and BE8010 anvil die from TOX® were used to join the specimens [19]. The subsequent fatigue tests on the clinched joints were carried out using the same frequency, force ratio, and load cycle boundaries as in the flat specimen tests.

Parameter Identification
To identify the damage strength S and the damage exponent s, the parameters used in the HCF model are listed in Table 1. The closure parameter h determines the weighting of the compression component of the stress tensor, which is utilized in computing the energy density release rate Y^{μ} [6]. The parameter D_c represents the critical damage value that determines the failure of the RVE.

Sheet Metal 2025
Materials Research Proceedings 52 (2025) 184-191

Materials Research Forum LLC
https://doi.org/10.21741/9781644903551-23

First, Wöhler curves for varying values of s were computed. Based on these curves, a diagram showing the log-log slope of the Wöhler curve k versus the damage exponent s was generated, as shown in Fig. 2. By using the slope $k = -0.158$ computed in the last section, the damage exponent s was determined to be 4.3 from this diagram. Subsequently, using the determined s, the Wöhler curves for varying damage strength S were simulated, leading to the identification of an appropriate value for S. It is important to note that the stress distribution in the flat specimens under tensile loading is not perfectly uniform, leading to stress concentrations in certain areas, as Fig. 3 shows. As a result, the actual stress amplitudes cannot be calculated as the nominal ones $\sigma_a = F_a/A$. Instead, the stress amplitudes obtained from numerical simulations will be used. Consequently, the identified damage strength was calibrated as $S = 0.4$ MPa. The simulation results in Fig. 1 illustrate that the simulated fatigue life with the identified parameters align well with the experimental curve.

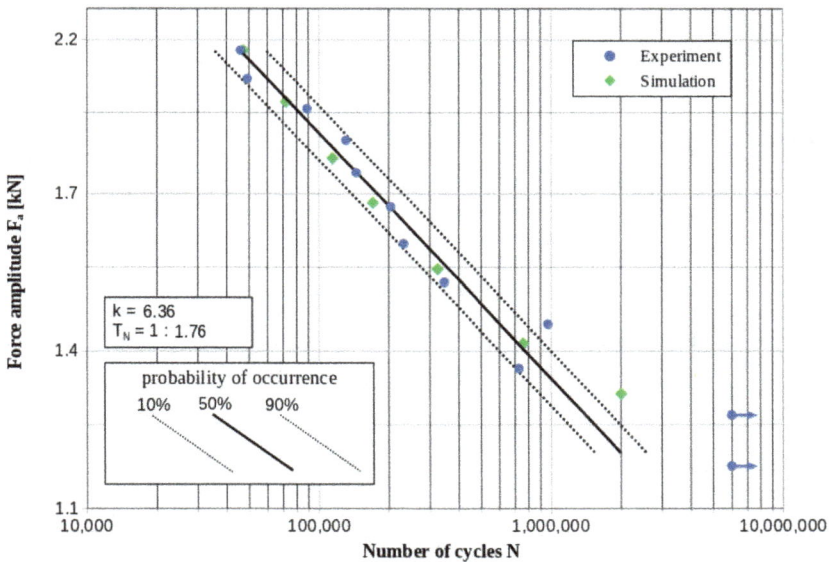

Figure 1: Experimental F-N curve of EN AW-6014 T4 with the testing and the simulation results.

Table 1: Parameters of EN AW-6014 T4 used for numerical simulation of the two-scale damage model.

Parameter	Physical meaning	Value	Ref.
E	Young's modulus	63 [GPa]	[20]
ν	Poisson ratio	0.34	[20]
σ_f	Fatigue strength	63 [MPa]	
σ_u	Maximum tensile strength	244.4 [MPa]	
ε_{pD}	Plastic strain threshold	0.19	[6]
C	Kinematic hardening parameter	220.6 [MPa]	[20]
h	Closure parameter	0.2	
D_c	Critical damage	0.5	

Sheet Metal 2025

Materials Research Forum LLC

Materials Research Proceedings 52 (2025) 184-191

https://doi.org/10.21741/9781644903551-23

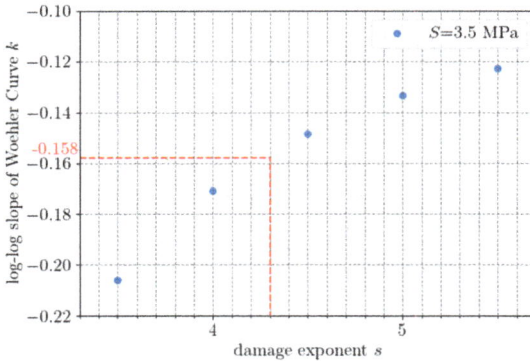

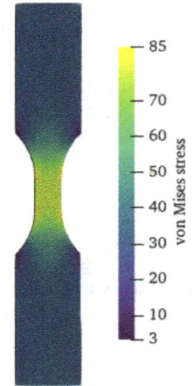

Figure 2: log-log slope of Wöhler curve k versus damage exponent s.
The red lines represent the determination of the damage exponent s.

Figure 3: Stress concentration
(in MPa) in the flat specimen.

Validation and Discussion

The parameters for the two-scale damage model were identified in the previous section. In this section, a numerical study on clinched joints is conducted to validate these parameters, as EN AW-6014 is commonly used in clinched joints, which are widely employed in car body structures. The simulated results were compared with the experimental ones for clinched joints of the same geometry and material, EN AW-6014 T4.

The numerical model of the clinched joint is shown in Fig. 4. The joint is formed by two 2 mm-thick sheets joined with a conical punch. The interface between the sheets is defined as a contact surface with the friction coefficient, assumed as $\mu = 1$. The model was implemented in the finite element tool LS-DYNA and solved implicitly using a nonlinear solver with BFGS updates and a time increment of 0.05 s. The geometry, the material model with the flow stress curve and the mesh were obtained from the clinching process simulation by Bielak et al. [21]. In the present numerical simulation, the applied mesh consists of 44908 eight-node hexahedron elements with reduced integration. A significantly denser mesh is used around the clinch point (see Fig. 4) than in other regions, as most failure modes occur in this critical area.

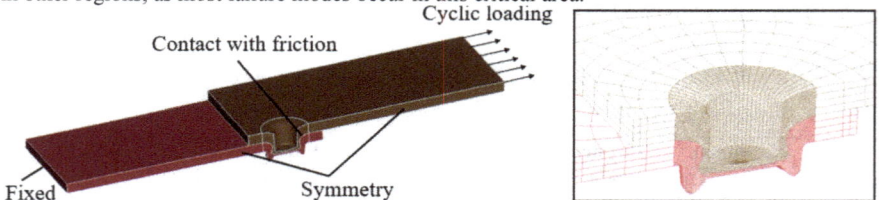

Figure 4: Geometry, boundary conditions and mesh of the clinched joint.

For simplification and due to the computational cost of simulating contact between metal sheets over millions of cycles, this aspect is not considered in the present work. Instead, the clinched joint is simulated for a single cycle to determine the stress state of the most critical element, which is assumed to be the failure location due to its largest von Mises stress amplitude [22]. The critical elements are shown in Fig. 5. This critical element is then treated as the RVE and computed until

Sheet Metal 2025

Materials Research Proceedings 52 (2025) 184-191

Materials Research Forum LLC

https://doi.org/10.21741/9781644903551-23

failure. While this approach allows for a fast and efficient prediction, it neglects effects such as load path shifting caused by increasing damage for a static friction coefficient.

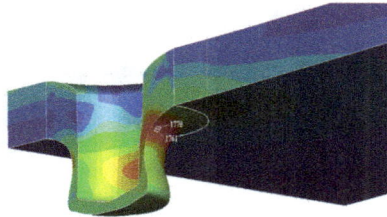

Figure 5: Punch-sided sheet of clinched joints and location of the critical elements (element 1770: for $F_a > 0.84$ kN; element 1761: for $F_a < 0.84$ kN).

Fig. 6 shows the comparison between the experimental results and the simulation. The figure illustrates that the two-scale damage model, using the identified parameters, predicts the HCF behavior of clinched joints quite well. However, the deviation between the simulated and experimental results may be attributed to several factors.

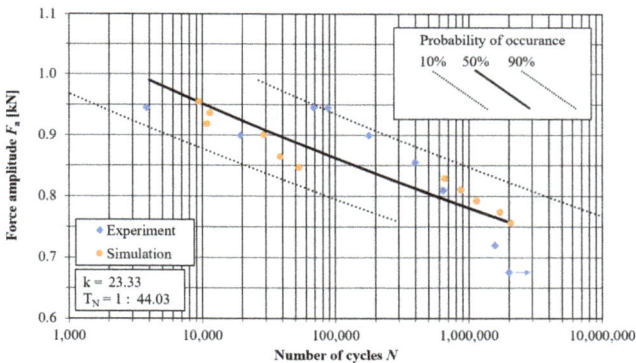

Figure 6: Comparison of experimental Wöhler curve of clinched joints and simulation results.

To improve fatigue life predictions, the following aspects should be considered:

(a) The geometry of the simulated clinched joint deviate slightly from that of the clinched joints used in the experiments. As a result, the simulated fatigue life may occasionally fall within the range of a 10% to 90% probability of occurrence.

(b) The jump between 0.846 kN and 0.828 kN occurs because the critical element, which experiences the highest von Mises stress, shifts to a neighboring element, as shown in Fig. 5. It is assumed that using a finer mesh would reduce this discrepancy. However, this would significantly increase the computational cost.

(c) As the force amplitude increases, the von Mises stress can easily exceed the material's yield stress due to the stress concentration in the neck area, which is inherent to the structure of clinched joints. This exceedance, leading to macroscopic plastic strain, is neglected by the two-scale damage model. Consequently, at higher force amplitudes, the predicted fatigue life tends to be longer than the experimental results and the two-scale damage model is not well-suited for such cases.

Sheet Metal 2025
Materials Research Proceedings 52 (2025) 184-191

Materials Research Forum LLC
https://doi.org/10.21741/9781644903551-23

(d) The friction between the sheets has to be defined accurately, as it depends on the surface characteristics, the lubricating conditions and the friction coefficient curve.

(e) The internal stress and pre-damage caused by the jointing process should be considered.

Summary

In the present work, the experimental results of fatigue testing on flat specimens and clinched joints made from aluminum alloy EN AW-6014 T4 were presented. Based on these results, parameter identification was carried out, determining the damage strength and damage exponent for the two-scale damage model used to predict high-cycle fatigue. Using the identified parameters, the HCF behavior of the clinched joint was numerically investigated. While the numerical results match the experimental results very well, accuracy is still limited by certain factors, such as a basic model for friction between the metal sheets. In the future, it will be valuable to identify the parameters for EN AW-6014 in the T6 condition to assess potential differences. Additionally, it is worthwhile to develop a more efficient model for predicting the HCF of clinched joints.

Acknowledgements

We gratefully acknowledge funding by the Deutsche Forschungsgemeinschaft (DFG, German Research Foundation) – TRR 285/2, subprojects B03 and A01– Project-ID 418701707.

References

[1] Gao, T., Sun, Z., Xue, H., & Retraint, D. (2020). Effect of surface mechanical attrition treatment on high cycle and very high cycle fatigue of a 7075-T6 aluminium alloy. International Journal of Fatigue, 139, 105798. https://doi.org/10.1016/j.ijfatigue.2020.105798

[2] Farfan, S., Rubio-Gonzalez, C., Cervantes-Hernandez, T., & Mesmacque, G. (2004). High cycle fatigue, low cycle fatigue and failure modes of a carburized steel. International Journal of Fatigue, 26(6), 673-678. https://doi.org/10.1016/j.ijfatigue.2003.08.022

[3] Cowles, B. A. (1996). High cycle fatigue in aircraft gas turbines-an industry perspective. International Journal of Fracture, 80, 147-163. https://doi.org/10.1007/BF00012667

[4] Wöhler, A. (1870). Über die Festigkeitsversuche mit Eisen und Stahl. Ernst & Korn.

[5] Schijve, J. (2003). Fatigue of structures and materials in the 20th century and the state of the art. International Journal of Fatigue, 25(8), 679-702. https://doi.org/10.1016/S0142-1123(03)00051-3

[6] Lemaitre, J., Sermage, J. P., & Desmorat, R. (1999). A two scale damage concept applied to fatigue. International Journal of Fracture, 97, 67-81. https://doi.org/10.1023/A:1018641414428

[7] Roters, F., Eisenlohr, P., Bieler, T. R., & Raabe, D. (2011). Crystal plasticity finite element methods: in materials science and engineering. John Wiley & Sons. https://doi.org/10.1002/9783527631483

[8] Cole, G. S., & Sherman, A. M. (1995). Light weight materials for automotive applications. Materials Characterization, 35(1), 3-9. https://doi.org/10.1016/1044-5803(95)00063-1

[9] Troeger, L. P., & Starke Jr, E. A. (2000). Microstructural and mechanical characterization of a superplastic 6xxx aluminum alloy. Materials Science and Engineering: A, 277(1-2), 102-113. https://doi.org/10.1016/S0921-5093(99)00543-2

[10] Kupfer, R., Köhler, D., Römisch, D., Wituschek, S., Ewenz, L., Kalich, J., ... & Troschitz, J. (2022). Clinching of aluminum materials-methods for the continuous characterization of process, microstructure and properties. Journal of Advanced Joining Processes, 5, 100108. https://doi.org/10.1016/j.jajp.2022.100108

[11] Harzheim, S., Hofmann, M., & Wallmersperger, T. (2023). Numerical fatigue life prediction of corroded and non-corroded clinched joints. Mechanics of Advanced Materials and Structures, 30(5), 961-966. https://doi.org/10.1080/15376494.2022.2140233

[12] Schramm, B., Harzheim, S., Weiß, D., Joy, T. D., Hofmann, M., Mergheim, J., & Wallmersperger, T. (2022). A review on the modeling of the clinching process chain-part III: operational phase. Journal of Advanced Joining Processes, 6, 100135. https://doi.org/10.1016/j.jajp.2022.100135

[13] OH, B. (1910). The exponential law of endurance tests. In Proc Am Soc Test Mater (Vol. 10, pp. 625-630).

[14] Martin, A., Hinkelmann, K., & Esderts, A. (2011). Zur Auswertung von Schwingfestigkeitsversuchen im Zeitfestigkeitsbereich-. Materials Testing, 53(9), 502-512. https://doi.org/10.3139/120.110255

[15] Kröner, E. (1961). Zur plastischen Verformung des Vielkristalls. Acta Metallurgica, 9(2), 155-161. https://doi.org/10.1016/0001-6160(61)90060-8

[16] Eshelby, J. D. (1957). The determination of the elastic field of an ellipsoidal inclusion, and related problems. Proceedings of the royal society of London. Series A. Mathematical and Physical Sciences, 241(1226), 376-396. https://doi.org/10.1098/rspa.1957.0133

[17] Lemaitre, J., & Doghri, I. (1994). Damage 90: a post processor for crack initiation. Computer Methods in Applied Mechanics and Engineering, 115(3-4), 197-232. https://doi.org/10.1016/0045-7825(94)90060-4

[18] VDEh, Stahlinstitut. "Prüf-und Dokumentationsrichtlinie für die experimentelle Ermittlung mechanischer Kennwerte von Feinblechen aus Stahl für die CAE-Berechnung." Stahl Eisen Prüfblatt SEP1240 1 (2006).

[19] DVS - Deutscher Verband für Schweißen und verwandte Verfahren e.V. Merkblatt DVS/EFB 3420: Clinching - basics. 2021.

[20] Friedlein, J., Wituschek, S., Lechner, M., Mergheim, J., & Steinmann, P. (2021, June). Inverse parameter identification of an anisotropic plasticity model for sheet metal. In IOP Conference Series: Materials Science and Engineering (Vol. 1157, No. 1, p. 012004). IOP Publishing. https://doi.org/10.1088/1757-899X/1157/1/012004

[21] Bielak, C. R., Böhnke, M., Beck, R., Bobbert, M., & Meschut, G. (2021). Numerical analysis of the robustness of clinching process considering the pre-forming of the parts. Journal of Advanced Joining Processes, 3, 100038. https://doi.org/10.1016/j.jajp.2020.100038

[22] Ewenz, L., Bielak, C. R., Otroshi, M., Bobbert, M., Meschut, G., & Zimmermann, M. (2022). Numerical and experimental identification of fatigue crack initiation sites in clinched joints. Production Engineering, 16(2), 305-313. https://doi.org/10.1007/s11740-022-01124-z

Characterization

SheMet 2025

Sheet Metal 2025

Materials Research Proceedings 52 (2025) 193-200

Materials Research Forum LLC

https://doi.org/10.21741/9781644903551-24

Evaluating the joinability of aluminum 2024 T351 for aerospace structures using aluminum solid self-piercing rivets

Felix Holleitner[1,a *], Knuth-Michael Henkel[2], and Normen Fuchs[3,b]

[1]Fraunhofer Institute for Large Structures in Production Engineering IGP, Albert-Einstein-Str. 30, 18059 Rostock, Germany

[2]University of Rostock, Faculty of Mechanical Engineering and Marine Technology, Chair of Joining Technology, Albert-Einstein-Str. 30, 18059 Rostock, Germany

[3]Stralsund University of Applied Science, School of Mechanical Engineering, Professorship for Quality Management, Production and Joining Technology, Zur Schwedenschanze 15, 18435 Stralsund, Germany

[a]felix.holleitner@igp.fraunhofer.de, [b]normen.fuchs@hochschule-stralsund.de

Keywords: Aluminum, Joining, Fatigue

Abstract. The load-bearing capacity of material equivalent aluminum joints with a developed solid self-piercing rivet (SSPR) was investigated. In particular, the effect of the rivet material on the joint strength under quasi-static and fatigue loading was analyzed systematically in order to evaluate the joinability. To demonstrate the load-bearing capacity of the Al-SSPR, it was also compared to a conventional cold-formed solid aluminum rivet (Al-SR). Under a quasi-static load, the load-bearing capacity of the developed aluminum SSPR is 10% lower than that of a geometrically identical SSPR made of steel (St-SSPR) and is approximately 27% lower than that of the Al-SR. However, the fatigue strength was found to be equivalent. Thus, the joinability of the aluminum alloy 2024 T351 up to a total sheet thickness of at least 3.0 mm using the developed Al-SSPR was demonstrated.

Introduction

The high-strength aluminum alloy (Al) 2024 T351 has excellent lightweight properties and is widely used in aircraft structures. For the installation of frame couplings, clips for cables and cladding and lockbolts made of titanium, as well as solid rivets made of aluminum, are primarily used. These manual joining technologies require time-consuming preparatory work, such as the insertion of the pilot holes, deburring, and the multiple positioning of the components [1]. A suitable and efficient process for joining structural components made of aluminum is solid self-piercing riveting. It combines the pre-hole operation with the rivet installation process and can be easily automated [2].

During the installation process, the rivet itself acts as a punching tool. This results in a high rivet stress. Therefore, solid self-piercing rivets (SSPRs) are typically made of steel, e.g., the corrosion-resistant martensitic steel alloy 1.4035 (X46CrS13). However, the joining of the dissimilar materials of the rivet and sheet does have disadvantages in terms of corrosion [3] and thermal expansion [4]. To solve this problem, an SSPR made of aluminum 7068 T651 was developed.

As the joint formation of Al-2024 T351 with an SSPR made of Al-7068 T651 has already been investigated in a previous study [5], the aim is to evaluate the SSPR joint strength in comparison to the state-of-the-art aerospace cold-formed aluminum solid rivets (Al-SRs) [6].

Experimental procedure

The plate material Al-2024 T351 had a thickness of 1.5 mm. To increase the corrosion resistance, the sheet material had a surface coating of pure aluminum. The percentage of the "Alclad layer"

Content from this work may be used under the terms of the Creative Commons Attribution 3.0 license. Any further distribution of this work must maintain attribution to the author(s) and the title of the work, journal citation and DOI. Published under license by Materials Research Forum LLC.

Sheet Metal 2025
Materials Research Proceedings 52 (2025) 193-200

Materials Research Forum LLC
https://doi.org/10.21741/9781644903551-24

in relation to the total sheet thickness was approximately 3.5%. The rivet material Al-7068 T651 was available as bar material, with a diameter of 8.0 mm.

For the mechanical characterization of the material, tensile tests were carried out under a quasi-static load. The mechanical properties and hardness of the rivet and sheet materials are summarized in Table 1.

The hardness of the Al-SSPR is only one third of the hardness of the SSPR made of steel. Thus, the application range of the aluminum rivet is limited [5].

Table 1 Mechanical properties and hardness of the sheet and rivet materials (* SR hardness at the head, ** SR hardness at the cold-formed locking head).

Material	Used as	Tensile strength [MPa]	Yield strength [MPa]	Elongation [%]	Hardness [HV1]
Al-2024 T351 (AlCu4Mg1)	Sheet	437 ± 5	345 ± 4.7	14 ± 1.6	143 ± 2
Al-7068 T651 (AlZn7.5Mg2.5Cu2)	SSPR	655 ± 5	603 ± 10	11 ± 2	207 ± 2
St-1.4035 (X46CrS13)	SSPR	–	–	–	625 ± 6
Al-2117 A-T42 (DIN EN 6081:2023)	SR	–	–	–	110* - 170**

To join Al-2024 T351 with a total sheet thickness of 3.0 mm, an SSPR with a shaft diameter of 4.80 mm was used. This ensures repair and replacement by a larger rivet and is a standard size for aerospace fasteners [7]. A drawing of the SSPR in superposition to the microsection of the rivet and the rivet groove geometry in detail are given in Figure 1(a). The geometries of the Al-SSPR and the St-SSPR were identical. Both rivets were in total 5.8 mm long and had a flat rivet head with a diameter of 5.9 mm.

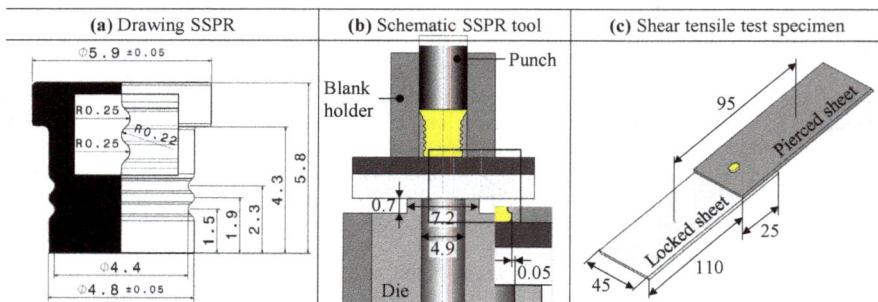

Fig. 1 Drawing of the SSPR (a), configuration of the riveting tool (b), and shear tensile test specimen for quasi-static and fatigue investigations (c) (all dimensions in mm).

The die had an inner diameter of 4.9 mm. As a result, the cutting clearance, which is the equidistant space between the surfaces of the punch (rivet) and the cutting plate (die) [8], was 0.05 mm. The die embossing ring had an outer diameter of 7.2 mm. Therefore, the ring area was approx. 22 mm² and reliably prevented the pre-embossing of the die-sided sheet during the piercing of the sheets. A schematic of the SSPR tool is given in Figure 1(b).

Single-lap shear specimens were fabricated for the quasi-static and fatigue tests, as illustrated in Figure 1(c). The overlap length was set to a uniform 25 mm and the free clamping length was set to 95 mm. The quasi-static tensile shear tests were performed with a universal testing machine (ZWICK/ROELL Z50) in force-controlled mode. The tests were carried out under quasi-static load

Sheet Metal 2025   Materials Research Forum LLC
Materials Research Proceedings 52 (2025) 193-200   https://doi.org/10.21741/9781644903551-24

conditions with a test speed of 10 mm/minute. The local displacement was recorded by a video extensometer (ZWICK/ROELL videoXtens) at the center of the joint overlap.

The fatigue tests were performed with a load ratio of R = 0.1 (tensile mode) and a frequency of f = 10 Hz using an electrodynamic testing machine (ZWICK/ROELL LTM10 T). The tests were stopped when more than 2 million cycles were reached or when the failure of the specimen resulted in a loss of the joint integrity.

Results and discussion
Joint formation
The cross-sections of the investigated joints are presented in Figure 2. Punching through the Al-2024 T351 sheets with a total sheet thickness of 3.0 mm resulted in a compression of the Al-SSPR of about 2%. However, this rivet compression was not critical, as the rivet grooves remained intact and were properly filled with material from the die-side sheet (see Figure 2(a)). Furthermore, the joint formation was gap-free and therefore fulfilled the quality standards. The form fit of Al-2024 T351 joints with the Al-SSPR is analyzed in more detail in [5].

In comparison to the Al-SSPR, the geometrically identical SSPR made of steel was not compressed due to its significantly higher hardness. Embossing with a maximum force of 45 kN resulted in a sufficient filling of the rivet groove. The average filling was 81% for the St-SSPR and 83% for the Al-SSPR.

Fig. 2 Cross-sections of joints with two 1.5-mm-thick Al-2024 T351 sheets with the Al-SSPR (a), St-SSPR (b), and Al-SR (c).

The Al-SR also had a shaft diameter of 4.8 mm. While forming the locking head at the bottom sheet, the rivet diameter increased slightly (see Figure 2(c)). With a diameter of 9.6 mm, the SR head was considerably larger than that of the SSPR. This might increase the load-bearing capacity of these joints under a shear load.

Quasi-static characteristics
The quasi-static test results are presented in Figure 3. The diagram on the left shows the load-displacement curves of all specimens. At the beginning, the curves' characteristics are almost linear, indicating an almost elastic joint behavior. At a tensile load of approximately 3,000 N, the SSPR begins to tilt irreversibly due to the moment created between the overlapping sheets. As a result, the gradient of the force-displacement curves decreases. After exceeding the maximum load, on average at 3,991 N for the joints with the Al-SSPR and at 4,256 N for the joints with the St-SSPR, the rivet is pulled out from the locked sheet. As expected, the large SR head and the die-side locking head almost completely prevented the rivet from tilting. This increased the load-bearing capacity of the joints with the Al-SR. The peak load was on average 5,498 N.

All specimens were tested until failure and evaluated based on the peak load, the load at a displacement of 0.05 mm, and the energy absorption according to [9]. The load at a displacement of 0.05 mm represents predominantly elastic joint characteristics and can be interpreted as the technical performance capability of the joint. Consequently, a shear-tensile load above this limit causes extensive and irreversible damage in the joint.

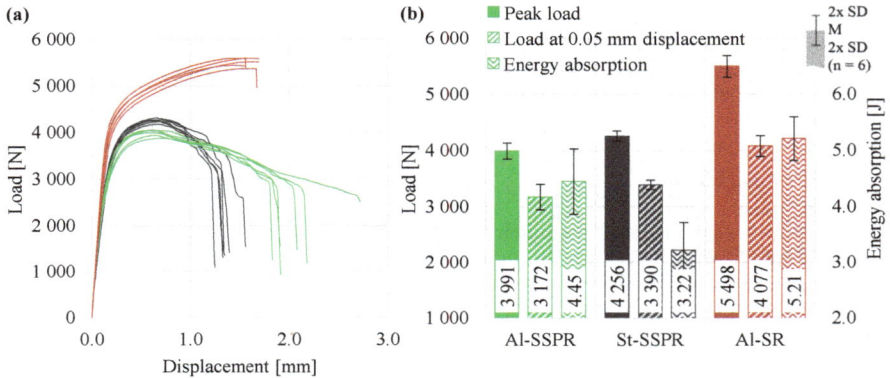

Fig. 3 Load-displacement curves from quasi-static tests (a) and average peak loads and joint energy absorption (b).

The average quasi-static test results are presented in Figure 3(b). The difference between the Al-SSPR and the St-SSPR in the peak load and the elastic limit is below 10%. Moreover, the Al-SSPR has a superior energy absorption capacity. The difference is about +25 %. For the application, the failure characteristics of the Al-SSPR under a quasi-static tensile shear load are satisfactory. However, the load capacity of the Al-SSPR is approximately 27% less than that of the Al-SR.

Static joint failure
The failure of the Al-SSPR and the failure of the St-SSPR were similar. The rivets tilted severely and the rivet holes widened. Furthermore, the rivet heads were cutting slightly into the top sheets of the specimens. All joints failed due to the extraction of the rivet foot from the locked sheet. Damage to the Al-SSPR and St-SSPR themselves was not detected. A typical failed specimen with the Al-SSPR is shown in Figure 4(a).

Joints with the Al-SR failed due to the fracture of the rivet itself (see Figure 4(b)). If the shear strength of the rivet material was exceeded, the rivet head sheared off completely. There was no evidence of the loss of the joint load-bearing capacity, indicating rivet failure. It is worth noting that all the failures occurred in the potentially weak area between the large rivet head and the rivet shaft.

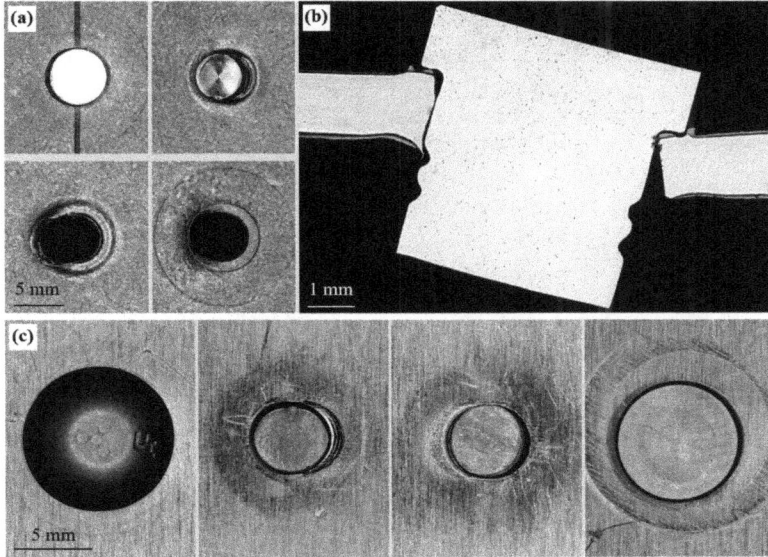

Fig. 4 Typical failure of an Al-SSPR joint (a), cross-section of a failed Al-SSPR (b), and failure of an Al-SR joint (c) in the quasi-static shear test.

Fatigue characteristics

The fatigue data of the investigated joints are summarized in Table 2. The different load levels were chosen with the aim of observing a failure between 10^4 and 2×10^6 cycles to determine the finite life fatigue curve (F-N curve) using linear regression in a double logarithmic representation.

Table 2 Fatigue data of the investigated joints (load amplitude and fatigue cycle).

Load amplitude [N]		1 395	1 305	1 215	1 125	1 035	945	900
	Al-SSPR	80 789	123 079	207 353	311 761	808 261	1 205 448	1 726 419
Cycles	St-SSPR	68 666	118 446	209 441	406 841	675 924	1 271 757	1 991 167
	Al-SR	77 886	132 023	256 426	404 772	596 150	1 171 741	1 752 619

In the logarithmic coordinate system, the linear equation can be expressed as follows:

$$\log(N) = a + k \cdot \log(F), \tag{1}$$

where N is the number of fatigue cycles, F is the fatigue load amplitude (in N), and a and k are the undetermined coefficients that can be calculated using the least squares method. Furthermore, the linear correlation coefficient R^2 for each F-N curve was determined. As Figure 5(a) shows, the number of cycles can be approximated well. The low variation resulted in a high correlation coefficient of $R^2 = 0.99$ for all three joint types. Finally, these curves indicate a 50% probability of joint failure.

The difference in the fatigue life between the two SSPR series is very small. The technical fatigue limit at 2 million load cycles is about 900 N for the SSPR made of steel and about 880 N for the joints with the Al-SSPR. The gradient exponent k of the F-N curves correlates with the

Sheet Metal 2025
Materials Research Proceedings 52 (2025) 193-200

Materials Research Forum LLC
https://doi.org/10.21741/9781644903551-24

notch effect of the joints. It is 7.5 for the joints with the St-SSPR and 7.1 for the joints with the Al-SSPR. The smaller the gradient exponent, the smaller the fatigue resistance. According to Issler et al. [10], such gradient exponents can be classified as representing a medium notch effect.

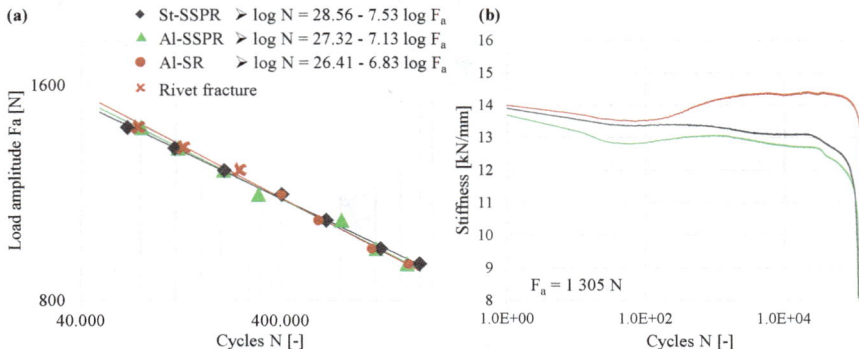

Fig. 5 F-N curves of the investigated joints (a) and characteristic joint-stiffness curves at a load amplitude of $F_a = 1,305$ N (b).

For the Al-SR joints, the fatigue limit is 880 N. Accordingly, the load-bearing capacity of the Al-SR joints under the fatigue load is much lower in comparison to the quasi-static properties. The gradient exponent of the F-N curve is 6.8. The fatigue resistance of the Al-SR is therefore slightly lower than that of the Al-SSPR. In fact, considering the high load amplitudes in relation to the technical performance capability (quasi-static), the Al-SSPR even has a superior fatigue strength.

In Figure 5(b), the stiffness curves of exemplary joints, tested at a load amplitude of 1,305 N, show the typical loss of the joint integrity. However, the failure was quite different for these joints.

Fatigue joint failure
Figure 6 shows the failed SSPR and SR specimens at three different load amplitudes. All SSPR joints failed due to the fracture of the sheet metal. At the highest load amplitude of 1,395 N, the SSPR tilted severely and fatigue cracks occurred in the pierced sheet, as well as in the locked sheet. The small cracks are indicated by arrows. Furthermore, oxidized friction particles colored the area around the rivet hole dark. As the test load decreased, the tilting of the rivet also decreased. In contrast, the area of friction increased. At the same time, the cracks originated at a larger distance from the rivet hole, almost outside of the friction contact (see Figure 6; $F_{a\_7} = 900$ N). Fatigue damage of the SSPR itself was not observed.

The failure mode of the SR joints depended on the load amplitude. At the three highest load levels, the rivet head sheared off (see Figure 6; $F_{a\_1} = 1,395$ N). This was similar to the failure in the quasi-static tests. In addition, small cracks were found in the sheet metal that originated from the rivet hole. The failure of the remaining specimens at lower test loads was due to cracks in the top sheet.

Materials Research Forum LLC
https://doi.org/10.21741/9781644903551-24

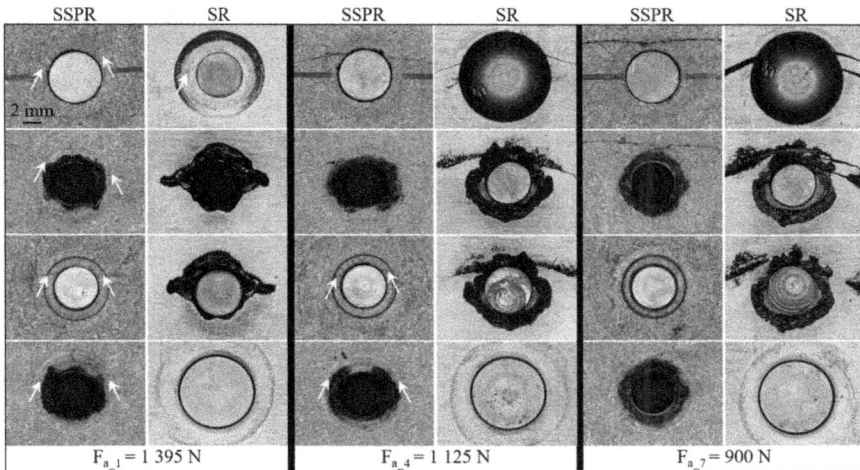

Fig. 6 Fatigue failure of SSPR and SR joints tested at different load amplitudes: front and back of the pierced sheet (a), and locked sheet (b). Small cracks are indicated by arrows.

Summary

In this study, the load-bearing capacity under quasi-static and fatigue shear loads of aluminum solid self-piercing rivets for the material equivalent joining of two 1.5-mm-thick Al-2024 T351 sheets was investigated. The main conclusions derived are the following:

Influence of the rivet material

- The load-bearing capacity of the St-SSPR joints under a quasi-static shear load is about 10% higher than the load-bearing capacity of the Al-SSPR joints.
- The energy absorption capacity of the material equivalent joints with the Al-SSPR significantly exceeds that of steel rivets (+25%).
- The fatigue strengths of joints with the Al-SSPR and joints with the St-SSPR are almost equal.
- The failure of the SSPR joints was due to sheet fracture.

Influence of the joining technology

- The load-bearing capacity (-27%) and the energy absorption of the Al-SSPR (-14%) under a quasi-static shear load are significantly lower than the load-bearing properties of the Al-SR.
- The fatigue strengths of joints with the Al-SSPR and joints with the Al-SR are almost equal. Moreover, the fatigue strength of the Al-SSPR is even superior in relation to the technical performance (quasi-static).
- The failure of the SR joints varied depending on the load level. The failure was due to rivet fracture and sheet fracture.

In summary, the load-bearing capacity of the developed Al-SSPR for the material equivalent joining of Al-2024 T351 has been evaluated. On the one hand, the Al-SSPR can substitute for the common SSPR made of steel for joining applications of thin lightweight sheet metals. On the other hand, the Al-SSPR can substitute for the Al-SR for secondary joining applications in the aircraft structure, such as clips for cables and cladding, due to its very good fatigue properties, and thus it can contribute to the joining efficiency. Future work will concentrate on the variation of the rivet

Materials Research Forum LLC
https://doi.org/10.21741/9781644903551-24

shaft diameter and the rivet material to extend the application limit of the Al-SSPR. In addition, the development of an SSPR made from the same aluminum alloy as the sheet material could benefit the recycling of a structure at the end of its life.

References

[1] M.-C. Wanner, A. Ebert, T. Strohbach, C. Blunk and N. Fuchs, Vollstanznieten kann Flugzeug-Endmontage optimieren, Ingenieurspiegel, 2012.

[2] DVS/EFB 3410: Self-piercing riveting - Overview, Technical Bulletin, DVS Media GmbH, 2018.

[3] L. Calabrese, L. Bonaccorsi, E. Proverbio, G. Di Bella and C. Borsellino, Durability on alternate immersion test of self-piercing riveting aluminium joint, Materials & Design, pp. 849-856, 2013. https://doi.org/10.1016/j.matdes.2012.11.016

[4] S. Sujatanond, Y. Miyashita, S. Hashimura, Y. Mutoh and Y. Otsuka, Bolt load loss behavior of magnesium alloy AZ91D bolted joints clamped with aluminum alloy A5056 bolt, Applied Mechanics and Materials, pp. 135-139, 2013. https://doi.org/10.4028/www.scientific.net/AMM.313-314.135

[5] F. Holleitner and N. Fuchs, Increasing the load-bearing capacity of aluminium solid self-piercing rivet joints of EN AW-2024 T351 sheets by geometric rivet modification, in International Symposium Research-Education-Technology XXVI, pp. 34-41, 2024.

[6] DIN EN 6081: Aerospace series - Rivet, universal head, close tolerance - Inch series, DIN Media GmbH, 2023.

[7] K. Engmann, Technologie des Flugzeuges, 7th ed. Vogel Communications Group GmbH, 2019.

[8] VDI 3349: Trim steels for large stamping dies, VDI-Gesellschaft Produktion und Logistik, 2013.

[9] DIN EN ISO 12996: Mechanical joining - Destructive testing of joints - Specimen dimensions and test procedure for tensile shear testing of single joints, DIN Media GmbH, 2013.

[10] L. Issler, H. Ruoß and P. Häfele, Festigkeitslehre - Grundlagen, 2nd ed. Springer-Lehrbuch, Springer-Verlag Berlin, 2003.

Sheet Metal 2025
Materials Research Proceedings 52 (2025) 201-209

Materials Research Forum LLC
https://doi.org/10.21741/9781644903551-25

Influence of the sampling procedure on the mechanical forming limits in the characterization of sheet metal foils

Jan Sommer[1,a*], Max Meerkamp[1,b], Martina Müller[1,c], Tim Herrig[1,d] and Thomas Bergs[1,2,e]

[1]Manufacturing Technology Institute MTI of RWTH Aachen University, Campus-Boulevard 30, 52074 Aachen, Germany

[2]Fraunhofer Institute for Production Technology IPT, Steinbachstrasse 17, 52074 Aachen, Germany

[a]j.sommer@mti.rwth-aachen.de, [b]m.meerkamp@mti.rwth-aachen.de, [c]m.mueller@mti.rwth-aachen.de, [d]t.herrig@mti.rwth-aachen.de, [e]t.bergs@mti.rwth-aachen.de

Keywords: Bipolar Plate, Material Characterization, Specimen Removal

Abstract. The aviation industry is committed to reducing its impact on the climate to meet political targets and enhance social acceptance. In particular, the ambitious climate protection targets of Flightpath 2050 require disruptive and revolutionary propulsion concepts. The use of fuel cell systems is one of these propulsion concepts and enables near carbon-neutral operation. A key component of the fuel cell is the bipolar plate, which is manufactured using sheet metal foil with thicknesses ≤ 100 µm. Failure-free forming of the sheet metal foil poses a particular challenge due to the high risk of cracking during forming. To evaluate the failure-free formability of the sheet metal foil, a numerical representation of the forming process is essential. A key parameter is the material-specific failure criterion (forming-limit-diagram (FLD)), which can be determined using Nakajima tests. Due to the low material thickness of the sheet metal foil, it is currently questionable what influence the sampling procedure of the required test specimens has on the resulting mechanical characteristic values. The aim of this paper is therefore to investigate the influence of the sampling procedure on the resulting failure limits in the Nakajima test, considering a corrosion-resistant austenitic stainless steel of type 1.4404. The results of this paper show that there is a strong correlation between the selected removal method, the resulting edge zone quality and thus also the determined material properties in the characterisation of foil materials. Machining and wire erosion of the foil using dummy sheets have been identified as promising removal methods.

Introduction

The Flightpath 2050 demands for a significant reduction in CO_2 and NO_x emissions as well as noise pollution in the aviation industry [1]. However, achieving the targets set out in Flightpath 2050 is only possible using innovative propulsion concepts. One such alternative propulsion concept is the hydrogen-powered polymer electrolyte membrane fuel cell (PEM fuel cell). [2] A key component of the PEM fuel cell is the metallic bipolar plate, which serves as a distributor of the operating media, electrical conductor and mechanical stabilizer [3]. Currently, thin sheet metal foils with thicknesses $s \leq 100$ µm made of stainless steel 1.4404 are mainly used to produce bipolar plates [4]. One of the greatest challenges in the production of these bipolar plates arises from the failure-free forming of the thin sheet metal foil, as there is an increased risk of cracking during the forming process due to the low sheet metal foil thickness [5]. To reduce the risk of cracking and avoid cost-intensive misproduction, the numerical design of the forming process is therefore essential. The numerical design of the forming process makes it possible to detect critical load conditions of the sheet metal foil at an early stage, to adjust the corresponding process parameters and thus to ensure the failure-free forming of the sheet metal foil. The forming limits

Content from this work may be used under the terms of the Creative Commons Attribution 3.0 license. Any further distribution of this work must maintain attribution to the author(s) and the title of the work, journal citation and DOI. Published under license by Materials Research Forum LLC.

Sheet Metal 2025
Materials Research Proceedings 52 (2025) 201-209

Materials Research Forum LLC
https://doi.org/10.21741/9781644903551-25

for sheet materials are usually determined using so-called forming limit diagrams, which indicate the failure limits of the material under multiaxial load. The standardized Nakajima test is used to determine such forming limit diagrams. However, sheets with thicknesses $s \leq 0.3$ mm are currently only insufficiently considered in the corresponding standard of the test norm. [6] This leads to eccentric and therefore non-standard failure patterns during the material test, which do not allow the subsequent determination of the failures [7]. One reason for the eccentric failure is the low sheet thickness, which limits the number of grains over the thickness of the sheet metal foil and thus the forming. In addition, inhomogeneities in the sheet metal foils are very significant due to the low semi-finished product thicknesses. Previous studies have therefore investigated the influence of scaling and modification of the test specimen and punch geometries on the eccentric failure [8, 9]. However, in addition to the evaluability of the failure pattern, the influence of the test specimen edge zones on the mechanical material characteristics due to the low semi-finished product thicknesses must be investigated. The edge zone is significantly influenced by the sampling of the test specimens for the respective mechanical test methods. Eroding, milling, water jet cutting, and laser cutting are known sampling procedures that also correspond to the standard for Nakajima tests. All these sampling procedures exert a mechanical and/or thermal load on the material during test specimen production [10]. Currently, it is unknown what effect the sampling procedure has on the determination of the failure characteristics of sheet metal foil. The aim of this work is therefore to investigate the influence of the sampling process in determining the failure characteristics of sheet metal foil and then to identify a suitable sampling process for the production of test specimens from sheet metal foils.

Materials and Methods
The tests were carried out on a corrosion-resistant, austenitic stainless steel of type 1.4404 (X2CrNiMo17-12-2) with a sheet metal foil thickness of $s = 100$ μm. The test specimens for the Nakajima tests were produced using different sampling procedures: water jet cutting, laser cutting, wire eroding, milling. Due to the low material thickness, direct removal from the semi-finished product is not possible without affecting the edge zone properties and the dimensional accuracy of the test specimens. The influence on and damage to the test specimen edge zone can occur during the processing of sheet metal foils due to vibrations during mechanical processing and surface damage caused by abrasive particles, for example. For this reason, the test specimens were packaged before investigating the sampling procedures. Only laser cutting was carried out without packaging, as the laser power is not sufficient to cut the total package thickness of the packaging. Packing was necessary to prevent the formation of wrinkles and cracks during the specimen sampling and to remove the specimens in the sense of fixation according to *Angeloni* [11]. The packaging process is shown in Fig. 1 a).

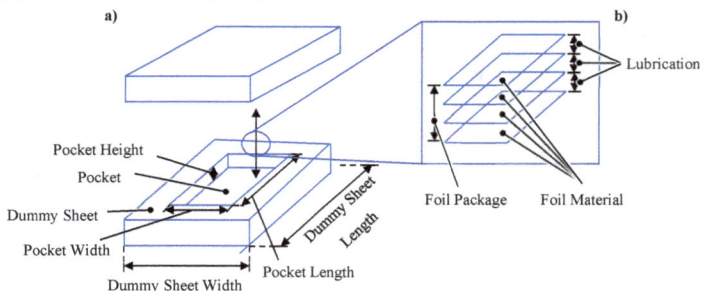

Fig. 1: a) Sampling using dummy sheets and b) Foil package

Sheet Metal 2025   Materials Research Forum LLC
Materials Research Proceedings 52 (2025) 201-209   https://doi.org/10.21741/9781644903551-25

After cutting, the metal foils were stacked, with a lubricating layer of special Vaseline from E-COLL being applied between the individual blanks. This lubricant was used to protect the metal foil from damage during the sampling by absorbing abrasive particles that are used during water jet cutting. The stacked blanks, the so-called foil package Fig. 1 b), were then placed in between dummy sheets. These aluminum dummy sheets were equipped with milled pockets whose dimensions are matched to the dimensions of the foil package. In laser cutting, the sheet metal foil was clamped as a long blank and each sample was cut out individually. According to *Angeloni*, the material should also be kept under as little tension as possible during this sampling procedure [11]. As there is no intermediate layer of the lubricant for this sampling procedure, the specimens were not cleaned after receipt.

The Nakajima test, which predicts material failure under a multiaxial stress state, was used to evaluate the sampling procedure. Six test specimen geometries were produced for the Nakajima tests for each sampling procedure. There are currently no standardized test specimen geometries for determining the forming limit diagram of sheet metal foils. However, the test specimen geometries used in this paper have already been defined as promising for sheet metal foil characterization in previous studies and are shown schematically Fig. 2 a) [8].

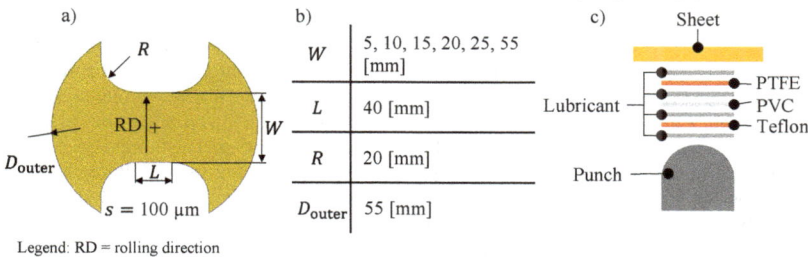

Legend: RD = rolling direction

Fig. 2: Nakajima test specimens b) Test specimens web width variation

The test specimens had an outer diameter $D_{outer} = 55$ mm, which is specified by the test setup. The fillet radius $R = 8$ mm and the parallel web length $L = 3$ mm were kept constant for all test specimens. Only the web width W was varied in the context of the Nakajima tests in accordance with the test schedule shown in Fig. 2 b). In the following, the variation of the web width is referred to as geometry $1 - 6$ (geometry 1 = Web width $W = 5\ mm$; geometry 6 = web width $W = 55\ mm$).

The Nakajima tests were carried out on the BUP200 sheet metal forming machine from *Zwick/Roell* The test setup comprised a die, a punch, the test specimens, a clamp and a lubrication system Fig. 2 c). The lubrication system used consists of a layer sequence of Polytetrafluorethylen (PTFE), PVC and graphite paste. After the lubrication system was applied to the test specimen, the clamps were lowered, and a defined clamping force F_N was applied. The clamping force F_N was kept constant at $F_N = 180$ kN to prevent the material from flowing out of the clamping area. The punch speed was set to $v_{St} = 0.5$ mm/s. In addition, only Nakajima tests at an angle ϕ of 90° relative to the web orientation was required for the stainless steel 1.4404 used in accordance with DIN EN ISO 12004. For optimal selection of the sampling procedure, the evaluation was based on a reproducible evaluation catalogue. The evaluation catalogue had the following requirements:

1. The Nakajima test specimens had to be evaluable, i.e. crack in the middle, during the test. Firstly, the specimens are checked visually by an operator for a proper crack and secondly, they must be evaluable using the Aramis software from Zeiss GOM. A proper crack profile ensured that the results were valid regarding the forming limit data.

Sheet Metal 2025   Materials Research Forum LLC
Materials Research Proceedings 52 (2025) 201-209   https://doi.org/10.21741/9781644903551-25

2. The test results had to show the lowest possible deviation of results. The statistical evaluation of multiple test runs using standard deviation was used to control the deviations in the measurement results.

3. Scanning electron microscope (SEM) images of the test specimens were also produced to evaluate the sampling procedure. A homogeneous, notch-free edge zone helped to avoid stress concentrations and ensured that no local weakening of the material was caused by the sampling procedure.

Results and Discussions

The evaluation of the sampling procedures is initially carried out by means of visual observation of the edge zone using SEMs. Fig. 3 shows the resulting edge zones after removal of the test specimens. As described in Materials and Methods, the target parameters of the edge zone are homogeneity and notch-free. Fig. 3 a) shows the resulting edge zone after sampling by milling. It can be seen that a very smooth edge zone is produced. Notches are not visible by visual inspection. Furthermore, the edge zone after machining shows a very homogeneous coloration, which indicates that no thermally activated processes such as oxidation occurred. In comparison to the milled specimen, Fig. 3 b) shows the edge zone after water jet cutting. Compared to the milled specimen, significantly deeper notches of approx. 10 µm are formed in the edge area. In addition to the notches in the edge area, marks can also be seen on the surface of the test specimen. One reason for this may be the use of abrasive particles during water jet cutting, which can cause small particles to penetrate the specimen package despite greasing. In addition, a white-coloured surface can be seen on the water jet cut specimen. However, oxidation of the material can be ruled out here, as no thermal load is applied to the material during water jet cutting. The eroded edge zone, which can be seen in Fig. 3 c), is difficult to evaluate as the metal spatter formed during eroding has deposited on the edge zone. The notch effect after eroding can therefore only be evaluated inadequately. Based on the images, however, it can be assumed that the mechanical properties are influenced by the metal spatter in the edge zone and behave in homogeneously due to locally varying amounts of metal spatter. However, it should be added that the amount of metal spatter can be reduced by optimizing the cutting parameters during eroding. Finally, Fig. 3 d) shows the edge zone after removal of the test specimen by laser cutting. In comparison with the eroded and water jet cut test specimens, the edge zone shows only minor notches. However, these are more pronounced than when the specimens were removed by milling. Furthermore, a white layer has formed on the surface of the edge zone during laser cutting. Due to the homogeneity, the coloration and the effective depth, it can be strongly assumed that a thermally induced oxide zone has formed because of laser processing. This can lead to altered mechanical material properties, which has a direct influence on the failure limits to be determined.

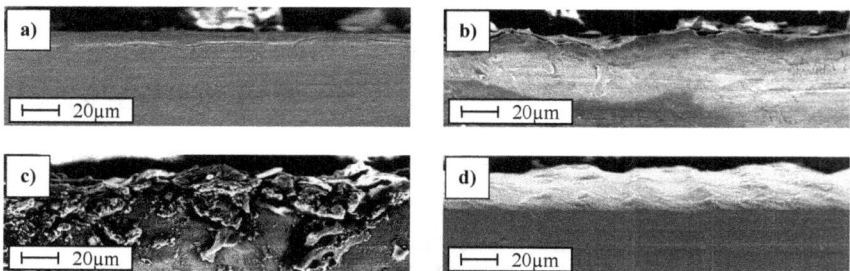

Fig. 3: Scanning-Electron-Microscope of cut edges after a) Milling b) Water jet cutting c) Eroding d) Laser cutting

Sheet Metal 2025
Materials Research Proceedings 52 (2025) 201-209

Materials Research Forum LLC
https://doi.org/10.21741/9781644903551-25

Based on the SEM images of the cut edges, it can be stated that machining leads to the least notch formation. Since notches in the edge zone serve as a nucleus for crack initiation, the formation of notches during sampling must be prevented. Furthermore, a homogeneous formation of the edge zone leads to homogeneous test results, as the deviation of the results is reduced due to a defined edge zone quality. In addition to the purely mechanical influences introduced by the sampling procedure via notches, the thermal influence of test specimen removal should not be neglected. The sampling procedure of eroding and laser cutting result in high temperatures in the cutting zone, which lead to the formation of metal spatter and oxide layers that influence the mechanical material behaviour accordingly. For a better evaluation of the thermal influence, microstructural images of the test specimens should be considered in later investigations.

As a further decision criterion in addition to the metallographically tested edge zone quality, material tests were carried out using Nakajima tests and evaluated according to the sampling procedure. The aim is to achieve the lowest possible deviation of the individual measurements of each test geometry to ensure a high repeatability of the measurements. Fig. 4 shows the forming limit diagrams of each sampling procedure at a punch speed of $v_P = 0.5$ mm/s.

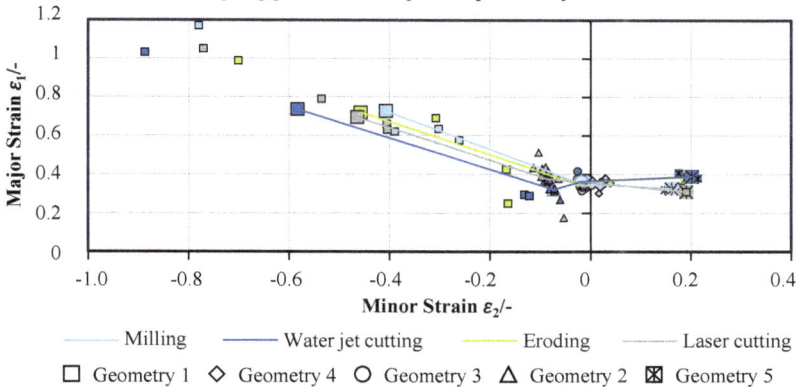

Fig. 4: Forming-Limit-Diagrams for specimens manufactured by milling, water jet cutting, eroding and laser cutting with a deformation rate of 0.5 mm/s

The deviating shapes of the markings represent a defined test specimen geometry as a function of the web width W. The small markings represent the values of the individual measurements, while the large markings show the mean values of a test specimen geometry as a function of the sampling procedure. The colored curves each mark one of the sampling procedures examined. To evaluate the sampling procedure, the number of evaluable samples, the position of the deformation diagram and the deviation of the individual measurements are considered. As can be seen in Fig. 4, the positions of the Forming Limit Diagrams (FLDs) differ slightly depending on the sampling procedure, but the tendency of the curves is very similar, as can be seen from the parallel curves of the FLDs.

The average major strain ε_1 of the first specimen geometry with a web width of $W = 5$ mm is $\varepsilon_1 = 0.75$ for each of the sampling procedures. The associated minor strain ε_2, on the other hand, differs greatly depending on the sampling procedure. For the water jet cut specimens, minor strain of $\varepsilon_2 = -0.58$ is achieved for geometry 1, while the milled specimens have a minor strain of $\varepsilon_2 = -0.41$ with identical major strain. The values of the eroded and laser-cut test specimen geometry 1 lie between those of the water jet cut and milled test specimens. The eroded and laser-cut specimens achieve a minor strain of $\varepsilon_2 = -0.47$ with a major strain of $\varepsilon_1 = 0.73$. The smallest major strain for all sampling procedures, except for water jet cutting, occurs for specimen geometry 3 with a web width of $W = 15$ mm. For the milled test specimens, major strain values of $\varepsilon_1 = 0.342$ and minor strain values of $\varepsilon_2 = -0.0121$ are achieved. The values of the eroded and laser-cut specimens ranged identically. Only the water jet cut test specimens achieved strain ratios of $\varepsilon_1 = 0.363$ to $\varepsilon_2 = -0.0179$. It is particularly noteworthy that the minimum of the main deformation in the water jet cut specimens occurs in specimen geometry 2 with a web width of $W = 10$ mm. Here, deformation ratios of $\varepsilon_1 = 0.322$ to $\varepsilon_2 = -0.0792$ are achieved. It is also noticeable that the curves of the milled and laser-cut specimens in the biaxial stress state (specimen geometry 6 with $W = 55$ mm) do not increase again but are constant to slightly decreasing. In contrast, the eroded and water jet cut test specimens show slightly increasing curves for the biaxial stress range. As the circular blanks of test specimen geometry 6 are fully clamped in the BUP200 clamping system during the test, the influence of the edge zone on the results obtained is negligible. The deviation of the results depending on the sampling procedure at the web width of $W = 55$ mm is accordingly due to material fluctuations. The deformation ratios of the biaxial test result in main deformations between $\varepsilon_1 = 0.388$ for the water jet cut specimens and $\varepsilon_1 = 0.31$ for the laser-cut specimens. The secondary shape change fluctuates between values of $\varepsilon_2 = 0.162$ for the milled specimens and $\varepsilon_2 = 0.202$ for the water jet cut specimens.

In summary, the highest strain ratios were achieved with the milled specimens in the area of the negative minor strain ε_2, the tensile-compression area. These flatten out in the area of the positive minor strain ε_2, the tensile-tensile load. The highest deformation ratios in this sub-range are achieved using water jet cutting and erosion. However, it is not possible to make a fundamental statement about the quality of the sampling procedure regarding the mechanical material properties. For this reason, the deviation of the individual measurements in relation to the mean value is also determined using the standard deviation to evaluate the sampling procedure regarding the repeatability of the material test. Table 1 shows the standard deviations and the number of validly tested specimens for each sampling procedure and specimen geometry. Looking first at the number of validly tested specimens, the largest number of accepted tests can be carried out using milling. Apart from the web width of $W = 25$ mm, all test specimens produced by milling can be tested and evaluated. Only in the case of geometry 6 do two test specimens crack in such a way that an evaluation is not possible. In contrast, only four of the six web widths W can be tested using water jet cutting and laser cutting. The web widths $W = 20$ mm and $W = 25$ mm do not provide any reference points for the FLD, so that the interpolation range of the curve is enlarged. The eroded test specimens, like the milled ones, can be fully tested except for the web width $W = 25$ mm. Compared to the milled specimens, only three of the six tested specimens can be evaluated due to invalid cracks at a web width of $W = 20$ mm. The milled and eroded test specimens therefore show the most reliable evaluation compared to water jet and laser cutting. The low volatility of the measurement results from the milled test specimens can also be seen from the standard deviation in the main and secondary shape change. The largest standard deviations result from testing the test specimens with a web width of $W = 5$ mm, as can be seen in Table 1. This can be attributed to the small web width W of the test specimens, which makes an optical evaluation of the deformation difficult and therefore also leads to a large deviation of the results. The smallest standard deviations are achieved with milled and laser-cut test specimens. Milling shows the smallest standard deviations of $s_{\varepsilon_1} = 0.244$ in major strain and $s_{\varepsilon_2} = 0.207$ in minor strain. The standard deviations of the other web widths W vary only marginally between the individual sampling procedures, so that it is not possible to make a well-founded statement about the repeatability based on the other test specimens. In summary, the test specimens produced by milling show the lowest standard deviation of the individual measurements and therefore lead to the most repeatable tests. The most reference points can also be generated with the milled specimens, as the milled specimens lead to the most validly tested geometries. The determined

Sheet Metal 2025
Materials Research Proceedings 52 (2025) 201-209

Materials Research Forum LLC
https://doi.org/10.21741/9781644903551-25

failure limits, the FLDs, provide only a slight indication of the influence of the sampling procedure. All FLD curves show a consistent tendency and vary in the determined limit values by a maximum of 0.17 in the minor strain ε_2 and 0.08 in the major strain ε_1.

Table 1: Deviation of mechanical material properties according to the sampling procedure

Milling Web width W [mm]	Standard deviation ε_1	Standard deviation ε_2	Valid Samples Tested
5	0.244	0.207	6
10	0.044	0.008	6
15	0.024	0.009	6
20	0.035	0.011	6
25	-	-	-
55	0.008	0.014	4
Water jet cutting Web width W [mm]	Standard deviation ε_1	Standard deviation ε_2	Valid Samples Tested
5	0.400	0.412	6
10	0.030	0.007	6
15	0.037	0.005	3
20	-	-	-
25	-	-	-
55	0.014	0.020	4
Eroding Web width W [mm]	Standard deviation ε_1	Standard deviation ε_2	Valid Samples Tested
5	0.385	0.327	6
10	0.024	0.004	6
15	0.007	0.001	6
20	0.007	0.007	3
25	-	-	-
55	0.012	0.018	4
Laser cutting Web width W [mm]	Standard deviation ε_1	Standard deviation ε_2	Valid Samples Tested
5	0.286	0.234	6
10	0.029	0.016	6
15	0.037	0.005	3
20	-	-	-
25	-	-	-
55	0.01	0.015	6

Summary and outlook

In this paper, the influence of the test specimen sampling procedure on the material characteristic values was investigated. For that matter, Nakajima tests of specimens with differing sampling procedures (water jet cutting, milling, laser cutting and wire eroding) were carried out and evaluated regarding SEM images of the edge zone, as well as the detected material properties. Initially, it was discovered that the direct removal of the test specimens from the semi-finished product resulted in damage to the test specimens. Consequently, packaging was initially conducted prior to the production of the test specimens. Subsequently, Nakajima tests were carried out. The forming limit diagrams exhibited minimal variation regardless of the sampling procedure used. In the tension-compression range, the specimens produced by milling exhibited the highest forming limit diagrams, while in the tension-tension range, the specimens manufactured using water jet cutting demonstrated the highest forming limit ratios. However, the milled specimens showed the lowest deviation in the determined forming limit ratios, as well as the most validly tested specimens. This is due to the higher edge zone quality of the milled specimens compared to the remaining sampling procedures. While testing, notches in the edge zone can lead to stress maxima, that result in local cracks before reaching the global forming limit. Therefore, it is suggested to use milling as the sampling procedure for the characterization of sheet metal foil, since it results in the highest edge zone quality as well as the lowest deviation of the measured failure values. However, further investigations on this topic must be rolled out. On the one hand, the influence of the thermal load on the microstructure formation in the edge zone of the test specimens must be investigated, as a microstructural transformation can lead to deviating material characteristics. In addition, the influence of the burr must be examined. While this is already significantly less pronounced in milling than in the other sampling procedures, a grinding finish of the edge zone could still be promising. In addition, the influence of the sampling procedure should be investigated at further forming speeds to further validate the causality between material characteristics and sampling procedure. It is also possible to transfer the investigated sampling procedures to other material tests such as the tensile test to further validate the findings of this paper.

Acknowledgements
This research was funded by Bundesministerium für Wirtschaft und Klimaschutz (BMWK,Federal Ministry for Economic Affairs and Climate Action;project number 20N2204D – KeyTech2GreenPower).

References
[1] Europäische Kommission: Generaldirektion Mobilität und, V.; Europäische Kommission: Generaldirektion Forschung und, I.: Flightpath 2050 : Europe's vision for aviation : maintaining global leadership and serving society's needs. Publications Office, 2011

[2] Töpler, J.; Lehmann, J.: Hydrogen and Fuel Cell Technologies and Market Perspectives. 1st 2016. Aufl., 2016 https://doi.org/10.1007/978-3-662-44972-1

[3] Schutzrecht. Schutzrecht DE 10 2021 104 821 A1 (01.09.2022). Hagel, M. G., M. Brugger, S. : Verfahren zum Herstellen eines Werkstücks, insbesondere einer Bipolarplatte

[4] Porstmann, S.; Wannemacher, T.; Drossel, W. G.: A comprehensive comparison of state-of-the-art manufacturing methods for fuel cell bipolar plates including anticipated future industry trends. In: Journal of Manufacturing Processes. 60 Jg., 2020. S. 366-383 https://doi.org/10.1016/j.jmapro.2020.10.041

[5] Bong, H. J.; Barlat, F.; Lee, M.-G.; Ahn, D. C.: The forming limit diagram of ferritic stainless steel sheets: Experiments and modeling. In: International Journal of Mechanical Sciences. 64 Jg., 2012, Nr. 1. S. 1-10 https://doi.org/10.1016/j.ijmecsci.2012.08.009

[6] Bauer, A.: Experimentelle und numerische Untersuchungen zur Analyse der umformtechnischen Herstellung metallischer Bipolarplatten. Verlag Wissenschaftliche Scripten, 2020

[7] 120004-2:2021, N. D. Bestimmung der Grenzformänderungskurve für Bleche und Bänder July 2021: DIN, July 2021

[8] Sommer, J.; Müller, M.; Herrig, T.; Bergs, T.: Simulative and Empirical Investigation of Test Specimen Geometries for the Determination of Forming Limit States in the Tensile-Compression Range for Austenitic Stainless Steel Foil Material. Cham: Springer Nature Switzerland, 2024 https://doi.org/10.1007/978-3-031-42093-1_25

[9] Briesenick, D.; Liewald, M.; Radonjic, R.; Karadogan, C.: Enhanced accuracy in springback prediction for multistage sheet metal forming processes. Berlin, Heidelberg: Springer Berlin Heidelberg, 2019 https://doi.org/10.1007/978-3-662-60417-5_11

Sheet Metal 2025

Materials Research Forum LLC

Materials Research Proceedings 52 (2025) 201-209

https://doi.org/10.21741/9781644903551-25

[10] Förster, R.; Förster, A.: Einteilung der Fertigungsverfahren nach DIN 8580 in Einführung in die Fertigungstechnik. Berlin, Heidelberg: Springer Berlin Heidelberg, 2018 https://doi.org/10.1007/978-3-662-54702-1_2

[11] Angeloni, C.; Liverani, E.; Ascari, A.; Fortunato, A.: Characterization and process optimization of remote laser cutting of current collectors for battery electrode production. In: Journal of Materials Processing Technology. 324 Jg., 2024. S. 118266 https://doi.org/10.1016/j.jmatprotec.2023.118266

Sheet Metal 2025

Materials Research Forum LLC

Materials Research Proceedings 52 (2025) 210-219

https://doi.org/10.21741/9781644903551-26

Processing of the hypoeutectic AlSi9 alloy with twin-roll casting by using copper shells

Moritz Neuser[1,a*], Kay-Peter Hoyer[1,b] and Mirko Schaper[1,c]

[1]Chair of Materials Science, Paderborn University, Warburger Straße 100, 33098 Paderborn, Germany

[a]neuser@lwk.upb.de, [b]hoyer@lwk.upb.de, [c]schaper@lwk.upb.de

Keywords: Aluminium, Microstructure, Twin-Roll Casting

Abstract. Lightweight design is one of the central topics of the automotive industry since reducing mass can save emissions over the entire life cycle of a component. Nowadays, vehicle structures usually consist of a multi-material design, which poses the additional challenge of joining these different materials. Mechanical joining is the most common way of joining different types of materials. Cast aluminium alloys of the AlSi-system have a low ductility, which causes cracks during the mechanical joining process in the joint. One research approach is to achieve a fine microstructure by influencing the solidification rate since this results in increased mechanical properties, specifically the elongation at fracture and yield strength. A very fine microstructure can be achieved by utilizing Twin-roll casting (TRC) which is a continuous casting process in which high solidification rates of more than 100 K/s occur. In this study, the hypoeutectic cast aluminium alloy AlSi9 is processed in the TRC process using copper rollers. The cast strips are investigated regarding the microstructure-property correlation. A variation of the roller materials and cooling conditions allows for an increase in the solidification rate, whereby a defined, fine microstructure can be achieved, which enhances the mechanical properties of the hypoeutectic aluminium casting alloys.

Introduction

Lightweight design represents a methodology aimed at weight reduction through utilizing materials in an application-specific manner, resulting in decreased emissions across the entire lifecycle - from production to end-of-life disposal. In automotive engineering, lightweight design is essential not only for vehicles with internal combustion engines (ICEV) but also for battery electric vehicles (BEV). In conventional ICEVs, reducing the weight by 100 kg can lower fuel consumption by approx. 0.3 litres/100 km [1]. A similar situation also affects BEVs. Although the latest developments are continuously increasing the energy density of batteries, the range challenge remains. This is because less mass needs to be accelerated, a factor particularly significant considering that three of the four driving resistances, which influence the vehicle's range, are mass-depending [2,3].

A vehicle body shell is usually manufactured using the multi-material construction method; different materials are set at the most suitable points. Particularly at the nodal points of a body, such as in the spaceframe design, cast components are used that are particularly suitable for absorbing compressive forces. Another example is the spring dome, which accommodates the spring-damper system of a vehicle. These components are joined to the bodywork. A mechanical process is selected for joining to create a suitable connection between two different materials. For lightweight construction reasons, an aluminium (Al) alloy from the aluminium-silicon (AlSi) system is mostly for these components. These alloys are divided into hypoeutectic, eutectic and hypereutectic alloys, whereby the eutectic composition is 12.5 wt% [4]. Typical examples are the naturally hard AlSi9 alloy and the age-hardenable AlSi10Mg alloy. Both alloys of the Al-Si system are characterised by excellent castability accompanied by very low solidification shrinkage and

Content from this work may be used under the terms of the Creative Commons Attribution 3.0 license. Any further distribution of this work must maintain attribution to the author(s) and the title of the work, journal citation and DOI. Published under license by Materials Research Forum LLC.

Sheet Metal 2025
Materials Research Proceedings 52 (2025) 210-219

Materials Research Forum LLC
https://doi.org/10.21741/9781644903551-26

good strength [5,6]. Nevertheless, the low elongation at fracture of both alloys poses a challenge, in particular for mechanical joining as shown in the studies Neuser et al. [7] and [8]. The high degrees of deformation that are achieved during the formation of the joint occur particularly in the closing head of the joint resulting in the initiation of cracks due to the high tensile stress applied during joining [9].

The brittle character of AlSi alloys is caused by the needle-shaped silicon (Si) that forms during secondary crystallisation as a result of solidification. This occurs in particular at slower solidification rates, which are typical for sand casting (1 to 4 K/s) [7]. To increase the mechanical properties, refining agents can be introduced into the melting phase. By adding the elements sodium (Na) or strontium (Sr) in the melting phase, the morphology of silicon can be directly influenced during solidification, so that no Si needles, but a lamellar Si, are formed. This is known as modification [4]. The fine morphology of the Si leads to a significant increase in mechanical properties such as elongation at break and tensile strength [4,6]. In addition, the microstructure can be significantly influenced by controlling the solidification rate [10]. According to Fredriksson et al. [10], by increasing the solidification rate to > 16.8 K/s, the morphology of the Si does not form as a needle-shaped microstructure, but instead crystallises as a fine-lamellar Si.

Additionally to the morphology of the Si, the secondary dendrite arm spacing (DAS) of the α-aluminium is also decisive for the mechanical properties of the AlSi alloys, as the following relationship exists: the faster the solidification rate, the smaller the DAS and, accordingly, the higher the mechanical properties [11]. The solidification rate can be achieved via the component geometry, the casting process, or active tempering [7,12].

One process for manufacturing semi-finished aluminium products is continuous casting (Twin-roll casting, TRC). The TRC process is divided into horizontal and vertical processes [13]. It combines a primary moulding process and an almost simultaneous reshaping process. The vertical process scheme is shown in Fig. 1.

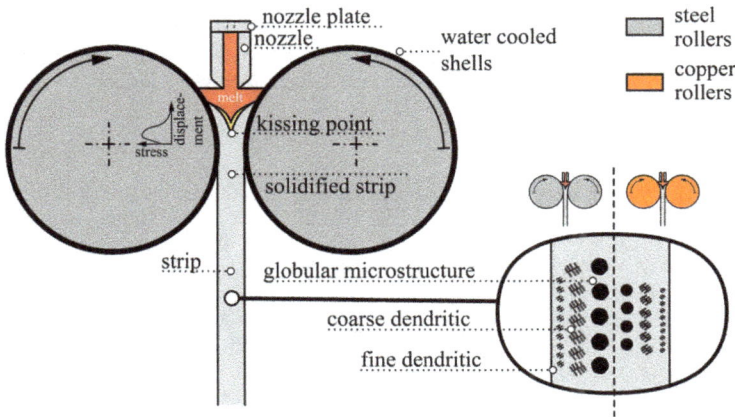

Figure 1: TRC-process with steel- or copper rollers; schematic comparison of microstructure, following Neuser et al. [9]

The melt contacts the rotating, internally water-cooled steel rollers, as a result of which the surface layer solidifies very quickly in contact with the rollers. Due to the continuous rotation of the rollers, the solidified material reaches the kissing point, where the two previously separate parts of the strip combine to form one solidifying strip. The subsequent forming of AlSi alloys is carried

Sheet Metal 2025   Materials Research Forum LLC
Materials Research Proceedings 52 (2025) 210-219   https://doi.org/10.21741/9781644903551-26

out with moderate forming forces to obtain a strip that is as texture-free as possible. The combined casting-rolling process enables skipping several intermediate steps in the production of semi-finished products: The forming process starts directly in the melting phase up to the final contour thickness and ultimately realises energy savings compared to a conventional process chain [13]. In particular, the processing of high manganese TWIP steel for crash-relevant applications in the TRC process has been extensively researched by the Institute of Metal Forming (IBF, Aachen) in studies such as Daamen et al. [14] and Daamen et al. [15]. In addition, various investigations have been carried out at the Osaka Institute of Technology using the example of the hypoeutectic AlSi7 (cf. Haga et al. [16]) and the hypereutectic AlSi alloys AlSi16 and AlSi20 (cf. Haga et al. [17]).

Steel rollers are usually used in TRC but there have also been investigations addressing copper rollers by processing hypoeutectic aluminium casting alloys, see the studies of Haga et al. [16] and Grydin et al. [18] as well as by processing the material 1.4301, stainless steel (Daamen et al. [19]). These have a higher heat capacity, which means that even higher casting speeds can be achieved on an industrial scale [20]. The use of copper shells also allows the TRC process to achieve solidification rates in excess of 200 K/s, as demonstrated in the study by Grydin et al. [18]. Concerning the processing of AlSi alloys, this means that the Al melt can solidify even faster and the mechanical properties, especially the elongation at fracture, of the hypoeutectic AlSi alloy can be further increased due to the higher heat capacity of the copper rollers [18]. This effect is in line with the study by Talaat et al. [21] that the higher the solidification rate, the finer the Si morphology. Ultimately, this enhances the mechanical properties [11].

A typical value for the solidification rate for sand casting is 1 K/s with a DAS of 25 μm and a component thickness of 2 mm [7]. In the study by Grydin et al. [18], it was shown that a solidification rate of approx. 100 to 200 K/s can be realised in the casting-rolling process and a hypoeutectic AlSi alloy with fine-lamellar Si can be successfully processed [18]. There is a DAS of 9 μm with a strip thickness of 2 mm. Due to their fine microstructure, the AlSi10Mg and AlSi9 processed in the TRC have high elongation at fracture and yield strengths compared to sand casting, which in turn is essential for their suitability for joining [9]. To address the effect of an increased solidification rate and the resulting mechanical properties influenced by applying copper rollers in the TRC process, AlSi9 is processed and comprehensively characterised, microscopically and mechanically.

Materials and Methods
Hypoeutectic Aluminium Casting Alloys EN AC-AlSi9.

In this study, two hypoeutectic aluminium casting alloys out of the Aluminium-Silicon system (AlSi), the naturally-aged aluminium alloy EN AC-AlSi9 (European Norm—aluminium cast product, according to DIN EN 1706) are investigated [4]. Both alloys are provided by TRIMET Aluminium SE (Germany, Essen). The AlSi9 alloy has been pre-modified by Trimet with an AlSr master alloy. Due to restrictions by Trimet, the alloy composition may not be published. Precipitation hardening is not applied since AlSi9 does not contain high additions of magnesium (Mg). In addition, a TA750SN thermal analyser from Ideco GmbH (Bocholt, Germany) was used to determine the liquidus and solidus temperatures of AlSi9, the difference between which provides the solidification interval.

Casting Process and Trails.

The aluminium casting alloy was processed using a vertical casting-rolling caster with water-cooled rollers on the inside [5]. Steel rollers made of X38CrMoV5-3 (1.2367) and copper rollers made of CuCr1Zr were used. The steel rollers have a width of 200 mm and a diameter of 370 mm, the copper rollers have a dimension of width of 200 mm, a diameter of 370 mm and a shell thickness of 40 mm. The caster is connected to the primary cooling water of the research building and the heat is extracted from the secondary cooling circuit via a heat exchanger. The secondary circuit of the caster is filled with a 50:50 glycol water mixture and dissipates the heat absorbed

Sheet Metal 2025
Materials Research Proceedings 52 (2025) 210-219

Materials Research Forum LLC
https://doi.org/10.21741/9781644903551-26

from the rollers. In general, the characteristics and the design features of this type of vertical caster are explained in Grydin et al. [22] and the TRC testing setup is shown in the authors' previous study [9] (cf. Fig. 3 in study [9]). Table 1 lists the process parameters applied during TRC.

Table 1: TRC-process parameters by using copper rollers

Melt temperature [°C]	720.0
Pouring temperature [°C]	715.0
Casting rate [m/min]	5.25
Thickness of strip [mm]	2.5
Roll force [kN]	70.0
Cooling water flow rate [m^3/h]	2.4

Measurement of Chemical Composition.
After casting, the chemical composition was determined using specimens taken from the cast strips via Optical Emission Spectrometer (OES). The examination was carried out on a Q4 Tasman from Bruker (Germany, Karlsruhe).

Heat Treatment of Aluminium Casting Alloys.
A completely different heat treatment process chain was chosen for the naturally hard AlSi9. Due to the lower Mg content, precipitation hardening is not possible. However, to increase the elongation at fracture, annealing at a temperature of 380 °C and a holding time of 2 hours was carried out. This heat treatment routine was determined in preliminary research by Neuser et al. [23]. It has been observed that after 2 hours of soaking, the hardness hardly decreases, and a kind of pleat is formed [23]. The annealing was carried out in the circulating air furnace Fresenberg POV125.600 (Germany, Wipperfürth).

Preparation and Examination of Microscopic Investigation.
During preparation for the microscopic analyses, the specimens were embedded in the two-component system CEM 1000 blue from Cloren Technology (Germany, Wegberg). A mixing ratio of 2 (hardener):0.8 (powder) was used. All samples were ground in stages, starting with a grain size of 500 up to a grain size of 2500. Afterwards, the specimens were polished with a polishing machine using the polishing agent Colloidal Silica Suspension from Cloeren Technology with a particle size of 50 nm and a pH value of 9 to 10 for 18 hours. Finally, the samples were cleaned with ethanol. For the determination of the DAS, details are given in the study by Neuser et al. [9], and for the calculation of the solidification rate based on the DAS, a description is given in [18].

Examination of mechanical investigation.
For the mechanical characterization of the cast strips, samples were taken for tensile- and hardness tests as well as for joining samplings. In advance, the samples for the tensile and hardness testing were ground with 1200-grit sandpaper.

The tensile tests were carried out following DIN EN ISO 6892-1 on the TableTop tensile testing machine from MTS Systems Corporation (USA, Eden Prairie). The tensile specimen geometry used is shown in the previous study by Neuser et al. [9] (cf. Fig. 5). The specimens were tested at a test rate of 1.5 mm/min and gripped with a clamping pressure of 8 MPa. Furthermore, the frequency at which the force and strain were recorded was 1024 Hz. At least 10 samples per condition were tested.

Brinell hardness tests were carried out on specimens following DIN EN ISO 6506-1. A 2.5 mm diameter tungsten carbide ball was used and a test load of 62.5 kp was applied for 10 to 15 seconds. At least five hardness marks were performed per condition.

Sheet Metal 2025

Materials Research Forum LLC

Materials Research Proceedings 52 (2025) 210-219

https://doi.org/10.21741/9781644903551-26

Results and Discussion - Microstructure characteristics and mechanical properties

The measured chemical composition of the AlSi9 is listed in Table 2. With a silicon content of 10.42 wt% and an Mg content of 0.014 wt%, the alloy used corresponds to the DIN EN 1706 and is, therefore, a hypoeutectic AlSi alloy. The standard does not provide for modification with the element Sr. The alloy used contains 0.049 wt% Sr and is, therefore, above a limit value of 200 ppm, which is required for modification [24]. At slow solidification rates, below 20 K/s, a refining element such as Sr is necessary to transform the Si morphology from a plate-like to a lamellar finely dispersed morphology. In particular, if the Sr content is between 0.04 wt% and 0.06 wt%, the elongation at fracture of the AlSi alloys can be improved [25].

Table 2: Chemical composition via OES of AlSi9, measurement after TRC-process

AlSi9								
Elements	Al	Si	Mg	Mn	Fe	Cu	Sr	other
Mean value [wt%]	88.640	10.420	0.014	0.470	0.098	0.031	0.049	0.458
Standard deviation [wt%]	0.105	0.097	0.0003	0.008	0.0007	0.0005	0.0002	

In Fig. 2, three images of the microstructure of AlSi9 are displayed, which was processed in the TRC process with copper rollers. In this context, reference is made to the authors' previous study [9] demonstrating the microstructure of an AlSi9 processed by TRC using steel rolls. The left and right images show the microstructure of the strip sections solidified directly on the copper rollers. Here, an orientated dendritic columnar solidification occurred, as illustrated in the process diagram in Fig. 1. Furthermore, the very small formation of the dendritic α-aluminium directly on the edge is noticeable, which is induced by the high heat dissipation of the copper rollers. A homogeneous microstructure distribution at the edges is also evident in Fig. 2. In comparison, in the middle image of Fig. 2, a coarser microstructure can be detected and a globular dendritic structure can be seen in the centre of the strip due to the all-around solidification. This observation confirms the graphical illustration of Fig 1. In addition, segregation has taken place in the centre of the strip, in particular the eutectic fraction consisting of the Si phase is higher at certain points. The occurrence of segregation in the centre of the strip supports the hypothesis that the solidification rate in the centre of the strip is lower than in the outer areas of the strip, as the Si phase is completely solidified at the eutectic temperature.

Sheet Metal 2025
Materials Research Proceedings 52 (2025) 210-219

Materials Research Forum LLC
https://doi.org/10.21741/9781644903551-26

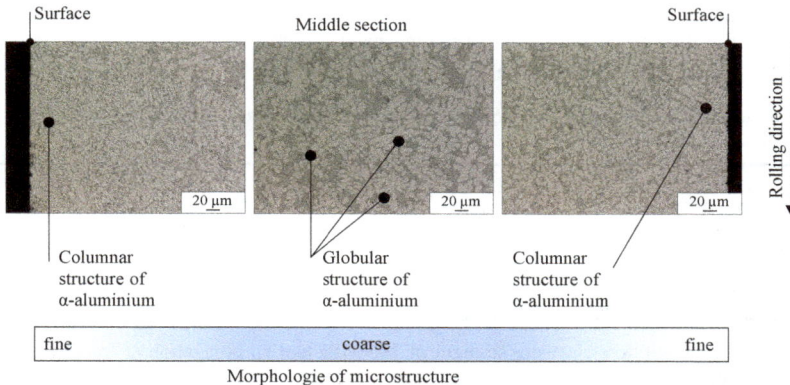

Morphologie of microstructure

Figure 2: Morphology of AlSi9 strip manufactured using TRC, divided into outer sections (left and right image) and middle section (middle image). Cross-section of the thickness of the strip, which is shown graphically in the spotlight of Fig. 1.

Depending on the solidification rate to be achieved, the microstructure and, therefore, the mechanical properties can be influenced. This can be further enhanced by additional heat treatment. Solution annealing at a temperature of 380 °C has allowed a further modification of the morphology, particularly, of the silicon morphology, so that a previously lamellar Si morphology can be transformed into a globular Si morphology. This in turn can improve the elongation at fracture of the alloy, as the round Si particles provide fewer starting points for possible crack initiation. Images of the microstructure in the as-cast and heat-treated states are shown in Fig. 3. In image a) it can be seen that the Si is very finely dispersed due to the high solidification rate of 100.83 K/s and that only very few coarser Si lamellae are present. To calculate the solidification rate based on the DAS, the solidification interval of the AlSi9 was determined at 20.4 K using thermal analysis. The subsequent solution annealing at 380 °C leads to embedded Si even more finely in the α-aluminium matrix (image b)). Furthermore, it can be recognized from these images, in particular from image a), that the high solidification speed reduces the distance between the Si lamellae to such an extent that the Si is also finely distributed.

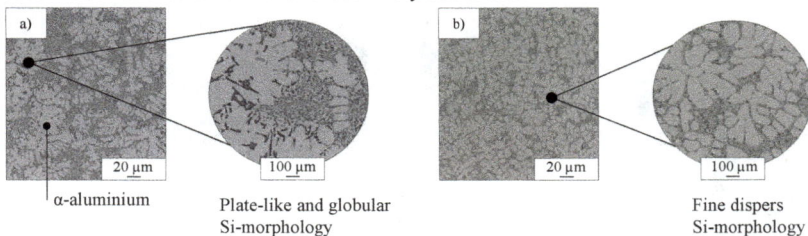

Figure 3: Comparison of Si morphology, a) as-cast state; b) heat-treated state 380 °C/2 h soaking time

Analogue to the preliminary study of Neuser et al. [9], the DAS was determined, and the procedure is also described in the mentioned publication. A DAS of 6.11 μm was obtained. In contrast, a DAS of 11.31 μm was obtained for AlSi9 in the study by Neuser et al. [9] using steel rollers. This difference already indicates that the heat dissipation in the solidification process could

Sheet Metal 2025
Materials Research Proceedings 52 (2025) 210-219

Materials Research Forum LLC
https://doi.org/10.21741/9781644903551-26

be significantly increased by using Cu rollers. Furthermore, the smaller the solidification rate, the smaller the resulting DAS and the higher the mechanical properties. The mechanical properties determined for the two conditions, as-cast and heat-treated (380 °C/2 h), are listed in Table 3.

Table 3: Mechanical properties of AlSi9 processed by TRC in the as-cast and heat-treated state (380 °C/2 h)

	State of AlSi9	
Mechanical properties	As-cast	Heat-treated (380 °C/2 h)
Hardness [HBW]	61.3 ± 0.4	57.8 ± 0.4
Tensile strength [MPa]	195.0 ± 9.8	162.2 ± 12.3
Yield strength [MPa]	93.3 ± 12.5	72.2 ± 8.5
Elongation at fracture [%]	17.0 ± 2.9	14.0 ± 3.8

In the as-cast condition, a tensile strength of 195 MPa and an elongation at fracture of 17 % were achieved. In contrast, the tensile strength was reduced by 17 % and the elongation at fracture by 14 % after the heat treatment. Compared to processing using steel rollers, the tensile strength was increased by 8 % and the elongation at fracture by 9 % [9]. Similar values were also determined by Haga et al. [16] for processing the hypoeutectic AlSi7, whereby a slightly higher elongation at fracture of 18 % was achieved. However, this is due to the slightly lower Si content of 7 wt%. An increase in elongation at fracture in the as-cast state is achieved due to the finely dispersed Si based on the higher solidification speed and the grain sizes can be minimised. One challenge identified was that the high solidification rate induced by the copper rollers used resulted in a high amount of solidification on the surface. In comparison, however, there are still semi-solidified areas in the centre of the strip. This results in cold runs and micro-solidification shrinkage cavities. During microscopic examinations, these small round microlunkers could be detected in the centre of the strip, which supports this hypothesis (see Fig. 4). However, the shrinkage cavities are not resolved due to the low process-specific rolling forces in this case and are also based on the small silicon content of 9.5 wt%. Applying an alloy with a higher silicon content of 12 wt%, which is near-eutectic, could minimise the formation of micro-lunkers. Nevertheless, such alloys have a narrower solidification interval, which can be assumed to avoid large cold runs during the solidification process on the rollers. This reduces the mechanical properties such as tensile strength.

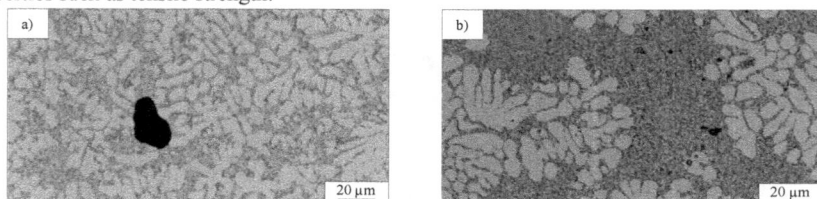

Figure 4: Microscopic images of the cross-section in the middle of the strip. a) macro-lunker; b) various micro-lunkers in eutectic phase

Conclusions

In this study, the hypoeutectic alloy AlSi9 was successfully processed in the TRC process using copper rollers and evaluated microscopically and mechanically. Due to the increased heat dissipation of the copper rollers, it was shown that the solidification rate in the TRC process can be further increased. This was reflected in the reduction of the DAS to 6.11 μm. The large solidification interval of 20.4 K, which is typical for AlSi casting alloys, poses a challenge when processing AlSi9 in the TRC with copper rollers. As a result, the alloy is in the semi-solidified phase during the TRC process in which the band formation takes place at the kissing point. The solidification cavities cannot be prevented either by the lack of replenishment nor do the low molding forces contribute to the cavities closing again. As a result, the elongation at break can be increased by 8 % compared to the use of steel rollers. Nevertheless, the other mechanical properties such as the tensile strength cannot be increased, as would be expected due to the lower DAS. It was also shown that the high solidification speed can further reduce the distance between the Si particles, which are even more finely dispersed than when using steel rollers. Although further heat treatment like solution annealing results in the Si being distributed even more finely in the α-aluminium matrix, the elongation at break cannot be increased any further.

Funding
This research was funded by the Deutsche Forschungsgemeinschaft (DFG, German Research Foundation) – TRR 285/2 – Project-ID 418701707. The authors thank the German Research Foundation for their organizational and financial support.

Acknowledgements
The authors would like to take this opportunity to thank Mr Moritz Klöckner as well as Mr Moritz Mann, both from Paderborn University, for their support in the preparation and conduction of casting trials as well as Mr Olexandr Grydin for contributing his detailed expertise while conducting and evaluating the TRC trails.

References
[1] httpLeichtbau in der Fahrzeugtechnik; Friedrich, H.E., Ed.; Springer Fachmedien Wiesbaden: Wiesbaden, 2017.

[2] httpMallick, P.K. Materials, Design and Manufacturing for Lightweight Vehicles, 2nd ed.; Elsevier Science & Technology: San Diego, 2021, ISBN 978-0-12-819029-6.

[3] httpLumley, R.N. Fundamentals of aluminium metallurgy: Production, processing and applications; Woodhead Publishing: Cambridge, MA, 2018, ISBN 978-0-08-102063-0.

[4] httpHatch, J.E. Aluminum: Properties and physical metallurgy; American Society for Metals: Metals Park, OH, 1984, ISBN 1615031693.

[5] httpD. Vojtech, J. Serak, O. Ekrt. Improving the casting properties of high-strength aluminium alloys. Materiali In Tehnologije 2004, 38, 99-102.

[6] httpAdvanced Light Alloys and Composites; Ciach, R., Ed.; Springer: Dordrecht, 1998, ISBN 9789401590686.

[7] httpNeuser, M.; Grydin, O.; Andreiev, A.; Schaper, M. Effect of solidification rates at sand casting on the mechanical joinability of a cast aluminium alloy. Metals 2021, 11, 1304, doi:10.3390/met11081304. https://doi.org/10.3390/met11081304

[8] httpNeuser, M.; Grydin, O.; Frolov, Y.; Schaper, M. Influence of solidification rates and heat treatment on the mechanical performance and joinability of the cast aluminium alloy AlSi10Mg. Prod. Eng. Res. Devel. 2022, 1-10. https://doi.org/10.1007/s11740-022-01106-1

[9] httpNeuser, M.; Kappe, F.; Ostermeier, J.; Krüger, J.T.; Bobbert, M.; Meschut, G.; Schaper, M.; Grydin, O. Mechanical Properties and Joinability of AlSi9 Alloy Manufactured by Twin-Roll Casting. Adv Eng Mater 2022, 24, 2200874. https://doi.org/10.1002/adem.202200874

[10] httpFredriksson, H.; Åkerlind, U. Solidification and crystallization processing in metals and alloys; John Wiley & Sons, Ltd: Chichester, UK, 2012. https://doi.org/10.1002/9781119975540

Materials Research Forum LLC

https://doi.org/10.21741/9781644903551-26

[11] httpKaufman, J.G.; Rooy, E.L. Aluminum alloy castings: Properties, processes, and applications; ASM International: Materials Park, OH, 2004, ISBN 1615030476. https://doi.org/10.31399/asm.tb.aacppa.9781627083355

[12] httpVossel, T.; Wolff, N.; Pustal, B.; Bührig-Polaczek, A. Influence of Die Temperature Control on Solidification and the Casting Process. Inter Metalcast 2020, 14, 907-925. https://doi.org/10.1007/s40962-019-00391-4

[13] httpBarekar, N.S.; Dhindaw, B.K. Twin-Roll Casting of Aluminum Alloys - An Overview. Materials and Manufacturing Processes 2014, 29, 651-661. https://doi.org/10.1080/10426914.2014.912307

[14] httpDaamen, M.; Güvenç, O.; Bambach, M.; Hirt, G. Development of efficient production routes based on strip casting for advanced high strength steels for crash-relevant parts. CIRP Annals 2014, 63, 265-268. https://doi.org/10.1016/j.cirp.2014.03.025

[15] httpDaamen, M.; Haase, C.; Dierdorf, J.; Molodov, D.A.; Hirt, G. Twin-roll strip casting: A competitive alternative for the production of high-manganese steels with advanced mechanical properties. Materials Science and Engineering: A 2015, 627, 72-81. https://doi.org/10.1016/j.msea.2014.12.069

[16] httpHaga, T.; Takahashi, K.; Ikawa, M.; Watari, H. A vertical-type twin roll caster for aluminum alloy strips. Journal of Materials Processing Technology 2003, 140, 610-615. https://doi.org/10.1016/S0924-0136(03)00835-5

[17] httpHaga, T.; Inui, H.; Watari, H.; Kumai, S. Casting of Al-Si hypereutectic aluminum alloy strip using an unequal diameter twin roll caster. Journal of Materials Processing Technology 2007, 191, 238-241. https://doi.org/10.1016/j.jmatprotec.2007.03.012

[18] httpGrydin, O.; Neuser, M.; Schaper, M. Influence of Shell Material on the Microstructure and Mechanical Properties of Twin-Roll Cast Al-Si-Mg Alloy. In Proceedings of the 14th International Conference on the Technology of Plasticity - Current Trends in the Technology of Plasticity: ICTP 2023 - Volume 3, 1st ed.; Mocellin, K., Ed.; Springer: Cham, 2024; pp 589-596, ISBN 978-3-031-41340-7. https://doi.org/10.1007/978-3-031-41341-4_61

[19] httpDaamen, M.; Förster, T.; Hirt, G. Experimental and numerical investigation of double roller casting of strip with profiled cross section. Special Edition: 10th International Conference on Technology of Plasticity, ICTP 2011 2011, 93-98.

[20] httpSpathis, D.; Clemente, A.; Tsiros, J.; Arvanitis, A.; Wobker, H.-G. The use of copper shells by Twin Roll Strip Casters. TMS Light Metals 2010.

[21] httpTalaat, E.-B.; Fredriksson, H. Solidification Mechanism of Unmodified and Strontium Modified Al-Si Alloys. Materials Transactions JIM 2000, 41, 507-515. https://doi.org/10.2320/matertrans1989.41.507

[22] httpGrydin, O., Yu.; Ogins'kyy, Y., K.; Danchenko, V.M.; Bach, F.-W. Experimental Twin-Roll Casting Equipment for Production of Thin Strips. Metallurgical and Mining Industry 2010, 2.

[23] httpNeuser, M.; Böhnke, M.; Grydin, O.; Bobbert, M.; Schaper, M.; Meschut, G. Influence of heat treatment on the suitability for clinching of the aluminium casting alloy AlSi9. Proceedings of the Institution of Mechanical Engineers, Part L: Journal of Materials: Design and Applications 2022, 146442072210758. https://doi.org/10.1177/14644207221075838

Sheet Metal 2025
Materials Research Proceedings 52 (2025) 210-219

Materials Research Forum LLC
https://doi.org/10.21741/9781644903551-26

[24] httpEbhota, W.S.; Jen, T.-C. Effects of modification techniques on mechanical properties of Al-Si cast alloys. In Aluminium Alloys - Recent Trends in Processing, Characterization, Mechanical Behavior and Applications; Sivasankaran, S., Ed.; InTechOpen: Rijeka, Croatia, 2017, ISBN 978-953-51-3697-2. https://doi.org/10.5772/intechopen.70391

[25] httpPezda, J. A R C H I V E S of 24/3 Effect of modifying process on mechanical properties of EN AC-43300 silumin cast into sand moulds. Archives of Foundry Engineering 2009, 9.

Sheet Metal 2025
Materials Research Proceedings 52 (2025) 220-227

Materials Research Forum LLC
https://doi.org/10.21741/9781644903551-27

The effect of height to diameter ratio at stack compression tests on biaxial yield stress

Martin L. Kölüs[1,2,a], Gábor Kalácska[2,b] and Gábor J. Béres[1,c] *

[1]Mechanical- and Laser Beam Technologies Research Group, John von Neumann University, 6000 Kecskemét, Izsáki út 10. Hungary

[2]Institute of Technology, Magyar Agrár- és Élettudományi Egyetem (MATE), 2100 Gödöllő, Páter K. u. 1. Hungary

[a]kolus.martin@nje.hu, [b]kalacska.gabor@uni-mate.hu, [c]beres.gabor@nje.hu

Keywords: Metal Forming, Sheet Metal, Compression Test

Abstract. This study presents experimental stack compression test results of samples with different height-to-diameter (aspect) ratios for a simplified calculation of the friction- and the stress state contribution on the flow stress compared to the case of simple tension. Standard uniaxial tension tests formed the basis of the investigations as reference curves, and the stack compression tests were carried out on the same material, on commercially available DC04 steel sheet. The aim was to experimentally observe and possibly separate the anisotropic, pure material behavior from the friction-forced reaction of the material to determine the equi-biaxial yield stress for proper yield surface election. Our results show that with proper lubrication conditions, the stress increase caused by friction can be kept at low value for high aspect ratios, as well as with the comparison of the high and low aspect ratios, the value of the Coulomb friction coefficient, and thus the friction contribution on the flow stress can be calculated. After a stress compensation based on the obtained friction coefficient, the calculated stress ratios led to the monitoring of the goodness of different yield surfaces.

Introduction

For the industrial practice of sheet metal forming, it is important to know the yield properties of the applied material both in the product design/development and the technology maintenance phases. Their usual goal is to apply the most suitable yield condition during finite element simulations, but especially those working in production want to achieve it with less time and money investment (especially for relatively low-value products). Knowing the equi-biaxial yield stress (σ_{eb}) could give them the chance to check if they have chosen correctly when using a tensile test calibrated yield surface model, such as Hill48 [1] and Yld89 [2]. Or, the calibration of advanced criteria (e.g. BBC2003 [3]) could become possible.

Stack-compression test could be a good alternative of hydraulic bulge test and biaxial tension test to evaluate σ_{eb}, but mainly the contribution of friction to the measured stress (load) value rises questions. This study shows experimental stack-compression test on samples with different aspect (height-to-diameter) ratios, for the investigation of this friction contribution.

A recently published comprehensive study emphasized the importance of stack compression test method on material characterization, and mention that the Coulomb friction coefficient (μ) falls between 0.02...0.07 when using Teflon sheet and grease [4]. Merklein and Kuppert [5] also used Teflon foils, but the exact value of μ is not discussed. Coppieters et al. used finite element based inverse method to determine friction coefficient in [6] and [7] and found that if μ=0.05, the model fits to the experiments well at oil lubrication. Kuwabara et al. [8] applied sintered PCD layer in 1 mm thickness on the dies surface to eliminate friction and they classified the resulting deformation uniform at stack compression tests. Steglich et al. [9] also proved that stack

Content from this work may be used under the terms of the Creative Commons Attribution 3.0 license. Any further distribution of this work must maintain attribution to the author(s) and the title of the work, journal citation and DOI. Published under license by Materials Research Forum LLC.

compression between polished dies treated with Teflon spray provides similar characteristics to the hydraulic bulge test and to the cruciform-specimen biaxial test. The same lubrication method was used by Rahmaan et al. [10], who evaluated the friction coefficient as 0.06…0.08. The relative error in flow stress caused by the friction is described by An and Vegter [11] for different friction coefficients, but the variation of friction [12] can make the situation even more complex.

Beside the above-mentioned phenomenon, friction can cause uncertainty in the in-plane deformation, which is also suitable for yield criterion calibration [13, 14]. Thus, authors think that the stack compression test is of great interest today for calculating mechanical properties of sheet metals, especially yield behavior, in which this paper focuses on equi-biaxial stress acquisition at different geometries, while the lubrication conditions assumed to be constant.

Applied material

Commercially available DC04 thin sheet metal, with nominal thickness (t) of 1.0 mm was used for the investigations. This material clearly exhibits plastic anisotropy due to the cold rolling process, which the material passes through in the last step of manufacturing. Besides, this sheet is known as a single-phase steel material, since the ferrite phase strongly dominates the microstructure.

The basic material properties in mechanical aspects of the tested sheet are presented in the following chapter. The chemical compound and other micro-structural properties are partly out of the topic of this manuscript, and thus these are not discussed here.

Experimental procedure

This section summarizes the most important details of the experimental tests. Tension and stack compression tests were carried out for characterizing the plastic behavior in this study.

Tensile tests

The DIN EN ISO 6892-1 standard served as the basis of the tensile tests. The applied A_{80} specimen geometry and the test conditions can be seen on Fig. 1. The tests were performed at room temperature, with a constant crosshead speed of 20 mm/min, which resulted in 0.0042 s^{-1} initial and approximately 0.0034 s^{-1} final strain rate, at the moment of neck occurring.

The force measurement was realized through Instron 100 kN force cell, and Instron AVE video extensometer was responsible for the strain measurements both in the longitudinal and in the transverse directions.

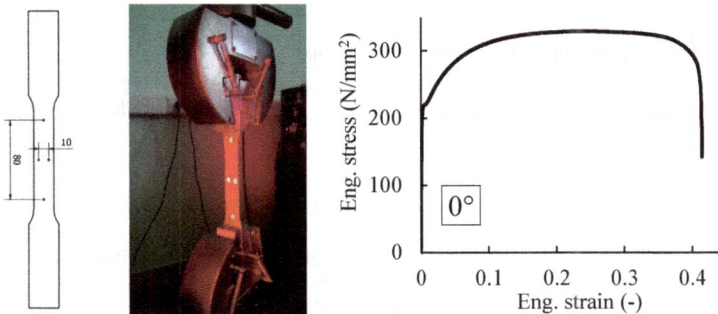

Fig. 1: Applied tension test specimen and the engineering stress – strain curve

The engineering stress-strain curve for one representative sample parallel to the rolling direction (0°) is also given on Fig. 1. The measurements were carried out with three times reduplication.

Sheet Metal 2025

Materials Research Proceedings 52 (2025) 220-227

Materials Research Forum LLC

https://doi.org/10.21741/9781644903551-27

The true stress – equivalent plastic strain curve (flow curve) and the relationship of the true transverse - thickness strains are visible in Fig. 2 for the same sample. In Fig. 2 right, the isotropic case ("ISO") is also indicated. This data is considered as the basis, to which the following compression tests relate later. The average mechanical properties and those deviation during uniaxial tension load are listed in Table 1. Here YS refers to yield strength and UTS to ultimate tensile strength. In terms of strain, A_{80} is the total, A_g is the uniform elongation of the material. The r-value is indicated in parallel (r_0), perpendicular (r_{90}) and in 45° to the rolling direction (RD) (r_{45}). The average r-value is marked by R.

Table 1: Mechanical properties obtained by uniaxial tension tests

YS_0 [N/mm^2]	UTS_0 [N/mm^2]	A_{80_0} [%]	A_{g_0} [%]	r_0 [-]	r_{45} [-]	r_{90} [-]	R [-]
220±6.0	328±11.0	41.0±1.5	21.4±2.0	1.823±0.1	1.311±0.1	2.380±0.1	1.706

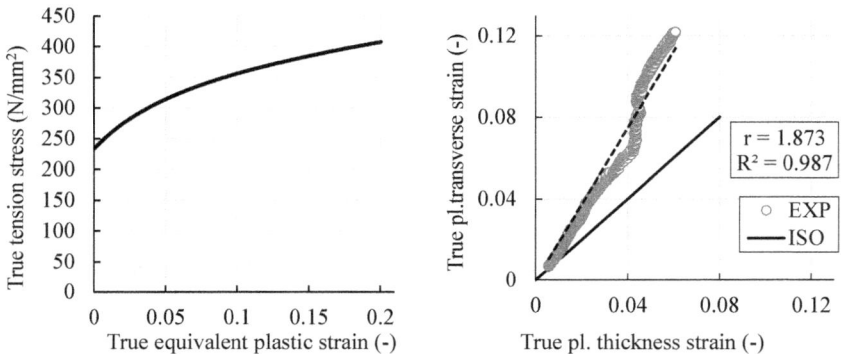

Fig. 2: True stress – strain curve (left) and true transverse - thickness strain curve (right) in 0° to the RD, measured by tensile test

Compression tests

The compression tests were carried out under almost identical conditions as described in the previous subchapter. The testing device (Instron 5900R) and the temperature were the same as the tension tests. The strain rate ($\dot\varepsilon$) has been set depending on the initial height of the stack (h_0) to ensure its constant and quasi-static value for all cases. The crosshead speed during the compressive motion (v_c) was adjusted according to Eq. 1,

$$v_c = \dot\varepsilon \cdot h_0 , \tag{1}$$

in which $\dot\varepsilon = 0.1$ min^{-1} (0.0017 s^{-1}) initial strain rate for all measurements. The repetition was three times per layout, again.

Three layouts were investigated with different initial heights (h_0). The initial, nominal diameter (d_{nom}) of each specimen was uniformly 10 mm. The applied height to nominal diameter ratios indicated on Fig. 3, are the following: 1x10 mm, 5x10 mm and 10x10 mm.

Sheet Metal 2025
Materials Research Proceedings 52 (2025) 220-227

Materials Research Forum LLC
https://doi.org/10.21741/9781644903551-27

1x10   5x10   10x10

Fig. 3: Layouts of the tested specimens

Although the 10x10 layout seems uneven in the picture, the samples were manufactured with great care, using laser beam cutting. The effect of the heat was monitored by micro-hardness tests and our observation is that microstructural change is negligible on the perimeter of the sample. After that a minimum of 50 individual specimens' geometry were measured by an optical microscope (Zeiss Discovery V8) and micrometer in order to make sure the exact, initial geometry of the disks. The obtained data are the initial diameter in rolling (d_0) and transverse direction (d_{90}) direction as well as the thickness (t), which count as $d_0 = 9.73\pm0.01$, $d_{90} = 9.88\pm0.03$ and t = 1.04 ± 0.02 mm.

To avoid as many experimental mistakes as possible, precisely machined dies with polished surface as well as self-developed positioning devices were used (Fig. 4). For minimizing the friction, Luba21® high-pressure oil, specially developed to compressive forming operations was applied on the dies' surface. The lubrication was renewed before every new test. The positioning device, besides, took care of keeping the stacks' elements in the same central axis, and adjusted the orientation aligned in the rolling direction that was painted on to the surface of the specimens.

Fig. 4: The photo of the compression dies, the positioning and the test setup

The measurements continued until the specimens reached half of their initial heights. The true compressive stress – strain curves as well as the true plastic flow stress - plastic work curves obtained by both tension and stack compression tests are shown by Fig. 5.

Discussion of the material behavior and the friction contribution
The tension, as reference curve and the stack-compression test results for the 1x10, 5x10 and 10x10 layouts can be seen in Fig. 5 right, with black and red curves respectively. We assume that the flow stress difference between the tension and compression methods lies in two things: the stress state (yield surface) and the friction (neglecting strength differential effect).

Sheet Metal 2025
Materials Research Proceedings 52 (2025) 220-227

Materials Research Forum LLC
https://doi.org/10.21741/9781644903551-27

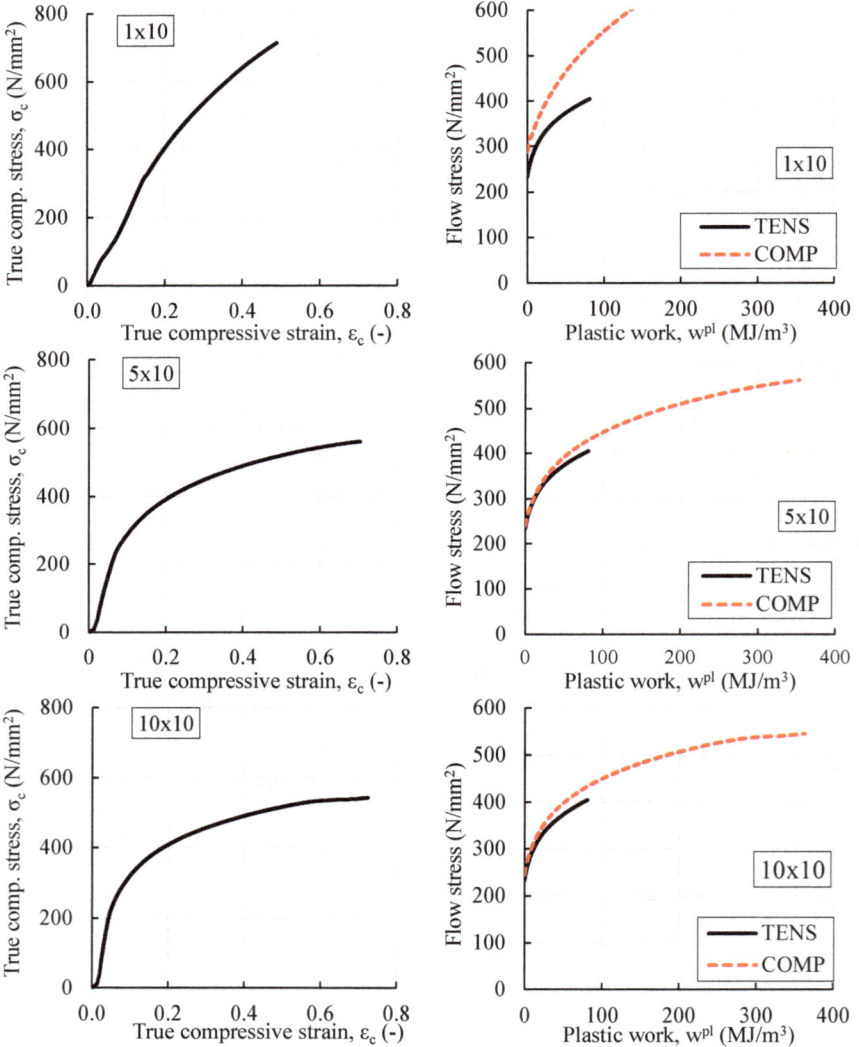

Fig. 5: Compressive stress – compressive strain curves (left) and plastic flow stress – plastic work curves (right)

Table 2 lists the theoretical flow stress increase, i.e. the calculated contribution of friction (Ucalc) for 10x10, 5x10 and 1x10 layouts for different μs using the Siebel equation [15]:

$$\sigma_c = \bar{\sigma} \cdot U_{calc} = \bar{\sigma} \cdot \left(1 + \frac{\mu}{3} \cdot \frac{d}{h}\right) , \tag{2}$$

Sheet Metal 2025   Materials Research Forum LLC
Materials Research Proceedings 52 (2025) 220-227   https://doi.org/10.21741/9781644903551-27

In the above equation, d and h refer to the instantaneous diameter and height of the stack as well as σ_c is the flow resistance and $\bar{\sigma}$ refers to the uniaxial flow stress. The instantaneous diameter and height of the stacks were calculated at the compressive strain equal to the neck occurring strain of the tensile specimens ($\varepsilon_{neck\_ave} = 0.201$), using the volume constancy. Therefore, for isotropic materials, the flow resistance should be equal with both the equi-biaxial and the uniaxial flow stresses at $\mu = 0$. But in case of anisotropy, equi-biaxial flow stress should differ according to the unknown yield surface [16], even if $\mu = 0$.

Table 2: Mechanical properties obtained by uniaxial tension tests

μ [-]	1x10 layout Ucalc.	5x10 layout Ucalc.	10x10 layout Ucalc.	$Ucalc_{1x10}$ / $Ucalc_{5x10}$	$Ucalc_{1x10}$ / $Ucalc_{10x10}$
0.02	1.088	1.018	1.009	1.069	1.079
0.04	1.177	1.035	1.018	1.137	1.156
0.06	1.265	1.053	1.027	1.201	1.232
0.08	1.353	1.071	1.035	1.264	1.307

Table 2 shows the ratios of the calculated friction contributions (Ucalc/Ucalc) of the different cases, too. The simple disk compression was set as reference to determine the $Ucalc_{1x10}/Ucalc_{5x10}$ and $Ucalc_{1x10}/Ucalc_{10x10}$ ratios. It is assumed that the ratio of the friction contribution only depends on the geometry, since both the material anisotropy and the lubrication method were the same for the different layouts. It leads to the possible comparison of the experimental stress increase ratios to the calculated ones, as a simplified friction coefficient determination.

The experimental stress increases between the uniaxial tension and compression (Uexp), arising from the friction and the unknown anisotropic behavior, count as 1.296 for the 1x10, 1.061 for the 5x10 and 1.070 for the 10x10 layouts. Thus, the stress increase ratios are $Uexp_{1x10}/Uexp_{5x10} = 1.221$ and $Uexp_{1x10}/Uexp_{10x10} = 1.211$. These results are close to the Ucalc/Ucalc ratios if $\mu \approx 0.6$. If we accept this value as the simplified-acquired friction coefficient, the remaining stress increase should be the stress state contributions of the unknown yield criterion. In this case, it is 3.1% for the 1x10, 0,8% for the 5x10 and 4.3% for the 10x10 layout, resulting in 2.7% average.

This stress state contribution can also be theoretically calculated for certain yield loci, and now it is possible to check which yield surface is closer to the already, purely stress state dependent flow stress contribution that results from the material anisotropy. Hill48 yield criterion leads to the following correlation,

$$\frac{\sigma_b}{\bar{\sigma}} = U_{exp} = \sqrt{\frac{1+R}{2}}, \tag{3}$$

in which constant R-value, independent of specimen orientation is used. According to it, the stress increase should be 19.0% (however neglecting planar anisotropy).

Calculating with the Yld89 yield surface, the stress increase is much smaller and gives 3.0%, which is closer to the average 2.7%. After some simplification on the original formula [2], Uexp can be calculated in this context as:

$$\left[\frac{a + a|h|^M + c|1-h|^M}{2}\right]^{\frac{1}{M}} = \frac{\bar{\sigma}}{\sigma_b} = U_{exp}. \tag{4}$$

It is still worth mentioning that Abspoel et al. [17] found that the value in question, the equi-biaxial uniaxial yield stress ratio can be calculated with statistical method, based on tension test data only. With this calculation Uexp gets 10.0%, which is moderately close to the average 2.7% specified here.

Conclusion

Stack compression test method on samples with different geometries was used to embody equibiaxial tests with the aim of proper yield surface selection for the industrial practice. Stress increase between the tension and the stack compression cases was investigated through considering both the friction- and the stress state contribution. Based on the comparison of the experimentally measured and the calculated stress increase ratios (Ucalc and Uexp), the friction coefficient was determined near to $\mu \approx 0.06$ with a simplified approach. This led to 26.5%, 5.3% and 2.7% friction contribution in stress for 1x10, 5x10 and 5x10 height-to-diameter ratios, respectively. The remaining stress increase, an average value of 2.7% was considered as the contribution of the anisotropic material behavior through the stress state, and was compared to different theories. It was observed that the Yld89 gives an appropriate description of the anisotropy for the investigated DC04 material.

Acknowledgement

„A Projekt a Kulturális és Innovációs Minisztérium Nemzeti Kutatási, Fejlesztési és Innovációs Alapból nyújtott támogatásával és az NKFI Hivatal által kibocsátott Támogatói Okirat alapján valósult meg. Támogató Okirat száma: MEC_R 149631."

References

[1] R. Hill, A theory of the yielding and plastic flow of anisotropic metals, The hydrodynamics of non-Newtonian fluids. I (1947) 281-297.

[2] F. Barlat, J. Lian, Plastic behavior and stretchability of sheet metals. Part I: A yield function for orthotropic sheets under plane stress conditions, International Journal of Plasticity Vol. 5 (1989) 51-66.

[3] D. Banabic, H. Aretz, D.S. Comsa, L. Paraianu, An improved analytical description of orthotropy in metallic materials, International Journal of Plasticity 21 (2005) 493-512.

[4] S. Coppieters, H. Traphöner, F. Stiebert, T. Balan, T. Kuwabara, A.E. Tekkaya, Large strain flow curve identification for sheet metal, J. Mat. Pro. Tec. 308 (2022) 117725. https://doi.org/10.1016/j.jmatprotec.2022.117725

[5] M. Merklein, A. Kuppert, A method for the layer compression test considering the anisotropic material behavior, Int. J. Mater. Form. Vol. 2 (2009) 483-486. doi:10.1007/s12289-009-0592-8

[6] S. Coppieters, P. Lava, H. Sol, A. Van Bael, P. Van Houtte, D. Debruyne, Determination of the flow stress and contact friction of sheet metal in a multi-layered upsetting test, J. Mat. Pro. Tec. 210 (2010) 1290-1296. https://doi.org/10.1016/j.jmatprotec.2010.03.017

[7] S. Coppieters, M. Jackel, C. Kraus, T. Kuwabara, F. Barlat, Influence of hydrostatic pressure shift on the flow stress in sheet metal, Proc. Manuf. 47 (2020) 1245-1249. https://doi.org/10.1016/j.promfg.2020.04.196

[8] T. Kuwabara, R. Tachibana, Y. Takada, T. Koizumi, S. Coppieters, F. Barlat, Effect of hydrostatic stress on the strength differential effect in low-carbon steel sheet, Int. J. Mater. Form. 15:13 (2022) https://doi.org/10.1007/s12289-022-01650-2

[9] D. Steglich, X. Tian, J. Bohlen, T. Kuwabara, Mechanical testing of thin sheet magnesium alloys in biaxial tension and uniaxial compression, Exp. Mechanics 54 (2014) 1247-1258. https://doi.org/10.1007/s11340-014-9892-0

[10] T. Rahmaan, J. Noder, A. Abedini, P. Zhou, C. Butcher, M.J. Worswick, Anisotropic plasticity characterization of 6000- and 7000-series aluminum sheet alloys at various strain rates, Int. J. of Impact Eng. 135 (2020) 103390. https://doi.org/10.1016/j.ijimpeng.2019.103390

[11] Y.G. An, H. Vegter, Analytical and experimental study of frictional behavior in through-thickness compression test, J. Mat. Pro. Tec. 160 (2005) 148-155. https://doi.org/10.1016/j.jmatprotec.2004.05.026

[12] M. Kraus, M. Lenzen, M. Merklein, Contact pressure-dependent friction characterization by using a single sheet metal compression test, Wear 476 (2021) 203679. https://doi.org/10.1016/j.wear.2021.203679

[13] F. Barlat, J.C. Brem, J.W. Yoon, K. Chung, R.E. Dick, D.J. Lege, F. Pourboghrat, S.-H. Choi, E. Chu, Plane stress yield function for aluminim alloy sheets – Part 1: theory, Int. J. Plast. (2003) 1297-1319

[14] S. Kim, J. Lee, F. Barlat, M-G. Lee, Formability prediction of advanced high strength steels using constitutive models characterized by uniaxial and biaxial experiments, J. Mat. Pro. Tec. 213 (2013) 1929-1942. http://dx.doi.org/10.1016/j.jmatprotec.2013.05.015

[15] E. Siebel, H. Beisswänger, Tiefziehen, Hanser, München, 1955.

[16] H. Mulder, Differential hardening of low carbon steel in sheet metal forming, Doctoral Dissertation, University of Twente, 2023.

[17] M. Abspoel, M.E. Scholting, M. Lansbergen, Y. An, H. Vegter, A new method for predicting advanced yield criteria input parameters from mechanical properties, J. Mat. Pro. Tec. (2017) http://dx.doi.org/doi:10.1016/j.jmatprotec.2017.05.006

Sheet Metal 2025
Materials Research Proceedings 52 (2025) 228-233

Materials Research Forum LLC
https://doi.org/10.21741/9781644903551-28

Consideration of residual stresses and damage in the fracture mechanical investigation of mechanically joined structures

Deborah Weiß[1,a] *, Tobias Duffe[1,b], Tintu David Joy[1,c] and Gunter Kullmer[1,d]

[1]Applied Mechanics (FAM), Paderborn University, Pohlweg 47-49, 33098 Paderborn, Germany

[a]weiss@fam.upb.de, [b]duffe@fam.upb.de, [c]joy@fam.upb.de, [d]kullmer@fam.upb.de

Keywords: Residual Stress, Damage, Fracture Mechanical Simulation

Abstract. The process of joining is used in numerous sectors of the manufacturing industry, where constructions composed of individual components or metal sheets are combined to form complex structures. A straightforward and pervasive approach for joining materials of disparate natures and coated surfaces is clinching. During the clinching process, plastic deformation, residual stresses and damage are introduced into the joint. Due to time-varying service loads cracks can initiate and propagate in the vicinity of the joint which limits the lifetime of the clinched structure. In order to prevent those damage cases, it is crucial to perform fracture mechanical evaluation of cracks in the joint region. Therefore, this publication deals with the question of how plastic deformation, residual stresses and damage need to be considered for the assessment of a crack. For this purpose, simple substitute models are employed to illustrate the principles based on the clinching application example.

Introduction

In the contemporary era, the term "joining" is employed in a multitude of manufacturing sectors, including the appliance and electrical industries. In these fields, the process of joining multiple components or metal sheets into compound structures is a common practice [1]. One method of combining dissimilar coated materials within a construction is clinching. In the initial phase of the clinching process, the sheets are secured between the blank holder and the die. Subsequently, the punch exerts pressure on the sheets, penetrating them. As the punch is inserted into the parts to be joined, the lower sheet is compressed and pressed against a rigid die. The material begins to flow radially, forming an undercut. This results in an inseparable, force-locked and form-fit connection with a high degree of plastic flow [2]. It should be emphasized that the joining process itself causes damage to the clinch joint and creates residual stresses. The clinching process gives rise to primarily compressive residual stresses within the clinch and in the surrounding area. As the distance from the clinch increases, the compressive residual stresses tend to be counterbalanced by tensile residual stresses, see Fig. 1a [3]. The substantial plastic deformations that occur during the clinching process can result in ductile damage to the material, which can be reduced by introducing variations in geometry or process parameters [4]. The findings of [5] indicate that in the absence of a plastic history, the strength of the joint is underestimated, predominantly due to the absence of plastic hardening. Conversely, in the absence of in-process damage, the strength is overestimated, underscoring the necessity for damage modelling throughout the joining process. Continuum mechanical models such as Lemaitre's damage model describe the damage quantitatively by reducing the modulus of elasticity, see Fig. 1b. Damage influences the failure of materials and components and leads to technical cracks [6]. As soon as there are technical cracks in components, fracture mechanics is used to assess the remaining service life of components [7]. The question thus arises as to how one might address the issue of plastically deformed components, pre-existing damage and residual stresses in the context of fracture mechanics. This paper addresses this question.

Content from this work may be used under the terms of the Creative Commons Attribution 3.0 license. Any further distribution of this work must maintain attribution to the author(s) and the title of the work, journal citation and DOI. Published under license by Materials Research Forum LLC.

Sheet Metal 2025   Materials Research Forum LLC
Materials Research Proceedings 52 (2025) 228-233   https://doi.org/10.21741/9781644903551-28

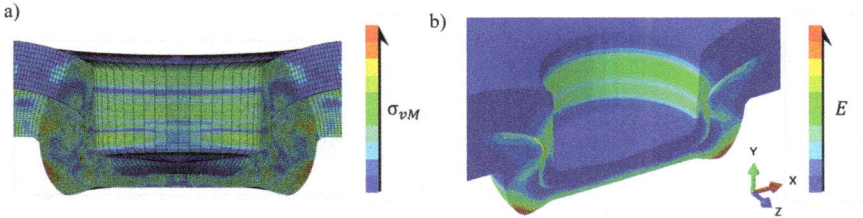

Fig. 1. Illustration of a clinched joint with (a) residual stresses and (b) reduced modulus of elasticity due to damage during the process

Stress Intensity Factor

In fracture mechanics, cracks are defined as local separations of the material a component or a structure is made of. They result in a local redirection of the flow of forces and consequently, a singular stress field in the vicinity of the crack. This field can be expressed by Eq. 1, utilising the polar coordinates r and φ and the angle-dependent functions f_{ij}. Because it is not possible to ascertain the potential danger of a crack from the infinite stresses at its tip, the stress intensity factor (SIF) K_I is the critical fracture mechanical stress parameter in a Mode I loading situation. K_{II} is employed for a Mode II loading situation, while K_{III} is derived from a Mode III loading situation [7]. IRWIN [8] proposed a classification of local crack modes based on the types of loading experienced by a crack and the directions of propagation observed. Mode I is applicable to all normal stresses σ that result in crack opening. Mode II is applicable to all plane shear stresses τ that result in the opposing sliding of the crack surfaces in the x-direction, thereby causing the crack to deflect with a kinking angle φ_0. Mode III is distinguished by the transverse sliding of the crack planes in the z-direction, which occurs due to a non-planar stress state. This leads to a twisting of the crack with a twist angle ψ_0.

$$\sigma_{ij} = \frac{1}{\sqrt{2\pi r}} \left[K_I \cdot f_{ij}^I(\varphi) + K_{II} \cdot f_{ij}^{II}(\varphi) + K_{III} \cdot f_{ij}^{III}(\varphi) \right]. \qquad (1)$$

The stress intensity factor is used to quantify the intensity of the stress field around the crack tip and can be calculated using elasticity theory, numerical or experimental methods. One method, that has been demonstrated to be an efficacious numerical procedure for the analysis of diverse crack problems in linear elasticity is the Modified Virtual Crack Closure Integral (MVCCI) and is also used for the determination of the stress intensity factors in this paper [9]. With the help of the finite element method, it is possible to calculate the nodal forces F_i at the crack tip and the relative displacements u_i of two opposite nodes on the cack flanks at a distance Δa from the crack tip in the different coordinate directions x, y and z and thus the energy release rate G_I by using the following Eq. 2 [7].

$$G_I(a, \Delta t_k)_k = \frac{1}{\Delta t_k \cdot \Delta a} \cdot W_k^y \qquad \text{with} \qquad W_k^y = \frac{1}{2} \left[F_{i,k}^y(a) \cdot \Delta u_{i-1,k}^y(a) \right]. \qquad (2)$$

For the plane stress state, the conversion from the energy release rate G_I into the stress intensity factor K_I can be done with the modulus of elasticity E and Eq. 3.

$$K_I = \sqrt{E \cdot G_I}. \qquad (3)$$

Sheet Metal 2025   Materials Research Forum LLC
Materials Research Proceedings 52 (2025) 228-233   https://doi.org/10.21741/9781644903551-28

Substitute Models

To gain a deeper understanding of the topics of deformed components, pre-damage and residual stresses in the context of fracture mechanics, these challenges are analyzed in greater detail using a simple, well-known problem, the so-called Kirsch problem [10]. The Kirsch problem concerns an infinite sheet with a hole under uniaxial tension. The dimensions of the sheet under examination are 200 x 200 mm² with a hole radius r_h of 10 mm and a finer meshed area with an outer radius r_o of 25 mm, see Fig 2a. Double symmetry is utilized, and only quarter models of the finer meshed region are shown in the following figures. The assumed ideal-plastic material has a modulus of elasticity of 210000 MPa, a Poisson´s ratio of 0.3, and a yield strength of 200 MPa.

In the first model, a tensile stress of 25 MPa is initially applied, which corresponds to a load within the linear elastic range, see Fig. 2b. The load results in maximum principal stress of 77.9 MPa at the notch which is typically somewhat higher than three times the applied stress since the sheet under examination is not an infinite sheet. To generate significant plastic deformation in the vicinity of the notch, the sheet is subjected to a stress of 100 MPa, see Fig. 2c.

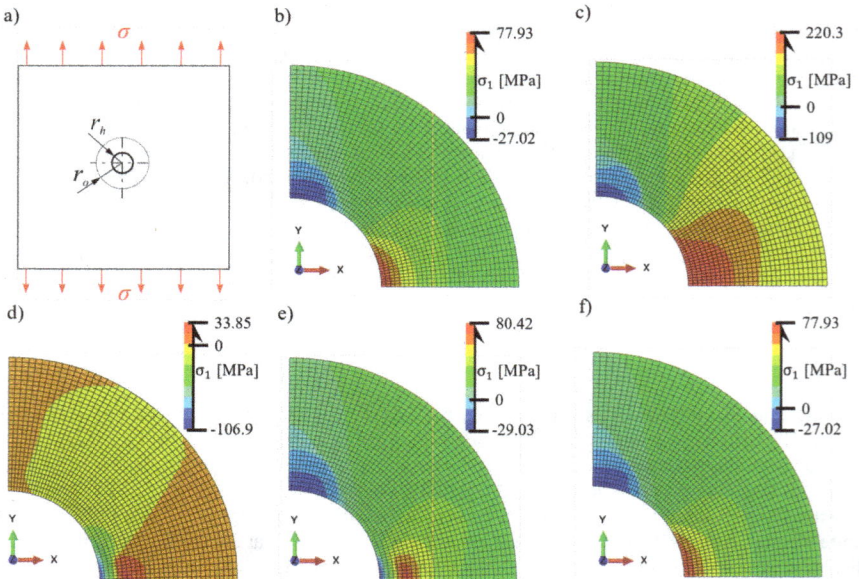

Fig. 2. (a) General model; (b) Linear elastic notch stresses after a load of 25 MPa; (c) Elastic plastic notch stresses after a load of 100 MPa; (d) Notch with residual stresses after unloading; (e) Notch stresses after reload of 25 MPa; (f) Resulting notch stresses after reload of 25 MPa subtracted by the residual stresses

The subsequent unloading of the specimen generates a maximum residual compressive stress of approximately −107 MPa in the notch base, see Fig. 2d. Subsequently, the deformed sheet is loaded again with 25 MPa, which simulates a moderate load during operation. This loading sequence is similar to the formation of a clinching joint, where residual stresses also occur with subsequent operational loading of the deformed structure. The resulting stress field (Fig. 2e) can be considered to correspond to the superposition of the elastic stress field (Fig. 2b) and the residual

Sheet Metal 2025   Materials Research Forum LLC
Materials Research Proceedings 52 (2025) 228-233   https://doi.org/10.21741/9781644903551-28

stress field after unloading (Fig. 2d). Once the residual stress field has been subtracted, the linear elastic stress field remains (Fig. 2f), which can be compared directly with the stress pattern shown in Fig. 2b. This applies since the external geometry of the structure and the boundary conditions do not change during the load sequence. Thus, the result is the same if the residual stress field due to plastic deformation is mapped on the deformed structure as prestress.

In a further model, the notch is subjected to a compressive overload of the same magnitude as the previous tensile overload. The resulting stress distribution at the notch is found to be virtually the opposite. In contrast to the residual compressive stresses of -107 MPa observed in the notch base following unloading of the plasticized specimen, the corresponding residual tensile stress of 107 MPa is evident with compressive overload. In the case of compressive overloads, the elastic stress field also overlaps with the residual stresses after unloading. However, it should be noted that the residual tensile stresses in the notch base must not exceed the yield strength, as they must not be too high. However, these residual tensile stresses dissipate as soon as a crack grows.

To investigate the influence of the residual stresses on a crack, a 5 mm long crack is inserted into the residual stress field of the notch in a third model, starting from the edge of the hole, see Fig. 3a. To generate the crack in the FE model and to avoid penetration of the crack flanks, the symmetry boundary conditions along the x-axis are removed in this section and replaced by contact boundary conditions against a rigid edge. This corresponds to a crack length $a = 15$ mm. The loads on the model still without the crack are selected according to the first model under tensile overload of 100 MPa and a residual compressive stress of -107 MPa in the notch base after unloading as shown in Fig. 2d. Since the residual stresses represent an overall state of equilibrium, residual compressive stresses near the edge cause tensile residual stresses at a certain distance from the edge, see Fig. 2e. The introduction of the crack in the residual stress field of the notch causes a redistribution of the residual stresses. The primary residual tensile stresses along the crack flanks cause a local opening of the crack and tensile stresses at the crack tip. Due to the compressive stresses near the notch edge, the crack remains closed there. Although the crack tip is constantly open during reloading, it only experiences an increase in stress intensity when the crack flanks also open at the notch edge. When the sheet is reloaded with 25 MPa, the crack is completely open. The comparison of the ligament stress curves in Figure 3b shows that in this case the stresses after reloading in the residual stress field are identical to the linear elastic stress curve for the same load. The stress intensity factors at the crack determined using the MVCCI method, for linear elastic material behavior and the reloaded specimen with elastic ideal plastic material behavior are the same, too. As Figure 3b also shows, the same result is achieved if the residual stresses at the notch after an overload are mapped onto the notch specimen without the crack and the same boundary conditions with linear elastic material behavior as prestress. When mapping residual stresses as prestresses, it must be ensured that the initial geometry and the target geometry as well as the boundary conditions are the same. Then the crack is introduced and a load of 25 MPa follows. However, during unloading, the crack flanks already come into contact with the notch edge due to the residual stresses before the external load becomes zero. As a result, the crack tip remains open and there is no further reduction in stress intensity at the crack. Thus, the amplitude of the stress intensity decreases, and the crack grows more slowly. In a crack growth simulation, the stress intensity at the crack must then be determined for both the maximum and the minimum external load, because the proportionality to the external load is no longer given. Nevertheless, linear elastic material behavior can be assumed for both load conditions for the fracture mechanics evaluation. For the consideration of residual stresses in the crack growth simulation at clinching points, this means that the residual stresses can be mapped onto the deformed geometry due to the clinching process without a crack. The crack is then inserted and linear elastic material behavior can be assumed for the subsequent crack growth simulation.

Sheet Metal 2025   Materials Research Forum LLC
Materials Research Proceedings 52 (2025) 228-233   https://doi.org/10.21741/9781644903551-28

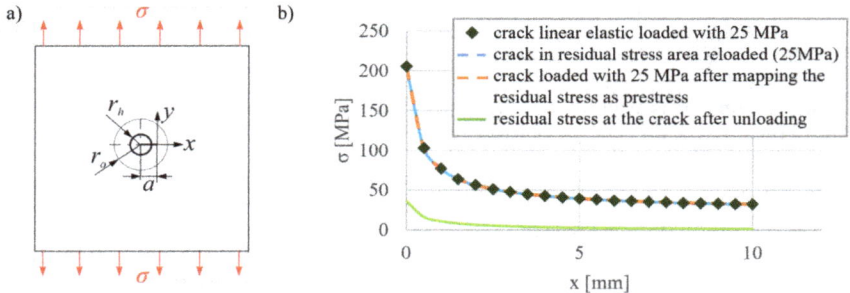

Fig. 3. (a) General model with crack; (b) Ligament stress curves at the crack after tensile overload at the notch

With regard to the clinch connection, it now remains to be clarified how pre-damage should be dealt with. For this purpose, a further model is set up in which, as typical in the damage modeling, a modulus of elasticity reduced by 50% is defined in a radius of 20 mm around the center of the hole, see Fig. 4a.

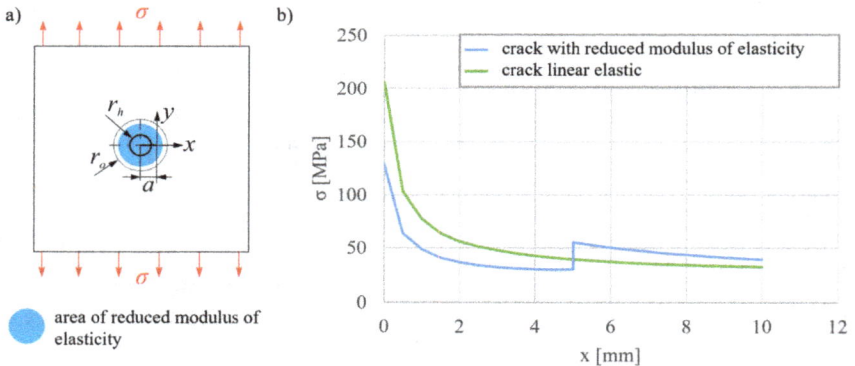

Fig. 4. (a) General model with crack and area of reduced modulus; (b) Stress curves at the crack with reduced modulus of elasticity

The calculation of the stress intensity factors shows that the reduction of the modulus of elasticity according to Eq. 3 also leads to a reduction of the stress intensity factor from 189 MPamm$^{0.5}$ to 116.77 MPamm$^{0.5}$, as the crack is shielded in the area of the reduced modulus of elasticity. The area with a higher modulus of elasticity ($x > 5$ mm) around the area with a reduced modulus of elasticity is subjected to higher stresses, which is what the damage is intended to achieve, see Fig. 4b. It can therefore be stated that for the crack growth simulation, the crack must be inserted into an area with full modulus of elasticity. This means that the damage calculation provides the position and orientation of the initial crack. In the crack growth simulation, the consideration of a reduced modulus of elasticity due to damage is not appropriate.

Summary

This paper shows in principle how residual stresses due to forming processes and damage due to fatigue loading should be considered in fracture mechanics analyses. Simulations using simplified models are used to show that residual stresses can be taken into account as initial stresses. Damage is usually expressed in damage modeling by a reduction in the modulus of elasticity. When calculating stress intensity factors, it is not appropriate to consider a locally reduced modulus of elasticity, as this reduces the stress intensity at the crack. Damage modeling provides information about the location and orientation of the initial crack. A reduced modulus of elasticity should not be assumed for a fracture mechanics analysis.

Acknowledgements

This project was funded by the Deutsche Forschungsgemeinschaft (DFG, German Research Foundation) – TRR 285/2 – 418701707.

References

[1] D. Geoffrey, Materials for Automobile Bodies, second ed., Butterworth-Heinemann Ltd, Oxford, 2012.

[2] DVS/EFB 3420, 2012: Merkblatt 3420: Clinchen - Überblick. Deutscher Verband für Schweißen und verwandte Verfahren e.V., Europäische Forschungsgesellschaft für Blechverarbeitung e.V.

[3] J. Gibmeier, R. Lin, M. Odén, B. Scholtes, Residual Stress Distributions around Clinched Joints, Materials Science Forum 404-407 (2002) 617-622. https://doi.org/10.4028/www.scientific.net/MSF.404-407.617

[4] Y. Abe, K. Mori, T. Kato, Joining of high strength steel and aluminium alloy sheets by mechanical clinching with dies for control of metal flow, Journal of Materials Processing Technology 212 (4) (2012) 884-889. https://doi.org/10.1016/j.jmatprotec.2011.11.015

[5] E. Roux, P.O. Bouchard, Kriging metamodel global optimization of clinching joining processes accounting for ductile damage. Journal of Materials Processing Technology 213 (7) (2013) 1038-1047. https://doi.org/10.1016/j.jmatprotec.2013.01.018

[6] J. Lemaitre, Local approach of fracture. Engineering Fracture Mechanics 25 (5-6) (1986) 523-537. https://doi.org/10.1016/0013-7944(86)90021-4

[7] H.A. Richard, M. Sander, Fatigue Crack Growth, Springer Verlag, Switzerland 2016. https://doi.org/10.1007/978-3-319-32534-7

[8] G.R. Irwin, Analysis of Stresses and Strains Near the End of a Crack Traversing a Plate, Journal of Applied Mechanics 24 (1957) 361-364. https://doi.org/10.1115/1.4011547

[9] E.F. Rybicki, M.F. Kanninen, A finite element calculation of stress intensity factors by a modified crack closure integral, Engineering Fracture Mechanics 9 (1977) 931-938. https://doi.org/10.1016/0013-7944(77)90013-3

[10] G. Kirsch, Die Theorie der Elastizität und die Bedürfnisse der Festigkeitslehre, Zentralblatt Verein Deutscher Ingenieure 42 (1898) 797-807.

Sheet Metal 2025   Materials Research Forum LLC
Materials Research Proceedings 52 (2025) 234-241   https://doi.org/10.21741/9781644903551-29

Inverse parameter identification for the delamination behaviour of metal-polymer-metal sandwich materials

Moritz Kuhtz[1,a *], Jonas Richter[1], Andreas Hornig[1,2,3] and Maik Gude[1]

[1]Institute of Lightweight Engineering and Polymer Technology (ILK), TUD Dresden University of Technology, Holbeinstr. 3, 01307 Dresden, Germany

[2]Center for Scalable Data Analytics and Artificial Intelligence Dresden/Leipzig (ScaDS.AI), TUD Dresden University of Technology, Chemnitzer Str. 46b, 01187 Dresden, Germany

[3]Department of Engineering Science, University of Oxford, Parks Road, OX1 3PJ Oxford, United Kingdom

[a]Moritz.Kuhtz@tu-dresden.de

Keywords: Composite, Delamination, Modelling

Abstract. The double cantilever beam (DCB) test is a standardised method for characterising delamination properties in composites, adapted here for sandwich structures with metal cover layers and a polymer core (MPM). An inverse parameter identification approach is used based solely on force-displacement data to characterise interface properties, eliminating the need for crack length measurements. DCB tests are conducted with varying pre-crack configurations, core thicknesses, and loading rates. The results show that critical strain energy release rates G_{Ic} increase with increasing core thicknesses, loading rates, and cohesive pre-cracks, while interface strengths R_3^+ are influenced less by these factors. Numerical simulations using a refined mesh based on a mesh study and a cohesive zone model are used to characterise the delamination behaviour. The inverse method of determining cohesive zone model parameters R_3^+ and G_{Ic} by fitting numerical force-displacement curves to experimental results provides significant improvements in accuracy and efficiency compared to manual re-calibration, reducing the pre- and post-processing time while providing robust parameter estimation for delamination behaviour in MPM composites.

Introduction

The DCB test is an established and standardised method for characterising interface properties for monolithic materials such as unidirectional fibre-reinforced epoxy composite materials [1] and can be also applied for sandwich material for characterising the delamination behaviour between cover and core layers [2]. This method is applicable when the total energy applied is entirely transformed into delamination work. In the case of high-strength interfaces in combination with ductile materials such as forming steel, the specimens must be stiffened to gain a valid delamination behaviour. This is achieved by using stiffening elements bonded to the cover layers [3]. These backing beams enable valid delamination failure for MPM composites and allow the characterisation of the interface properties by means of the interface tensile strength R_3^+ and the Mode I critical strain energy release rate G_{Ic}.

G_{Ic} describes the integral under the stress-crack tip opening curve, which is usually represented as a bilinear relationship. A linearly increasing tensile force causes a linear increase in stress σ_3 until the interface strength R_3^+ is reached. With further loading, irreversible damage to the interface is assumed, which is reflected in a linearly decreasing stress. When the interface is completely separated, no more force can be transmitted and the resulting interfacial tensile stress is zero [4].

The parameter identification for sandwich materials is complex due to the link between material and structural phenomena. Here, an inverse material parameter identification method can be used to efficiently characterise the material properties [5]. It is also being investigated whether this

Content from this work may be used under the terms of the Creative Commons Attribution 3.0 license. Any further distribution of this work must maintain attribution to the author(s) and the title of the work, journal citation and DOI. Published under license by Materials Research Forum LLC.

Sheet Metal 2025
Materials Research Proceedings 52 (2025) 234-241

Materials Research Forum LLC
https://doi.org/10.21741/9781644903551-29

method can replace the use of time-consuming crack length measurement during the test. Therefore, the inverse parameter identification method is used in the present study to describe the delamination behaviour of MPM composites. The model parameters describing the delamination behaviour are determined by fitting the model results to the experimental results.

DCB experiments

The investigated material system is a MPM sandwich consisting of two high strength formable, electrolytic galvanised steel cover sheets DPK 30/50+ZE (HCT500X, thyssenkrupp Steel Europe AG, grade number: 1.0939) with a thickness of 0.48 mm and a PP/PE core of 0.3 mm thickness. It was manufactured by roll-bonding with an adhesive agent (one-component epoxy resin Köratac FL201, Kömmerling Chemische Fabrik GmbH) of a thickness of approx. 0.01 mm [6]. As recommend in [3], the specimens are stiffened with 15 mm thick backing beams. In addition to these prior investigations, the influences of the pre-crack configuration, the core layer thickness and the loading rate on the delamination behaviour in the DCB test are investigated here (Table 1). In the case of an adhesive pre-crack, a separating film (PTFE) is inserted between a metallic cover layer and the polymer core material. In the case of cohesive pre-crack, the separating film is inserted between two PE/PP core layers. In addition, three different core layer thicknesses (0.3 mm, 0.6 mm and 1.2 mm) are used. Finally, the loading rate is increased from a quasi-static (0.167 mm/s) to a moderately fast rate (10 mm/s).

Table 1: Investigated parameters of the DCB test

Configuration	Pre-crack	Core layer thickness [mm]	Loading rate [mm/s]	Number of specimens
#1	adhesive	0.3	0.167	3
#2	adhesive	0.6	0.167	3
#3	cohesive	0.6	0.167	3
#4	cohesive	1.2	0.167	3
#5	adhesive	0.3	10	4
#6	adhesive	0.6	10	3
#7*	cohesive	0.6	10	-
#8	cohesive	1.2	10	4

* Due to failed data recording, this data is not available for analysis.

The width and the thickness of the DCB specimens were measured before conditioning for 24 h at 23 °C and 50 % humidity prior to testing. The initial delamination length is measured separately for all specimens as well. The DCB experiments were carried out using a Zwick 1475 testing machine and a digital image correlation system ARAMIS 5 M from GOM. The deformation measurements are based on a frame size of 4096 pixel × 1000 pixel with facet sizes of 20 pixel × 20 pixel, facet distances of 12 pixel × 12 pixel at a frame rate of 5 frames per second for the quasi-static loading and 2000 frames for the moderate fast tests.

In contrast to prior work, G_{Ic} is not evaluated from the given calculations that base on the force-displacement curve and the measurement of the delamination front. Here, the force-displacement curve and the initial pre-crack length is used for the inverse parameter identification approach.

Sheet Metal 2025   Materials Research Forum LLC
Materials Research Proceedings 52 (2025) 234-241   https://doi.org/10.21741/9781644903551-29

Numerical model

The numerical model initially corresponds to the work presented so far [3]. The nodes relating to the support are fixed translationally in each direction. The nodes related to the load application are fixed in the x and y directions and are moved in the z direction by a function in time. The load is applied by a constant acceleration of $60000 \, \text{mm}/_{\text{s}^2}$ to achieve a compromise between computational effort and the influence of the kinetic energy of the models.

A mesh dependency has been identified in [7], which is why a mesh study is carried out here too. Starting from the reference [3], the mesh density of the MPM composite is increased by factors of 2 and 4 in each direction. Fig. 1 shows the reference model and the model used for inverse parameter identification, which has twice as many elements in the x-direction and four times more elements in the y- and z-directions.

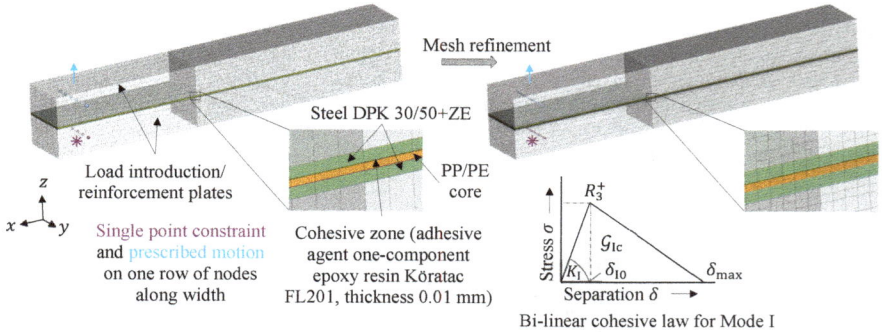

Fig. 1: Reference mesh [3] and refined mesh

The cohesive zone is represented by a bilinear tension-separation law (*MAT_138), the polymer is modelled by an elastic-plastic model (*MAT_24), as are the steel cover layers. Since the stiffness of the steel is considerably higher than that of the polymer, only the steel is modelled as a function of strain rate. The support beams are modelled as elastic bodies. In order to take manufacturing deviations into account, the simulation model is adapted for each configuration with regard to the dimensions and the initial delamination length.

Inverse parameter identification

Fig. 2a) shows the impact of the mesh densities on the F-u results. Based on prior work [3], the reference mesh (1x1y1z) is refined in length (x), width (y) and thickness direction (z) by the factors 2 and 4. All simulations with twice as many elements in z-direction (2z) show an unstable behaviour resulting in an unrealistic failure behaviour. One possible explanation for this could be a specific resonance behaviour. Hence, these simulations are dismissed for further analysis. Incidentally, the remaining simulations show the same qualitative behaviour, whereby the noise of the signal decreases with increasing mesh fineness. As a compromise between computation time and result quality, the variant (2x4y4z) is selected for further analyses in the framework of inverse parameter identification.

The influence of the strength R_3^+ and energy release rate G_{Ic} is analysed also. Fig. 2b) shows the influence of the strength and Fig. 2c) the influence of the energy release rate on the force-displacement behaviour. Upper and lower limits for the values were defined for this purpose: $10 \le R_3^+ \le 220 \, \text{N}/_{\text{mm}^2}$ and $0.5 \le G_{Ic} \le 8.0 \, \text{N}/_{\text{mm}}$. R_3^+ essentially influences the force increase of the force-displacement curve, whereby low strengths result in a highly non-linear deformation

Sheet Metal 2025
Materials Research Proceedings 52 (2025) 234-241

Materials Research Forum LLC
https://doi.org/10.21741/9781644903551-29

behaviour. $\mathcal{G}_{Ic}$ determines the course after the force peak is reached, whereby higher force levels are achieved with higher values for the energy release rate. The upper and lower limits of this parameter study ($R_3^+ = [10, 220]$ N/mm^2 and $\mathcal{G}_{Ic}=[0.5, 8.0]$ N/mm) are also used to set the limits for the inverse parameter identification.

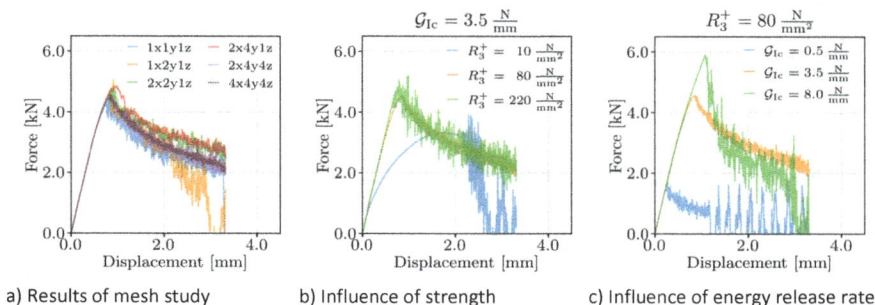

a) Results of mesh study   b) Influence of strength   c) Influence of energy release rate

Fig. 2: Force-displacement curves of DCB model

Fig. 3 shows the procedure for inverse parameter identification, which is carried out with LS-OPT 7.0. Firstly, the parameter spaces for R_3^+ and $\mathcal{G}_{Ic}$ are defined based on the values from the preliminary work. 100 value pairs are generated on the basis of a space-filling design (SFD). These 100 models are then calculated with LS-DYNA R13.1 and the resulting F-u curves are analysed. Subsequently, they are compared with one representative experimental force-displacement curve, and the resulting deviations are calculated using the mean squared error method (MSE). Finally, the parameter combination with the smallest difference between simulation and experiment is selected.

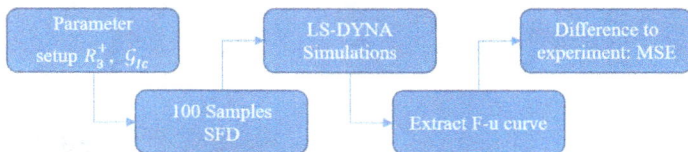

Fig. 3: Inverse parameter identification flow chart

Results and discussion

Fig. 4 shows a representative specimen for each configuration after the test (side view) and the fracture surface. Not all specimens are fully delaminated during the test, e.g. configuration #3, but rather separated after the test to analyse the fracture surfaces.

All specimens exhibit adhesive fracture close to the interface at one (#1, 3, 4, 5, 8) or both interfaces (# 2, 6). In cases of a cohesive pre-crack (#3, 4, 8), the crack spreads into one of the two interfaces within a few millimetres. In these samples, the crack does not propagate to the other side. The test specimens with 0.6 mm core layer thickness and adhesive pre-cracking (# 2, 6) show a pronounced two-sided adhesive failure, which subsequently leads to large plastic deformations and failure of the polymer core. An influence of the loading rate on the appearance of the fracture

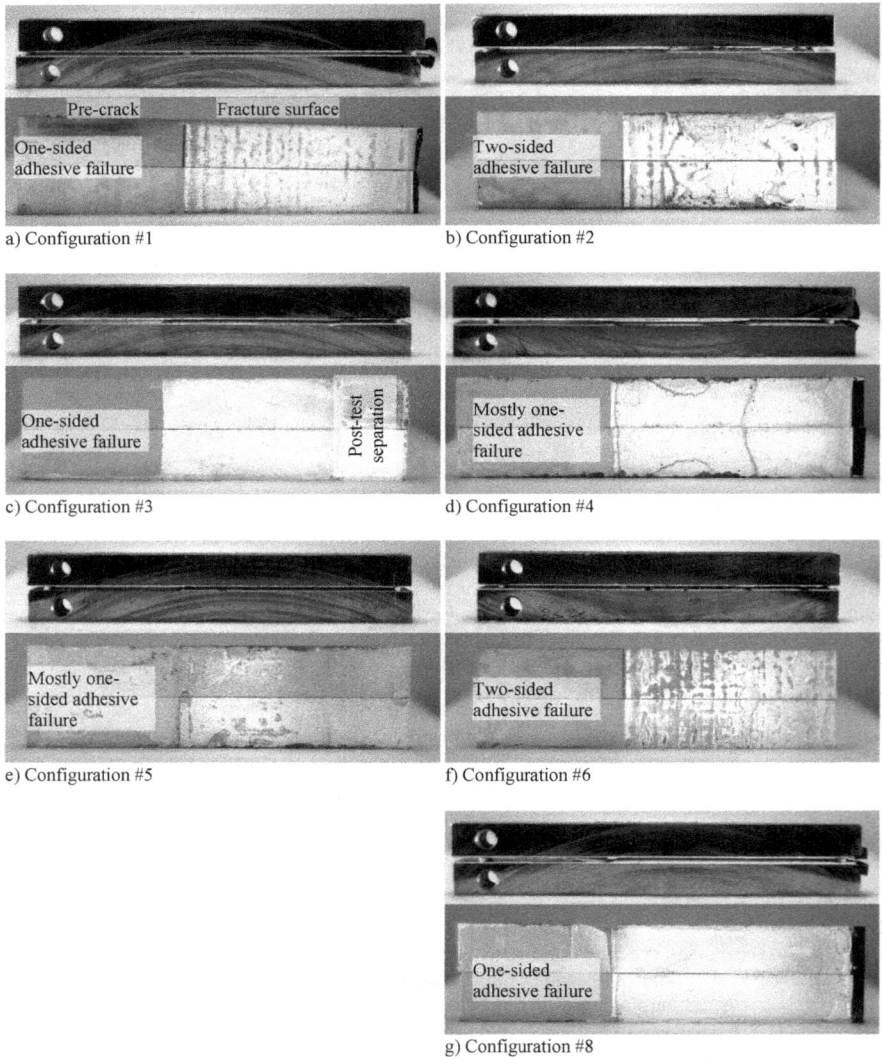

a) Configuration #1

b) Configuration #2

c) Configuration #3

d) Configuration #4

e) Configuration #5

f) Configuration #6

g) Configuration #8

Fig. 4: DCB specimens after test

surfaces cannot be determined. For comparison with the simulated F-u curves, the mean force curve is used for each configuration (Fig. 5).

Fig. 5a-g) shows the results of the inverse parameter identification as force-displacement curves with the values for the parameters R_3^+ and G_{Ic} for which the difference between the simulated and experimental curves is least. On the one hand, it can be stated that the method leads to better results

Sheet Metal 2025   Materials Research Forum LLC
Materials Research Proceedings 52 (2025) 234-241   https://doi.org/10.21741/9781644903551-29

than manual re-calibration [3]. In particular, the force maxima and the non-linear behaviour are well obtained for most configurations and are mostly within the limits of the experimental results. On the other hand, this approach saves a lot of time in the pre-and post-processing as well as the iterative re-calibration of the models at the expense of higher computational time, as more models are analysed.

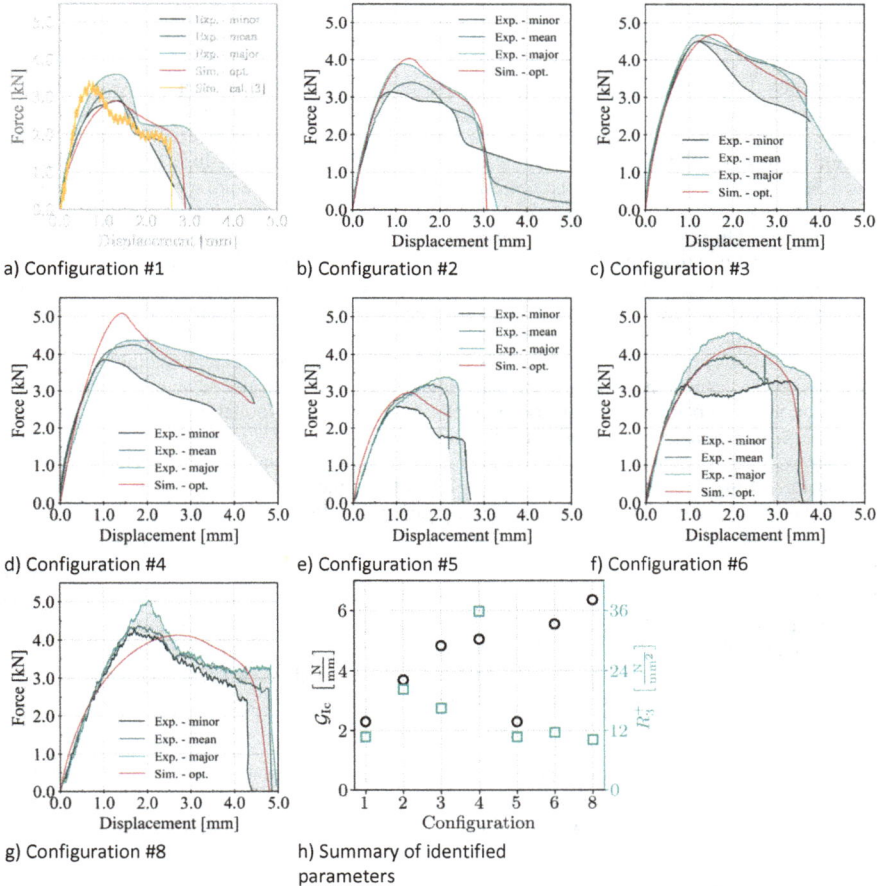

Fig. 5: Results of the mesh study and the parameter identification

Furthermore, clear tendencies can be recognised when analysing the force-displacement curves. On the one hand, the force maxima and the maximum displacements increased with increasing core density, which can be attributed to a higher bending stiffness and a higher plastic deformation in the core material. On the other hand, the forces increase with increasing load rate, whereby the maximum displacements decrease slightly, which is typical behaviour for the materials used. An influence of the pre-crack configuration on the structural behaviour can be determined. The specimens with a cohesive pre-crack show a slightly higher peak force and a slightly higher maximum displacement.

The results of the parameter identification, which are shown in Fig. 5h), confirm the following: the higher the core layer thickness and the higher the loading rate, the higher the critical strain energy release rate. The samples with cohesive pre-cracking also show higher values for $\mathcal{G}_{Ic}$. A trend for the strength cannot be clearly identified. Only a decrease in strength can be observed with increased loading rate.

Summary
The present study shows the potential of inverse parameter identification in characterising the interface properties of MPM composites. A significantly better agreement between simulation and experimental results can be achieved in comparison to manual re-calibration. In addition, this method dispenses with the time-consuming crack length determination, so that a high experimental effort, such as that associated with an optical measuring system, can be reduced. Furthermore, the method reduces personnel effort for the preparation and evaluation of simulation models to the detriment of slightly higher calculation times, as more models are solved. Through this evaluation it can be determined that the critical strain energy release rate increases with increasing core layer thickness and loading speed and that a cohesive pre-crack can also be described with a higher strain energy release rate. The interface strength is less dependent on the analysed parameters.

The modelling quality can be improved in the future through extended cohesive laws. For example, [9, 10], on the basis of the work of [8], show that a strain-rate-dependent elastic-plastic cohesive law represents the delamination behaviour well. The models use more parameters, which is why an iterative parameter identification method based on metamodel optimisation should be used to keep the number of simulations at a reasonable level.

Moreover, the tests and models have shown that the deformation and failure behaviour of the core material also influences the failure behaviour of the interface. Therefore, extended models have to be utilised in the future with regard to the strain rate-dependent, plastic behaviour.

Acknowledgements
The authors would like to thank the German Research Foundation (DFG project number 407352905) for the financial support. Moreover, the authors gratefully acknowledge the computing time provided by them on the high-performance computers Barnard at the NHR Centre NHR@TUD. This is funded by the Federal Ministry of Education and Research and the state governments participating on the basis of the resolutions of the GWK for the national high-performance computing at universities (www.nhr-verein.de/unsere-partner).

References
[1] ISO 15024:2023, Fibre-reinforced plastic composites - Determination of mode I interlaminar fracture toughness, GIC, for unidirectionally reinforced materials

[2] Manikandan, P.; Chai, G. B. Mode-I Metal-Composite Interface Fracture Testing for Fibre Metal Laminates, Advances in Materials Science and Engineering, 2018, 4572989, doi: https://doi.org/10.1155/2018/4572989

[3] Richter, J.; Kuhtz, M.; Hornig, A.; Harhash, M.; Palkowski, H.; Gude, M. A Mixed Numerical-Experimental Method to Characterize Metal-Polymer Interfaces for Crash Applications. Metals 2021, 11, 818. https://doi.org/10.3390/met11050818

[4] Cornec, A.; Scheider, I.; Schwalbe, K.-H. On the practical application of the cohesive model, Engineering Fracture Mechanics 70 (14), 2003, pp. 1963-1987. https://doi.org/10.1016/S0013-7944(03)00134-6

Sheet Metal 2025
Materials Research Proceedings 52 (2025) 234-241

Materials Research Forum LLC
https://doi.org/10.21741/9781644903551-29

[5] Shi, Y., Sol, H., Hua, H. Material parameter identification of sandwich beams by an inverse method, Journal of Sound and Vibration 290(3-5), 2006,pp. 1234-1255. https://doi.org/10.1016/j.jsv.2005.05.026

[6] Harhash, M.; Kuhtz, M.; Richter, J.; Hornig, A.; Gude, M.; Palkowski, H. Trigger geometry influencing the failure modes in steel/polymer/steel sandwich crashboxes: experimental and numerical evaluation. Compos. Struct. 2021, 262, 113619. https://doi.org/10.1016/j.compstruct.2021.113619

[7] Turón, A.; Dávila, C.G.; Camanho, P.P.; Costa, J. An engineering solution for mesh size effects in the simulation of delamination using cohesive zone models. Engineering Fracture Mechanics 2007, 74, pp. 1665-1682. https://doi.org/10.1016/j.engfracmech.2006.08.025

[8] Marzi, S.; Hesebeck, O.; Brede, M.; Kleiner, F. A Rate-Dependent, Elasto-Plastic Cohesive Zone Mixed-Mode Model for Crash Analysis of Adhesively Bonded Joints. 7th European LS-DYNA Conference, 2009, Salzburg. https://lsdyna.ansys.com/wp-content/uploads/attachments/B-VI-02.pdf

[9] Lißner, M.; Alabort, E.; Erice, B.; Cui, H.; Petrinic, N.A rate dependent experimental and numerical analysis of adhesive joints under different loading directions. Eur. Phys. J. Spec. Top. 227, 85–97 (2018). https://doi.org/10.1140/epjst/e2018-00070-x

[10] Sønstabø, J. K.; Morin, D.; Langseth, M. A Cohesive Element Model for Large-Scale Crash Analyses in LS-DYNA. 14th International LS-DYNA User Conference, 2016, Detroit, https://lsdyna.ansys.com/wp-content/uploads/2022/12/a-cohesive-element-model-for-large-scale-crash-analyses-in-ls-dyna-r.pdf

Sheet Metal 2025
Materials Research Proceedings 52 (2025) 242-249

Materials Research Forum LLC
https://doi.org/10.21741/9781644903551-30

A dieless Nakajima test for additively deposited materials

Rui F.V. Sampaio[1,a*], Pedro M.S. Rosado[1,b], João P.M. Pragana[1,c],
Ivo M.F. Bragança[2,d], Chris V. Nielsen[3,e], Carlos M.A. Silva[1,f] and
Paulo A.F. Martins[1,g]

[1]IDMEC, Instituto Superior Técnico, Universidade de Lisboa, Portugal

[2]CIMOSM, Instituto Superior de Engenharia de Lisboa, Instituto Politécnico de Lisboa, Portugal

[3]Department of Civil and Mechanical Engineering, Technical University of Denmark

[a]rui.f.sampaio@tecnico.ulisboa.pt, [b]pedro.s.rosado@tecnico.ulisboa.pt,
[c]joao.pragana@tecnico.ulisboa.pt, [d]ibraganca@dem.isel.ipl.pt, [e]cvni@dtu.dk,
[f]carlos.alves.silva@tecnico.ulisboa.pt, [g]pmartins.tecnico.ulisboa.pt

Keywords: Additive Manufacturing, Forming, Nakajima Test

Abstract. This paper focuses on a novel Nakajima test for characterizing the formability of additively deposited materials without extracting sheet blanks. Material is deposited by wire-arc additive manufacturing in the form of 'Π-shaped' vertical walls, and the testing regions are then machined to the desired thickness and shapes for subsequent deformation with a hemispherical punch in a press. Digital image correlation is used to obtain the strain loading paths of the deposited material up to failure. The novel test eliminates the need for the special-purpose tool setup of conventional Nakajima tests, simplifying the overall procedure and reducing costs.

Introduction

Combining metal additive manufacturing (MAM) with metal forming gives rise to new hybrid manufacturing routes that explore the advantages of both technologies to create high-quality parts with complex designs. Hybridization of MAM with metal forming can be used to introduce reinforcements [1] and functional elements [2] in parts, which would have been very difficult or impossible to obtain through traditional forming processes alone or, alternately, to produce preforms that are subsequently subjected to traditional sheet, bulk, and sheet-bulk forming processes to fabricate net-shape or near-net-shape parts in small to medium batch sizes [3].

However, the hybridization of MAM with metal forming suffers from anisotropy and formability problems [4] caused by the metallurgical structure of the deposited materials that often restrict its domain of applicability to low plastic deformation levels. This is particularly relevant to material deposited by wire-arc additive manufacturing (WAAM) rather than other MAM processes, and it justifies the paper's focus on WAAM-deposited materials.

WAAM is a form of wire-directed energy deposition (wire-DED) that utilizes an electric arc to provide the thermal energy necessary for melting and depositing the wire feedstock layer by layer and shares working principles with well-established gas metal arc, gas tungsten arc, or plasma arc welding processes. Its growing interest stems from its ability to create large-scale parts with low to medium quality due to its material efficiency and high deposition rates (5-6 kg/h) and due to the widespread availability of arc welding-based machines and motion systems in metalworking companies.

However, WAAM still faces difficulties meeting the parts' geometrical, surface, and metallurgical requirements due to the large melt pools created by the electric arc. These pools give rise to differential expansion and contraction across the deposited material during the heating and cooling cycles, as well as to porosity, precipitation reactions, and formation of dendritic-based

Content from this work may be used under the terms of the Creative Commons Attribution 3.0 license. Any further distribution of this work must maintain attribution to the author(s) and the title of the work, journal citation and DOI. Published under license by Materials Research Forum LLC.

Sheet Metal 2025
Materials Research Proceedings 52 (2025) 242-249

Materials Research Forum LLC
https://doi.org/10.21741/9781644903551-30

columnar grain structures [5]. The latter is responsible for the above-mentioned reported difficulties related to anisotropy and formability limits.

This paper aims to characterize the formability of WAAM-deposited stainless steel by focusing on a novel dieless Nakajima testing procedure that was recently proposed by the authors [6]. In contrast to the conventional Nakajima setup shown in Fig. 1, the novel testing procedure is carried out directly on the deposited material without the need to extract sheet blanks or use the special purpose tool setup of conventional Nakajima tests, which typically involves a punch, a die, and a blank holder [7]. The design of the dieless Nakajima test is supported by finite element simulations, utilizing in-house software, and by experimentation employing digital image correlation (DIC) to obtain the strain loading paths up to failure in principal strain space. Material deposition by WAAM is simulated by finite elements using new modelling features of the in-house software.

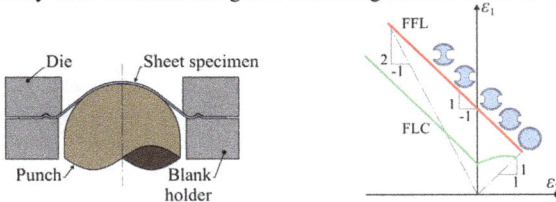

Fig. 1 Schematic representation of the conventional Nakajima test with illustration of the forming limit curve (FLC) and fracture forming line (FFL) in principal strain space obtained from different sheet blank geometries (adapted from [6]).

Experimentation

Material Deposition. The work was carried out in AISI 316L stainless steel deposited by WAAM using an ESAB Luc Aristo 400 gas metal arc welding power source coupled with a 3-axis CNC router table. The AISI 316L wire feedstock, with a diameter of 1.0 mm, was supplied through the welding torch and melted with a spray transfer mode onto hot rolled AISI 316L baseplates having 16 mm thickness.

Table 1 summarizes the main WAAM parameters used to construct the 'Π-shaped' vertical walls (hereafter referred to as 'deposited walls' or simply 'walls'). These parameters were retrieved from the authors' previous work [8], which used the same wire and baseplate materials to construct vertical walls under a stable electric arc.

Table 1 Summary of the WAAM processing parameters utilized in the deposition of AISI 316L.

Current [A]	Voltage [V]	Wire feed speed [m/min]	Travel speed [m/min]	Shielding gas	Gas flow rate [l/min]	Layer height [mm]
100	16.5	6	0.5	Argon 99.9%	10	1.8

The deposited walls with main dimensions shown in Fig. 2a were constructed through single-bead deposition layers along sequences with variable starting points at each layer to balance the heating and cooling cycles.

Dieless Nakajima Test. The dieless Nakajima test requires machining the web plates of the deposited walls to obtain a good-quality circular testing region of uniform thickness. In the current investigation, this region had a 1 mm thickness and 73.5 mm diameter at the center. Waterjet cutting was employed to modify the original circular geometry, which was aimed at providing equibiaxial strain loading paths, into an arc-shaped geometry, which was aimed at delivering near uniaxial strain loading paths. The overall test geometries were more compact, with a 0.7 size ratio compared to that recommended in the ISO 12004-2 standard.

Sheet Metal 2025   Materials Research Forum LLC
Materials Research Proceedings 52 (2025) 242-249   https://doi.org/10.21741/9781644903551-30

(a)

(b)

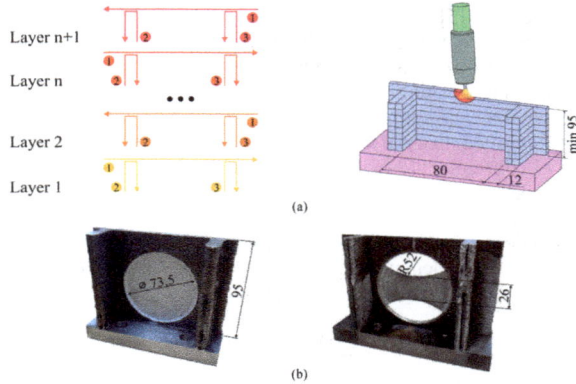

Fig. 2. (a) Schematic representation of the deposition strategy utilized in WAAM and (b) photographs of the deposited walls after machining the upper end and the central region to obtain circular and arc-shaped Nakajima test geometries (adapted from [6]).

The edges of the deposited walls were also machined perpendicularly to the building direction with good surface quality for subsequent fixing in the multidirectional tool, where the dieless Nakajima tests were performed (Fig. 3a). The tool converts the vertical crosshead movement of the hydraulic testing machine into the horizontal movement of a hemispherical punch with 70 mm diameter using a cam slide unit with a working angle $\alpha = 30°$ that consists of a sliding wedge actuator attached to the upper bolster. To minimize friction, a 0.5 mm Teflon sheet and molybdenum disulfide grease were added between the hemispherical punch and the testing region.

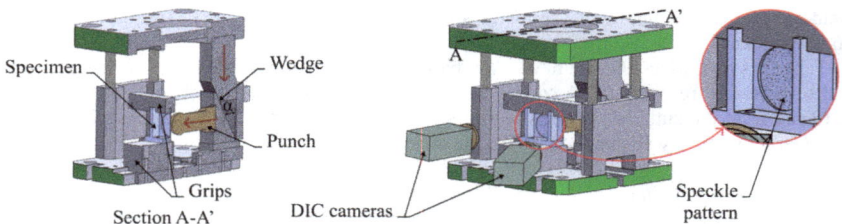

Fig. 3 Schematic representation of the dieless Nakajima testing setup in the multidirectional tool with detail of the setup used by the digital image correlation (DIC) system (adapted from [6]).

The side of the region of interest opposite to the hemispherical punch was painted white and, subsequently, with a stochastic black dot pattern (Fig. 3b) to ensure a reliable collection of the evolution of the in-plane strains over time $\varepsilon_1(t)$, $\varepsilon_2(t)$, using a commercial digital image correlation (DIC) system from Dantec Dynamics (model Q-400 3D). This system featured two 6-megapixel resolution cameras equipped with 50.2 mm focal length lenses and f/8 aperture (Fig. 3b), and the region of interest was illuminated by a spotlight during testing. Images were captured by the cameras at a shutter frequency of 10 Hz.

The authors' software [9] was utilized to merge the evolutions of in-plane strains over time $\varepsilon_1(t)$, $\varepsilon_2(t)$ into strain loading paths $\varepsilon_1 = f(\varepsilon_2)$ in principal strain space by removing the time dependency.

Sheet Metal 2025
Materials Research Proceedings 52 (2025) 242-249

Materials Research Forum LLC
https://doi.org/10.21741/9781644903551-30

Numerical Simulation

The numerical simulation of material deposition by WAAM and Nakajima testing was carried out using the in-house computer program I-FORM. In the case of WAAM, the approach involved the transformation of the non-linear thermo-mechanical macro-scale model into a staggered sequential solution of the heat transfer (1) and the quasi-static equilibrium (2) equations (under the absence of body forces) within a time increment Δt,

$$k\nabla^2 T - \rho c \frac{\partial T}{\partial t} + \dot{q} = 0 \,. \tag{1}$$

$$\frac{\partial \sigma_{ij}}{\partial x_j} = 0 \,. \tag{2}$$

In the above equations, k represents the thermal conductivity, ρ stands for the mass density, c denotes the specific heat capacity, $\dot{q}$ is the total input power density, σ_{ij} represents the Cauchy stress tensor, and x_j denotes an arbitrary position within the baseplate and deposited material at the current instant of time t. The total power density $\dot{q}$ in (1) is approximated through a heat source model, which in the current software implementation consists of the asymmetric Gaussian moving double-ellipsoid proposed by Goldak et al. [10], because it effectively accounts for the total power transferred to both filler and baseplate materials.

The thermal loads induced by temperature changes within a time increment Δt are included in the solution of the quasi-static equilibrium equations (2) through a thermal strain increment $\Delta \varepsilon^t = \alpha(T)\Delta T$, where the symbol α denotes the coefficient of thermal expansion. The thermal strain increment $\Delta \varepsilon^t$ is then combined with the elastic strain increment $\Delta \varepsilon^e$, calculated from the elasticity modulus and Poisson ratio, and the plastic strain increment $\Delta \varepsilon^p$ determined from the material flow curve to obtain the total increment of strain $\Delta \varepsilon = \Delta \varepsilon^e + \Delta \varepsilon^p + \Delta \varepsilon^t$ and the distribution of stresses, through the constitutive equation,

$$\sigma_{ij} = C_{ijkl}(\varepsilon_{kl} - \varepsilon_{kl}^p - \varepsilon_{kl}^t) \,. \tag{3}$$

where the symbol C_{ijkl} denotes the fourth order material stiffness tensor.

Fig. 4a provides a detailed view of the model used in the numerical simulation of material deposition by WAAM. The baseplates were clamped to a table to minimize distortion, and the clamps were included in the finite element model using single node constraints of the baseplates' left and right bottom surface edges. The remaining surfaces of the model were assigned with free convection boundary conditions with convection coefficient values equal to 15 (W/m²K) and radiation to the environment. The model consisted of approximately 70,000 hexahedral elements active at the end of the deposition, with mechanical and physical properties assumed to be temperature dependent. A 'born-dead element' variant of the 'element birth' technique, where new mesh elements are gradually activated as the heat source moves to simulate the formation, detachment, and impingement of droplets from the filler material into the molten pool was utilized.

Macro-scale modeling approach of WAAM was validated in a previous work of the authors [11], which used the same material and revealed an excellent agreement between numerically predicted and experimentally measured temperatures, and between numerically predicted and experimentally measured distortions after cooling and unclamping the baseplates.

Sheet Metal 2025

Materials Research Proceedings 52 (2025) 242-249

Materials Research Forum LLC

https://doi.org/10.21741/9781644903551-30

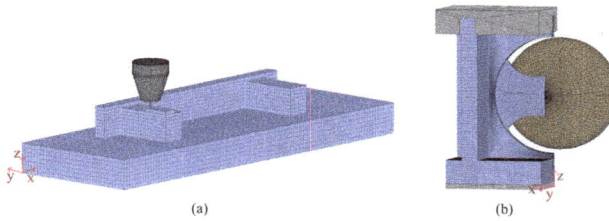

Fig. 4 Finite element model of the (a) material deposition by WAAM and (b) dieless Nakajima test performed with an arc-shape geometry after some hemispherical punch displacement.

The numerical simulation of the Nakajima test made use of the finite element flow formulation built upon the quasi-static equilibrium equations (2). The model treated the deposited walls as deformable objects, and the material was assumed as isotropic, following the Levy-Mises constitutive equations. The flow stress (4) was determined by averaging the values obtained for the transverse, longitudinal, and inclined directions in a previous study of the authors [8],

$$\sigma = 1200\varepsilon^{0.35} \text{ (MPa)} . \tag{4}$$

Fig 4b shows a typical model considering a symmetrical plane and discretizing the deposited walls with approximately 50,000 three-dimensional hexahedral elements. The discretization is more refined at the center of the region of interest, where the hemispherical punch first contacts the deposited material.

The hemispherical punch and the grips holding the upper end of the deposited walls were assumed to be rigid objects and discretized using spatial triangular elements. The Prandtl law of constant friction was utilized between the rigid objects and the deposited walls, with the friction factor at the contact with the punch and the lower bolster of the multidirectional tool being respectively equal to 0.1 and 1.0. The first value accounted for the Teflon sheet placed in the region of interest. In contrast, the second value allowed the model to accurately replicate the fixing of the odd-sized deposited walls to the lower bolster through bolts.

Results and Discussion

Temperature and Temperature Gradients in WAAM. Figs. 5a and 5b show the finite element predicted distribution of temperature and temperature gradient at an instant of material deposition corresponding to the tenth layer. The maximum temperature is capped at the melting temperature (approximately equal to 1400°C) to facilitate the identification of the molten pool, and the temperature distribution discloses contours that gradually increase at the front and rear of the molten pool, in agreement with the asymmetric Gaussian distribution of heat flux.

Reheating caused by deposition of the different layers combined with the heat flow into the baseplate, which remains cooler, gives rise to a temperature gradient that is maximum where material is being deposited and aligned with the building height. This alignment closely mirrors the orientation of the primary arms of dendritic growth along the building direction (Fig. 5c), and it is crucial for understanding the formation of columnar grain-based microstructures and the highly anisotropic behavior of the material deposited by WAAM.

Sheet Metal 2025   Materials Research Forum LLC
Materials Research Proceedings 52 (2025) 242-249   https://doi.org/10.21741/9781644903551-30

(a)   (b)

(c)

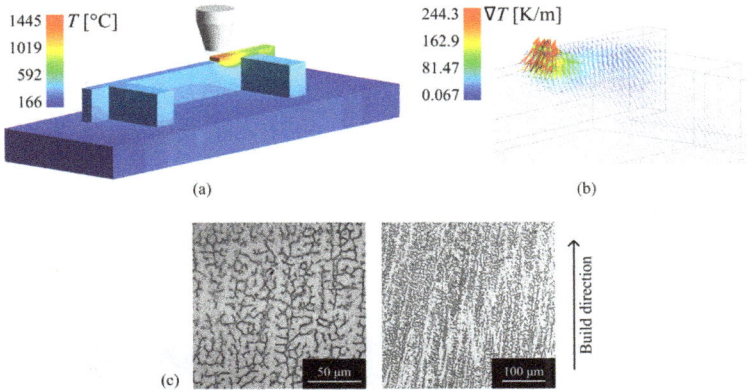

Fig. 5 Finite element predicted distribution of (a) temperature and (b) temperature gradient during deposition of the tenth layer of the main wall. (c) Microstructure observation of a sample extracted from the main wall.

Structural Integrity of the Dieless Nakajima Setup. The finite element predicted effective strain distributions after 20 mm punch displacement for the circular and arc-shaped test geometries shown in Fig. 6 demonstrate that the deposited walls outside the area of interest possess sufficient strength to maintain their original structural integrity. Their function is akin to the die and blank holder of the blank sheet, which is mimicked by the thinner machined region of interest undergoing plastic deformation due to contact with the hemispherical punch.

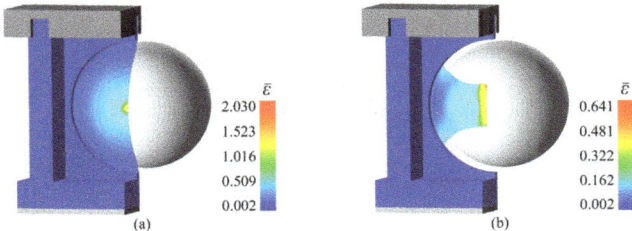

(a)   (b)

Fig. 6 Finite element predicted distribution of effective strain after 20 mm punch displacement for the (a) circular and (b) arc-shaped test geometries.

Strain Loading Paths and Failure. Fig. 7a shows the finite element predicted accumulation of ductile damage according to the McClintock ductile fracture criterion [12] for the circular and arc-shaped test geometries in case of assuming the material as isotropic. Higher damage values accumulate at the center of the region of interest, in contact with the hemispherical punch pole, validating the new procedure as an alternative to conventional Nakajima tests using toolsets with a punch, a die, and a blank holder.

The utilization of the previously mentioned software [9] to obtain the strain loading paths $\varepsilon_1 = f(\varepsilon_2)$ in principal strain space from DIC measurements gave the results shown in Fig. 7b In both test geometries, the open markers represent the final DIC measurements on the area where the crack was triggered, whereas the solid markers denote the in-plane strains at the fracture. The latter were obtained from thickness measurements along the crack to obtain the thickness strains given

Sheet Metal 2025   Materials Research Forum LLC
Materials Research Proceedings 52 (2025) 242-249   https://doi.org/10.21741/9781644903551-30

by $\varepsilon_3 = ln(t_f/t_o)$, where t_o is the initial thickness and t_f is the thickness at fracture, assuming no changes in the minor strain $\Delta\varepsilon_2 = 0$ after the last DIC measurement.

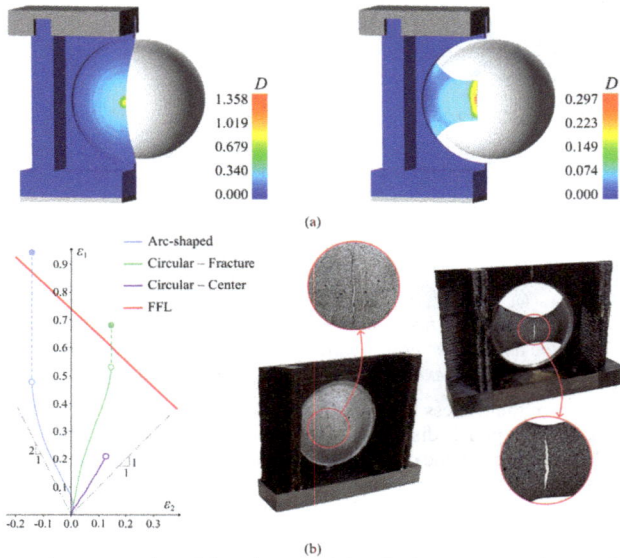

Fig. 7 (a) Finite element predicted distribution of ductile damage according to the McClintock fracture criterion after 20 mm punch displacement for the circular and arc-shaped test geometries. (b) Strain loading path evolutions with pictures of both test samples after failure.

In the case of the circular test geometry, two locations were chosen, corresponding to the center and the zone where the crack was triggered. The strain loading paths corresponding to these two locations are different and confirm the position of the crack slightly shifted away from the center (Fig. 7c) and in close agreement with the fracture forming limit (FFL) of this material that the authors previously determined [8] (refer to the red line in Fig. 7c).

The crack shift in the circular test geometry is attributed to the combined effect of friction and anisotropy. Even with a Teflon sheet between the hemispherical punch and the region of interest, friction carries part of the in-plane loads and allows cracks to initiate at the edge of the tool contact, away from the center. Anisotropy, caused by the dendritic-based columnar grain microstructure of the deposited walls, contributes to the cracks' alignment with the building direction.

Conclusions

The dieless Nakajima test enables direct characterization of the formability limits of deposited materials without extracting sheet blanks. The results obtained with circular and arc-shaped geometries confirm the validity of the proposed design, which utilizes the surrounding material of the region of interest to replicate the die and blank holder, ensuring strain loading paths similar to those obtained in conventional Nakajima sheet formability tests.

Macro-scale thermo-mechanical modeling of WAAM, performed independently from the formability test modelling, predicts vertical temperature gradients that closely mirror the orientation of dendritic growth along the building direction, which leads to a general anisotropic behavior of the deposited materials. This behavior, together with good lubrication leads to a shift of cracks away from the center of the region of interest in the case of circular geometry.

Acknowledgements

The authors would like to acknowledge the support provided by Fundação para a Ciência e a Tecnologia (FCT) and IDMEC for its financial support via the project LAETA Base Funding (DOI: 10.54499/UIDB/50022/2020).

References

[1] M. Bambach, A, Sviridov, A. Weisheit, J.H. Schleifenbaum, Case studies on local reinforcement of sheet metal components by laser additive manufacturing, Metals 7 (2017) 113. https://doi.org/10.3390/met7040113

[2] M. Merklein, D. Junker, A. Schaub, F. Neubauer, Hybrid additive manufacturing technologies - An analysis regarding potentials and applications, Phys. Procedia 83 (2016) 549-559. https://doi.org/10.1016/j.phpro.2016.08.057

[3] J.P.M. Pragana, R.F.V. Sampaio, I.M.F. Bragança, C.M.A. Silva, P.A.F. Martins, Hybrid metal additive manufacturing: A state-of-the-art review, Adv. Ind. Manuf. Eng. 2 (2021) 100032. https://doi.org/10.1016/j.aime.2021.100032

[4] C. López, A. Elías-Zúñiga, I. Jiménez, O. Martínez-Romero, H.R. Siller, J.M. Diabb, Experimental determination of residual stresses generated by single point incremental forming of AlSi10Mg sheets produced using SLM additive manufacturing process, Materials 11 (2018) 2542. https://doi.org/10.3390/ma11122542

[5] A Shah, R. Aliyev, H. Zeidler, S. Krinke, A review of the recent developments and challenges in wire arc additive manufacturing (WAAM) process, J. Manuf. Mater. Process. 7 (2023) 97. https://doi.org/10.3390/jmmp7030097

[6] R.F.V. Sampaio, P.M.S. Rosado, J.P.M. Pragana, I.M.F. Bragança, C.M.A. Silva, L.G. Rosa, P.A.F. Martins, Formability assessment of additively manufactured materials via dieless Nakajima testing, J. Manuf. Mater. Process. 8 (2024) 180. https://doi.org/10.3390/jmmp8040180

[7] K. Nakajima, T. Kikuma, K. Hasuka, Study on the Formability of Steel Sheets, Yamata Technical Report, Japan, 1968.

[8] J.P.M. Pragana, I.M.F. Bragança, L. Reis, C.M.A. Silva, P.A.F. Martins, Formability of wire-arc deposited AISI 316L sheets for hybrid additive manufacturing applications, Proc. Inst. Mech. Eng. Part L J. Mater. Des. Appl. 235 (2021) 2839-2850. https://doi.org/10.1177/14644207211037033

[9] R.F.V. Sampaio, N.S.M. Alexandre, J.P.M. Pragana, I.M.F. Bragança, C.M.A. Silva, P.A.F. Martins, A software for research and education in ductile damage, Adv. Ind. Manuf. Eng. 7 (2023) 100127. https://doi.org/10.1016/j.aime.2023.100127

[10] J. Goldak, A. Chakravarti, M.A. Bibby, A new finite element model for welding heat sources, Metall. Trans. B 15 (1984) 299-305. https://doi.org/10.1007/BF02667333

[11] J.P.M. Pragana, R.F.V. Sampaio, I.M.F. Bragança, C.M.A. Silva, C.V. Nielsen, P.A.F. Martins, Macro-scale finite element simulation of wire-arc additive manufacturing, Proc. Inst. Mech. Eng. Part L J. Mater. Des. Appl. (2024) (in press). https://doi.org/10.1177/14644207241272840

[12] F.A. McClintock, A criterion for ductile fracture by the growth of holes, J. Appl. Mech. 35 (1968) 363-371. https://doi.org/10.1115/1.3601204

Polymers and composites

SheMet 2025

Sheet Metal 2025   Materials Research Forum LLC
Materials Research Proceedings 52 (2025) 251-259   https://doi.org/10.21741/9781644903551-31

Joining process for fiber-reinforced thermoplastics and sheetmetal without additional adhesion promoter

Bernd-Arno Behrens[1], Annika Raatz[2], Sven Hübner[1],
Christoph Schumann[2], Jörn Wehmeyer[1, a*]

[1]Leibniz University Hannover, Institute of Forming Technology and Machines, An der Universität 2, 30823, Garbsen, Germany

[2]Leibniz University Hannover, Institute of Assembly Technology, An der Universität 2, 30823, Garbsen, Germany

[a]wehmeyer@ifum.uni-hannover.de

Keywords: Tool, Manufacturing, Polymer

Abstract. Thanks to their excellent mechanical properties and low structural weight, multi-material structures offer a promising solution for lightweight design, body construction, and functionalisation in the automotive industry. A common approach is combining metal and plastic to enhance the performance of the final component compared to single-material structures. This paper presents the development of a manufacturing cell for joint forming and heat-assisted press joining of steel and continuous fiber-reinforced thermoplastics, specifically using unidirectional carbon-fiber tapes. To achieve shorter cycle times and ensure cost-effective production, the manufacturing cell was equipped with two robots for automated handling and utilised an isothermal two-stage forming tool concept. The produced composite components were evaluated regarding their mechanical performance, confirming the feasibility of the process. All composite parts demonstrated higher specific load capacity compared to pure steel components. Cycle times of less than 60 s were consistently achieved, marking a significant reduction in process time compared to variothermal tool concepts.

Introduction

The efficient use of resources is becoming increasingly crucial, both in response to environmental changes and from a competitive standpoint. Several strategies can enhance resource efficiency, such as designing products more efficiently, reducing raw material consumption in manufacturing, or enabling the separation and recycling of materials.

By combining metals with fiber-reinforced plastics (frp) to create hybrid components, the benefits of both material groups can be combined, allowing for synergies and optimal utilisation of lightweight construction potential. However, one major challenge in implementing such material concepts is often the lack of cost-effectiveness in the associated manufacturing processes. Producing fiber-reinforced plastic-metal hybrid components require tailored handling and forming processes that accommodate the differing material behaviours, along with the creation of a large-area bond between the two components. Research and industry focus on two main approaches for producing these joints:

One approach is to bond the two materials using a suitable adhesive. While this method can significantly enhance the bond strength, it is also costly and presents challenges in manufacturing. Adhesive bonding adds extra process steps, requires additional safety precautions, and extends processing times due to adhesive curing.

A more efficient method for creating a large-area bond between metals and fiber-reinforced plastics is press joining using variothermic tools, which alternate between heating and cooling. In this process, the materials to be joined are heated within the tool, causing the thermoplastic to melt

Content from this work may be used under the terms of the Creative Commons Attribution 3.0 license. Any further distribution of this work must maintain attribution to the author(s) and the title of the work, journal citation and DOI. Published under license by Materials Research Forum LLC.

Sheet Metal 2025   Materials Research Forum LLC
Materials Research Proceedings 52 (2025) 251-259   https://doi.org/10.21741/9781644903551-31

and bond with the metal component under pressure. Afterwards, the tool is cooled, allowing the thermoplastic to solidify. However, the duration of cooling phase and the thermal cycling involved make this approach impractical for industrial mass production. The extended process times and the high thermal stresses on the tools make it neither economically viable nor energy-efficient for large-scale use [1].

Most of the process variants developed so far have focused primarily on proving basic feasibility or meeting specific requirements, with initial results confirming success in these areas. The next crucial step is to prioritise cost reduction, specifically by reducing process complexity and cycle times. The challenge lies in developing processes that ensure full functionality while also offering economic advantages for mass production.

To address this, a process chain based on a two-stage, isothermal tool concept with automated handling technology was developed and implemented in this research for producing load-optimized components. The tool consists of two units: a heating and impregnation unit, and a forming and consolidation unit, both maintained at different constant temperatures. Robots are employed to stack the semi-finished products, load them into the first tool stage, transfer them between stages, and remove the finished components. Once the system was commissioned, a suitable process window for fiber impregnation and component consolidation was identified. The effects of varying process parameters, such as forming force, holding time, and tool temperature, on the stability and quality of the components were analysed. A comparison between pure steel components and load-optimized hybrid components will demonstrate the lightweight potential of the final products.

State of the Art
Fiber-reinforced thermoplastics
Fiber-reinforced thermoplastics (FRP) offer a promising solution to address the challenges where light weight and high specific modulus and strength are critical issues [2]. To meet future lightweight construction goals, the use of FRP has grown significantly in recent years, becoming an essential factor in modern design in generell. Semi-finished FRP products can be categorized into several types, including long fiber-reinforced thermoplastics (LFT), glass mat reinforced thermoplastics (GMT), unidirectional fiber-reinforced tapes (UD tapes), and organic sheets [3].

Thermoplastics can be moulded through stamp forming processes when subjected to high temperatures, enabling large-scale production depending on the type of matrix material used [4, 5]. In addition to consolidating the matrix materials, other joining methods, such as clinching, can also be utilized at elevated temperatures. [6].

Although fiber-reinforced plastics (FRPs) are regarded as a distinct class of materials, they are essentially composites made up of multiple materials. In this composite structure, the components have well-defined roles: the high-strength fibers handle mechanical loads and compressive forces, while the matrix, in which the fibers are embedded, secures their position, provides support, and transfers mechanical stresses to the fibers [7].

Heat-assisted press joining of hybrid FRP steel components
Heat-assisted press joining allows for the connection of thermoplastic FRPs and metals, offering the advantage of reduced process times and streamlined production chains. During this process, the thermoplastic matrix material of the fiber composite is melted, facilitating a fusion bond with the metallic joining partner [8]. By utilizing heat-assisted press joining, the bonding step is eliminated, allowing for the achievement of structural strengths through effective process control and surface pre-treatment [9]. Within BRECHER, a plant technology for heat-assisted press joining of metal-plastic hybrid joints was developed. Here, the thermoplastic is melted by heat conduction due to the contact between the metallic workpiece and the tool in the joining zone [10]. The contact temperature at which the molten plastic wets the metal surface is critical for bond formation. For

Sheet Metal 2025
Materials Research Proceedings 52 (2025) 251-259

Materials Research Forum LLC
https://doi.org/10.21741/9781644903551-31

instance, when combining polyamide 66 and steel, the optimal joining temperature is found to be between 260 °C and 270 °C [11]. In contrast, a joining temperature of 250 °C is regarded as promising for achieving a bond between polyamide 6 and steel [12]. SCHEIK et al. dealt with inductive heating of the tool [13] and conductive heating of the metallic joining component with the aim of reducing process time [14]. The heat-assisted press joining process consists of several steps. Initially, both materials need to be heated to the specified joining temperature, which is monitored using thermocouples. For instance, the thermoplastic can be heated using an infrared field [15]. The heating occurs at the surface of the joining zone, eliminating the need for consolidation of the organic sheet. A pyrometer is used to monitor the temperature of the organic sheet. Once both materials have attained the required joining temperatures, the metal sheet is pressed onto the organic sheet with a specified force. During this process, the molten thermoplastic flows into the undercut features of the metal surface, creating a form-fit joint. Additionally, material bonds are established between the thermoplastic and the metal surface [16]. Following this, the cooling process commences, often facilitated by compressed air. Once the thermoplastic in the organic sheet has solidified, the test specimen can be extracted. The essential process parameters for heat-assisted press joining include the joining temperatures of both materials and the joining force, which have been extensively studied [15, 17].

So far, a wide variety of pretreatment processes have been considered, such as mechanical pretreatment by sandblasting [10, 18], imprinting of structures [19, 20], corundum blasting [21, 22] or different chemical pretreatments [23], which create an activated and partly new disordered surface structure.

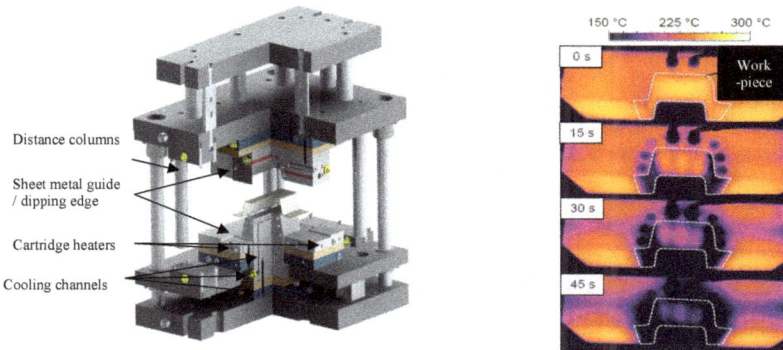

Fig. 1. *Sectional view of the CAD model of the variothermal forming tool (left), temperature profile of the punch during cooling (right) [24].*

DRÖDER et al. manufactured a variothermal forming tool for the production of hat sections consisting of carbon fiber reinforced thermoplastic tapes and metal sheets with different frame angles [25]. After the tool was put into operation, hybrid hat profiles were produced and component properties were determined. The forming tool has a maximum heating power of 12.4 kW and is heated to 250 °C. The tool can be cooled down to a removal temperature of 150 °C within 45 s due to cooling channels, filled with permanently circulating water, running close to the contour.

The CAD model of the forming tool is displayed in a sectional view in Figure 1. However, the tool requires several minutes for reheating, which was the primary motivation behind the work presented in this paper. The objective is to develop a process chain that enables economically

Sheet Metal 2025   Materials Research Forum LLC
Materials Research Proceedings 52 (2025) 251-259   https://doi.org/10.21741/9781644903551-31

viable production times for FRP-metal hybrids. Additionally, another aim is to conserve energy by eliminating the need to reheat the die after a part has been produced.

Experimental setup
Work piece
The component developed here is a hybrid sandwich component with specially reinforced areas. It consists of metallic cover sheets (TH340 (MagiZinc ZM70)) on which PA6 (Polyamid 6) films are located as adhesion promoters. The PA6 can connect to the steel due a surface roughness R_a of 0.57 µm, which was experimentally measured. No other adhesives are needed to create a permanent bond. The core of the component consists of a full-surface layer of UD-CFRP tape (unidirectional carbonfiber reinforced plastic tape). The areas of the reinforcing structures are supported with further narrow strips of UD-CFRP tape. Figure 2 shows the schematic layer structure of the semi-finished product and a manufactured component.

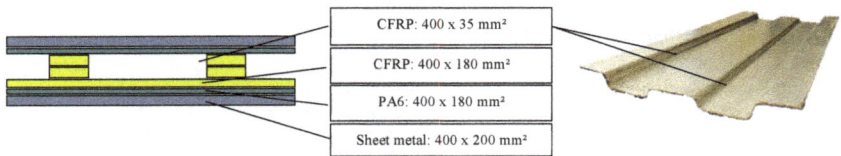

CFRP: 400 x 35 mm²

CFRP: 400 x 180 mm²

PA6: 400 x 180 mm²

Sheet metal: 400 x 200 mm²

Fig. 2. *Layer structure of the semi-finished products (left), manufactured component (right) [26].*

Two-stage isothermal heating / impregnating and consolidation / forming tool
To manufacture sandwich components that can withstand bending loads while saving energy costs and significantly reducing production time, a two-stage isothermal heating/impregnation tool and a consolidation/forming tool were developed, ensuring compatibility with automated handling technology. Unlike the variothermal tools described in the literature, this concept features both stages heated continuously at different temperatures using heating coils. This design eliminates the energy-intensive heating process required for each new component, and cooling channels are also rendered unnecessary.

In the first tooling stage, the stacked semi-finished products are heated, and the loose carbon fibers are impregnated into the PA6 plastic under a force of 325 kN, resulting in a thermal bond. The temperature for this stage is set at 270 °C, which is 10 °C below the decomposition temperature of PA6. This temperature is chosen to ensure that the PA6 melts, creating an optimal heat cushion for transferring to the second tooling stage, while preventing the plastic fromsolidifying before the actual forming process. Contours in the die of the mold help prevent any slipping of the additional layers of CFRP-UD tape used for reinforcement.

In the second tooling stage, forming occurs at a temperature of 120 °C, which accommodates the deeper contours of the die. This temperature is selected to ensure that the plastic remains in a molten state until the forming is completed. Cooling takes place during an average holding time of 25 s in the mold. After this period, the consolidation of the composite is finalized, and the component, which is still warm but completely solidified, can be removed.

Manufacturing cell with automated handling technology
To further decrease production time and enhance the repeatability of tests, an automated stacking system for the semi-finished products and an automated transfer mechanism into the tools were developed, utilising two robots. A Kuka KR6 robot was used for stacking the semi-finished products. The challenge was ensuring that the robot's gripper could reliably handle and place both rigid metal sheets and CFRP UD tapes of various sizes, as well as flexible PA6 film in its

Sheet Metal 2025
Materials Research Proceedings 52 (2025) 251-259

Materials Research Forum LLC
https://doi.org/10.21741/9781644903551-31

unprocessed state. Once the stacking process is complete, the fixed stacks of semi-finished products are deposited at a designated location for further handling. A Kuka KR60 robot was responsible for transferring the stacked semi-finished products between the heating stage, forming stage, and the final deposit location. The stacking and transfer sequences were synchronised to optimise production times. With a heating or forming time of 20 s (+12 s for opening and closing) and a transfer time of 38 s, the total production time per component amounts to 70 s. Since both stages of the tool are loaded in each cycle, this value can be effectively halved, leading to a theoretical cycle time of 35 s. Figure 3 illustrates the complete manufacturing cell.

Fig. 3. *Production cell consisting of Kuka KR60, Kuka KR6, two-stage isothermal tool and Dunkes 250 hydraulic press [26].*

Conducting experiments

The testing of the double-hat profiles, manufactured using the test setup described in chapter 3, was carried out by means of three-point bending tests following DIN 14130. The test components were inserted in such a way that the areas with the UD tape reinforcements were maximally far off the compression fin of the test unit and were thus loaded in tension. The specific load capacity of the composites can be calculated from the weight of the composites and the maximum average forces per test series from the three-point bending tests. Depending on the combination, different weights are obtained. The specific load value can be determined from the maximum force values using the equ. 1:

$$Max.\,spec.\,load\ \left[\frac{N}{g}\right] = \frac{Force}{Weight\ of\ the\ unit} \qquad [1]$$

Results

The results of the 3-point Bending tests are displayed in Figure 4. To determine the optimal manufacturing parameters, the temperature of the forming tool (100 °C and 140 °C) as well as the forming force (250 kN – 400 kN) were varied out through the test series.

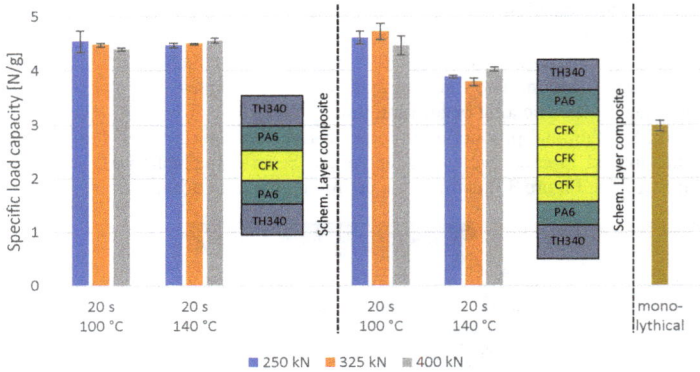

Fig. 4. *Comparison of the specific load capacity with different manufacturing parameters and layer structure: TH340 without UD tape reinforcement (left), TH340 with UD tape reinforcement (right) [27].*

The comparison shows components with single and multi-layer UD-Tape reinforcements, as well as a monolythical component as a reference to check the effects of the UD-Tape reinforcements. It is obvious that all hybrid components can withstand higher specific loads than the monolithic component. The largest difference here was 58.39 % [3-layer structure (manufactured with a tool temperature of 100 °C and a forming force of 325 kN) compared to monolithic component]. Furthermore, it can be seen that lower temperatures in the forming tool are advantageous for the production of components with several layers of UD-Tape reinforcements, as the performance of those specimens were higher. The reason for this is that the larger accumulation of PA6 material does not solidify quickly enough at higher temperatures, which leads to swelling of the matrix material and slippage of the fibre layers. However, for specimens with only one UD tape layer, the die temperature was not as important.

Summary and Conclusion

This paper presents a novel two-stage isothermal tool concept integrated with automated handling technology, utilising two robots. This tooling system is designed to produce stacks of semi-finished products made from steel, PA6 film, and unidirectionally reinforced carbon fiber tapes. In the final process, both tool stages are continuously occupied, with the first robot consistently stacking new semi-finished products. With a heating or forming time of 20 s (+12 s for opening and closing) and a transfer time of 38 s, the production time per component totals 70 s. Since both stages of the tool are always loaded during each cycle, this time can be effectively halved, resulting in a theoretical cycle time of 35 s. Compared to a variothermic tool, this process time represents an 8.5-fold reduction for this component, highlighting the significant time-saving potential of the two-stage isothermal tool concept.

Furthermore, a thorough evaluation of all manufactured components was conducted by means of 3-point bending tests. Various test series were analyzed, focusing on different manufacturing parameters and layer structures. All test series demonstrated an increase in specific load capacities of up to 58.4% when compared to a single 1 mm thick sheet.

As part of this project, a process was developed that allows for the production of components from loosely stacked semi-finished products composed of sheet metal and UD tapes, utilizing a two-part isothermal heating and impregnation stage followed by a consolidation and forming stage. The individual assembly of components from distinct layers, combined with the relatively short cycle time, positions this production technology as a viable solution for economically manufacturing load-bearing components in series production.

Outlook

In current research project a manufacturing method is being developed with which a metal insert is integrated into the FRP, which is here glass matt reinforced thermoplastic (GMT) with a tangled arrangement of the fibres, by means of extrusion moulding. This doubles the available joining surface compared to bonding the metal reinforcement. The project is based on the hypothesis that the integrated metal insert reinforces the component itself and improves the connection to surrounding structures and thus the maximum force transmission. A possible demonstrator component is shown in Figure 5. Initial tensile shear tests showed an adhesive and cohesive fracture behaviour, which indicates a good bond between metal (22MnB5) and the GMT.

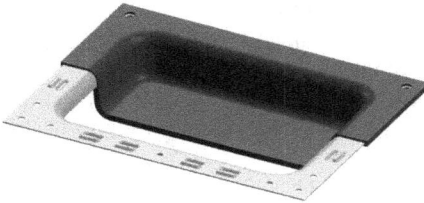

Fig. 5. *CAD model of an FRP component with integrated metallic reinforcement structure with functional elements for improved bonding.*

Acknowledgements

The authors thank the Industrial Collaborative Research (IGF) and the European Research Association for Sheet Metal Working (EFB) for the financial support of the research project "Economical production of load-compliant FRP/metal composites" (IGF-Code: 19560) and "Reinforcement of FRP components using metal inserts" (IGF-Code: 01IF22941N (22941)).

References

[1] A. Neumann, et. al. Ermittlung der Prozessgrenzen bei der Umformung von Musterbauteilen aus Faser-Kunststoff/MetallVerbunden; Schlussbericht zu dem IGF-Vorhaben (2016)

[2] K. Friedrich, A.A. Almajid, Manufacturing of Advanced Polymer Composites for Automotive Applications. In: Applied Composite Materials (2013) https://doi.org/10.1007/s10443-012-9258-7

[3] S. Gebai, A. Hallal, Composite Materials Types and Applications, in: Mechanical Properties of Natural Fiber Reinforced Polymers. Emerging Research and Opportunities (2018) https://doi.org/10.4018/978-1-5225-4837-9

[4] B.-A. Behrens, et. al., Development of a combined Process of Organic Sheet forming and GMT Compression Molding. In: Procedia Engineering 207, Cambridge (2017), pp. 101-106 https://doi.org/10.1016/j.proeng.2017.10.745

[5] B.-A. Behrens, S. Hübner, A. Neumann, Forming sheets of metal and fibre-reinforced plastics to hybrid parts in one deep drawing process. In: Procedia Engineering (2014) 81, pp. 1608-1613 https://doi.org/10.1016/j.proeng.2014.10.198

[6] B.-A. Behrens, et al., Forming and Joining of Carbon-Fiber-Reinforced Thermoplastics and Sheet Metal in One Step. In: Procedia Engineering 183 (2017), pp. 227-232 https://doi.org/10.1016/j.proeng.2017.04.026

[7] H. Schürmann, Konstruieren mit Faser-Kunststoff-Verbunden. Springer-Verlag, (2007) https://doi.org/10.1007/978-3-540-72190-1

[8] C. Ageorges,L. Ye, M. Hou, Advances in fusion bonding techniques for joining thermoplastic matrix composites: a review, Composites: Part A, Bd. 32 (2001), pp. 839-857 https://doi.org/10.1016/S1359-835X(00)00166-4

[9] A. Heckert, M.F. Zaeh, Induction Heated Joining of Aluminum and Carbon Fiber Reinforced Nylon 66, Journal of Laser Applications, Bd. 27 (2015), Nr. S2 https://doi.org/10.2351/1.4906380

[10] C. Brecher, Integrative Produktionstechnik für Hochlohnländer, Springer-Verlag (2011) https://doi.org/10.1007/978-3-642-20693-1

[11] D. Flock, Wärmeleitungsfügen hybrider Kunststoff-Metall-Verbindungen

[12] F. Reig, R. Steinhilper, Handbuch Konstruktion, 2. Auflage, München. Hanserverlag (2018) https://doi.org/10.3139/9783446456198.fm

[13] S. Scheik, M. Schleser, U. Reisgen, Thermisches Direktfügen von Metall und Kunststoff - Eine Alternative zur Klebtechnik?, Adhäsion (2012) https://doi.org/10.1007/978-3-658-04025-3_16

[14] J. Schoene, B. Marx, et. al, Fügen von Metall-Kunststoff-Verbunden, ISF Direkt 49 (2014)

[15] K. Lippky,S. Hartwig, D. Blass, K. Dilger, Bonding performance after aging of fusion bonded hybrid joints. International Journal of Adhesion and Adhesives (2019) https://doi.org/10.1016/j.ijadhadh.2019.01.025

[16] M. Üzüm, Metal polymer hybrids - Multiscale adhesion behavior and polymer dynamics. Technische Universität Berlin, Dissertation (2015)

[17] K. Lippky, M. Mund, D. Blass, K. Dilger, Investigation of hybrid fusion bonds under varying manufacturing and operating procedures. Composite Structures (2018) https://doi.org/10.1016/j.compstruct.2018.01.078

[18] J. P. Bergmann, M. Stambke, Potential of laser-manufactured polymer-metal hybrid joints, Physics Procedia 39 (2012), pp. 84-91 https://doi.org/10.1016/j.phpro.2012.10.017

[19] C. Ageorges, L. Ye, M. Hou, Advances in fusion bonding techniques for joining thermoplastic matrix composites: a review, Composites: Part A, Bd. 32 (2001), pp. 839-857 https://doi.org/10.1016/S1359-835X(00)00166-4

[20] R. Matsuzaki, N. Tsukamoto, J. Taniguchi, Mechanical Interlocking by im-printing micropatterns for improving adhesive strength of polypropylene. International Journal of Adhesion & Adhesives 68 (2016), pp. 124-132 https://doi.org/10.1016/j.ijadhadh.2016.03.002

[21] K. Mittal, A. Pizzi, Adhesion Promotion Technique. Technological Applications. Basel: Marcel Dekker. ISBN: 0-8247-0239-5 (1999), pp. 2, 4-5, 19-20

[22] R. Velthuis, LAUDATIO - Fügetechnik - Verfahrensvergleich: Leichtbau aus Metall und Faser-Kunststoff-Verbunden, in: Kunststoffe: Werkstoffe, Verarbeitung, Anwendung (2007)

[23] M. Didi, P. Mitschang, Diskontinuierliches Induktionsschweißen von CF/PEEK und CF/PA66 mit Aluminium, CCeV (2011)

[24] S. Bräunling, E. Staiger, Textil-Blech-Verbund Hybride auf Basis von Kohlenstofffasern und Thermoplast durch umformende Verbundherstellung, laufendes EFB Projekt, 3. Zwischenbericht AK-Sitzung "Hybride Strukturen" (2015)

Materials Research Forum LLC
https://doi.org/10.21741/9781644903551-31

[25] K. Dröder, M. Brand, A. Gerdes, T. Grosse, H. Grefe, K. Lippky, F. Fischer, K. Dilger, An Innovative Approach for Joining of Hybrid CFRP-Metal Parts by Mechanical Undercuts: Proceedings, Euro Hybrid Materials and Structures (2014)

[26] J. Wehmeyer, et. al: Process chain for forming and consolidating fiber-reinforced thermoplastics and metallic sheets in a two-stage isothermal tool, 42nd Conference of the International Deep Drawing Research Group (IDDRG 2023)

[27] B.-A. Behrens, A. Raatz, J. Wehmeyer, F. Bohne, R. Lorenz, C. Schumann, Wirtschaftliche Fertigung belastungsgerechter FVK/Metall-Verbunde: IGF-Vorhaben 19603 N (2021)

Sheet Metal 2025
Materials Research Proceedings 52 (2025) 260-267

Materials Research Forum LLC
https://doi.org/10.21741/9781644903551-32

Efficient failure information propagation under complex stress states in fiber reinforced polymers: From micro- to meso-scale using machine learning

Johannes Gerritzen[1,a*], Andreas Hornig[1,2,3,b] and Maik Gude[1,c]

[1]Institute of Lightweight Engineering and Polymer Technology, TUD Dresden University of Technology, Holbeinstr. 3, 01307 Dresden, Germany

[2]Center for Scalable Data Analytics and Artificial Intelligence (ScaDS.AI) Dresden/Leipzig, TUD Dresden University of Technology, Chemnitzer Straße 46b, 01187, Dresden, Germany

[3]Department of Engineering Science, Solid Mechanics and Materials Engineering, University of Oxford, OX1 3PJ, Oxford, United Kingdom

[a]johannes.gerritzen@tu-dresden.de, [b]andreas.hornig@tu-dresden.de, [c]maik.gude@tu-dresden.de

Keywords: Fiber Reinforced Plastic, Machine Learning, Failure

Abstract The failure behavior of fiber reinforced polymers (FRP) is strongly influenced by their microstructure, i.e. fiber arrangement or local fiber volume content. However, this information cannot be directly used for structural analyses, since it requires a discretization on micrometer level. Therefore, current failure theories do not directly account for such effects, but describe the behavior averaged over an entire specimen. This foundation in experimentally accessible loading conditions leads to purely theory based extension to more complex stress states without direct validation possibilities. This work aims at leveraging micro-scale simulations to obtain failure information under arbitrary loading conditions. The results are propagated to the meso-scale, enabling efficient structural analyses, by means of machine learning (ML). It is shown that the ML model is capable of correctly assessing previously unseen stress states and therefore poses an efficient tool of exploiting information from the micro-scale in larger simulations.

Introduction

Fiber reinforced polymers (FRP) play a crucial role in lightweight applications due to their excellent specific properties [1]. This allows significant weight reduction and thus improvement of energy efficiency in the mobility sector and of frequently accelerated parts in general [2]. However, one major obstacle for the wider application of FRP is the challenge of joining FRP with dissimilar materials [3]. So far, adhesive bonding has been the widest spread solution [4]. This however is undesirable when considering end of life and recycling, because dejoining is impossible or requires a significant amount of effort. Hence, material preserving recycling has yet to be widely adopted for FRP [5].

One possible solution to improve the dejoinability, and thus the recyclability, of FRP is the implementation of mechanical joining technologies. However, these lead to significant changes in the local material structure and therefore the joints load bearing behavior [6]. Established failure criteria for FRP originate in thin walled materials with almost plane stress conditions [7]. Even though many theories have been extended to consider all stress components, even biaxial stress states still pose a significant challenge [8]. One significant contributing factor are the microscopic inhomogeneities of FRP, which strongly influence the failure behavior [9].

With rising computing power, the study of failure behavior by finite element analyses (FEA) on micro-scale has received more attention in research. The effect of fiber distribution as well as yarn alignment and waviness on elastic properties of entire plies has been studied with a special

Content from this work may be used under the terms of the Creative Commons Attribution 3.0 license. Any further distribution of this work must maintain attribution to the author(s) and the title of the work, journal citation and DOI. Published under license by Materials Research Forum LLC.

Sheet Metal 2025
Materials Research Proceedings 52 (2025) 260-267

Materials Research Forum LLC
https://doi.org/10.21741/9781644903551-32

focus on draping effects in [10]. Pulungan et al. investigated the effect of local microstructure and RVE size on the failure behavior under transverse tension [11].

Such micro-scale models allow deep insight into the local behavior. However, their computational cost remains prohibitive for real world structures [12]. For elastic properties, homogenization approaches are well established to efficiently take information from the micro-scale into account on the meso-scale. This is enabled by the well accepted models on meso-scale. Given the lack thereof for FRP specific failure criteria under arbitrary stress states, the homogenization approach cannot directly be applied to failure behavior.

Fueled by the growing application of machine learning (ML) techniques, some approaches have been presented, obtaining data from micro-scale simulations and training various ML models on these data. Chen et al. used an RVE subjected to three stress components to predict critical loading conditions and trained these into an NN with all used data points being close to the failure envelope [13]. Wan et al. added the aspect of failure probability by analyzing multiple RVEs and taking all their results into account for the subsequent training [14]. None of these approaches take the simultaneous superposition of all stress components into account. Since this is crucial to accurately describe FRP behavior in the zone of a mechanical joint, the aim of this work is to establish a dataset from simulations on micro-scale representative volume elements (RVEs) under arbitrary loading conditions and subsequently train an ML model to assess criticality of a full stress state. This ML model is intended as alternative to established failure criteria, which can be incorporated into meso-scale FEA in future works.

Modeling approach
Material structure. Data regarding constituent behavior and uniaxial strengths are taken from the glass fiber-epoxy with Silenka E-Glass fibers and MY750/HY917/DY063 epoxy resin [15]. Here, data for a thermoset are used because of the amount of reliable data on its failure behavior and its status as the de facto standard for modeling FRP failure behavior. The different matrix system does not affect the presented development of a methodology for data driven failure determination and its propagation across length scales.

Based on the published data, constitutive models for the constituents are chosen. The glass fibers are modeled using an isotropic linear elastic model, with Young's modulus and Poisson ratio taken directly from [15]. Failure is modeled as element deletion by a minimum/maximum principal strain criterion with corresponding strain values taken as uniaxial ones. For the epoxy resin, inelasticities have to be considered. These are modeled as purely plastic up to triaxiality dependent damage initiation, using LS-DYNA keywords *MAT_PIECEWISE_LINEAR_PLASTICITY and *MAT_ADD_DAMAGE_GISSMO as commonly described in literature, i.e. [16]. Under uniaxial tension and compression, this leads to excellent agreement with the published strength and failure strain: Under tension a maximum stress of 79 MPa occurs in the simulation and 80 MPa are given in the paper, with a corresponding failure strain of 4.9 % and 5 % respectively. Under compression, the constitutive model leads to a strength of 117 MPa, the paper states 120 MPa.

The constituents are assembled in a representative volume element (RVE) as shown in Fig. 1. Given the structured placement of individual fibers, identical edge lengths in x and y direction are highly important to ensure transversal isotropy in the RVE's failure behavior. For this study, no fiber-matrix debonding is considered.

Sheet Metal 2025

Materials Research Proceedings 52 (2025) 260-267

Materials Research Forum LLC

https://doi.org/10.21741/9781644903551-32

Figure 1 Geometrical arrangement of fibers in the RVE

Data generation. As database for subsequent ML model training a total of 5000 loadcases triggering all six stress components in the RVE are analyzed. To define these, 5000 unit vectors $\vec{v}$ in six dimensional stress space are taken such that the minimum distance of their endpoints is maximized, using the implementation of the maximin algorithm [17] in the open source Python package diversipy [18]. This ensures a good sampling of the full dimensional design space [19] and therefore yields ample information on the failure surface. An estimate for failure along each of the unit vectors is obtained by calculating the reserve factor f_{res} through the failure criterion proposed in [20] with the further adaptions from [21]; the best performing criteria on general stress states tested in [7]. Given the homogeneous formulation of the failure criterion, the estimates can be obtained from $\sigma_{fail} \approx \vec{v} / f_{res}$. To ensure simulations leading to failure, a termination time is chosen that is expected to lead to $2\sigma_{fail}$ when extrapolating from the initial RVE stiffness. For the analysis, periodic boundary conditions are applied to the RVE using the *INCLUDE_UNITCELL keyword with 3 additional nodes as control points. To apply the intended stress states to the RVE, they are first transformed to their respective strain equivalent using the RVEs mesoscopic compliance matrix and from that to nodal displacements taking the RVE geometry into account. For details of the latter transformation, the reader is referred to [22]. The resulting values are applied as displacement boundary conditions to the control points.

Simulations are carried out on the high performance computing cluster Barnard at the NHR Center of TU Dresden. Given that the objective of this study is the modeling of the onset of failure, simulations are terminated once one element has been deleted. To reduce the data's dependency on integration step width, simulations are restarted from the last state before failure with the step size decreased and sampling frequency increased by a factor of 100.

From the simulation results, homogenized stress as well as failure information on the RVE are extracted, aligned by the simulation time, using the open source Python library lasso-python [23]. Here, information on failure mode is determined by the first material to fail, allowing for the differentiation between fiber failure (FF) and inter fiber failure (IFF). While neither failure model is triggered, a proxy mode "no failure" (nF) is added. This data is aggregated across all loadcases and loaded into three distinct databases with a unique identifier (UID) per loadcase.

For the training of the ML model, the database is split into training, test and validation sets, enforcing stratification by the loadcase UID to prevent cross contamination of the datasets. A ratio of 70:20:10 is chosen for this.

Machine learning. In this study, failure detection is treated as classification task. This is addressed by a fully connected neural network (NN), taking the six independent stress components as input vector in the form $[\sigma_1, \sigma_2, \sigma_3, \tau_{12}, \tau_{23}, \tau_{13}]$ and mapping them to the considered failure modes. To achieve the highly non-linear mapping necessary, three hidden layers with 58, 65, 21 neurons respectively are used. The parameters were obtained by hyperparameter optimization,

Sheet Metal 2025
Materials Research Proceedings 52 (2025) 260-267

Materials Research Forum LLC
https://doi.org/10.21741/9781644903551-32

using the tool OmniOpt [24]. During this, the number of neurons per layer, batch size and learning rate were modified by the underlying Bayesian optimization algorithm until no improvement on the test loss occurred for 25 sequential iterations. For the output layer, the softmax activation function is used, leading to a probability prediction for the failure modes; all other layers have "Swish" [25] as activation function. The training is conducted using the Adam optimizer [26] for up to 200 epochs with categorical crossentropy loss, a batch size of 221 and a learning rate of $10^{-2.033}$. Early stopping is activated to avoid overfitting if no improvement of the loss, evaluated on the test set, occurs for 50 epochs.

Results and discussion

Simulation. The RVE setup is first used to simulate uniaxial stress states. Strengths obtained from the simulations are given alongside experimental values in Table 1. From the comparison of the values, it becomes clear that the chosen setup is capturing the FRP's uniaxial failure behavior well in all cases except for transverse tension. This deviation can be attributed to the idealized fiber-matrix interface, since failure under transverse tension is typically initiated by fiber matrix debonding [27]. Therefore, results obtained by the presented RVE with the constitutive models for its constituents are considered representative for FRP and constitute a valid basis for developing a methodology for a data driven failure criterion.

Table 1 Strength values from experiment and RVE simulation

	$R_\parallel^+$ [MPa]	$R_\parallel^-$ [MPa]	$R_\perp^+$ [MPa]	$R_\perp^-$ [MPa]	$R_{\perp\parallel}$ [MPa]
WWFE [15]	1280	800	40	145	73
RVE	1292	799	77	139	67

From the 5000 loadcases, 4577 lead to IFF and 94 to FF. The remaining 329 cases did not yield usable results, 209 due to numerical problems and 120 since termination time was reached in the simulation before the RVE failed. To alleviate the substantial imbalance of the failure modes, an additional 100 loadcases with $\sigma_\parallel > 0.9\,R_\parallel^+$ or $\sigma_\parallel < -0.9\,R_\parallel^-$ were simulated, all leading to FF.

Machine Learning. During the training, of the ML model, loss and accuracy improve quickly for 35 epochs. Afterwards, model performance plateaus before signs of overfitting occur after 57 epochs. This is caught by the early stopping algorithm and training is terminated after 107 epochs with restoration to the state after 57 epochs. The development is shown in Fig. 2, with the loss on the left and accuracy on the right. The final model achieves an accuracy of 87 % on the training data and 84 % both on test and validation data.

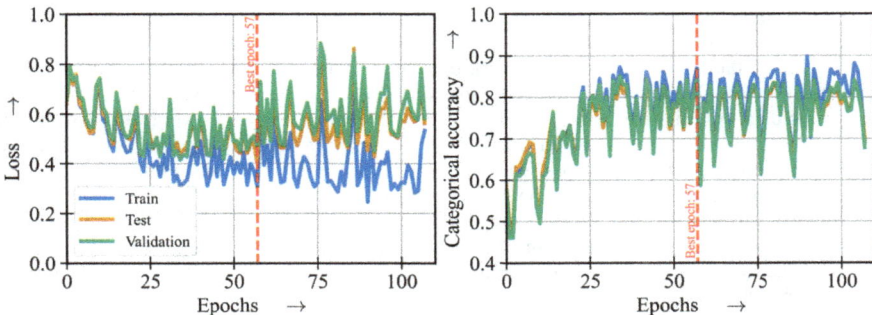

Figure 2 Development of performance metrics throughout model training

Sheet Metal 2025

Materials Research Proceedings 52 (2025) 260-267

Materials Research Forum LLC

https://doi.org/10.21741/9781644903551-32

Additional insight into model performance is given confusion matrices. In this, each row represents the actual label and each column the model predictions. In each cell, the ratio of predictions for the respective true class is given. Hence, on the main diagonal represents correctly classified datapoints for each failure mode. On the left of Fig. 3, the confusion matrix for the validation dataset is shown. From this it becomes clear, that detecting IFF poses the highest challenge to the model, especially differentiating from nF. Similarly, a confusion of nF and IFF occurs, though significantly less frequently. This challenge can be attributed to the very high sampling frequency in the simulation when approaching the failure point. Therefore, the ternary decision has to be changed completely based on miniscule changes in the stress state.

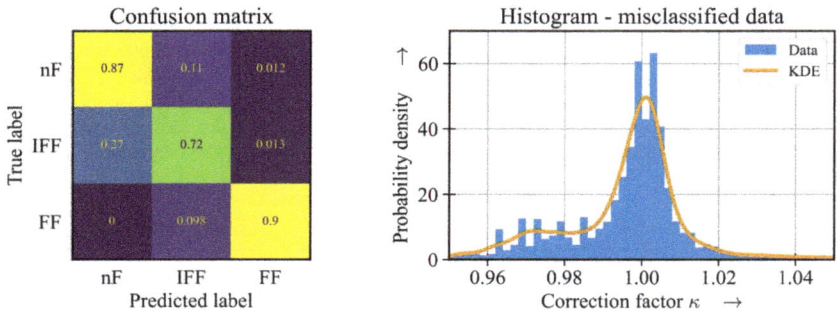

Figure 3 Confusion matrix for the validation data set (left) and correction factor κ for stress state necessary to obtain correct failure mode (right)

To quantify how far off the model predictions are in cases of misclassification, the stress state of misclassified data points is multiplied with a correction factor κ, which is smaller than 1 in cases where failure was predicted for uncritical stress states and greater than 1 otherwise. On the right in Fig. 3, a histogram of κ in the range 0.95 to 1.05 is shown alongside a kernel density estimate (KDE). Based on the KDE, an estimation of the increase in accuracy with percentage of acceptable deviation is possible. 77 % of misclassifications can be corrected κ in ±5 %, leading to an increase in accuracy of 12.3 %. With κ in ±1 %, still 49 % correction rate can be achieved, improving the accuracy by 8 %. This allows to take additional engineering judgement into account

Conclusion

A methodology for the development of a data driven failure criterion for FRP under arbitrary stress states has been established. Based on results of RVE simulations, a database of stress state and corresponding failure model was created. This database was used to train a simple NN on the posed classification task. It could be shown, that the NN achieves high accuracy on training, testing and validation data. Additionally, a large share of misclassifications is based on very narrow margins. Useable accuracy can therefore be further increased by incorporating engineering judgement. The current version of the model can be used for post hoc analyses of FRP parts, efficiently using the failure information from micro-scale on the meso-scale.

The strong oscillations of loss and accuracy during training could be indicators of a suboptimal network architecture or the demand for additional data. Therefore, generating additional data and focusing the hyperparameter optimization on modulating the architecture are expected to further improve model performance. Additionally, transferring the trained model to a user material routine would further enhance its capabilities and allow the usage during simulations to trigger element deletion.

To further take effects from the heterogeneity in the microstructure observable in mechanical joints of FRP into account, the data generation can directly be extended to statistical volume elements. This allows to capture the influence that local variations in fiber volume content have on failure initiation.

Funding

This research was funded by the Deutsche Forschungsgemeinschaft (DFG, German Research Foundation) – TRR 285/2 – 418701707 – sub-project A03.

Data availability

Data is available upon request from https://trr285.uni-paderborn.de/en/publications.

References

[1] D. Rajak, D. Pagar, P. Menezes, and E. Linul, "Fiber-Reinforced Polymer Composites: Manufacturing, Properties, and Applications," Polymers, vol. 11, no. 10. MDPI AG, p. 1667, Oct. 12, 2019. https://doi.org/10.3390/polym11101667.

[2] Prashanth S, Subbaya KM, Nithin K, Sachhidananda S, "Fiber Reinforced Composites - A Review," Journal of Material Science & Engineering, vol. 06, no. 03. OMICS Publishing Group, 2017. https://doi.org/10.4172/2169-0022.1000341.

[3] Y. Chen, X. Yang, M. Li, K. Wei, and S. Li, "Mechanical behavior and progressive failure analysis of riveted, bonded and hybrid joints with CFRP-aluminum dissimilar materials," Thin-Walled Structures, vol. 139. Elsevier BV, pp. 271–280, Jun. 2019. https://doi.org/10.1016/j.tws.2019.03.007.

[4] B. Ravichandran and M. Balasubramanian, "Joining methods for Fiber Reinforced Polymer (FRP) composites – A critical review," Composites Part A: Applied Science and Manufacturing, vol. 186. Elsevier BV, p. 108394, Nov. 2024. https://doi.org/10.1016/j.compositesa.2024.108394.

[5] R. Kupfer, L. Schilling, S. Spitzer, M. Zichner, and M. Gude, "Neutral lightweight engineering: a holistic approach towards sustainability driven engineering," Discover Sustainability, vol. 3, no. 1. Springer Science and Business Media LLC, May 27, 2022. https://doi.org/10.1007/s43621-022-00084-9.

[6] B. Gröger, J. Gerritzen, A. Hornig and M. Gude, „Developing a numerical modelling strategy for metallic pin pressing processes in fibre reinforced thermoplastics to investigate fibre rearrangement mechanisms during joining" Proc IMechE Part L: J Materials: Design and Applications (in press)

[7] A. Kaddour and M. Hinton, "Maturity of 3D failure criteria for fibre-reinforced composites: Comparison between theories and experiments: Part B of WWFE-II," Journal of Composite Materials, vol. 47, no. 6–7. SAGE Publications, pp. 925–966, Mar. 2013. https://doi.org/10.1177/0021998313478710.

[8] L. Wan, Z. Ullah, D. Yang, and B. G. Falzon, "Comprehensive inter-fibre failure analysis and failure criteria comparison for composite materials using micromechanical modelling under biaxial loading," Journal of Composite Materials, vol. 57, no. 18. SAGE Publications, pp. 2919–2932, May 31, 2023. https://doi.org/10.1177/00219983231176609.

[9] F. Naya, C. González, C. S. Lopes, S. Van der Veen, and F. Pons, "Computational micromechanics of the transverse and shear behavior of unidirectional fiber reinforced polymers including environmental effects," Composites Part A: Applied Science and Manufacturing, vol. 92. Elsevier BV, pp. 146–157, Jan. 2017. https://doi.org/10.1016/j.compositesa.2016.06.018.

[10] B. Liang et al., "Multi-scale modeling of mechanical behavior of cured woven textile composites accounting for the influence of yarn angle variation," Composites Part A: Applied Science and Manufacturing, vol. 124. Elsevier BV, p. 105460, Sep. 2019. https://doi.org/10.1016/j.compositesa.2019.05.028.

[11] D. Pulungan, G. Lubineau, A. Yudhanto, R. Yaldiz, and W. Schijve, "Identifying design parameters controlling damage behaviors of continuous fiber-reinforced thermoplastic composites using micromechanics as a virtual testing tool," International Journal of Solids and Structures, vol. 117. Elsevier BV, pp. 177–190, Jun. 2017. https://doi.org/10.1016/j.ijsolstr.2017.03.026.

[12] G. Balokas, S. Czichon, and R. Rolfes, "Neural network assisted multiscale analysis for the elastic properties prediction of 3D braided composites under uncertainty," Composite Structures, vol. 183. Elsevier BV, pp. 550–562, Jan. 2018. https://doi.org/10.1016/j.compstruct.2017.06.037.

[13] J. Chen, L. Wan, Y. Ismail, J. Ye, and D. Yang, "A micromechanics and machine learning coupled approach for failure prediction of unidirectional CFRP composites under triaxial loading: A preliminary study," Composite Structures, vol. 267. Elsevier BV, p. 113876, Jul. 2021. https://doi.org/10.1016/j.compstruct.2021.113876.

[14] L. Wan, Z. Ullah, D. Yang, and B. G. Falzon, "Probability embedded failure prediction of unidirectional composites under biaxial loadings combining machine learning and micromechanical modelling," Composite Structures, vol. 312. Elsevier BV, p. 116837, May 2023. https://doi.org/10.1016/j.compstruct.2023.116837.

[15] P. Soden, M. Hinton, A. Kaddour, "Lamina properties, lay-up configurations and loading conditions for a range of fibre-reinforced composite laminates," Composites Science and Technology, vol. 58, no. 7. Elsevier BV, pp. 1011–1022, Jul. 1998. https://doi.org/10.1016/s0266-3538(98)00078-5.

[16] F. Rickhey, T. Park, and S. Hong, "Damage prediction in thermoplastics under impact loading using a strain rate-dependent GISSMO," Engineering Failure Analysis, vol. 149. Elsevier BV, p. 107246, Jul. 2023. https://doi.org/10.1016/j.engfailanal.2023.107246.

[17] M. E. Johnson, L. M. Moore, and D. Ylvisaker, "Minimax and maximin distance designs," Journal of Statistical Planning and Inference, vol. 26, no. 2. Elsevier BV, pp. 131–148, Oct. 1990. https://doi.org/10.1016/0378-3758(90)90122-b.

[18] S. Wessing, "Two-stage methods for multimodal optimization," Technische Universität Dortmund, 2015, doi: 10.17877/DE290R-7804.

[19] V. R. Joseph, "Space-filling designs for computer experiments: A review," Quality Engineering, vol. 28, no. 1. Informa UK Limited, pp. 28–35, Jan. 02, 2016. https://doi.org/10.1080/08982112.2015.1100447.

[20] R. G. Cuntze and A. Freund, "The predictive capability of failure mode concept-based strength criteria for multidirectional laminates," Composites Science and Technology, vol. 64, no. 3–4. Elsevier BV, pp. 343–377, Mar. 2004. https://doi.org/10.1016/s0266-3538(03)00218-5.

[21] Proceedings of the European Conference on Spacecraft Structures, Materials and Mechanical Testing 2005 (ESA SP-581). 10-12 May 2005, Noordwijk, The Netherlands. Edited by Karen Fletcher. Published on CD-Rom, id.138.1

[22] W. Tian, L. Qi, X. Chao, J. Liang, and M. Fu, "Periodic boundary condition and its numerical implementation algorithm for the evaluation of effective mechanical properties of the

composites with complicated micro-structures," Composites Part B: Engineering, vol. 162. Elsevier BV, pp. 1–10, Apr. 2019. https://doi.org/10.1016/j.compositesb.2018.10.053.

[23] Diez, C., Ballal, N. & "rao014" Open-lasso-python/lasso-python: Home of the open-source CAE Library Lasso-Python. GitHub Repository., https://github.com/open-lasso-python/lasso-python

[24] P. Winkler, N. Koch, A. Hornig, and J. Gerritzen, "OmniOpt – A Tool for Hyperparameter Optimization on HPC," Lecture Notes in Computer Science. Springer International Publishing, pp. 285–296, 2021. https://doi.org/10.1007/978-3-030-90539-2_19.

[25] P. Ramachandran, B. Zoph, and Q. V. Le, "Searching for Activation Functions," 2017, arXiv. https://doi.org/10.48550/ARXIV.1710.05941.

[26] D. P. Kingma and J. Ba, "Adam: A Method for Stochastic Optimization," 2014, arXiv. https://doi.org/10.48550/ARXIV.1412.6980.

[27] L. Yang, Y. Yan, Y. Liu, and Z. Ran, "Microscopic failure mechanisms of fiber-reinforced polymer composites under transverse tension and compression," Composites Science and Technology, vol. 72, no. 15. Elsevier BV, pp. 1818–1825, Oct. 2012. https://doi.org/10.1016/j.compscitech.2012.08.001.

Sheet Metal 2025
Materials Research Proceedings 52 (2025) 268-275

Materials Research Forum LLC
https://doi.org/10.21741/9781644903551-33

Modeling approaches for the decomposition behavior of preconsolidated rovings throughout local deformation processes

Benjamin Gröger[1,a*], Johannes Gerritzen[1,b],
Andreas Hornig[1,2,3,c] and Maik Gude[1,d]

[1]Institute of Lightweight Engineering and Polymer Technology (ILK), TUD Dresden University of Technology, Holbeinstr. 3, 01307 Dresden, Germany

[2]Center for Scalable Data Analytics and Artificial Intelligence Dresden/Leipzig (ScaDS.AI), TUD Dresden University of Technology, Strehlener Straße 12-14, 01069 Dresden, Germany

[3]Department of Engineering Science, University of Oxford, Parks Road, OX1 3PJ Oxford, United Kingdom

[a]benjamin.groeger@tu-dresden.de, [b]johannes.gerritzen@tu-dresden.de,
[c]andreas.hornig@tu-dresden.de, [d]maik.gude@tu-dresden.de

Keywords: Process, Thermoplastic Fiber Reinforced Plastic, Finite Element Method (FEM)

Abstract. This paper investigates two modeling approaches for the simulation of the deformation and decomposition behavior of preconsolidated rovings above the thermoplastic matrix' melting temperature. This is crucial for capturing the local material structure after processes introducing highly localized deformation such as mechanical joining processes between metal and fiber reinforced thermoplastics (FRTP). A generic finite element (FE) model is developed, incorporating interfaces discretized through either cohesive zone (CZ) elements or Coulomb friction-based contacts. The material parameters for the FE elements are derived from the initial stiffness of a statistical volume element (SVE) at micro scale modelled with an Arbitrary-Lagrange-Eulerian method for three load cases. The CZ properties calculated are based on the shear viscosity of the composite. The CZ and contact modelling approaches are evaluated using three load cases of the SVE, comparing force-displacement curves. Under simple loading conditions, such as normal pressure tension and bending, both methods produce similar results; however, in complex load cases, the CZ approach shows clear advantages in handling interface interactions and shows robust simulations. The CZ approach thus presents a promising method for simulating roving decomposition in FRTP-metal joining applications above the matrix' melting temperature.

Introduction

The industrial application of continuous fiber reinforced thermoplastics (FRTP) requires reliable manufacturing processes and consistent resulting mechanical properties. These properties are determined by the inner material structure, which is significantly influenced by the manufacturing process. During forming process like joining sheet metal and FRTP, the initial material structure – characterized by a uniform fiber alignment and predictable mechanical behavior – is transformed into a more complex structure with three-dimensional (3D) fiber orientations, varying fiber volume content f (FVC) [2] and potentially defects [3]. As a result, the joint's mechanical properties are often difficult to predict and typically require experimental validation. The deformation mechanisms in the forming process of FRTP are based on specific mechanisms on meso scale [5] and matrix flow processes like percolation or squeeze flow [6] as well as inter- and intraply shear [7–9].

The ongoing developments in the technologie sector require a robust design strategy to predict and quantify the resultant material structure of the FRTP and the resultant load bearing behavior

Content from this work may be used under the terms of the Creative Commons Attribution 3.0 license. Any further distribution of this work must maintain attribution to the author(s) and the title of the work, journal citation and DOI. Published under license by Materials Research Forum LLC.

Sheet Metal 2025
Materials Research Proceedings 52 (2025) 268-275

Materials Research Forum LLC
https://doi.org/10.21741/9781644903551-33

of the joint. To simulate such joining operations and the occurring mechanisms the Arbitrary-Lagrange-Eulerian (ALE) method in combination with an established modeling strategy of fiber bundles in between micro and meso scale [10,11] is used. Due to the considered fluid-structure-interaction (FSI), contact between the fiber bundles and the ALE method the simulation on full scale parts is computationally prohibitive. Therefore, it has to be adapted to meso scale, which allows both accurate predictions of the material structure and computationally efficient simulations. As shown by [12], this requires the simulation to cover the deformation and decomposition of individual rovings within the textile architecture. Throughout this, separation is caused by intraply shear, both parallel and transverse to the fiber direction.

Considering roving decomposition due to intraply shear while forming can be accomplished intrinsically in the constitutive model [13] or explicitly by a CZ modeling approach [14] or within contact. When using the intrinsic approach with a constitutive model, computationally intensive descriptions such as crack propagation, element splitting, or extremely fine meshes within the continuum mesh must be employed. In contrast, the CZ modeling approach enables simplified material modeling and predefined paths for the deletion of CZ elements, facilitating the separation of roving elements with reduced computational effort. Within the contact modeling approach the FE elements already separated, but due to the friction, the separation is impeded.

In the presented paper, the two approaches with explicit separation zone are compared for modeling the decomposition of an FRTP roving above its matrix's melting temperature. A generic model using 3D solid elements is employed to represent the cross-section of the roving. The coupling of the solid elements is achieved using CZ modeling and friction based contact modeling approaches. Therefore, a homogenized elastic description of the material behavior must be employed for the FE solid elements and separation for the CZ elements or contact definitions. The required material data are based on numerical simulations on micro scale. The final results are compared across deformation as well as force-displacement curves.

Material modeling

The determination of material parameters for FRTP above their melting temperature presents a challenging task due to the lack of standardized testing methods. Therefore, a numerical approach on micro scale to simulate the shear behavior of continuous glass fibers (GF) and polypropylene (PP) is used. For this, a statistical volume element (SVE) containing 430 fibers with a diameter of 14 μm and a resultant FVC $f = 63$ % is set up. The cube model with an edge length of 0.32 mm is shown in Fig. 1. The ALE method is used to model the viscous melt's flow through the ALE mesh domain (not shown) and accounts for FSI as a coupling between fibers, plates, and the melt. The capability to simulate such problems is demonstrated in [15]. The initial slopes of the resultant force-displacement-curves are used to calibrate a constitutive model for the homogenized roving, the further data to validate the presented modeling strategy.

The material parameters for GF are given in Table 1 and the molten PP Borealis BJ100-HP is described with a shear rate dependent viscosity at 200 °C according to [16]. Three distinct load cases with an average shear rate $\dot{\gamma}$ of $1000\,\mathrm{s}^{-1}$ are simulated: shear in fiber direction τ_{31}(longitudinal), shear transverse to fiber direction τ_{32} (transversal) and normal compaction σ_{33} (normal).

Table 1: Mechanical properties of GF

Property	ρ [to/mm^3]	E [MPa]	ν [-]
Value	2.54e-9	69500	0.23

Materials Research Forum LLC

https://doi.org/10.21741/9781644903551-33

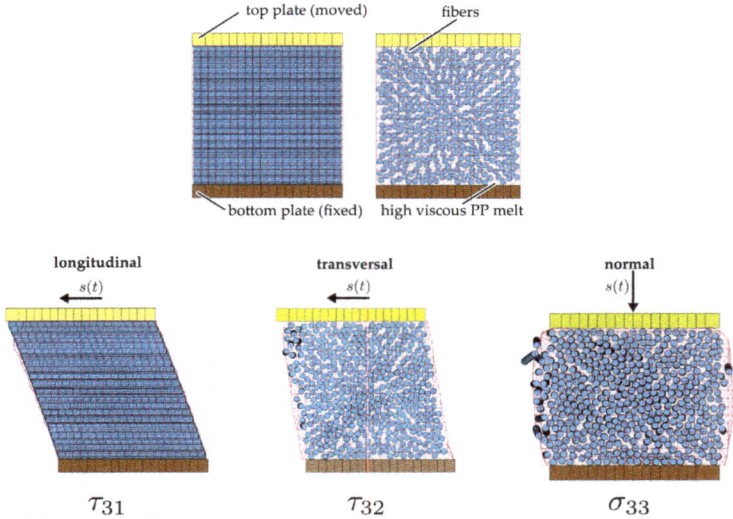

Fig. 1:SVE with 430 fibers in viscous melt evaluates shear and pressure resistance to predict FRTP elastic parameters above melting temperature

The respective deformed states are shown in Fig. 1. The motion of the top plate $s(t)$ is defined as displacement controlled boundary condition (BC). Due to the motion of the top plate, both the matrix and the embedded fibers move in the direction of displacement. From the initial slope of the respective force-displacement curves, elastic constants for the homogenized roving model are determined, with the exception of Young's modulus in fiber direction (E_1), which is calculated by rule of mixture. Poisson ratios are chosen close to 0.5, given the melt's almost incompressible behavior. The resulting parameters are given Table 2.

Table 2: Elastic material properties of the homogenized GF/PP material based on initial stiffness of the SVE model

Property	E_1 [MPa]	E_2 [MPa]	E_3 [MPa]	G_{12} [MPa]	G_{13} [MPa]	G_{23} [MPa]	ν_{12} [-]	ν_{13} [-]	ν_{23} [-]
Value	44,380	10.8	10.8	3.91	3.91	3.84	0.49	0.49	0.49

Cohesive Zone Modeling. Modeling the matrix rich layer behavior between the fibers for intraply shear for the CZ elements follows the work of Roberts [17]. The longitudinal viscosity η_L can be calculated by the equation (1)

$$\frac{\eta_L}{\eta_M} = \frac{1-f}{(1-\sqrt{f})^2},$$ (1)

where η_M is the dynamic viscosity of pure matrix and f as FVG. For the presented investigation, the ratio η_L/η_T is assumed to be 1. The limit shear viscosity η_∞ is assumed to be the matrix viscosity η_M with 2.68 Pa s [16]. These assumptions and material parameters lead to

$$\eta_L = \eta_T = 308 \text{ Pa s.}$$ (2)

For the CZ modeling a trilinear traction separation law in LS-Dyna is used (*MAT_185) [18]. The maximum stress τ_{max} is calculated with the shear rate $\dot{\gamma}$ of $1000s^{-1}$ and $\eta_{L/T}$, by

$$\tau_{max} = \sigma_{max} = \eta_L \dot{\gamma}. \tag{3}$$

The modeling parameters for CZ material model are given in Table 3 with the density ρ of the thermoplastic melt, a maximum stress σ_{max}, maximum separation in normal (NLS) and tangential direction (TLS) in length scale and the scaled distance of peak traction (λ_1), beginning of softening (λ_2) and failure (λ_{fail}).

Table 3: Mechanical properties for the trilinear traction separation law of the CZ modeling

Property	ρ	σ_{max}	NLS	TLS	λ_1	λ_2	λ_{fail}
	[to/m^3]	[MPa]	[mm]	[mm]	[-]	[-]	[-]
Value	7.5e-10	0.308	0.1	1.0	1e-3	0.95	1.0

Contact based modeling. The contact-based model, which involves Coulomb friction contact, requires the HERSEY number H [19], based on:

$$H = \frac{\eta_L \cdot v_{relative}}{p}, \tag{4}$$

with the relative velocity $v_{relative}$ between the sliding parts, the normal pressure p and the intraply shear viscosity η_L. The normal pressure is set to 0.1 MPa. Due to the fiber contact the intraply shear viscosity η_L is chosen instead of the pure matrix viscosity η. In accordance with the assumption of equation 3, the relative velocity $v_{relative}$ is determined by

$$\dot{\gamma}_{inf} = v_{relative}/d_{gap}, \tag{5}$$

and the assumed shear gap d_{gap} between the fibers of 7e-3 mm, which is half of the fiber diameter. The resultant friction coefficient μ for this numerical study is based on [19] with

$$\mu = 6.12 \cdot H + 0.27. \tag{6}$$

This leads to a friction coefficient $\mu = 0.29$.

Numerical setup
An initial roving cross section as a single ply is simplified to an element set containing 18 solid elements with spatial dimensions of 0.32 x 0.32 x 0.32 mm (Fig. 2- blue). Rows of elements along the fiber direction (1-direction) share nodes, resulting in a consistent mesh. Interfaces between each row, both along and transverse to the fiber direction, are modeled using CZ elements with a thickness of 0.001 mm or through contact definition (Fig. 2- orange).

Sheet Metal 2025
Materials Research Proceedings 52 (2025) 268-275

Materials Research Forum LLC
https://doi.org/10.21741/9781644903551-33

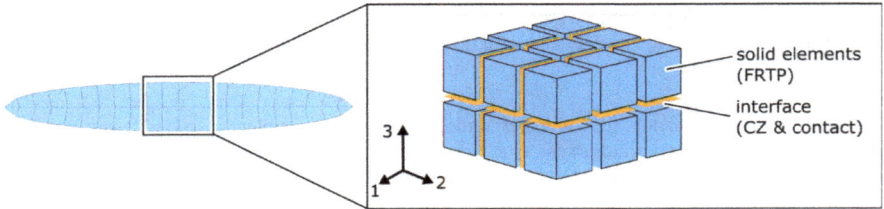

Fig. 2: Model setup of 18 cubic solid elements (blue) with interface modeling via CZ elements or contact definition in fiber direction (orange).

Due to the preliminary shear simulations and occurring phenomena in forming processes four load cases are investigated (Fig. 3). The displacement boundary condition is defined as a constant acceleration of $a = 55555 \ mm/s^2$ which avoids initial reaction forces and leads to more stable simulations. For each stimulation a total time t of $6 \cdot 10^{-3}$ s is simulated, leading to a total displacement of 1mm. The force displacement curves allow a comparison between the approaches.

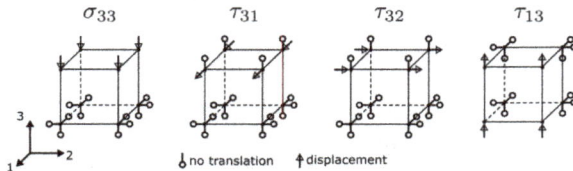

Fig. 3 Load cases to investigate roving separation with proposed modeling strategies

Results

With the proposed modeling strategy the material response of a roving is composed by the elastic stiffness of the FE elements and behavior the CZ elements or contact definition. The resultant force-displacement-curves of each load case for the SVE, the CZ model (CZM) and the contact model are given in Fig. 4.

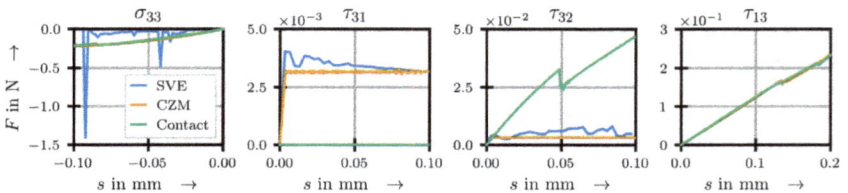

Fig. 4: Force displacement curves of each load case for the SVE, CZ and contact model

The results of the SVE serve as the basis for comparing the ability of the CZ and contact models to simulate the intraply shear phenomena. Due to the ALE modeling approach used in the SVE, the τ_{13} load case cannot be simulated. For the load case σ_{33}, the downward displacement of the top plate should be interpreted from right to left in the graph. As s decreases, the force increases due to the compression (cf. Fig. 1). The load cases of the SVE show an initial stiffness and a fluctuating force as the top plate movement increases. While the normal compaction force (σ_{33}), fluctuates significantly, the other two curves exhibit a force plateau (τ_{31}, τ_{32}).

The resultant deformation after 0.1 mm of deformation for CZ and contact model is presented in Fig. 5. For σ_{33}, both CZ and contact model are similar, predicting an increasing compression

Sheet Metal 2025
Materials Research Proceedings 52 (2025) 268-275

Materials Research Forum LLC
https://doi.org/10.21741/9781644903551-33

force, with deformation also being quite similar, though none can capture the low force level observed in the SVE and the Barreling effect (cf. Fig. 1 normal). Furthermore, the contact model exhibits penetration, resulting in unrealistic stress states.

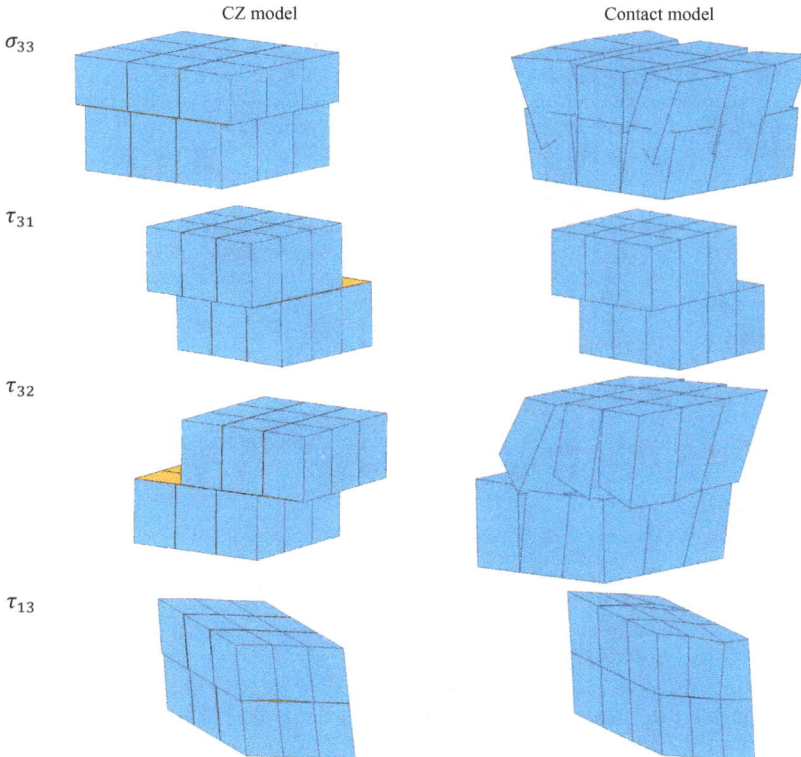

CZ model · Contact model

σ_{33}

τ_{31}

τ_{32}

τ_{13}

Fig. 5: Resultant deformation of both models for each load case

For τ_{31}, the sliding of the elements, interpreted as the interaction between the fibers in longitudinal direction, is underestimated in the contact model, whereas for the CZ modelling, the predicted force aligns well with the converged force. But the deformation of both models show similar deformation. For τ_{32}, where the FE solid elements become entangled in the contact model, the sliding force is overestimated. This contrasts with the CZ model, where the deformation between both rows of elements is accounted by the CZ, resulting in a constant reaction force that aligns with the SVE model. For the τ_{13} load case, there is no difference between the two approaches in the graph, as only a bending load is applied. However, in the contact model, the deformation is handled by the solid elements, which can lead to unrealistic stress states. In contrast, the deformation in the CZ model is handled by the CZ elements and a more homogenous deformation of the whole model.

It can be seen, that the deformation behavior for both approaches is similar for simple test cases (σ_{33}, τ_{13}). At more superimposed stress states the contact approach leads to unrealistic deformation and penetration. The resultant reaction forces underestimated for τ_{31} and overestimated for τ_{32} compared to the CZ approach. The CZ approach leads to expected deformation and show force-displacement-curves which are in agreement with numerical results of the SVE.

Conclusion

A simplified model to evaluate two different approaches for predicting process induced roving decomposition was developed. Such decomposition can lead to a change of the initial shape of the cross section. To address this, both a CZ model and a contact model were applied. Using material data based on a micro scale model, simulations indicate that the CZ approach is better suited to predict roving decomposition in forming processes of unidirectional FRTP. Given the promising results of the CZ modeling, this approach will be employed in further investigations to simulate the entire roving during the joining process. Future investigations should consider computational efficiency, incorporate more accurate material parameters based on experimental test and explore enhancements to the CZ material model which consider compression forces in thickness direction [20]. In addition to numerical tasks, experimental testing is necessary to obtain material data for unidirectional materials above melting temperature and to provide experimental validation for the simulation results.

Acknowledgements
The authors gratefully acknowledge the GWK support for funding this project by providing computing time through the Center for Information Services and HPC (ZIH) at TU Dresden.

Funding
This research was funded by the Deutsche Forschungsgemeinschaft (DFG, German Research Foundation) – TRR 285/2 – 418701707 – sub-project A03.

Data availability
Data is available upon request from https://trr285.uni-paderborn.de/en/publications.

References

[1] Troschitz J, Gröger B, Würfel V, Kupfer R, Gude M. Joining Processes for Fibre-Reinforced Thermoplastics: Phenomena and Characterisation. Materials (Basel) 2022;15(15). https://doi.org/10.3390/ma15155454

[2] Seidlitz H, Gerstenberger C, Osiecki T, Simon S, Kroll L. High-performance lightweight structures with Fiber Reinforced Thermoplastics and Structured Metal Thin Sheets. JMSR 2014;4(1). https://doi.org/10.5539/jmsr.v4n1p28

[3] Kunze E, Galkin S, Böhm R, Gude M, Kärger L. The Impact of Draping Effects on the Stiffness and Failure Behavior of Unidirectional Non-Crimp Fabric Fiber Reinforced Composites. Materials (Basel) 2020;13(13). https://doi.org/10.3390/ma13132959

[4] Azzouz R, Allaoui S, Moulart R. Composite preforming defects: a review and a classification. Int J Mater Form 2021;14(6):1259–78. https://doi.org/10.1007/s12289-021-01643-7

[5] Remko Akkerman, Sebastiaan Haanappel, Ulrich Sachs, Bert Rietman. Complex stamp forming of advanced thermoplastic composites. In: R. Rolfes, E.L. Jansen, editors. 3rd ECCOMAS Thematic Cconference on Mechnical Response of Composites 2011. Leibniz Universität Hannover; 2011, p. 27–34

[6] Shuler SF, Advani SG. Transverse squeeze flow of concentrated aligned fibers in viscous fluids. Journal of Non-Newtonian Fluid Mechanics 1996;65(1):47–74. https://doi.org/10.1016/0377-0257(96)01440-1

[7] Barnes JA, Cogswell FN. Transverse flow processes in continuous fibre-reinforced thermoplastic composites. Composites 1989;20(1):38–42. https://doi.org/10.1016/0010-4361(89)90680-0

[8] Sheet Metal 2023. Materials Research Forum LLC; 2023. https://doi.org/10.21741/9781644902417

[9] Murtagh AM, Mallon PJ. Chapter 5 Characterisation of shearing and frictional behaviour during sheet forming. In: Composite Sheet Forming. Elsevier; 1997, p. 163–216.

[10] Gröger B, Würfel V, Hornig A, Gude M. Forming process induced material structure of fibre-reinforced thermoplastics - Experimental and numerical investigation of a bladder-assisted moulding process. Journal of Advanced Joining Processes 2022;5(7):100100. https://doi.org/10.1016/j.jajp.2022.100100

[11] Gröger B. Numerical and experimental investigations of piercing fibre-reinforced thermoplastics. In: Sheet Metal 2023. Materials Research Forum LLC; 2023, p. 171–178.

[12] Borowski A, Gröger B, Füßel R, Gude M. Characterisation of Fibre Bundle Deformation Behaviour—Test Rig, Results and Conclusions. JMMP 2022;6(6):146. https://doi.org/10.3390/jmmp6060146

[13] Dörr D, Henning F, Kärger L. Nonlinear hyperviscoelastic modelling of intra-ply deformation behaviour in finite element forming simulation of continuously fibre-reinforced thermoplastics. Composites Part A: Applied Science and Manufacturing 2018;109(0):585–96. https://doi.org/10.1016/j.compositesa.2018.03.037

[14] Kapshammer A, Miron MC, Dangl L, Major Z. Interface Characterization of Consolidated PPGF Tapes on PPGF Mat Material. Polymers 2023;15(4):935. https://doi.org/10.3390/polym15040935

[15] Gröger B, Wang J, Bätzel T, Hornig A, Gude M. Modelling and Simulation Strategies for Fluid-Structure-Interactions of Highly Viscous Thermoplastic Melt and Single Fibres-A Numerical Study. Materials (Basel) 2022;15(20). https://doi.org/10.3390/ma15207241

[16] Gröger B, Gerritzen J, Hornig A, Gude M. Developing a numerical modelling strategy for metallic pin pressing processes in fibre reinforced thermoplastics to investigate fibre rearrangement mechanisms during joining. Proceedings of the Institution of Mechanical Engineers, Part L: Journal of Materials: Design and Applications 2024;238(12):2286–98. https://doi.org/10.1177/14644207241280035

[17] Roberts RW, Jones RS. Rheological characterization of continuous fibre composites in oscillatory shear flow. Composites Manufacturing 1995;6(3-4):161–7. https://doi.org/10.1016/0956-7143(95)95007-L

[18] Manual L-DKU's. vol. II. Livermore Software Technology Corporation (LSTC) 2013.

[19] Gorczyca-Cole JL, Sherwood JA, Chen J. A friction model for thermostamping commingled glass–polypropylene woven fabrics. Composites Part A: Applied Science and Manufacturing 2007;38(2):393–406. https://doi.org/10.1016/j.compositesa.2006.03.006

[20] M. Kuhtz, J. Gerritzen, J. Wiegand, A. Hornig, M. Gude. Modelling delamination in fibre-reinforced composites subjected to through-thickness compression by an adapted cohesive law. [November 01, 2024.627Z].

Sheet Metal 2025   Materials Research Forum LLC
Materials Research Proceedings 52 (2025) 276-283   https://doi.org/10.21741/9781644903551-34

Combination of metal forming and injection moulding in one tool

Juliane Troschitz[1,a]*, Sven Bräunling[2,b], Matthias Kahl[3,c], Frank Schneider[3,d], Thomas Krampitz[4,e], Robert Kupfer[1,f], Maik Gude[1,g] and Alexander Brosius[2,h]

[1]Institute of Lightweight Engineering and Polymer Technology, TUD Dresden University of Technology, Germany

[2]Chair of Forming and Machining Processes, TUD Dresden University of Technology, Germany

[3]FEP Fahrzeugelektrik Pirna GmbH & Co. KG, Germany

[4]Institute for Mineral Processing Machines and Recycling Systems Technology, TU Bergakademie Freiberg, Germany

[a]juliane.troschitz@tu-dresden.de, [b]sven.braeunling@tu-dresden.de, [c]matthias.kahl@fepz.de, [d]frank.schneider@fepz.de, [e]thomas.krampitz@iart.tu-freiberg.de, [f]robert.kupfer@tu-dresden.de, [g]maik.gude@tu-dresden.de, [h]alexander.brosius@tu-dresden.de

Keywords: Deep Drawing, Injection Moulding, Electric Architecture Casing

Abstract. In electric vehicles, the mechanical, media and thermal loads are significantly reduced compared to conventional vehicles, whereas at the same time the challenges in terms of electromagnetic compatibility, lightweight construction and acoustics are increasing. For this purpose, economically and technically competitive, recyclable electric architecture casings are to be developed. These will be realized as hybrid components in which a thin metal sheet provides the electromagnetic shielding and the polymer component transfers the mechanical loads. For the production, a hybrid process is developed in which deep drawing and injection moulding are combined in one tool. This paper presents the results of component, process and tool development, manufacturing studies, component analysis and recycling tests using a scaled generic test structure.

Introduction

The transition from conventional to electrified powertrains opens up new possibilities for the design of electric vehicles and their components. The installed components are changing and so are their requirement profiles. The challenges in terms of electromagnetic compatibility, lightweight construction and acoustics are increasing significantly for electric vehicles. A new approach is to develop recyclable, ecologically, economically and technically competitive electric architecture casings that meet the high electromagnetic requirements and have a high potential to be used in a wide range of industries. For this purpose, the housing structure is to be realized as a hybrid component in which the polymer component transfers the mechanical loads and the electromagnetic shielding is provided by a thin metal sheet.

The production of metal-polymer hybrid parts has so far required multi-stage manufacturing processes consisting of high-precision sheet metal forming, complex sheet metal preparation and final injection moulding [1]. This leads to double tool and machine costs for sheet metal forming and injection moulding. In addition, complex sheet metal preparation is necessary in order to achieve sufficient adhesion between the joining partners [2]. An alternative approach is to use the polymer melt pressure during injection moulding for the hydroforming of the metal sheet, so called polymer injection forming (PIF) [3,4]. The PIF has been the subject of extensive research. Experimental investigations have already been carried out on simple test structures [5-9] and numerical [8,9] and analytical [10] models have been developed to describe the process. In addition, deep drawing and injection moulding can be combined in one tool, which has already been successfully demonstrated on metal cup geometries with internal polymer ribbing [11,12],

Content from this work may be used under the terms of the Creative Commons Attribution 3.0 license. Any further distribution of this work must maintain attribution to the author(s) and the title of the work, journal citation and DOI. Published under license by Materials Research Forum LLC.

Sheet Metal 2025

Materials Research Forum LLC

Materials Research Proceedings 52 (2025) 276-283

https://doi.org/10.21741/9781644903551-34

but has not yet been transferred to more complex structures. The approach of integrating the deep drawing and injection moulding processes in one mould offers the potential to significantly shorten the established process route and thus achieving considerable economic advantages [13].

In preliminary studies, the authors already demonstrated the feasibility of the technology for combining deep drawing and injection moulding in one tool for a simple and small tub geometry [14]. The sheet metal is first deep drawn by the closing movement of the mould and then finally formed by the polymer melt pressure. In order to achieve recyclability at the end of the product life cycle, the materials are not joined by adhesion but rather by form-fit. This article presents work on proving the technology for the production of complex housing structures. For this purpose, a downscaled generic demonstrator was developed.

Generic Test Structure

The scaled generic test structure for an electric architecture casing is shown in Fig. 1. The drawing depth is 40 mm and the polymer wall thickness is 2.5 mm. The generic test structure contains typical design elements of a housing structure. These include ribs, clearance holes for contacting, metallic bushes for screw joints and mechanical anchoring points to provide a form fit between metal sheet and polymer component. The flange has a circumferential groove into which an O-ring sealing is inserted when assembling the casing. As a challenge, the metal sheets of the two halves of the housing need to be in contact over a large area at the flange in order to ensure sufficient electromagnetic shielding. This requires a defined metallic residual flange that is as flat as possible. At the same time, however, the flange has to be drawn far enough to ensure complete overmoulding of the outer edge of the sheet metal, as no trimming is planned after deep drawing.

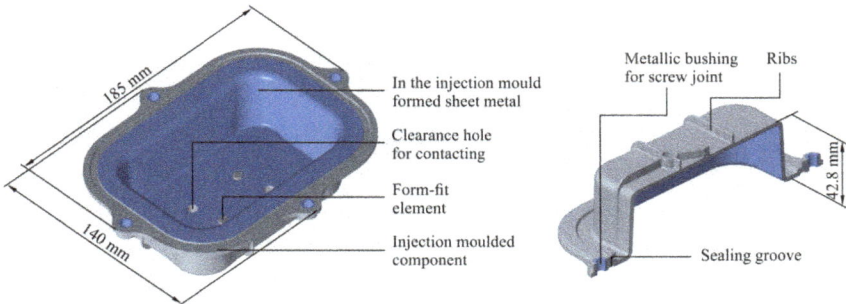

Fig. 1: CAD of the generic test structure

Materials

The housing structure is to be realized as a hybrid component in which the polymer component transfers the mechanical loads and the electromagnetic shielding is provided by a thin metal sheet. The metal sheet thickness was chosen to be as thin as possible in order to increase the weight as little as possible and to meet the sustainability requirements. At the same time, however, it must be thick enough to be deep-drawn. The metallic component is only intended for EMC and not for transferring loads. The materials used for the studies were a 0.3 mm aluminium sheet (Al99.5 EN-AW 1050A) for the metallic component and glass fibre-reinforced polypropylene (Mafill® CR XG 5544 H, 30 % glass fibres) for the injection moulding compound. In further studies, other sheet thicknesses and steel are also investigated. The metallic bushings are compression limiters for M5 screws made of aluminium.

Development of the combined forming-injection moulding process

The process for combined deep drawing and injection moulding in one tool is shown schematically in Fig. 3. First, the sheet metal (not preheated) is inserted on pins on the ejector side (ES) into the open injection mould as well as the bushings on the nozzle side (NS) (a). The blank holder (on the ES) and the inner die section (on the NS) are then moved forward hydraulically. The tool is then closed (b) pressing the metal sheet against the die by the blank holder. As the tool continues to close, the sheet metal is deep drawn by the punch, whereby the hold-down pressure is kept constant hydraulically (c). Finally, the inner part of the die retracts to open the injection mould cavity (d). When the polymer melt is injected, the metal sheet is finally formed by the polymer melt pressure (d). Once the cooling time has elapsed, the hybrid component can be ejected. Since deep drawing takes place with the closing movement of the mould, the process time corresponds to that of a standard injection moulding process (excluding insertion time of the metal sheet). For the selection of a suitable injection moulding machine, not only the clamping force is relevant, but in particular the force-displacement characteristic during closing.

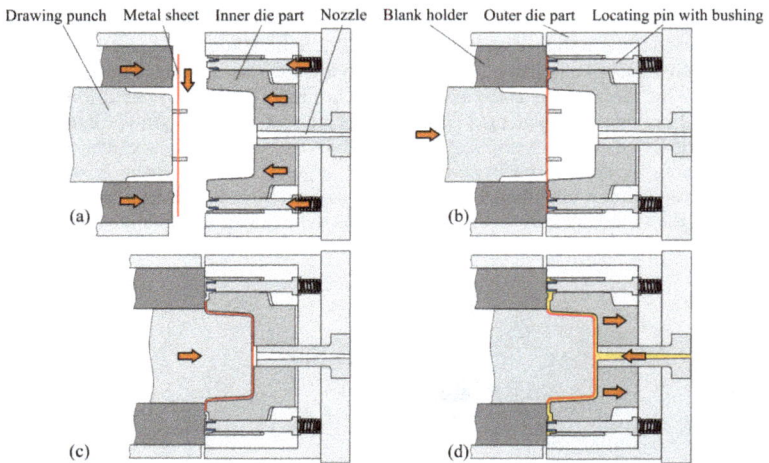

Drawing punch   Metal sheet   Inner die part   Nozzle   Blank holder   Outer die part   Locating pin with bushing

Fig. 2: Schematic process sequence for combined deep drawing and injection moulding in one tool

To design and validate the metal forming process, forming simulations were carried out using PAM-STAMP 2021.0.1 (ESI-GROUP). For the process and tool design, both geometric parameters of the tool were varied and the influence of typical process parameters such as blank holder pressure and friction ratios on the forming result were investigated. A two-stage forming process was simulated: 1) deep drawing of the sheet metal and 2) the final forming by the polymer melt pressure during injection moulding using a substitute model, so-called external high-pressure forming. The entire process was divided into five stages for the simulation:

- Closing the tool, placing and embossing the blank holder
- Path-controlled deep drawing (explicit simulation)
- Retracting inner die to the injection position, stress relief of the sheet (implicit simulation)
- External high-pressure forming (explicit simulation)
- Stress relief / springback of the sheet (implicit simulation)

The element type was set as shell, the active parts as rigid bodies (min. element size 0.2 mm) and the sheet metal as elastic-plastic (min. element size 0.1 mm). The material model chosen for the aluminium was Hill 48 / 90. For the final forming of the metal sheet by the polymer melt pressure during injection moulding (external high-pressure forming), mould filling simulations were previously carried out to determine the level of polymer melt pressure (max. 12 MPa).

Sheet Metal 2025   Materials Research Forum LLC
Materials Research Proceedings 52 (2025) 276-283   https://doi.org/10.21741/9781644903551-34

The generic test structure contains many construction elements, particularly in the flange area (Fig. 1), which represent a particular challenge for sheet metal forming in order to explore the limits of the process. The sealing groove, the ribs and the bushings cause significant resistance to the flow of the sheet metal material into the direction of the cup. Therefore, each of these additional interfering contours increases the force to be transmitted (which is limited by the tensile strength of the sheet metal) and thus reduces the actual (global) forming, i.e. in particular in this case the drawing depth. Numerous numerical studies were conducted on different design variants in order to configure and arrange the interfering contours in a manner that at least reduces their negative influence on the overall deep drawing behaviour. As an example, Fig. 3 shows the results of two FE simulations (after step 1: deep drawing) in comparison: with and without bushings. In (a), cracking can clearly be identified during the deep drawing process due to excessive sheet thinning caused by the resistance of the interfering contour of the tangential rib. In variant (b), at least the sheet metal could be drawn without cracking, even if the flange end position still needs to be optimized. For the process design, this means that the pins with the bushings should be pushed forward during deep drawing and not pulled back in order to support the material flow (cf. Fig. 2).

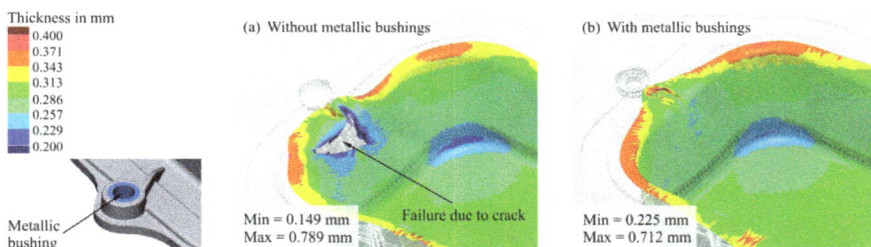

Fig. 3: Comparison of deep drawing simulations with and without bushings

In addition, the numerical studies have shown that the influence of friction on deep drawing is very high due to the large ratio of component size to sheet thickness. In particular, it should be noted that deep drawing is a process with indirect force application (via the bottom – main forming area is the flange). This means that the entire drawing force has to be transferred via the small ring cross-section in the wall area, which is then at risk of cracking. All additional interfering contours increase the risk of cracks. Exemplarily, Fig. 4 shows that even a minimal increase in the friction coefficient from 0.11 to 0.13 leads to the formation of cracks in the corner radius.

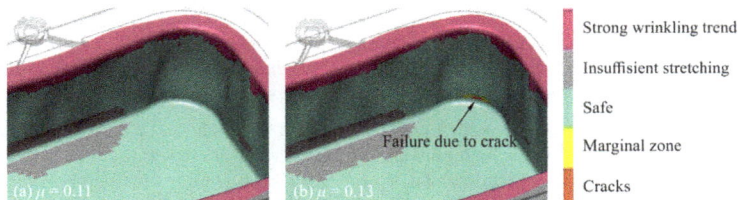

Fig. 4: Comparison of deep drawing simulations with different friction coefficients

Determining the exact design of the end position of the residual flange of the metal sheet proved to be a complex task, since the respective end shape is highly dependent on friction conditions, geometry and material properties, as well as on the injection moulding process. The forming simulation results for the final design of the generic test structure (Fig. 1) show that a uniform residual flange is formed (Fig. 5). Fig. 5b presents the subsequent forming of the sheet metal as a result of the injection moulding. A typical wrinkle formation in the corner area can be observed,

Sheet Metal 2025   Materials Research Forum LLC
Materials Research Proceedings 52 (2025) 276-283   https://doi.org/10.21741/9781644903551-34

which results from the reduction of the ring cross-section (component circumference) during subsequent forming by the polymer melt pressure.

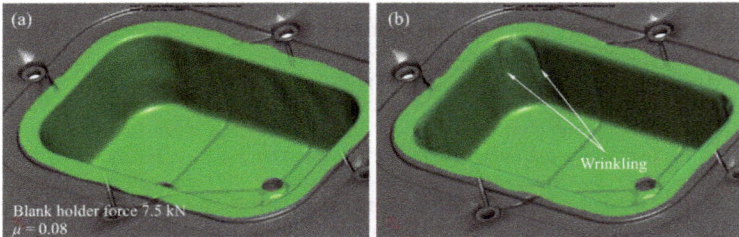

Fig. 5: Results of the forming simulations: (a) after the deep-drawing process with uniform flange formation and (b) after injection moulding (external high pressure forming in substitute model)

Manufacturing studies

Deep drawing. The manufacturing tests were carried out with an Arburg ALLROUNDER 370 H injection moulding machine with plasticizing unit 290. Initially, pure deep drawing tests were carried out in order to iteratively optimize the blank cutting and its positioning and to validate the forming simulation. It was found that no crack-free deep drawing was possible either with or without forming lubricants like polyethylene dispersion (Fig. 6b), even with varying blank holder forces between 4.5 kN and 10 kN. As the simulation has already shown, the process is very sensitive with regard to the friction coefficient due to the many interfering contours in the flange area. Only with an additional foil as a drawing aid it was possible to deep-draw the sheet without cracks. As a result, a circumferentially uniform flange area was created and the edges only showed minimal wrinkling (Fig. 6c), which is important as no trimming is carried out after deep drawing.

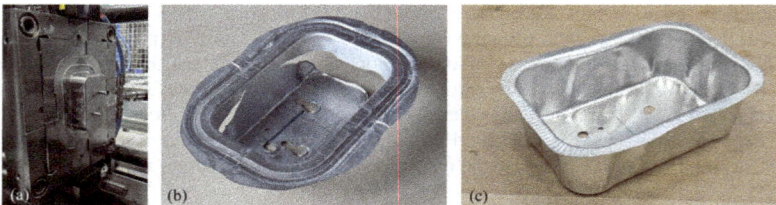

Fig. 6: (a) Ejector side of the mould with retracted blank holder and deep drawn sheet (b) with polyethylene dispersion with solid lubricant Raziol Drylub WA 03 T and (c) with additional Polifilm PFT 25/60 P PV4 L.Blue foil as lubricant (blank holder force 7 kN)

Injection moulding. As it is not expedient to overmould the metal sheet part that is covered with foil, the foil was removed outside the mould and the deep drawn sheet was then reinserted into the mould for overmoulding. The deep drawn sheets were successfully completely overmoulded (Fig. 7). The comparison with the results of the forming simulation in Fig. 4b shows very good concordance. The characteristic wrinkles can be found in all four corner areas, both in the simulation and in the real component. A sufficiently good, uniform residual flange formation can be observed (Fig. 7), as required in casing structures as contacting surface for electromagnetic shielding. Despite the slight waviness of the sheet metal, there is hardly any overmoulding of the metal sheet on the edge of the residual flange. Electromagnetic tests on two screwed housing halves will be conducted in further studies to assess the influence of housing geometry and the flange. Previous measurements on planar material demonstrated good attenuation in the range of 50 to 80 dB (at frequencies from 30 MHz to 1.5 GHz) [15]. An upscaling of the hybrid structure will also be tested. It should be noted that multi-stage deep drawing is not possible with this process, which leads to restrictions in the forming of the sheet metal (e.g. drawing ratio) and must be taken into account in the component design.

Specifications of the injection moulding process	
Melt temperature	250 °C
Metal sheet	Not preheated (Ambient temperature)
Mould temperature	50 °C
Injection rate	35 cm³/s
Clamping force	600 kN
Cooling time	20 s

Fig. 7: Generic test structure manufactured by deep drawing and injection moulding in one tool

Experimental investigations

Computed tomography. Computed tomography (CT) analyses were carried out to analyse the component structure in detail. In Fig. 8a, it can be observed that the metal sheet detaches from the polymer component in some areas. This is due to the different shrinkages of the materials, which causes the metal sheet to bulge partially. This bulging can be reduced or prevented by using appropriate form-fit elements, where holes are provided in the metal, for example, through which polymer is injected. This was exemplarily tested in the ground area of the generic test structure (Fig. 8b).

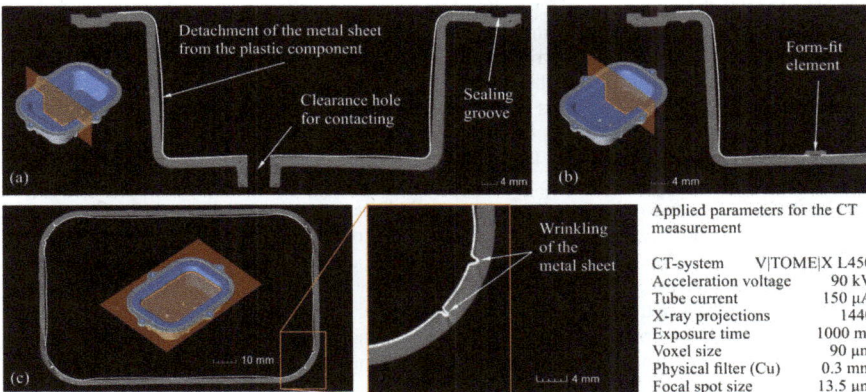

Applied parameters for the CT measurement	
CT-system	V|TOME|X L450
Acceleration voltage	90 kV
Tube current	150 µA
X-ray projections	1440
Exposure time	1000 ms
Voxel size	90 µm
Physical filter (Cu)	0.3 mm
Focal spot size	13.5 µm

Fig. 8: CT analysis of the generic test structure manufactured

The typical wrinkles in the corner areas are shown in Fig. 8c. It can be observed that there is sufficient residual wall thickness of the polymer part. Since the thin metal sheet is only intended to provide electromagnetic shielding in the housing structure and the polymer component transmits the mechanical loads, the wrinkles and detachment (local gap formation between polymer and metal) are not considered critical for the intended application. In order to reduce wrinkling and bulging due to residual stresses, the polymer melt could possibly be injected during the deep drawing process, although this would increase the demands on the process and the tool.

Recycling. Good recyclability mainly requires a good composite liberation of the structure, a sufficient separation characteristic for sorting and recyclable materials. The hybrid structure should be easy to disintegrate due to its design. The materials aluminium and thermoplastic have high characteristic differences in the range of electrical conductivity. This is used as a separation

Sheet Metal 2025
Materials Research Proceedings 52 (2025) 276-283

Materials Research Forum LLC
https://doi.org/10.21741/9781644903551-34

feature to achieve good sorting of the mixture of materials. The materials are then recycled to a high standard in the respective recycling routes.

The preferred variant of the mechanical recycling route was a (A) twin-shaft axial-split rotory shear with a stressing speed of approx. 0.5 m/s and a (B) single-shaft rotor shear (stressing speed approx. 10 m/s). The hybrid component was successfully shredded with both machines under shear load (Fig. 9) and liberated for sorting. With (A), 1.7 % residual mixed material remained and the fines content (< 2 mm) amounted to 0.8 %. In contrast, the more complex stress with a longer residence time in the process chamber in (B) led to complete liberation without residual non-liberated material. However, the amount of fines increased to 1.2 %.

Fig. 9: Unsorted and sorted shredded products from the generic test structure

Summary

The developments and investigations presented confirme the feasibility of combining the deep drawing of metal sheets and the injection moulding of polymers in one tool for the production of hybrid casing structures. This was demonstrated on a scaled generic test structure containing many construction elements of a housing structure. These include ribs, clearance holes, metallic bushes and a circumferential groove in the flange, which represent a particular challenge for sheet metal forming in order to explore the limits of the process. As a challenge, the metal sheets of the two halves of the housing need to be in contact over a large area at the flange in order to ensure sufficient electromagnetic shielding. This requires a defined metallic residual flange that is as flat as possible, with no trimming after deep drawing. For process development, FE simulations of the forming and injection moulding processes were carried out and validated. These form an important basis for the future design of complex components. Due to the challenging interfering contours in the flange area, crack-free deep drawing was only possible with a foil as a lubricant, which was removed before injection moulding. In order to combine deep drawing and injection moulding directly in one process in the future, the interfering contours in the flange area or the drawing depth should be reduced. The production tests showed that the residual flange of the metal sheet is sufficiently well and evenly formed, as required in casing structures as a contacting surface for electromagnetic shielding. The metal sheet and the polymer component are not joined by adhesion but rather by form fit enabling composite liberation at the end of the product life cycle, as demonstrated by mechanical recycling tests.

Acknowledgements

The authors would like to thank all project partners of the EAC+ project. The EAC+ research and development project is funded by the Federal Ministry of Economic Affairs and Climate Action (BMWK) and supervised by Project Management Jülich (PTJ), Funding ref.: 03LB3022.

References

[1] A. Al-Sheyyab, Light-Weight Hybrid Structures - Process Integration and Optimized Performance, Dissertation, Friedrich-Alexander-Universität Erlangen-Nürnberg (2008).

[2] W. Koshukow, A. Liebsch, J. Wippermann, B. Kolbe, R. Kupfer, J. Troschitz, M. Buske, G. Meschut, M. Gude, Influence of Plasma Coating Pretreatment on the Adhesion of

Thermoplastics to Metals, Proceedings of the Munich Symposium on Lightweight Design (2022) 85-96. https://doi.org/10.1007/978-3-031-33758-1_7

[3] S. Farahani, Polymer Injection Forming: A New Age Technology for Manufacturing Polymer-Metal Hybrids, Dissertation, Clemson University (2018).

[4] S. Farahani, V. A. Yerraa, S. Pilla, Analysis of a hybrid process for manufacturing sheet metal-polymer structures using a conceptual tool design and an analytical-numerical modelling, J. Mater. Process. Technol. 279 (2020) 116533. https://doi.org/10.1016/j.jmatprotec.2019.116533

[5] M.M. Hussain, B. Rauscher, M. Trompeter, A. E. Tekkaya, Potential of Melted Polymer as Pressure Medium in Sheet Metal Forming, Key Engineering Materials 410-411 (2009) 493-501. https://doi.org/10.4028/www.scientific.net/KEM.410-411.493

[6] G. Lucchetta, R. Baesso, Polymer Injection Forming (PIF) Of Thin-Walled Sheet Metal Parts - Preliminary Experimental Results, AIP Conf. Proc. 907 (2007) 1046-1051. https://doi.org/10.1063/1.2729652

[7] A.E. Tekkaya, M.M. Hussain, J. Witulski, The non-hydrostatic response of polymer melts as a pressure medium in sheet metal forming, Prod. Eng. Res. Devel. 6 (2012) 385-394. https://doi.org/10.1007/s11740-012-0392-8

[8] M.M. Hussain, M. Trompeter, J. Witulski, A.E. Tekkaya, An Experimental and Numerical Investigation on Polymer Melt Injected Sheet Metal Forming, ASME. J. Manuf. Sci. Eng. 134(3) (2012) 031005. https://doi.org/10.1115/1.4006117

[9] W. Michaeli, R. Maesing, Injection Moulding and Metal Forming in One Process Step, Progress in Rubber, Plastics and Recycling Technology, 26(4) (2010) 155-166. https://doi.org/10.1177/147776061002600401

[10] S. Farahani, A. F., Arezoodar, B. M., Dariani, S. Pilla, An Analytical Model for Nonhydrostatic Sheet Metal Bulging Process by Means of Polymer Melt Pressure, J. Manuf. Sci. Eng. September 140(9) (2018) 091010. https://doi.org/10.1115/1.4040429

[11] W.-G. Drossel, C. Lies, A. Albert, R. Haase, R. Müller, P. Scholz, Process combinations for the manufacturing of metal-plastic hybrid parts. IOP Con-ference Series: Materials Science and Engineering 118 (2016) 1-10. https://doi.org/10.1088/1757-899X/118/1/012042

[12] A. Albert, W. Zorn, M. Layer, W.-G. Drossel, D. Landgrebe, L. Kroll, W. Nendel, Smart Process Combination for Aluminum/Plastic Hybrid Components, Technologies for Lightweight Structures 1(2) (2017) 44-53. https://doi.org/10.21935/tls.v1i2.91

[13] D. Landgrebe, V. Kräusel, A. Rautenstrauch, A. Albert, R. Wertheim, Energy-efficiency in a hybrid process of sheet metal forming and polymer injection moulding, Procedia CIRP 40 (2016) 109-114. https://doi.org/10.1016/j.procir.2016.01.068

[14] J. Troschitz, S. Bräunling, M. Kahl, F. Schneider, F. Folprecht, L. Schilling, M. Gude, A. Brosius, Kombination von Spritzgießen und Metallumformung in einem Prozess, 28. Fachtagung über Verarbeitung und Anwendung von Polymeren (2023) 1.11.

[15] M. Gruber, M. Beltle, S. Tenbohlen, Measurement and Simulation of the Shielding Effectiveness of Planar Material with Apertures using a ASTM D4935 TEM Cell, International Symposium on Electromagnetic Compatibility - EMC Europe, Krakow, Poland (2023) 1-5. https://doi.org/10.1109/EMCEurope57790.2023.10274278

Machine learning

Sheet Metal 2025
Materials Research Proceedings 52 (2025) 285-292

Materials Research Forum LLC
https://doi.org/10.21741/9781644903551-35

Impact of the parameter distribution on the predictive quality of metamodels for clinch joint properties

Jonathan-Markus Einwag[1,a]*, Yannik Mayer[1,b], Stefan Goetz[1,c] and Sandro Wartzack[1,d]

[1]Friedrich-Alexander Universität Erlangen-Nürnberg, Engineering Design, 91058 Erlangen, Germany

[a]einwag@mfk.fau.de, [b]yannik.mayer@fau.de, [c]goetz@mfk.fau.de, [d]wartzack@mfk.fau.de

Keywords: Uncertainty, Machine Learning, Mechanical Joining

Abstract. The growing significance of lightweight design, reveals drawbacks of conventional joining processes such as welding, which are known to consume a considerable amount of energy. This fosters the use of mechanical joining processes including clinching. However, the lack of universally applicable design methods results in a cost- and time-intensive design process. The utilization of machine learning methods can overcome these drawbacks. To ensure a reliable clinch joint design, inherent uncertainties of the design parameter such as tool deviations need to be considered in the design process. Varying distributions of design parameters, due to changes in the manufacturing process, can lead to high-computational effort in recalculating the resulting clinch joint properties with numerical simulations. Current metamodel-based methods for consideration of inherent uncertainties within the design parameters do not investigate the transferability of metamodels to different distributions of design parameters, which can lead to incorrect predictions. Therefore, this contribution investigates the performance of several metamodels on differently distributed design parameters. The obtained results indicate that metamodels demonstrate the best performance when training and evaluation distributions are identical and that polynomial regression models perform best on disparate distributions, when trained on uniform distributions.

Introduction

Climate change, a major driver of technological innovation, is contributing to the growing importance of lightweight design [1]. Conventional joining processes such as welding, are not suitable for new, resource-efficient lightweight designs, due to their high energy consumption and their inability to join dissimilar materials [2]. This has led to the emergence of the use of mechanical joining processes such as clinching. To design high strength clinch joints, it is essential to comprehend the relationship between the tool geometry and the resulting clinch joint properties [3]. These are primarily defined by three geometrical properties, the neck thickness, the interlock and the bottom thickness [4] as well as the shear- and tensile strength. Consequently, numerous researches have examined the relationship between the tool geometry and the resulting joint properties, employing experimental studies and numerical simulations [3]. The influence of the geometry of the clinching tools on the neck thickness and the interlock is examined in [5–7], while [6] additionally investigates the tensile strength of the clinch joint. It can be generally concluded, that when subjected to tensile load, a small interlock results in the separation of the upper sheet from the lower sheet, whereas a thin neck can lead to a fracture in the neck area [3]. These phenomena both reduce the strength of the joint [3], which leads to a simultaneous optimization problem of maximizing the neck thickness and the interlock. This optimization problem is solved in [8] based on the results of [6] by using a response surface methodology (RSM) to describe the relationship between the tool geometry and the clinch joint properties and an automatic optimization procedure. Based on this results, four design parameters were further optimized in [9] with the RSM and a Kriging interpolation to achieve a higher tensile strength of

Content from this work may be used under the terms of the Creative Commons Attribution 3.0 license. Any further distribution of this work must maintain attribution to the author(s) and the title of the work, journal citation and DOI. Published under license by Materials Research Forum LLC.

Sheet Metal 2025   Materials Research Forum LLC
Materials Research Proceedings 52 (2025) 285-292   https://doi.org/10.21741/9781644903551-35

the joint. Another approach to model the relationship between the tool geometry and the clinch joint properties, namely the interlock, the neck thickness and the tensile strength, with an optimization of the latter is described in [10]. To investigate the influence of upstream manufacturing processes, the influence of the pre-straining of the sheets on the tensile- and shear strength of the clinch joint is investigated in [11] by employing linear and quadratic regression models. Since these approaches mainly focus on the optimization of existing clinch joints, they cannot be used to design joints for a new joining task. Therefore, in [3] an analytical method for calculating the needed geometric properties for the required tensile strength of the clinch joint was derived. This method is used to iteratively vary the most influential parameters of the tool geometry, followed by a numerical simulation to achieve the required geometric properties and determine the parameters of the tool geometry [3]. In [12] an assistance system for the design of joining parts is proposed, where several metamodels are evaluated and employed. In combination with parameterized CAD-parts and manufacturing knowledge, it ensures a high-quality initial design. However, these methods only focus on a nominal clinch joint design and are missing the consideration of uncertainties of the design parameters, such as inevitable variations of the tool geometry or process parameters. This can result in incorrect estimations of the real clinch joint properties and therefore should be considered in the design process [13, 14]. To analyze the influence of varying design parameters, in [15, 16] a sensitivity analysis of uncertain parameters, such as varying material and geometric tool parameters, on geometric clinch joint properties is executed. Subsequently, the failure probability of the clinch joint is reduced with a robustness analysis with optimized tool parameters. To allow for a more realistic estimation of clinch joint properties with an additional confidence interval, a gaussian process regression model is employed in [13], to consider uncertainties of the material and friction conditions. While this method shows a high potential, it completely neglects uncertainties of the tool geometry. These are considered in [17], where the transferability of metamodels trained on nominal data for first estimations of deviations of clinch joint properties is investigated by generating a database with a variation of the punch diameter and the die depth according to general tolerances [18]. The calculated and clustered uncertainty values can then be assigned to the nominal metamodels as a first estimation of deviations of clinch joint properties. In [14], the impact of the distributions of uncertain process parameters, including friction between tools and sheets and between sheets, is examined. A metamodel trained on uniformly distributed design parameters is employed to assess the influence of uniform and gaussian distributed design parameters on the geometric clinch joint properties.

As a change of the manufacturing process can lead to a change of the actual distributions [19], it can result in high-computational effort in recalculating the distribution of the resulting clinch joint properties. Overall, it would be useful for the designer to know the actual underlying distributions of the design parameters to predict the actual distributions of the clinch joint properties. However, this would require extensive and time-consuming measurements of a considerable number of samples of the design parameters, e.g. of the sheet thicknesses and the tool geometry, and may lead to disparate distributions due to varying batch sizes [20]. As a consequence, the usage of a metamodel, that can be used to sufficiently estimate the distribution of clinch joint properties, can result in cost and time savings and allows for the definition of safe process windows, such as the allowed deviations of the tool geometry. For the metamodel to be applicable to disparate distributions, it is essential that it exhibits a high degree of generalizability. Although the training and evaluation distributions differ in [14], the sufficiency of the metamodel for considering different distributions is not investigated. Due to this, this publication aims to identify the optimal training distribution and metamodel when uncertainties of design parameters are investigated. This in turn gives rise to the following research question:

What is the optimal training distribution of training data for metamodels predicting the clinch joint properties, considering the inherent uncertainties within the design parameters?

Sheet Metal 2025
Materials Research Proceedings 52 (2025) 285-292

Materials Research Forum LLC
https://doi.org/10.21741/9781644903551-35

Methodical approach for the data generation and metamodeling

As shown in Fig. 1, the approach for the investigation of the influence of the design parameter distribution on the performance of metamodels can be classified in two parts, the generation of data (left) and the training as well as the evaluation of the metamodels (right).

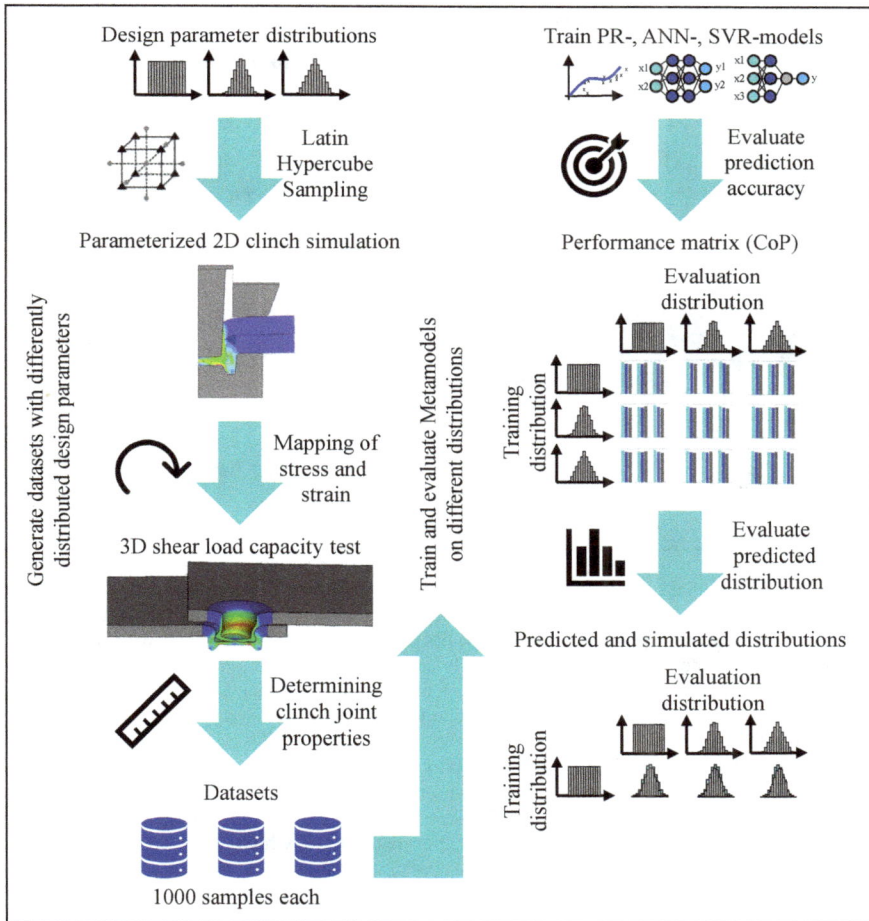

Figure 1: Methodical approach for data generation and metamodeling

In the data generation step, parametric studies with 9 design parameters in range of general tolerances [18] for the tools, sheet thickness tolerances according to [21] and further assumptions are executed, as illustrated in Table 1. In these ranges, datasets with 3 distributions are generated, namely uniform, normal and a realistic one, combining triangular distribution of the geometric tool parameters, as it is common for small series, a normal distribution for the sheet thickness, as it can be assumed for series production and a uniform distribution of the process parameter punch velocity *pvel* and remaining sheet thickness *rst* [20]. For the normal distribution, it is estimated that 99.73% (6σ) of parts are of a satisfactory quality, whereas all samples with uniform and triangular distributed design parameters fall within the specification limit. The Latin Hypercube Sampling method is employed to guarantee comprehensive coverage of the parameter space [22].

Table 1: Design parameters of the 2D-clinch process simulation

Design Parameter	Minimum Value	Maximum Value
Punch diameter pd	4.95 mm	5.05 mm
Die depth hd	1.25 mm	1.35 mm
Die bottom diameter db	4.95 mm	5.05 mm
Die groove diameter dg	6.1 mm	6.3 mm
Die diameter dd	7.9 mm	8.1 mm
Punch-sided sheet thickness st1	1.95 mm	2.05 mm
Die-sided sheet thickness st2	1.95 mm	2.05 mm
Remaining sheet thickness rst	0.75 mm	0.85 mm
Punch velocity pvel	1.95 mms⁻¹	2.05 mms⁻¹

The simulation chain consists of an adapted, parameterized 2D-Simulation of the clinching process from [13, 23], followed by a 3D-shear load capacity test with a mapping of the stress and strain from the clinching process analogous to [24]. All simulations are solved with the implicit LS-Dyna solver in the version R9. Subsequently an automatic evaluation of the geometric clinch joint properties, such as the neck thickness (NE) and the interlock (IL), similar to [25] and of the maximum shear load capacity (SF) is carried out. In the second step (Fig. 1, right), suitable metamodels for the prediction of clinch joint properties, an artificial neural network (ANN), a polynomial regression model (PR) and a support vector regression model (SVR), identified in earlier studies [12, 26], are trained and evaluated on these datasets. In this contribution, the models from the scikit-learn library are utilized [27]. Each metamodel algorithm is trained and evaluated for each clinch joint property and design parameter distribution, which results in 3^3 metamodels. A training and test split is applied to the data, whereby the training set comprises 80% of the data and the test set 20%. For the ANN and SVR models the input data is normalized and a preceding optimization step ensures optimal arguments.

Results and Discussion
In order to evaluate the performance of the metamodels on all datasets, a variety of evaluation techniques are employed. The coefficient of prognosis (CoP) [28], calculated according to Eq. 1., is employed to estimate the prediction accuracy of the metamodels as illustrated in Fig. 2. The CoP value is expressed on a scale from 0 to 1, with a value of at least 0.8 being considered indicative of sufficient precision [28].

$$\text{CoP}(y_p, y_t) = \left(\frac{\sum_{i=1}^{N} (y_p - \bar{y}_p) \times (y_t - \bar{y}_t)}{(N-1) \times \sigma_p \times \sigma_t} \right)^2. \tag{1}$$

According to the performance matrix in Fig. 2, the results demonstrate that most metamodels perform well on equal training and evaluation distributions, as only the PR and SVR models attain values marginally below the CoP threshold for the SF on normal and real distributed design parameters. Conversely, metamodels exhibit the lowest accuracy when trained on normal distributions and evaluated on uniform or real distributed design parameters. In particular, the PR model demonstrate suboptimal performance in this regard, whereas only the ANN model exhibits a satisfactory performance across all clinch joint properties. A similar, but less pronounced trend is observed in the case of real distributed design parameters. The metamodels trained on uniform parameters exhibit the best performance, when differently distributed design parameters were assumed, especially the PR model exhibits superior performance when trained with uniformly distributed design parameters in comparison to when the same distribution is employed for both training and evaluation. In this regard, the ANN and SVR models exhibit a performance that is relatively consistent, or slightly less inferior.

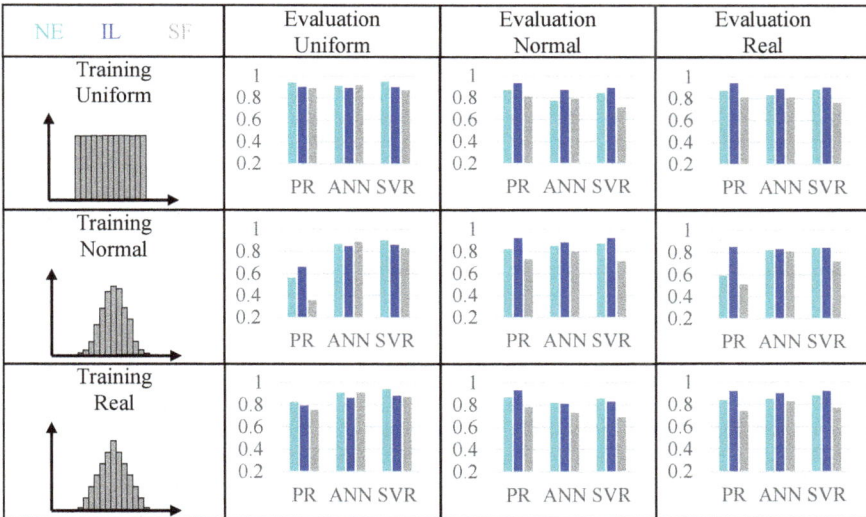

Figure 2: Performance matrix (CoP-Values) of all metamodels trained and evaluated on each distribution, NE (green), IL (blue), SF (grey).

As the CoP calculates the prediction accuracy of the metamodel on the entire test data, it is unable to evaluate whether the metamodel performs equally across the entire parameter space. This may result in a satisfactory overall performance, with a notable decline in the peripheral regions and can lead to incorrect predictions of the distribution of the clinch joint properties. Consequently, it is evident that further investigation of the predicted and simulated distributions of the clinch joint properties is required. Given its performance on diverse distributed datasets, the PR model trained on the uniform dataset is employed in this investigation. Initially, a Kolmogorov-Smirnov test is employed to ascertain the presence of a normal distribution for the predicted and simulated clinch joint properties, which yielded p-values above 0.05. This permits the supposition of normal distributions [29]. Subsequently, representative values, such as the mean value (μ) and three times the standard deviation (3σ), are employed to assess the discrepancy between the prediction of the metamodel and the simulation. Figure 3 compares the predicted and simulated values of μ and 3σ.

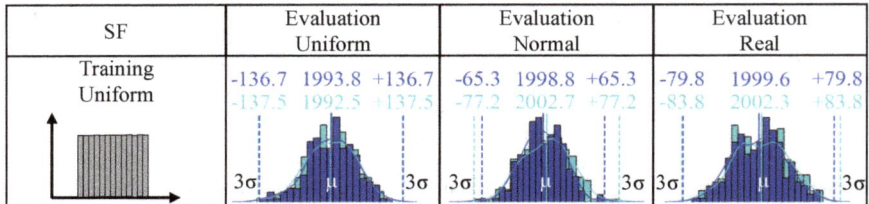

Figure 3: Predicted distributions of SF of PR-model for uniform, normal and real distributed design parameters, when trained on the uniform dataset

Overall, the investigation reveals a high degree of concordance of the predicted and simulated distributions, as a minor maximum discrepancy is observed for the μ (0.19 %) and the 3σ values (16.7 %) for the evaluation on normal distributed design parameters. Furthermore, it can be observed, that even with a similar CoP, the predicted and simulated 3σ values exhibit disparate degrees of variation, which underlines the necessity of this investigation. When trained and evaluated on normally distributed design parameters, the PR model shows a marginal improvement for the μ and for the 3σ values, while for real distributed training and evaluation parameters, the differences mainly remain the same. For identical training and evaluation distributions, the more complex models demonstrate superior performance. Reflecting the research question, Fig. 2 depicts, that most metamodels exhibit a high ability to estimate the clinch joint properties with uncertainties of 9 design parameters. The superior performance of ANN and SVR on normal and real distributions may stems from their adaptability and optimization algorithms, while PR's lower adaptability enhances its generalizability and transferability across diverse distributions.

Conclusion and Outlook
In this contribution, the impact of the training distribution on the predictive quality of metamodels is investigated. Therefore, three datasets with differently distributed design parameters are generated with a process simulation chain, consisting of a 2D-process simulation, followed by a 3D-shear load capacity test with a mapping of the stress and strain. Subsequently, three metamodeling algorithms (PR, ANN, SVR) are trained and evaluated on each dataset. The results show that metamodels can be used for the prediction of clinch joint properties, when uncertainties of the design parameters need to be considered. While all metamodels demonstrate the best performance (CoP) on the trained distribution, the use of a uniform distribution and a PR model is recommended when the actual distribution of design parameters remains uncertain. In other instances, it would be more beneficial to utilize the actual distribution and a more complex model, such as an ANN or SVR. This outcome provides a definitive answer to the research question. As demonstrated in [14], the impact of uncertainties of the design parameters varies across different nominal designs. Consequently, further investigations of the transferability of the metamodels presented in this contribution to different nominal designs should be conducted. Therefore, new datasets with different nominal designs should be generated, which can then be used for evaluation. This would permit an investigation into whether new simulations and metamodels are required in the event of a change of the nominal design. Moreover, the current databases are comprised solely of numerical generated data. Consequently, additional experiments are required to ascertain the

sufficiency of the metamodels, which would allow for the identification of current deficiencies of the metamodels and enable their refinement through a comparison with real observations.

Acknowledgement

Funded by the Deutsche Forschungsgemeinschaft (DFG, German Research Foundation) – TRR – 285/2 418701707

References

[1] Ewenz, Lars; Kuczyk, Martin; Zimmermann, Martina: Effect of the tool geometry on microstructure and geometrical features of clinched aluminum. In: Journal of Advanced Joining Processes Vol. 5 (2022), S. 100091. https://doi.org/10.1016/j.jajp.2021.100091

[2] Kascak, L.; Spisak, E.; Majernikova, J.: Clinching and Clinch-Riveting as a Green Alternative to Resistance Spot Welding. In: 2019 International Council on Technologies of Environmental Protection (ICTEP). Starý Smokovec, Slovakia: IEEE, 2019, S. 138-142. https://doi.org/10.1109/ICTEP48662.2019.8968973

[3] Lee, Chan-Joo, et al.: Design of mechanical clinching tools for joining of aluminium alloy sheets. In: Materials & Design Vol. 31 (2010), Nr. 4, S. 1854-1861. https://doi.org/10.1016/j.matdes.2009.10.064

[4] DVS-Merkblätter und -Richtlinien, Mechanisches Fügen (2009).

[5] De Paula, A.A., et al.: Finite element simulations of the clinch joining of metallic sheets. In: Journal of Materials Processing Technology Vol. 182 (2007), Nr. 1-3, S. 352-357. https://doi.org/10.1016/j.jmatprotec.2006.08.014

[6] Oudjene, M.; Ben-Ayed, L.: On the parametrical study of clinch joining of metallic sheets using the Taguchi method. In: Engineering Structures Vol. 30 (2008), Nr. 6, S. 1782-1788. https://doi.org/10.1016/j.engstruct.2007.10.017

[7] Lambiase, Francesco: Influence of process parameters in mechanical clinching with extensible dies. In: The International Journal of Advanced Manufacturing Technology Vol. 66 (2013), Nr. 9-12, S. 2123-2131. https://doi.org/10.1007/s00170-012-4486-4

[8] Oudjene, M.; Ben-Ayed, L.; Batoz, J.-L.: Geometrical Optimization Of Clinch Forming Process Using The Response Surface Method. In: AIP Conference Proceedings. Vol. 908. Porto (Portugal): AIP, 2007, S. 531-536. https://doi.org/10.1063/1.2740865

[9] Lebaal, Nadhir; Oudjene, Marc; Roth, Sébastien: The optimal design of sheet metal forming processes: application to the clinching of thin sheets. In: International Journal of Computer Applications in Technology Vol. 43 (2012), Nr. 2, S. 110. https://doi.org/10.1504/IJCAT.2012.046041

[10] Lambiase, F.; Di Ilio, A.: Optimization of the Clinching Tools by Means of Integrated FE Modeling and Artificial Intelligence Techniques. In: Procedia CIRP Vol. 12 (2013), S. 163-168. https://doi.org/10.1016/j.procir.2013.09.029

[11] Martin, Sven, et al.: Numerical investigation of the clinched joint loadings considering the initial pre-strain in the joining area. In: Production Engineering Vol. 16 (2022), Nr. 2-3, S. 261-273. https://doi.org/10.1007/s11740-021-01103-w

[12] Zirngibl, Christoph, et al.: Knowledge and Data-Based Design and Dimensioning of Mechanical Joining Connections. In: Volume 2: 42nd Computers and Information in Engineering Conference (CIE). St. Louis, Missouri, USA: American Society of Mechanical Engineers, 2022. https://doi.org/10.1115/DETC2022-89172

[13] Zirngibl, Christoph; Schleich, Benjamin; Wartzack, Sandro: Robust estimation of clinch joint characteristics based on data-driven methods. In: The International Journal of Advanced

Manufacturing Technology Vol. 124 (2023), Nr. 3-4, S. 833-845.
https://doi.org/10.1007/s00170-022-10441-7

[14] Zirngibl, C; Goetz, S; Wartzack, S: Influence of process variations on clinch joint
characteristics considering the effect of the nominal tool design. In: Proceedings of the
Institution of Mechanical Engineers, Part E: Journal of Process Mechanical Engineering (2024).
https://doi.org/10.1177/09544089241259347

[15] Drossel, Welf-Guntram; Israel, Markus; Falk, Tobias: Robustness evaluation and tool
optimization in forming applications.

[16] Drossel, Welf-Guntram, et al.: Unerring Planning of Clinching Processes through the Use of
Mathematical Methods. In: Key Engineering Materials Vol. 611-612 (2014), S. 1437-1444.
https://doi.org/10.4028/www.scientific.net/KEM.611-612.1437

[17] Bode, Christoph; Goetz, Stefan; Wartzack, Sandro: On the transferability of nominal
surrogate models to uncertainty consideration of clinch joint characteristics. In: Procedia CIRP
(2024). https://doi.org/10.1016/j.procir.2024.10.027

[18] DIN ISO 2768-1:1991-06, Allgemeintoleranzen; Toleranzen für Längen- und Winkelmaße
ohne einzelne Toleranzeintragung; Identisch mit ISO 2768-1:1989.

[19] Nigam, Swami D; Turner, Joshua U: Review of statistical approaches to tolerance analysis.
In: Computer-Aided Design Vol. 27 (1995), Nr. 1, S. 6-15. https://doi.org/10.1016/0010-
4485(95)90748-5

[20] Klein, Bernd: Prozessorientierte statistische Tolerierung: mathematische Grundlagen -
Toleranzverknüpfungen - Prozesskontrolle - Maßkettenrechnung - Praktische Anwendung; mit
61 Tabellen, Haus der Technik - Fachbuch. Renningen: expert-Verl, 2007.

[21] DIN EN 485-4:2019-05, Aluminium und Aluminiumlegierungen - Bänder, Bleche und
Platten - Teil 4: Grenzabmaße und Formtoleranzen für kaltgewalzte Erzeugnisse; Deutsche
Fassung EN 485-4:1993.

[22] Fisher, R. A.: The design of experiments. 9. ed. New York: Hafner Press, 1974.

[23] Bielak, C. R., et al.: Numerical analysis of the robustness of clinching process considering
the pre-forming of the parts. In: Journal of Advanced Joining Processes Vol. 3 (2021), S.
100038. https://doi.org/10.1016/j.jajp.2020.100038

[24] Bielak, Christian Roman, et al.: Further development of a numerical method for analyzing
the load capacity of clinched joints in versatile process chains. In: ESAFORM 2021 (2021).
https://doi.org/10.25518/esaform21.4298

[25] Zirngibl, Christoph; Schleich, Benjamin: Approach for the Automated Analysis of
Geometrical Clinch Joint Characteristics. In: Key Engineering Materials Vol. 883 (2021), S. 105-
110. https://doi.org/10.4028/www.scientific.net/KEM.883.105

[26] Zirngibl, Christoph; Schleich, Benjamin; Wartzack, Sandro: APPROACH FOR THE
AUTOMATED AND DATA-BASED DESIGN OF MECHANICAL JOINTS. In: Proceedings
of the Design Society Vol. 1 (2021), S. 521-530. https://doi.org/10.1017/pds.2021.52

[27] Pedregosa, Fabian, et al.: Scikit-learn: Machine Learning in Python. In: MACHINE
LEARNING IN PYTHON.

[28] Most, Thomas; Will, Johannes: Meta-model of Optimal Prognosis - An automatic approach
for variable reduction and optimal meta-model selection (2008).

[29] Massey, Frank J.: The Kolmogorov-Smirnov Test for Goodness of Fit. In: Journal of the
American Statistical Association Vol. 46 (1951), Nr. 253, S. 68-78.
https://doi.org/10.1080/01621459.1951.10500769

Sheet Metal 2025
Materials Research Proceedings 52 (2025) 293-300

Materials Research Forum LLC
https://doi.org/10.21741/9781644903551-36

Transient dynamic analysis: Performance evaluation of tactile measurement

Gregor Reschke[1,a] and Alexander Brosius[1,b]

[1]Chair of Forming and Machining Technology, TUD Dresden University of Technology, Germany

[a]gregor.reschke@tu-dresden.de, [b]alexander.brosius@tu-dresden.de

Keywords: Joining, Machine Learning, Transient Dynamic Analysis

Abstract. The assessment of mechanically joined connections, such as clinched connections, is usually conducted destructively. Applicable non-destructive testing methods like computed tomography are time-consuming and costly, or, like electrical resistance measurement, provide only a limited amount of information. A fast, non-destructive evaluation of the joints condition shall be made possible by using transient dynamic analysis (TDA). It is based on the introduction of sound waves and the evaluation of the response behavior after passing through the structure. This study focuses the application of TDA to clinched shear connections to evaluate the performance of the tactile measuring setup. Twenty-one series were investigated, covering variations in joining task, manufacturing and defect. The evaluation was carried out using machine learning to determine for which series characteristic signals may be detected. It was shown that a classification of the investigated specimens is possible, whereby the classification accuracy depends on the examined variation. Furthermore, the accuracy was evaluated as a function of frequency and results were concluded to identify the limits of the used measuring setup.

Introduction

Mechanical joining processes are based on the binding mechanisms form and force closure, which makes it particularly possible to join mixed compounds [1]. The persistent trend towards mixed-material construction in the automotive sector has led to significant development of this group of joining processes [2]. Mechanically joined connections can be designed not only to transfer mechanical loads, but also to conduct electricity [3]. This paper refers to clinching as a representative of this process group for the investigations. It is defined as joining by forming process utilizing a punch and die. For a comprehensive overview of this joining process, please refer to [1].

Typically, the characterization of clinched joints is performed destructively, for example, utilizing micrographs and mechanical tests [4]. Non-destructive characterizations are possible through computer tomography (CT) or electrical resistance measurements. The latter enables a rapid evaluation of conductivity, but provides only limited information, for example regarding the form closure. The CT provides detailed insights into the joining process and damage progress, but only enables a time- and cost-intensive analysis of discrete states [5]. Non-destructive testing of clinching joints with high temporal resolution is to become possible by means of transient dynamic analysis (TDA). This involves introducing elastic deformation waves into the structure and evaluating the damped response behavior of said structure as described in [6]. The overarching goal is to establish a conclusive correlation between signals and the underlying characteristics of the joint. Previous experimental and numerical investigations show qualitatively significant signal changes of the TDA signal due to emerging damages [7] or changes in the joining process [8]. The current research focuses the evaluation of this analysis, whereby the degrees of freedom of the system and the existing data volumes pose particular challenges.

Content from this work may be used under the terms of the Creative Commons Attribution 3.0 license. Any further distribution of this work must maintain attribution to the author(s) and the title of the work, journal citation and DOI. Published under license by Materials Research Forum LLC.

Sheet Metal 2025
Materials Research Proceedings 52 (2025) 293-300

Materials Research Forum LLC
https://doi.org/10.21741/9781644903551-36

One approach to analyzing complex data is machine learning. It allows analyzing large, multidimensional data with regard to existing structures or making predictions based on known input variables [9]. In the branch of supervised learning, a model is trained to describe a given data set. This model is then validated and evaluated on a basis of previously unconsidered input data [10]. To evaluate classification models, the accuracy can be used to evaluate all correct classifications or the F1 score to take incorrect classifications into account. These characteristic values are widely used for machine learning tasks. Please refer to [11] for the exact calculation of these and other characteristic values.

Methods

The experiments are carried out on clinched shear tensile joints. Clinch points with a nominal diameter of 8 mm were used with punch diameters ranging from 4.8 mm to 5.6 mm. The geometry of these specimens is based on DVS/EFB 3480-1 [4] with an additional pre-hole (Ø 3.3 mm) per joining partner for fixing the actuators, see Fig. 1. The aluminum components are joined in the T4 condition and are then, unless otherwise specified, artificially aged at 185 °C for 20 minutes to achieve the long-term stable T6 condition for the measurements. In industrial application the T6 condition is as standard for the product service life. A minimum of 10 specimens are examined for each series. All specimens show a defined tool offset of 0.2 mm, due to misalignment of the joining system.

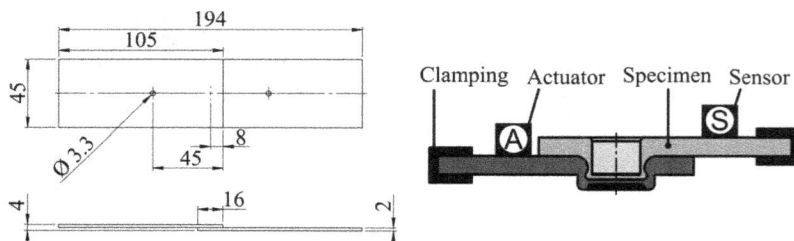

Fig. 1: Geometric dimensions of the specimens (left) and measuring setup for TDA

The measurement setup for the TDA includes a clamping for the specimen, the actuator and the sensor as displayed in Fig. 1 right. Piezoelectric stack actuators type P-016.00h (PI Ceramic GmbH, Lederhose, Germany) are used as actuator and sensor. The stack actuators are additionally equipped with a seismic mass (m = 9.5 g) and are axially preloaded (M_A = 1.5 Nm) using a M3 bolted connection to ensure reliable transmission of the vibrations.

During the measurement, elastic deformation waves are introduced into the specimen via the actuator. After passing through the clinching point, the response behavior is recorded at the sensor. For each specimen, the frequency band from 0.1 to 20 kHz is examined. A sinusoidal oscillation with a defined excitation frequency is introduced into the specimen for 2 seconds. After the settling process, the response behavior is recorded in the last 500 ms of the excitation. The sampling rate equals to eight times the excitation. This procedure is repeated in 100 Hz steps over the analyzed frequency band. In the evaluation of the measurement data, both time- and frequency-discrete characteristic values are determined. For this purpose, a fast Fourier transformation is carried out with the maximum evaluated frequency limited to 30 kHz or four times the excitation frequency. This takes into account the Nyquist–Shannon sampling theorem and avoids therefore aliasing effects.

A total of 21 series are joined and analyzed. These are divided into three sections and differ in terms of the joining task, the manufacturing process and the presence of defects. Table 1 contains an overview on the test series.

Table 1: Overview on test series

	Series	Combination	Characteristics
Joining Task	496-506	1	
	518-528	3	T6
	529-539	4	T6
	540-550	5	T6
	551-561	7	T6
	562-572	8	T6
Manufactering	573-583	7	T6, bottom thickness 0.8 mm
	584-594	7	T6, bottom thickness 0.6 mm
	595-605	7	T6, punch A48100
	606-616	7	T6, punch AB46100
	617-627	7	T6, punch AC48100
	726-736	7	T6, sheets degreased
	737-747	7	T6, sheets degreased & defined oiled
Defects	628-638	7	T4
	639-649	7	Heat treatment: 20 min, 220 °C
	650-660	7	T6, torsion 5 ° & turn back
	661-671	7	T6, torsion 30 ° & turn back
	694-704	7	T6, tension s = 0.3 mm
	672-682	7	T6, tension s = 0.5 mm
	683-693	7	T6, tension s = 1.0 mm
	705-725	7	T6, positional deviation on flange

The joining tasks, including the tools and the residual bottom thickness, are presented in Table 2. These combinations and interrelated joining parameters correspond to the joints sampled in Transregional Collaborative Research Centre 285/2. The internal numbering of the combination was adopted. The labels on the punches and dies are as specified by the system manufacturer TOX Pressotechnik. In the "manufacturing" group, the formation of the clinching point is modified by varying the bottom thickness, the punch used and the tribological properties between the joining partners. The defects investigated include errors in heat treatment, plastic mechanical stress (tension and torsion) and incorrect positioning of the clinched joints on the flange. The investigations on manufacturing and defects are based on combination 7.

In total, TDA measurements from 222 samples are subjected to evaluation here. Based on the TDA measurement methodology described, 200 individual measurements are present for each examined specimen. This results in a total of 44,400 data sets.

Sheet Metal 2025
Materials Research Proceedings 52 (2025) 293-300

Materials Research Forum LLC
https://doi.org/10.21741/9781644903551-36

Table 2: Materials and joining parameters of different combinations

Combi-nation	Punch sided sheet	Die sided sheet	Punch	Die	Bottom thickness [mm]
1	HCT590X 1.50 mm	HCT590X 1.50 mm	A50100	BD8016	0.75
3	EN AW-6014 2.00 mm	HCT590X 1.50 mm	AC56100	BD8016	0.80
4	EN AW-6014 1.00 mm	HCT590X 1.50 mm	AC56100	BC8014	0.70
5	HCT590X 1.50 mm	EN AW-6014 2.00 mm	A48100	BB8008	0.85
7	EN AW-6014 2.00 mm	EN AW-6014 2.00 mm	A50100	BE8012	0.70
8	EN AW-6014 1.00 mm	EN AW-6014 2.00 mm	A48100	BE8010	0.90

Results and Discussion

The aim of these investigations is to determine the performance of the tactile measurement setup and to identify the frequency ranges relevant for evaluation. The hypotheses of this study will be:

1. Different joining tasks show different TDA signals.
2. Different joining tasks can be clearly distinguished.
3. TDA signals of the "manufacturing" and "defects" sections are challenging to distinguish using the applied measurement setup.

Hypotheses 1 and 2 are based on the change of the vibrating masses with variation of the joining task. It is assumed that these differences are significant and can be detected in the considered frequency band. Investigations show that TDA signals change significantly with increasing tensional loading [7]. Köhler et al. also show numerically that changes in clinch point formation leads to changes in the TDA signal. However, these changes only significant in frequency ranges above 2.5 kHz [12]. It can therefore be assumed that the TDA signals also differ in the "manufacturing" and "defects" sections. The significance of these differences is presently unresolved. However, since small-volume defects in particular can be detected in frequency ranges above 20 kHz [13], it is assumed that the classification accuracy is reduced. The proof of the distinctiveness of the TDA signals shall be implicitly demonstrated by classifying the TDA signals using machine learning. With a high classification accuracy, it can be assumed that the signals show characteristics that allow a clear distinction.

In the first step of the evaluation, the time- and frequency-discrete parameters listed in Table 3 were calculated from the measuring data. These were calculated for the sensor signal as well as for the difference between actuator and sensor. As Heaton shows, classification algorithms can recognize mathematical relationships between features only to a limited extent, especially divisional relationships [14]. Therefore, additional parameters were calculated and used as features. These include multiplications and divisions of the frequency-discrete parameters. The matrix thus contains 44,400 data sets (rows) described by 33 features (columns) each.

The Random Forest Classifier from sklearn [15] is used for the machine learning model because of its robustness [10]. The objective of this study is not to optimize the machine learning task and the prediction accuracy of the model. Therefore, no other classifiers are taken into consideration. The hyperparameter tuning was carried out on basis of Bayesian Optimization using BayesSearchCV by scikit-optimize [16] in 50 runs with 5-fold cross-validation. The advantages of this algorithm for model optimization lie in its time efficiency, since the optimization is targeted [17]. For training and validation, the data was divided in an 80/20 ratio.

Sheet Metal 2025
Materials Research Proceedings 52 (2025) 293-300

Materials Research Forum LLC
https://doi.org/10.21741/9781644903551-36

Table 3: Calculated time- and frequency-discrete parameters

Time-discrete	Frequency-discrete
Maximum	Maximum amplitude
Minimum	Maximum frequency
Average	Amplitude excitation
RMS	Frequency excitation
Peak-to-peak	Phase shift
Standard deviation	
Average rectified value	
Crest factor	
Form factor	

The accuracy achieved for all data after hyperparameter tuning is 59.25 %. There is only a minimal difference between accuracy and F1 score (58.48 %), therefore only accuracy is used for further analysis. When evaluating the overall accuracy, it is necessary to clarify which accuracies are required and what is considered the reference. Here, only a random assignment (50 % accuracy) can be used as a benchmark for this application due to the preliminary nature of the study. The ultimate goal should be to achieve a significantly better accuracy, so that the threshold of 70 % is set as the target value for the remainder of the study. Concerning the engineered features can be stated, that 8 of the 10 engineered features are within the 20 most important features for the classification. The inclusion of these features is therefore relevant for the classification.

Fig. 2 shows the accuracy achieved for the individual series. The bars indicate the average accuracy of the series, while the error bars show the minimum and maximum specimen accuracy within the series. The limited number of specimens should be taken into account when interpreting the error bars. Significant differences between the series are apparent. Partly, very high accuracies are achieved, for example in series 650-660, reflecting the quality of the trained model. Due to these high achievable accuracies, low accuracies of other series are attributable to the available data and not the quality of the trained model.

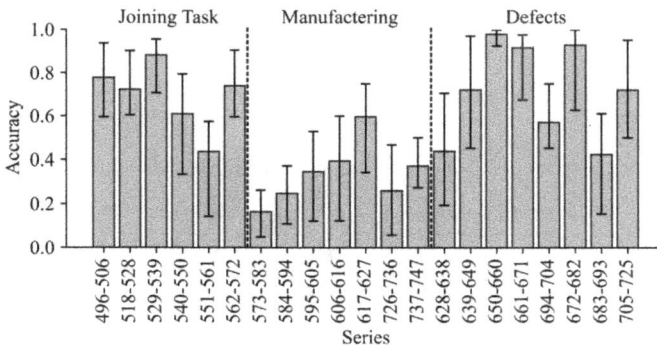

Fig. 2: Average classification accuracy with min. and max. specimen average

Results show that the series from the "joining task" and "defects" sections perform significantly better than the "manufacturing" section. Among the joining tasks, only two series are below 70% accuracy on average. These are series 551-561 (combination 7 in condition T6) and 540-550 (combination 5 in condition T6). Incorrect classifications of the series 540-550 primarily affect other joining tasks, as well as the series 705-725, which are characterized by position deviations on the flange. Incorrect classification of the series 551-561 particularly affects series in the "manufacturing" section. Considering the low accuracy achieved in this section and the similarity of the connections, this appears reasonable.

In the "manufacturing" section, the average accuracy of a series is never above 70%. This raises the question whether the examined variations cause significant changes in the properties of the clinched joints. For the variation in bottom thickness (series 573-583 and 584-594), Köhler et al. show numerical differences in the vibrational behavior at 20 kHz [8]. With regard to the variation of the punches (series 595-605, 606-616 and 617-627), Ewenz et. al. demonstrate changes in the mechanical strength with changes in the punches [18]. The performed torsion test is an indicator of the force closure component of the clinched connection, which has a significant influence on the TDA [13]. Changes in the tribological properties between the joining partners, as in the series 726-736 and 737-747, were examined in [19]. It was found that the connection resistance changed, which can also be attributed to altered binding mechanisms. In the "defects" section average accuracies below 70% are observed in three series as well. The significant influence of the joining properties through heat treatment (series 628-638) is shown in [19]. As shown in [20], defects due to tensile stress (series 694-704 and 683-693) lead to changes in the TDA signals.

The example of the sections "manufacturing" and "defects" demonstrates that a large variety of evidence for altered properties due to the introduced changes, can be found in the literature. At the same time, the classification accuracy of these series is below expectations. This suggests that the experimental setup could not detect unambiguous TDA signals for these series. This indicates the limitations of the applied measuring setup. On the one hand, it is possible that a change in the sensor placement could provide unambiguous signals. This can only be analyzed with a spatially resolved measurement. On the other hand, the examined frequencies can also be a reason for the poor accuracies. In this case, higher excitation frequency shall be analyzed.

It is known that the information content of the TDA signal depends decisively on the excitation frequency. This raises the question of whether frequency ranges can be identified on basis of these investigations that result in particularly high classification accuracies. Especially when using machine learning, it is essential to use data with a high information content to enable accurate classification. Therefore, Fig. 3 shows the accuracy depending on the excitation frequency for the different sections. In agreement with the results shown above, the accuracies of the sections "joining task" and "defects" are significantly better than those of the section "manufacturing". For the first two sections mentioned, it can be observed that better accuracies tend to be achieved at frequencies above 5 kHz. The "manufacturing" section does not show this. However, it can be seen that the best accuracies are achievable at frequencies > 15 kHz. This supports the thesis that limiting the excitation frequency limits the possibility of accurate classification.

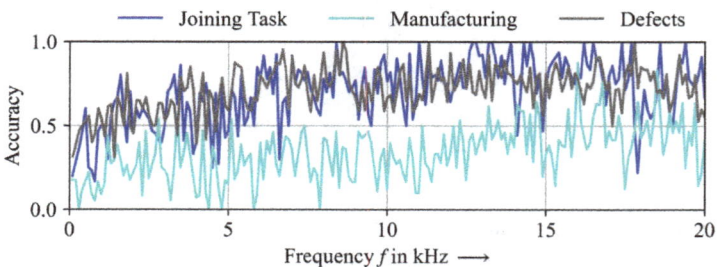

Fig. 3: Classification Accuracy depending on excitation frequency and section

Summary

In the present work, TDA measurements were performed on clinched shear tensile joints with the aim of determining the performance of the tactile measurement setup and identifying the frequency ranges relevant for the evaluation. Machine learning methods were used for this purpose.

It can be summarized that the different joining tasks can be reliably distinguished from each other and thus exhibit significantly different TDA signals. Likewise, introduced defects can be detected. However, differences due to changes in the manufacturing process cannot be adequately detected currently. In particular, these variations due to changes in the bottom thickness and the punches, as well as the tribological properties, can be well represented numerically. Therefore, the aim of future investigations should be to use numerical investigations to determine the reasons for the insufficient accuracy in the present work. On the basis of these investigations, strategies for the effective use of a scanning laser Doppler vibrometer can be developed.

With regard to the machine learning algorithm used, it can be stated that feature engineering can be used to calculate parameters that are significant for classification. Based on this knowledge, the algorithms used for this task can be further improved. In particular, the combination of dimension reduction and the use of synthetic data should be considered.

Acknowledgements

This research was funded by the German Research Foundation (DFG) within the project Transregional Collaborative Research Centre 285/2 (TRR 285/2, project number 418701707), sub-project C04 (project number 426959879).

References

[1] DIN Deutsches Institut für Normung e. V. Fertigungsverfahren Fügen: Teil 5: Fügen durch Umformen - Einordnung, Unterteilung, Begriffe 2003-09. Berlin: Beuth Verlag.

[2] Ostermann F. Anwendungstechnologie Aluminium. 3rd ed. Berlin: Springer Vieweg; 2014.

[3] Meschut G, Merklein M, Brosius A, Drummer D, Fratini L, Füssel U, et al. Review on mechanical joining by plastic deformation. Journal of Advanced Joining Processes. 2022;5:100113. https://doi.org/10.1016/j.jajp.2022.100113

[4] Europäische Forschungsgesellschaft Blechverarbeitung e.V. und DVS – Deutscher Verband für Schweißen und verwandte Verfahren e.V. Prüfung von Verbindungseigenschaften: Prüfung der Eigenschaften mechanisch und kombiniert mittels Kleben gefertigter Verbindungen 2021-06. Düsseldorf: DVS Media.

[5] Köhler D, Kupfer R, Troschitz J, Gude M. In Situ Computed Tomography-Analysis of a Single-Lap Shear Test with Clinch Points. Materials (Basel) 2021. https://doi.org/10.3390/ma14081859

[6] Lafarge R, Wolf A, Guilleaume C, Brosius A. A New Non-destructive Testing Method Applied to Clinching. In: Daehn G, Cao J, Kinsey B, Tekkaya E, Vivek A, Yoshida Y, editors. FORMING THE FUTURE: Proceedings of the 13th international conference on the. [S.l.]: Springer; 2021. p. 1461–1468. https://doi.org/10.1007/978-3-030-75381-8_121

[7] Köhler D, Stephan R, Kupfer R, Troschitz J, Brosius A, Gude M. In-situ Computed Tomography and Transient Dynamic Analysis of a Single-Lap Shear Test with a Composite-Metal Clinch Point. In: Liewald M, Verl A, Bauernhansl T, Möhring H-C, editors. Production at the Leading Edge of Technology. Cham: Springer International Publishing; 2023. p. 265–275. https://doi.org/10.1007/978-3-031-18318-8_28

[8] Köhler D, Sadeghian B, Kupfer R, Troschitz J, Gude M, Brosius A. A Method for Characterization of Geometric Deviations in Clinch Points with Computed Tomography and

Transient Dynamic Analysis. KEM. 2021;883:89–96.
https://doi.org/10.4028/www.scientific.net/KEM.883.89

[9] Alpaydin E. Maschinelles Lernen. 3rd ed. Berlin, Boston: De Gruyter; 2022.

[10] Sarker IH. Machine Learning: Algorithms, Real-World Applications and Research Directions. SN Comput Sci. 2021;2:160. https://doi.org/10.1007/s42979-021-00592-x.

[11] Hossin M, Sulaiman M. A Review on Evaluation Metrics for Data Classification Evaluations. IJDKP. 2015;5:1–11. https://doi.org/10.5121/ijdkp.2015.5201

[12] Köhler D, Sadeghian B, Troschitz J, Kupfer R, Gude M, Brosius A. Characterisation of lateral offsets in clinch points with computed tomography and transient dynamic analysis. Journal of Advanced Joining Processes. 2022;5:100089.
https://doi.org/10.1016/j.jajp.2021.100089

[13] Stephan R, Ewenz L, Brosius A, Zimmermann M. Anrisserkennung an geclinchten Proben während einer zyklischen Belastung unter Nutzung eines Scanning Laser-Doppler- Vibrometers: Crack Detection on Clinched Samples during Cyclic Loading using a Scanning Laser-Doppler-Vibrometer.

[14] Heaton J. An empirical analysis of feature engineering for predictive modeling. In: SoutheastCon 2016; 3/30/2016 - 4/3/2016; Norfolk, VA, USA: IEEE; 3/30/2016 - 4/3/2016. p. 1–6. https://doi.org/10.1109/SECON.2016.7506650

[15] N. N. scikit-learn Documentation RandomForestClassifier. https://scikit-learn.org/stable/modules/generated/sklearn.ensemble.RandomForestClassifier.html. Accessed 9 Aug 2024.

[16] N. N. Documentation BayesSearchCV. https://scikit-optimize.github.io/stable/modules/generated/skopt.BayesSearchCV.html. Accessed 2 Oct 2024.

[17] Bischl B, Binder M, Lang M, Pielok T, Richter J, Coors S, et al. Hyperparameter optimization: Foundations, algorithms, best practices, and open challenges. WIREs Data Min & Knowl 2023. https://doi.org/10.1002/widm.1484

[18] Ewenz L, Kalich J, Zimmermann M, Füssel U. Effect of Different Tool Geometries on the Mechanical Properties of Al-Al Clinch Joints. Key Engineering Materials. 2021;883:65–72. https://doi.org/10.4028/www.scientific.net/KEM.883.65

[19] Kalich J, Füssel U. Influence of the Production Process on the Binding Mechanism of Clinched Aluminum Steel Mixed Compounds. JMMP. 2021;5:105.
https://doi.org/10.3390/jmmp5040105

[20] Reschke G, Brosius A. Transiente Dynamische Analyse – Vergleich zeit- und frequenzdiskreter Auswertemethoden anhand geclinchter Aluminiumverbindungen. Proceedings der Tagung Werkstoffprüfung 2024. 2024. To be published after Dec. 6, 2024

Sheet Metal 2025
Materials Research Proceedings 52 (2025) 301-308

Materials Research Forum LLC
https://doi.org/10.21741/9781644903551-37

Predicting and identifying factors affecting sheet metal bending times using explainable AI

Alp Bayar[1,a], Johan Joubert[1,b] and Joost R. Duflou[1,c]

[1]KU Leuven, Department of Mechanical Engineering, Celestijnenlaan 300, 3001, Leuven, Belgium

[a]alp.bayar@kuleuven.be, [b]johan.joubert@kuleuven.be, [c]joost.duflou@kuleuven.be

Keywords: Man-Machine System, Machine Learning, Learning Curve Effect

Abstract. This study investigates the prediction of bending times in a sheet metal workshop using geometric information of parts without a bending sequence computation. Logs from two different sizes of bending machines from an industrial workshop are used, and data-centric machine-learning approaches are applied to achieve accurate predictions. Furthermore, the reduction in bending time with increasing production volume is analyzed as the learning effect in human-operated air bending. Explainable AI techniques, such as Partial Dependence Plots (PDP) and SHAP values, are used to interpret how model predictions are made. The learning effect is examined alongside geometric factors representing bending complexity: the number of bends and the total cutting length. The performances of XGBoost, random forest, and support vector machines (SVM) are compared in terms of bending time prediction, thereby demonstrating the approach's effectiveness.

Introduction

Sheet metal bending is a mature manufacturing process applied in highly customized labor-intensive production environments. Based on a technical drawing with the bending lines shown on a 2D unfolded sheet metal blank, the blank is gradually bent into a 3D geometry. Among different types of bending, air bending is widely used due to its flexibility, allowing the production of various geometries with one set of punches and dies, reducing tooling costs. However, these flexibilities require process planning to allocate tools considering material properties, minimize redundant operations, and reduce manufacturing times and costs. A bending sequence can be defined as the order of bending each line while considering the producibility of intermediate geometries without colliding with the punch or die. Although commercial software for bending process planning is available in the market, operators still often decide the bending sequences interactively depending on previous experience.

Estimating accurate manufacturing processing times allows workshops to efficiently allocate resources such as machines, tools, and operators, allowing precise calculation of labor and machine costs, optimization of production schedules, and increased responsiveness to changing market demands. This study aims to explore how bending times in a sheet metal workshop can be predicted using geometric information of the parts, without the need to compute a bending sequence. Understanding these relationships is crucial for improving production planning and efficiency in the context of Industry 4.0 and smart manufacturing. Additionally, the learning effect in human-operated air bending of different geometries has been studied using explainable AI approaches.

Explainable AI in this study refers to methods to enhance the interpretability of complex machine learning methods. It aims for practitioners to investigate the factors that affect the duration of the metal bending process more in detail, making it easier to adopt these technologies in sheet metal workshops since they are already lagging behind in digital transformation [1].

Content from this work may be used under the terms of the Creative Commons Attribution 3.0 license. Any further distribution of this work must maintain attribution to the author(s) and the title of the work, journal citation and DOI. Published under license by Materials Research Forum LLC.

Sheet Metal 2025   Materials Research Forum LLC
Materials Research Proceedings 52 (2025) 301-308   https://doi.org/10.21741/9781644903551-37

Related Work

In a similar study, sheet metal production costs were estimated using artificial neural networks and explained using multiple regression [2]. However, production cost as the prediction target is a composite metric calculated post-hoc, integrating labor and machine costs, excluding material costs. These cost elements are derived from secondary calculations and involve numerous assumptions, failing to capture the temporal dynamics of the manufacturing process. In contrast, our study focuses on predicting bending time, a directly measurable variable rooted in the physical characteristics and dynamics of production. Unlike production cost, which can be influenced by accounting practices, production time directly measures performance and can be used immediately to optimize manufacturing schedules, resource allocation, and machine utilization.

In another study trying to predict bending times, a dataset was created interactively by asking the operator about the tool setup, bending time, and additional features regarding the bending process such as the number of flips, whether additional setup tools were needed, etc. [3] The total bending time in [3] includes tool setup time and all the quality inspections, while in our study tool setup, configuration adjustments, quality inspections, and any other interruptions are detected through k-means clustering, allowing us to exclude these from the bending time measurements.

Finally, the learning effect in manufacturing, where workers become more efficient as they produce more units, was first introduced in [4] and is well-documented in the manufacturing literature, including applications in the semi-conductors [5] and other industries [6] within operations management. Our study examines how the learning effect in sheet metal bending varies with part complexity in terms of the number of bends and the total cutting length.

Data Collection, Pre-Processing, and Exploratory Analysis

This study uses data collected from an industrial partner and includes the production of parts on both a small and a medium-sized press brake, over a period of 6 months. The dataset contains 700 unique parts in different batch sizes. A batch is defined as the individual instances of the same part, produced sequentially under similar conditions within a single day. The parts used in this study have a wide range of geometric characteristics, with thicknesses ranging from 1mm to 4mm. The 2D bounding box widths -representing the smallest rectangular area that can fully enclose the part's 2D shape- span from 12mm to 1417mm, with an average of 290mm, while the heights range from 33mm to 2888mm, averaging around 877mm. The number of bends per part ranges from 1 to 20, with an average of 4,75 bends and with 3 and 4 bends being the most frequent. These statistics highlight the diversity in part complexity, making them representative of practical manufacturing scenarios. Features acquired from files in STEP format:

- Thickness
- 2D and 3D Minimum Bounding Box Dimensions
- Netto and Bruto Area of the Unfolded Sheet
- Total Cutting Length
- Number of Contours
- Number of Bending Lines
- Length of the Longest Bend

Fig. 1. 3D and 2D representation of a part with 5 parallel bends and 2 perpendicular bends

Sheet Metal 2025
Materials Research Proceedings 52 (2025) 301-308

Materials Research Forum LLC
https://doi.org/10.21741/9781644903551-37

Since the dataset has been mined through the machine logs, it required outlier detection and further pre-processing due to the high variance inherent in the bending time measurements. To manage this, the Z-score method was employed to identify outliers at individual part level. A Z-score threshold of 1,5 was used to filter out extreme time measurements for each part, which corresponds to 6% of the total dataset size. The processing time measurements tend to have a right-skewed distribution, because of the nature of measurement time. Therefore, the median processing time of a part is selected as a target to predict, rather than the mean, since it is more robust to outliers and skewness [7].

Clustering for identifying parts with quality inspections: Each workpiece's bending time is calculated by summing the intervals between pedal strokes for all bends, from machine logs. Some of the measured times include not only bending operations but also manual quality inspections and other processes. We applied an automated unsupervised approach to distinguish between measured bending times that include inspections and those that do not. K-means clustering proved particularly effective in identifying distinct groups within a dataset, revealing significant variations in production times that may be attributed to the presence or absence of quality inspections. Measurements have been clustered into inspected and non-inspected parts. Quality inspections are more often made at the start of a batch, as the operator gets more familiar and the process gets standard, fewer inspections are made (see Table 1). After clustering for each workpiece, 5000 measurements were identified as inspected while 23.000 were non-inspected. This suggests that a quality inspection or some form of interruption occurred once every five workpieces were produced.

Table 1. Distribution of duration (bending time) between normal and quality inspected parts

			min	25th	median	mean	75th	max
Before Clustering		Duration [s]	6	29	55	77	96	5215
		Batch Size	1	5	14	22	27	398
After Clustering	Normal Parts	Order in Batch	1	6	15	27	330	386
		Duration [s]	1	26	46	60	790	648
	Inspected Parts	Order in Batch	1	2	3	5	70	50
		Duration [s]	14	69	114	148	182	5215

In Table 1, the order in batch values for inspected parts is low, and the duration for producing those parts is higher than the normal parts. Order in batch refers to the sequence in which each instance of a part is produced within a batch. This implies that the high variance in time measurements is usually at the *beginning* of the production of each batch, likely due to the complex nature of the part requiring adjustments in the tool setup or machine settings. K-means clustering successfully identifies parts that were subjected to additional inspections or adjustments.

Experiments and Predictive Modeling
Three experiments were conducted to isolate different aspects of prediction bending time. In Experiment 1, predictions are made per batch, however, many parts had only a single batch, limiting the model's ability to capture the learning effect associated with varying batch sizes. To address this, Experiment 2 filtered out parts with only one batch, allowing the model to better focus on the impact of multiple batches. This filtering aimed to evaluate the impact of the batch size (quantity) on parts with multiple batches, ensuring the model could learn from diverse batch size information. Finally, in Experiment 3, only the geometric features are used to predict the bending time for each part, excluding the batch size information. The goal is to compare the predictive power of including batch sizes and assess the learning effect of human-operated bending. The

Sheet Metal 2025   Materials Research Forum LLC
Materials Research Proceedings 52 (2025) 301-308   https://doi.org/10.21741/9781644903551-37

overall process data flow diagram, as depicted in Fig. 2, involves multiple stages including data collection, pre-processing, and predictive modeling experiments.

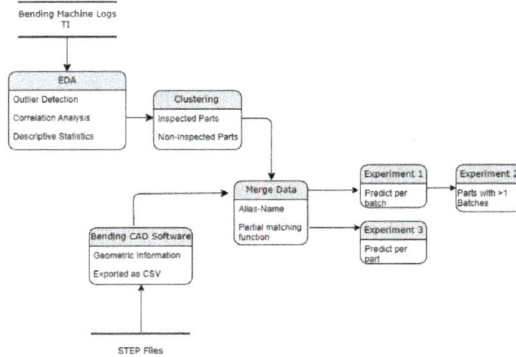

Fig. 2. Data Flow Diagram of the Application

Feature Importance and Explainability

XGBoost is a decision tree based gradient-boosting algorithm that is widely used for predictive modeling across various domains. In gradient boosting, decision trees are built sequentially with each tree aiming to correct the errors made by the previous trees. This differs from Random Forest, a bagging-ensemble algorithm, where multiple decision trees are built in parallel and trained with a different random subset of the data. The final predictions then are made by (weighted) averaging those results or individual trees. Both these algorithms can provide insights and a level of in-modeling explainability by analyzing the utilization of the features in their inner decision trees. While Random Forest calculates values such as Gini importance (mean decrease in impurity) and permutation importance (mean decrease in accuracy); XGBoost provides measures such as feature weight, gain, and cover [8]. In Fig. 3, relative feature importances from random forest are shown for Experiment 1:

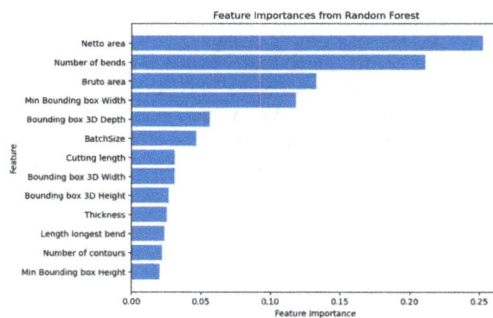

Fig. 3. Feature Gini Importance from Random Forest

For more comprehensive post-modelling explainability to provide detailed insights on how features influence model predictions, SHAP (Shapley Additive exPlanations) values, a recent game theory-based method, are used. SHAP values calculate the contribution of each feature to

Sheet Metal 2025
Materials Research Proceedings 52 (2025) 301-308

Materials Research Forum LLC
https://doi.org/10.21741/9781644903551-37

the prediction by evaluating its impact in combination with other features [9]. It provides a consistent and interpretable measure of how each feature influences the model's output, helping to explain and understand complex machine-learning models. In Fig. 4, relative gains of the best-performed XGBoost with the best-performed feature subset are presented. For most features, including netto area, red dots (representing higher feature values) show a positive correlation with bending time, meaning that larger values, such as larger sheets, take longer to bend. Blue dots (representing lower feature values) generally correspond to shorter bending times (negative values on the horizontal axis). However, batch size is the exception. In the 'BatchSize' row, smaller batch sizes (blue dots) are mostly on the positive side, indicating an increase in predicted bending time, while larger batch sizes (red dots) reduce the predicted time, demonstrating the learning effect in the production process.

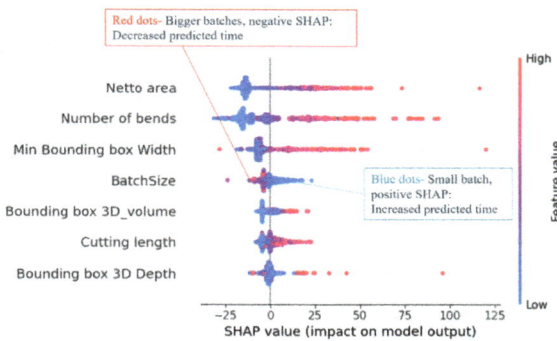

Fig. 4. Relative Feature Importances from XGBoost using SHAP Values

Results and Discussion

Initial exploratory analysis from bending machine logs revealed varying strengths of negative correlations between the order in batch and the bending time for each part. This suggests that as more units of a part are produced, the time taken per unit tends to decrease. There are two potential explanations for this trend: the learning effect and the decreasing variance over time. The learning effect phenomenon, as discussed in the related studies section, is documented in various fields of manufacturing. The decrease in variance reflects a stabilization in the process that the later workpieces in a batch are produced more consistently, with processing times converging toward a lower value than at the start. This stabilizing effect could be misinterpreted as purely a learning effect, but it also reflects the reduced uncertainty in the process. The experiments outlined in Fig. 2 were designed to incorporate the learning effect observed in the data.

To further analyze the effect of batch size on the bending time, a one-way Partial Dependence Plot (PDP) is created by running the model multiple times with different batch sizes while keeping all other factors in the test dataset constant [10]. The Y-axis shows partial dependence, which is in our case the predicted bending time in seconds.

Sheet Metal 2025   Materials Research Forum LLC
Materials Research Proceedings 52 (2025) 301-308   https://doi.org/10.21741/9781644903551-37

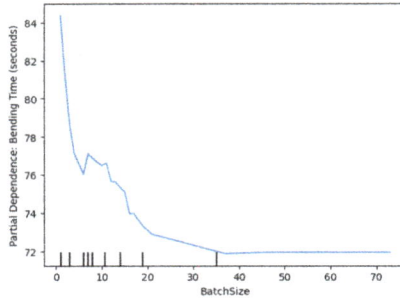

Fig. 5. Partial Dependence Plot Showing Effect of Batch Size on the Predicted Value

Fig. 5 illustrates that once the batch size increases to approximately 20-30 parts, the processing time stabilizes and no longer decreases significantly with further increases in batch size. Overall, as more parts are produced in the batch, the predicted bending time declines to an average reduction of 15% in the test dataset. To gain deeper insights into this learning effect, two-way partial dependence plots are used to investigate how part complexity, indicated by the *number of bends* and *cutting length*, influences the bending time. Figures 6 and 7 demonstrate that for parts with higher complexity—characterized by a large number of bends and an extensive cut length— the learning effect can lead to reductions in processing time of up to 22%.

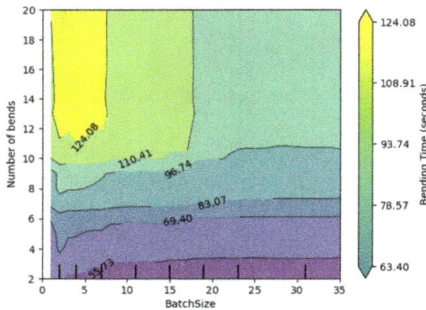

Fig. 6. Two-Way PDP: Batch Size and Number of Bends

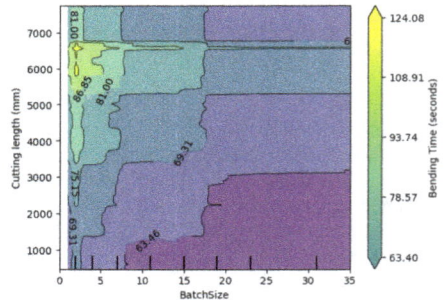

Fig. 7. Two-Way PDP: Batch Size and Cutting Length

For parts with a *larger number of bends (+12 bends)*, the predicted bending time is around 124 seconds and decreases sharply as batch size increases, as shown in Fig.6. For the *number of bends (6-12 bends)*, the rate of reduction in processing time is still visible but less dramatic compared to highly complex parts. The predicted time is lower, around 110–96 seconds, and the decrease is more gradual. This suggests that parts in this category still benefit from the learning effect, but the improvement is less substantial than for parts with a very high number of bends. And for *low number of bends (2–6 bends)*, the processing time starts around 83–69 seconds and shows only a smooth decrease.

Similarly, the cutting length (the total contour length for all internal and external contour) serves as another key indicator of part complexity. The PDP in Fig. 7 shows the relationship between the *cutting length, batch size*, and *the predicted bending time*. The rate of decrease in bending times varies across different cutting lengths, reflecting how the learning effect is influenced by the

Sheet Metal 2025
Materials Research Proceedings 52 (2025) 301-308

Materials Research Forum LLC
https://doi.org/10.21741/9781644903551-37

complexity of the part, as represented by the cutting length. For parts having a low cutting length (below 2000 mm), the bending times start around 69 seconds and the decrease with the batch size is more gradual. The color transitions show that the rate of improvement is relatively small compared to higher cutting lengths. For parts with longer cutting lengths, the reduction in processing time is substantial. As the batch size increases, the color changes from lighter (with processing times around 81 seconds) to darker shades (with times between 69–63 seconds), indicating a steeper learning curve due to the higher complexity of these parts, which likely require more operator adjustments.

Predicting the median bending time of a batch (experiments 1 and 2) is a more specific task compared to predicting the median bending time of each part (experiment 3), as it involves more features and requires the model to account for batch size related effects. Another thing to consider; filtering the dataset to exclude parts having only a single batch, reduces the variance of the dataset, improving the model's ability to generalization and resulting in higher accuracy. After applying this filter, datasets consisted of 500 parts and 880 batches for Experiments 1 and 3, while Experiment 2 worked with a reduced set of 200 parts and 580 batches. All experiments were performed using 5-fold cross-validation; 80% of the data was used for model training and 20% for testing. Grid search cross-validation is used for hyperparameter tuning of the algorithms.

Table 2. Experiment results

Experiment	Unit	Filter	XGB		RF		SVM	
			MAPE	R^2	MAPE	R^2	MAPE	R^2
1	Batch	No	29%	0.75	32%	0.74	32%	0.65
2	Batch	>1	20%	0.88	22%	0.88	24%	0.82
3	Part	N/A	58%	0.62	59%	0.65	50%	0.67

Table 2 shows the performances of each model in the three experiments. Filtering data to include only parts with multiple batches revealed that using Batch Size in the feature set resulted in a significant improvement in model performance, particularly for XGBoost, which saw an increase in R^2 by 0.15. XGBoost, better at capturing non-linear relationships and interactions between features, outperforms the other models in Experiment 2 when batch size is included in the feature set. XGBoost achieves the highest R^2 (0.88) and lowest Mean Absolute Percentage Error (MAPE) (20%) after filtering for multiple batches. In contrast, Random Forest and SVM show a more competitive performance when predicting processing times for individual parts (Experiment 3). This suggests that when batch effects and the learning curve are less critical, simpler models like SVM can sometimes offer better generalization, especially for tasks with less variance, such as part-level predictions.

Conclusion

This study demonstrates a data-centric machine learning application to predict bending times using data from bending machine logs. After cleaning the outliers, the high variance observed in exploratory data analysis can be attributed to factors such as varying operator experience and the impact of manually created bending sequences on total bending time. Notably, multiple valid bending sequences for a part result in various bending times. Because there are also quality inspections or tool adjustments and other time-consuming side activities of operators that have no available meta-data, a k-means clustering approach is used to find the predictions that are considered as inspected and non-inspected parts.

Experiments were designed to evaluate the impact of batch size and the learning effect on predicting bending times. XGBoost achieved an R^2 of 0.88 after filtering for multiple batches, demonstrating the importance of incorporating batch size information for improved performance.

Sheet Metal 2025   Materials Research Forum LLC
Materials Research Proceedings 52 (2025) 301-308   https://doi.org/10.21741/9781644903551-37

The use of explainable AI techniques, including SHAP values and partial dependence plots, not only improved model interpretability but also offered valuable insights into how batch size and part complexity influence bending times. These visualizations, combined with model performance metrics, underline that the machine learning model can capture the learning effect from the data. Also, the learning effect becomes more pronounced as the variety in production batch sizes increases, with complex parts showing a steeper reduction in bending times compared to simpler parts. This aligns with the hypothesis that operators gain efficiency as more parts are processed within a batch and the machine learning model can capture this natural phenomenon in the data.

The findings of this study offer a practical contribution to the manufacturing field by providing a data-driven methodology that can be adapted to similar scenarios in other job-shop-like environments, improving time estimates. For future work, in addition to the number of bends, information about their positions, such as the number of perpendicular bends, proximity to cutting contours, or the number of varying bending angles could be used. Perpendicular bends for instance require at least one flip operation, and likely have an impact on the bending time. These features could be integrated when geometric feature recognition tools are capable of processing 3D design files in bulk automatically and producing this information with this level of detail.

References

[1] World Economic Forum, "The digital revolution will transform the steel industry." Accessed: Oct. 01, 2024, Available: https://www.weforum.org/agenda/2019/06/the-digital-revolution-will-transform-steel-and-metals-companies/

[2] B. Verlinden, J. R. Duflou, P. Collin, and D. Cattrysse, "Cost estimation for sheet metal parts using multiple regression and artificial neural networks: A case study," Int. J. Production Economics, vol. 111, pp. 484-492, 2008. https://doi.org/10.1016/j.ijpe.2007.02.004

[3] Milingos, Spyridon-Romanos, "Prediction of Bending Process Time", International Hellenic University Jan. 2021.

[4] Wright T P, "Factors Affecting the Cost of Airplanes," Journal of the Aeronautical Sciences, vol. 3, Feb. 1936. https://doi.org/10.2514/8.155

[5] I. Tirkel, "Yield learning curve models in semiconductor manufacturing," IEEE Transactions on Semiconductor Manufacturing, vol. 26, no. 4, pp. 564-571, 2013, https://doi.org/10.1109/TSM.2013.2272017

[6] A. Tamás and T. Koltai, "Application of Learning Curves in Operations Management Decisions," Periodica Polytechnica Social and Management Sciences, vol. 28, no. 1, pp. 81-90, 2020. https://doi.org/10.3311/PPso.14136

[7] F. Figueiredo and M. I. Gomes, "The total median statistic to monitor contaminated normal data," Qual Technol Quant Manag, vol. 13, no. 1, pp. 78-87, Jan. 2016. https://doi.org/10.1080/16843703.2016.1139840

[8] X. Shi, Y. D. Wong, M. Z. F. Li, C. Palanisamy, and C. Chai, "A feature learning approach based on XGBoost for driving assessment and risk prediction," Accid Anal Prev, vol. 129, pp. 170-179, Aug. 2019. https://doi.org/10.1016/j.aap.2019.05.005

[9] S. M. Lundberg and S.-I. Lee, "A Unified Approach to Interpreting Model Predictions," Neural Information Processing Systems, 2017.

[10] T. Hastie, J. Friedman, and R. Tibshirani, "The Elements of Statistical Learning," 2001. https://doi.org/10.1007/978-0-387-21606-5

Sheet Metal 2025

Materials Research Proceedings 52 (2025) 309-317

Materials Research Forum LLC

https://doi.org/10.21741/9781644903551-38

Machine learning modeling of a deep drawing process for predicting resulting component properties after springback

Jonas Neumann[1,a] *, Umang Bharatkumar Ramaiya[1,b] and Marion Merklein[1,c]

[1]Institute of Manufacturing Technology, Friedrich-Alexander-Universität Erlangen-Nürnberg, Egerlandstraße 13, 91058 Erlangen, Germany

[a]jonas.jn.neumann@fau.de, [b]umang.ramaiya@fau.de, [c]marion.merklein@fau.de

Keywords: Sheet Metal, Deep Drawing, Machine Learning, ML Modeling

Abstract. Batch and process fluctuations in the fabrication of sheet metal components can lead to variations in product properties, potentially impacting subsequent manufacturing stages and increasing rejection rates. Identifying and analyzing these deviations is essential to maintain consistent product quality. To counteract such variations, a metamodel for a process chain, including deep drawing, clamping, and clinching, is proposed. This work presents a machine learning (ML) modeling approach for a deep drawing process with an s-rail tool to predict key properties such as effective plastic strain, maximum principal strain, von Mises stress, and residual sheet thickness after springback. The goal was to present a feasible approach for the ML modeling of the investigated deep drawing process and assess prediction qualities across different component sections. It was demonstrated that, with a moderate number of simulations, the predictive capabilities of the models for the forecasted parameters, as measured by the normalized root mean square error, averaged no more than just below 12 percent. However, certain regions, such as critical ends of the flange areas and transitions or straight sections near curved areas, had worse prediction quality compared to other sections. These regions were highlighted for future research, which should focus on incorporating additional features into the ML training process to improve predictions in these identified sections. In a broader scope, this ML modeling for the deep drawing process could be combined with ML modeling of subsequent processes, such as clamping and clinching, to develop a holistic metamodel for the entire process chain.

Introduction

Forming processes are subject to batch variations of semi-finished products and process fluctuations, which have a negative impact on the desired component quality, affect downstream process steps inside a process chain, and subsequently can lead to production rejects [1]. Especially the use of high-strength materials and the often-associated reduction in component wall thicknesses increase the sensitivity to fluctuations in the properties of the semi-finished products [2]. In order to meet high demands on product quality and reduce rejection rates, both reliable detection of deviations and a comprehensive understanding of the effects of fluctuations is necessary.

In this context, data-driven machine-learning (ML) models have great potential for forecasting the results of forming processes [1]. In recent years, machine learning has seen increasing application in forming processes. ML can be helpful in identifying patterns in material behavior and strain data, for instance, enabling more accurate predictions of forming limit curves [3]. Furthermore, ML can be helpful in predicting deep drawing defects using various classification methods [4] and predicting springback in sheet metal forming applications using various regression and artificial neural network methods [5]. Datasets for training various ML models for sheet metal forming processes can be generated by conducting experiments [6] and also by performing FE simulations to have a larger dataset [7]. Machine learning [8] and artificial neural networks are also useful for the optimization of FE simulations for better springback

Content from this work may be used under the terms of the Creative Commons Attribution 3.0 license. Any further distribution of this work must maintain attribution to the author(s) and the title of the work, journal citation and DOI. Published under license by Materials Research Forum LLC.

predictions for deep drawing applications [9]. Deep neural networks guided by theoretical material models can be used for a more sophisticated and accurate prediction of springback in sheet metal applications [10].

The forming process investigated is a deep drawing process to produce a s-rail geometry that is part of a three-stage process chain. Following the deep drawing process the s-rail-part is clamped with a counterpart and then both components are joined by clinching at multiple joining points. Deep drawing is a manufacturing process for producing sheet metal parts in which a flat sheet is formed into a three-dimensional hollow shape through a combination of compression and tension. A deep-drawing tool typically consists of a punch, a die and a blank holder. [11] Draw beads are often integrated into the tool as well in order to control the material flow in the forming zone [12]. The s-rail-geometry is ideally suited as a demonstrator, as it is similar to the components used in car bodies and the s-shape results in different stress states at different location throughout the component. [13]

To train ML algorithms to predict certain qualities of deep drawn components after springback of an s-rail, the most relevant process parameters and material-related parameters as well as resulting component properties are considered. A numerical setup is used to perform variant simulations of the deep drawing process including the springback. These variant simulations are later evaluated to determine the properties of the resulting component in the flange areas of the s-rails. The investigations focus on the flange areas in which the components are later clamped and joined with a counterpart. The goal of this work is to present a viable approach for the ML modeling of a deep drawing process for predicting component properties after springback in the flange areas of an s-rail part. Furthermore, the resulting prediction qualities of trained ML models for the investigated component properties at different locations of the flange areas will be investigated to identify regions of interest for improvements.

Numerical setup, feature extraction and ML modeling
For generating the necessary data pool for the ML modeling of the deep drawing process a numerical model in the software LS-DYNA was created and validated by Vallaster et. al. [13]. To validate the numerical model, comparisons between the data generated from experiments and data generated from numerical investigations were drawn. Specifically, the force-displacement-curves as well as the resulting part geometry were compared. Consequently, the numerical model was assessed as validated within the investigated parameter spaces. This numerical model was subsequently used to perform variant simulations. The setup of the simulation model can be seen in Fig. 1a). For the model the material HC340LA with a nominal sheet thickness of 2 mm was characterized by uniaxial tensile tests and layer compression test. The numerical model has a two-part structure, as illustrated in Fig. 1b). The first part of the model simulates the forming process using an explicit solver, while the second part addresses the springback behavior and the trimming of the part which are calculated with an implicit solver. [13] On the one hand, an explicit solver was chosen for the forming operation, as it is conditionally stable and preferable for dynamic models involving contact. On the other hand, an implicit solver was chosen for the springback and cutting operation, as it is unconditionally stable and commonly preferred in the solution of quasi-static problems. [14] To reduce the computational time for the numerical simulation of the forming process the velocity of the relative motion between the punch and the die is artificially increased by scaling the parameters, that are dependent on the velocity, so that 10 mm/s in a real experiment would be the equivalent to 250 mm/s in the simulation. [13]

For the variant simulations, a design of experiments (DOE) with variance-based sensitivity analysis (Sobol' method) was implemented by varying the input parameters of the deep drawing process within the range in which the numerical model was validated. The input parameters that are subject to variation are listed in Table 1. The scaling factors for the flow curve abscissa and

Sheet Metal 2025

Materials Research Forum LLC

Materials Research Proceedings 52 (2025) 309-317

https://doi.org/10.21741/9781644903551-38

ordinate as well as initial sheet thickness and young's modulus are considered to be able to cover batch fluctuations in the material properties to a certain degree.

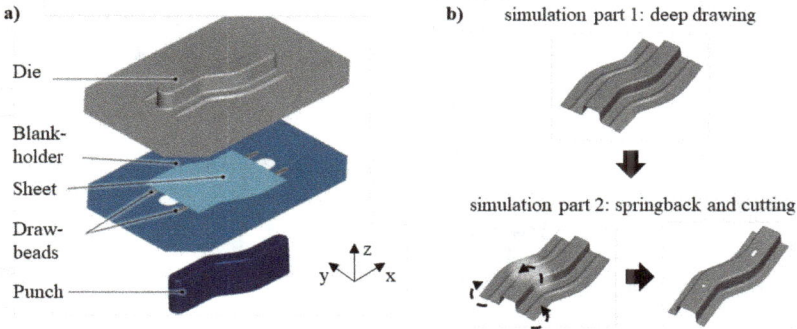

Fig. 1: a) Numerical model setup [13] and b) schematic display of the two-part model

To create the specific numerical models for the variant simulations with the desired parameter values predetermined by the DOE plan the keyword-files of the LS-DYNA model were parameterized with a Python-script-based workflow. Accordingly, 284 variant simulations were created, subsequently carried out with LS-DYNA and completed successfully. The average runtime for the simulation of the forming operation (simulation part 1) was around 4.5 hours, while the average runtime for the simulation of the springback and the cutting of the component (simulation part 2) was around 5 minutes. Those simulations are evaluated for data acquisition according to the procedures described in the following section.

Table 1: Input parameters for variant simulations and their respective limits

Parameter	Lower limit	Upper limit
Blank holder force	300 [kN]	650 [kN]
Friction coefficient	0.01	0.1
Draw bead height	3.1 [mm]	4.1 [mm]
Die velocity	10 [mm/s]	40 [mm/s]
Drawing depth	20 [mm]	40 [mm]
Initial sheet thickness	1.6 [mm]	2.1 [mm]
Young's modulus	200 [GPa]	220 [GPa]
Scaling factor flow curve abscissa	0.9	1.1
Scaling factor flow curve ordinate	0.9	1.1

As initially described, the investigation is focused on the component following the springback and trimming. Therefore, 50 evaluation points (EVPs) in the critical flange area of the part, where clamping and clinching operations are to be performed, were defined with specific coordinates. The location of the evaluation points is displayed in Fig. 2a) and the schematic location of the EVPs on the flange of the component is shown in Fig. 2b). In order to automatically find the nearest neighboring node to each EVP coordinate and the four elements surrounding it, a developed Python-script is used, as displayed in Fig. 2c). At each four elements related to the closest node to each evaluation point the effective plastic strain, the maximum principial strain, the von Mises stress and the residual sheet thickness are evaluated and the arithmetic mean value of the four elements is calculated for each parameter. The resulting values for the parameters describe properties of the component after springback and are referred to as output parameters in the following.

Sheet Metal 2025
Materials Research Proceedings 52 (2025) 309-317

Materials Research Forum LLC
https://doi.org/10.21741/9781644903551-38

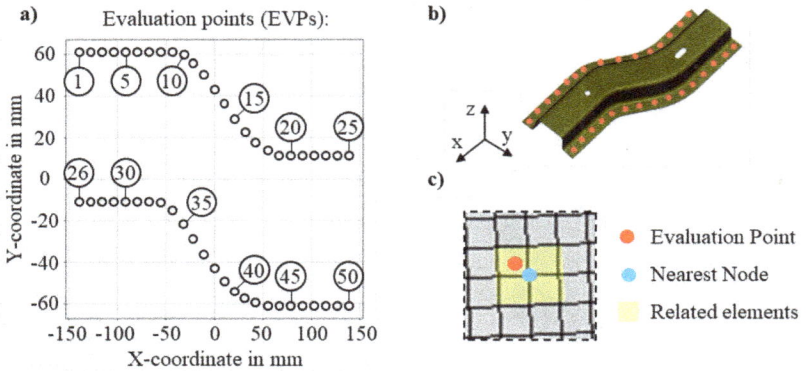

Fig. 2: a) Location of the EVPs, b) Schematic location of the EVPs on the component and c) Scheme for identifying relevant nods and elements

The software ClearVu Analytics of the company divis intelligent solutions GmbH is used for ML modeling to analyze the relationships between the parameters. Among other things the software provides the automated selection and adaptation of the learning methods according to the specific datasets and the comparison of different modeling algorithms using a statistical test to analyze the distribution of residuals. Furthermore, the software automatically carries out a generalization test through ten-fold-cross-validation. The best available ML algorithm for each application is determined on the basis of the minimum validation error and a statistical test. The currently available ML algorithms in ClearVu Analytics are gradient boosting model, linear model, neural network, random forest, gaussian process, support vector machine, kernel quantile regression, partial least squares regression, principal component regression, radial basis function model, decision tree and fuzzy model. [15]

For the ML modeling every single output parameter is individually linked with all available input parameters inside the dataset from the same variant simulation. The four output parameters which are to be predicted at the 50 EVPs using the ML algorithms, result in a total of 200 ML models. Thus, the optimal ML algorithm for each ML model is identified and considered for the investigations in the context of this work. Accordingly, the ML modeling of the deep drawing process is performed by using ClearVu Analytics with consideration of all available ML algorithms.

To comprehensively visualize the results of the ML modeling heatmap diagrams are used, where the normalized root mean square error (nRMSE), normalized by the difference between maximum value (maxA) and minimum value (minA) of the actual data, for the specific ML model with the best prediction quality for the respective output parameter at each evaluation point is shown. The nRMSE is a statistical metric for evaluating model performance. It measures the difference between the predicted and actual values. A lower nRMSE indicates a better model performance. The nRMSE is calculated with Eq. 1. [16]

$$nRMSE = \frac{RMSE}{maxA - minA}. \tag{1}$$

Sheet Metal 2025
Materials Research Proceedings 52 (2025) 309-317

Materials Research Forum LLC
https://doi.org/10.21741/9781644903551-38

Results

The results of the ML modeling for predicting the investigated component properties are shown in Fig. 3, where the EVP with the lowest and the highest nRMSE are highlighted for each output parameter.

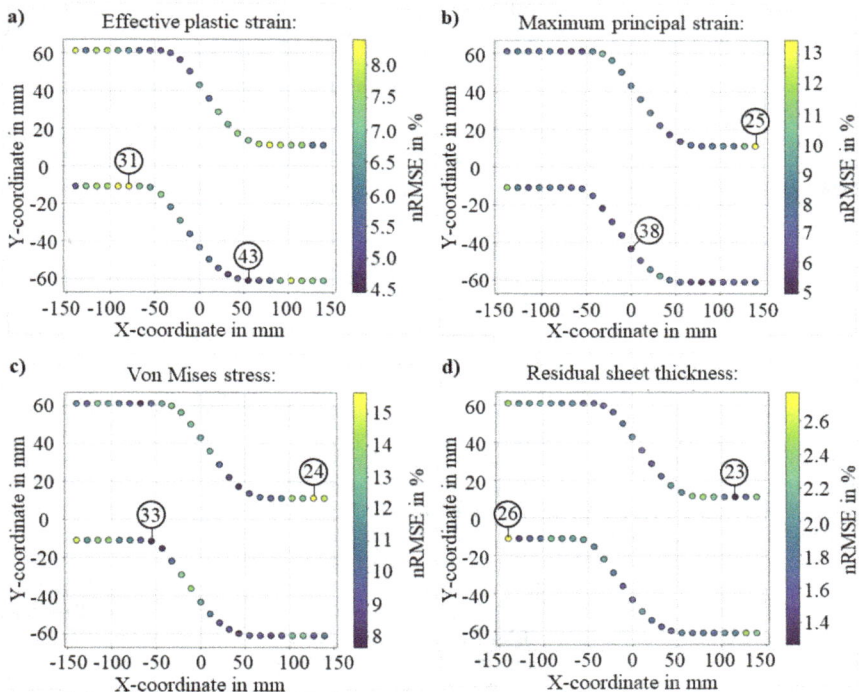

Fig. 3: ML Modeling result for a) effective plastic strain, b) maximum principal strain, c) von Mises stress and d) residual sheet thickness

The subsequent subchapters provide a detailed analysis of the results for the individual output parameters at various sections of the flanges of the s-rail. A comprehensive examination of the nRMSE values will offer insights into the error distribution and the stability of the predictions along the contour. This analysis is followed by an overall assessment of the results.

First, the modeling results for the effective plastic strain are examined. The average nRMSE for the effective plastic strain is 6.55 %, while the worst prediction quality with an nRMSE of 8.39 % was reached at EVP 31, which is located in the straight area on left side of the lower flange, and the best prediction quality with an nRMSE of 4.47 % was achieved at EVP 43, which is located right before the transition from the straight area on the right side of the lower flange towards the curved section. The nRMSE values for the upper flange are relatively stable on a low level around the curved section on the left side from EVP 8 to EVP 12 averaging 5.14 %. Apart from this stable section the prediction quality on the upper flange is worse, with a peaking nRMSE of 8.21 % at EVP 20 which is located just at the beginning of the curved section on the right side of the upper flange. In the surrounding region of the curved section on the right side (EVP 15 to EVP 23) the nRMSE is relatively high with an average of 7.44 %. Next to this region, the nRMSE decreases again towards the right end of the upper contour to an average of 5.79 % at EVP 24 and EVP 25 respectively. Contrastingly, towards the left end of the upper contour the nRMSE increases to an average of 7.37 % (EVP 1 to EVP 4), next to the before mentioned stable region with better prediction quality from EVP 8 to EVP 12. The EVP 26 located just left to this segment at the left end of the contour has a low nRMSE of 5,16 %. Furthermore, in the straight area just before the curved section on the left side of the lower flange from EVP 27 to EVP 31 has a relatively bad prediction quality with an average of 7,82 %. Similar to the upper flange, the nRMSE values for the lower flange are relatively stable on a low level in the curved section on the right side from EVP 40 to EVP 44 averaging 4.99 %. Towards the right end of the lower flange, the nRMSE increases to an average of 7.34 % from EVP 46 to EVP 50. Looking at the original dataset of the effective plastic strain used for training the models, a strong correlation between the prediction quality and the fluctuations in the dataset is noticeable.

Secondly, the modeling results for maximum principal strain are discussed. For the maximum principal strain an average nRMSE of 7.58 % was accomplished. Hereby, the best prediction quality was achieved with an nRMSE of 4.94 % at the EVP 38, which is located in the center of the straight middle section of the lower flange. The worst prediction quality was reached at EVP 25, which is located at the right end of the contour of the lower flange, with an nRMSE of 13.46 %. Regarding the upper flange, regions with relatively stable prediction qualities on different levels can also be identified. Starting on the left end of the contour until the turning point of the curved section (EVP 1 to EVP 8) an average nRMSE of 7.01 % signifies a relatively good prediction quality. Next to that from EVP 9 to EVP 15, specifically from the turning point curved section throughout the straight section of the middle part of the flange, the prediction quality abruptly decreases to an average of 8.59 %. This is in line with the observation that the fluctuations for the maximum principal strain in the original dataset. Hereby the average maximum percentage deviation from the mean value of the datasets for EVP 9 to EVP 15 is approximately 98.98 %, while the average maximum percentage deviation for the datasets for EVP 1 to EVP 8 is 94.00 %. Then all over the curved section on the right side of the flange (EVP 16 to EVP 20) the average nRMSE decreases again to 6.61 %, only to increase again towards the right end of the flange starting with an nRMSE of 8.02 % at EVP 21 rising to an nRMSE of 13.47 % at EV P25. On the lower flange at the left end, the prediction quality at EVP 26 with an nRMSE of 11.58 % is comparatively poor and then steadily increases over the curved section until EVP 34 with an nRMSE of 6.86 %. Following those four sections with relatively stable prediction qualities can be noticed. In the straight middle section (EVP 35 to EVP 39) the nRMSE decreases again and is relatively stable around an average of 5.98 %. Afterwards in the curved section on the right side of the flange (EVP 40 to EVP 43) the nRMSE increases again to an average of 8.29 %, followed by a decreased average nRMSE of in the following straight section from EVP 44 to 47 of 5.76 %. Finally, the prediction quality gets worse again towards the right end of the flange (EVP 48 to EVP 50) where the nRMSE averages 7.38 %.

Furthermore, the modeling results for von Mises stress are analyzed. For the von Mises stress, the overall worst prediction results were accomplished. In line with that, strong fluctuations of the parameter can be observed in the original dataset. The average nRMSE for all 50 EVPs is 11.02 %. Hereby, the worst prediction quality was achieved with an nRMSE of 15.67 % at the EVP 24, which is located close to the right end of the contour of the upper flange. The best prediction quality was reached at EVP 33, which is located at the transition point from the straight area of the left side of the lower flange to the curved area, with an nRMSE of 7.65 %. The nRMSE values for the upper flange are stable on a comparatively low level beginning on the left end of the flange from EVP 1 to EVP 8, with an average of 11.51 %. Throughout the curved section on the left side and the straight middle section (EVP 9 to EVP 14) the nRMSE decreases to an average of 12.16 %.

Sheet Metal 2025
Materials Research Proceedings 52 (2025) 309-317

Materials Research Forum LLC
https://doi.org/10.21741/9781644903551-38

Followed by a stable and good prediction quality in the area around the curved section on the right side of the flange from EVP 15 to EVP 20 with an average nRMSE of 9.23 %. While in the straight section towards the end of the right end of the flange (EVP 21 to EVP 25) the prediction quality clearly decreases again marked by an average nRMSE of 13.66 %. Regarding the lower flange, the EVP 26 on the left end of the contour has the lowest prediction quality with an nRMSE of 14.90 %. From EVP 27 onwards until close to the beginning of the curved section at EVP 31 the prediction quality is consistently better with an average nRMSE of 12.42 %. Throughout the curved section on the left side of the flange (EVP 32 to EVP 35) the nRMSE decreases even more and is stable around an average nRMSE of 8.86 %, only to abruptly rise to 13.20 % and 13.94 % in the center of the straight middle section at EVP 36 and EVP 37, respectively. It must be emphasized again, that this correlates with significant fluctuations of the dataset for those EVPs with values ranging from 275 MPa to 521 MPa at EVP 36 and from 344 MPa to 686 MPa at EVP 37. Following that spike, the average nRMSE decreases again to a stable level of 9.38 % in the area of the curved section on the right side of the flange and areas close by (EVP 38 to EVP 46). Onwards at the following two EVPs 47 and 48 the prediction quality gets worse with an average nRMSE of 12.65 % and then slightly improves again at the two final EVPs 49 and 50 towards the right end of the contour, where the nRMSE averages 10.03 %.

Finally, the modeling results for the residual sheet thickness are evaluated. Overall, the residual sheet thickness is the variable with the best prediction results. This clearly correlates with a low fluctuation of the initial dataset for this parameter. The dataset was, as expected, by far the most stable, as the deep drawing process itself does not result in significant deviations in the resulting sheet thickness. The average nRMSE for all 50 EVPs is 1.89 %. The best prediction quality was achieved with an nRMSE of 1.27 % at the EVP 23, which is located in the straight area of the right side of the upper flange. The worst prediction quality was reached at EVP 26, which is located at the left end of the contour of the lower flange, with an nRMSE of 2.77 %. On the upper flange of the component the nRMSE is relatively stable throughout most of the contour around an average of 1.81 %. Only at both ends of the flange (EVP 1 and EVP 25) and the curved section on the right side of the flange (EVP 19 to EVP 21) the prediction quality clearly drops off and the nRMSE values are stable around a combined average of 2.37 %. Likewise, to the upper flange at the ends of the lower flange, the nRMSE is the highest with a combined average of 2.52 %. In the straight section on the left side of the contour from EVP 27 to EVP 29 the prediction quality is the better than in the rest of the contour with an average nRMSE of 1.63 %. Moving into the curved section on the left side of the flange until the start of the straight middle section (EVP 30 to EVP 35), the nRMSE increases slightly to an average of 2.05 %. Throughout the rest of the lower flange from EVP 36 to EVP 48 the prediction quality is stable around an average nRMSE of 1.74 %.

Considering 284 variant simulations, the prediction quality of the ML models is deemed satisfactory for an initial ML modeling approach. However, this evaluation depends on the specific purpose and required tolerances for parameter prediction. Corresponding to the level of fluctuation in the training data, the ML models for residual sheet thickness exhibited the highest performance, while those for von Mises stress performed the worst. Fluctuations in the dataset may arise from the inherent characteristics of the parameters themselves, as well as from limitations in the validation of the simulation model, such as complex contact conditions at the ends of the component. An in-depth analysis of the prediction quality in different sections of the flange areas of the s-rail components identified regions with varying levels of susceptibility to reduced prediction quality. Overall, the findings highlight the need for targeted analyses and potential adjustments in modeling strategies, particularly around transitions, straight sections adjacent to curved areas, and critical ends of the flange, to enhance prediction stability along the contours.

Conclusion and outlook

The aim of this contribution was to present a viable approach for the ML modeling of a deep drawing process and to evaluate the resulting prediction qualities of trained ML models for the investigated component properties, to identify regions of interest for improvements. Therefore, it was shown, that with a moderate number of simulations relatively satisfactory prediction qualities were reachable. Furthermore, in certain regions like critical ends of the flange areas and transitions and straight sections adjacent to curved areas the prediction qualities get worse than in other flange sections of the component. Since those regions were identified, future research work should focus on the inclusion of additional features for the training of the ML algorithms to improve the prediction quality in the identified sections. The consideration of force-displacement-curve showed potential in preliminary examinations for targeted improvements. Additionally, a comprehensive analysis of the relationships between parameters and ML algorithms is required to better understand potential correlations. Moreover, the ML modeling of this specific process should be integrated into a metamodel of the overall process, which includes the main steps of deep drawing, clamping, and clinching. The metamodel would enable in-line process adaptations by applying adjusted parameters to the process steps, in response to batch and process fluctuations, in order to minimize production rejects.

Acknowledgement

This pre-competitive project was funded by the German Federal Ministry for Economic Affairs and Climate Protection with IGF funds.

References

[1] M. Liewald, T. Bergs, P. Groche, B. Behrens, D. Briesenick, M. Müller, P. Niemitz, C. Kubik, F. Müller, Perspectives on data-driven models and its potentials in metal forming and blanking technologies, Prod. Eng. Res. Devel. 16(5) (2022) 607-625. https://doi.org/10.1007/s11740-022-01115-0

[2] W. Drossel, F. Jesche, M. Israel, T. Falk, Sensitivitätsanalyse und Robustheitsbewertung beim mechanischen Fügen, Tagungsband T-037 des 3. Fügetechnischen Gemeinschaftskolloquiums (2013) 93-97.

[3] A. Thamm, F. Thamm, A. Sawodny, S. Zeitler, M. Merklein, A. Maier, Unsupervised Deep Learning for Advanced Forming Limit Analysis in Sheet Metal: A Tensile Test-Based Approach, Materials 16(21) (2023). https://doi.org/10.3390/ma16217001

[4] M. A. Dib, N. J. Oliveira, A. E. Marques, Single and ensemble classifiers for defect prediction in sheet metal forming under variability, Neural Comput. & Applic. 32(16) (2020) 12335-12349. https://doi.org/10.1007/s00521-019-04651-6

[5] R. Narayanasamy, P. Padmanabhan, Comparison of regression and artificial neural network model for the prediction of springback during air bending process of interstitial free steel sheet, J. Intell. Manuf. 23(3) (2012) 357-364. https://doi.org/10.1007/s10845-009-0375-6

[6] M. Wasif, M. Rababah, A. Fatima, S. Baig, Prediction of Springback using the Machine Learning Technique in high-tensile strength sheet metal during the V-Bending Process, Jordan J. Mech. Ind. Eng. 17(4) (2023) 481-488. https://doi.org/10.59038/jjmie/170403

[7] A. E. Marques, P. A. Prates, A. F. G. Pereira, M. C. Oliveira, J. V. Fernandes, B. M Ribeiro, Performance Comparison of Parametric and Non-Parametric Regression Models for Uncertainty Analysis of Sheet Metal Forming Processes, Metals 10(4) (2020) 457. https://doi.org/10.3390/met10040457

[8] M. Dezelak, I. Pahole, M. Ficko, M. Brezocnik, Machine learning for the improvement of springback modelling, APEM 7(1) (2012) 17-26. https://doi.org/10.14743/apem2012.1.127

[9] M. R. Jamli, N. M. Farid, The sustainability of neural network applications within finite element analysis in sheet metal forming: A review, Measurement 138 (2019) 446-460. https://doi.org/10.1016/j.measurement.2019.02.034

[10] S. Liu, Y. Xia, Z. Shi, H. Yu, Z. Li, J. Lin, Deep Learning in Sheet Metal Bending With a Novel Theory-Guided Deep Neural Network, IEEE/CAA J. Autom. Sin. 8(3) (2021) 565-581. https://doi.org/10.1109/JAS.2021.1003871

[11] E. Doege, B. Behrens, Handbuch Umformtechnik, Springer-Verlag Berlin Heidelberg, 2007.

[12] H. Schmid, Ganzheitliche Erarbeitung eines Prozessverständnisses von Tiefziehprozessen mit Ziehsicken auf Basis mechanischer und tribologischer Analysen, Dissertation, University of Erlangen-Nuremberg, 2022.

[13] E. Vallaster, S. Wiesenmayer, M. Merklein, Effect of a strain rate dependent material modeling of a steel on the prediction quality of a numerical deep drawing process, Prod. Eng. Res. Devel. 18(1) (2024) 47-60. https://doi.org/10.1007/s11740-023-01222-6

[14] F. J. Harewood, P. E. McHugh, Comparison of the implicit and explicit finite element methods using crystal plasticity, Comput. Mater. Sci. 39(2) (2007) 481-494. https://doi.org/10.1016/j.commatsci.2006.08.002

[15] divis GmbH, ClearVu Analytics (CVA) - Effiziente Datenanalyse, Modellierung, Vorhersage und Optimierung, 2023.

[16] A. Botchkarev, Performance Metrics (Error Measures) in Machine Learning Regression, Forecasting and Prognostics: Properties and Typology. IJIKM 14 (2019) 45-79. https://doi.org/10.28945/4184

Sustainability

Sheet Metal 2025
Materials Research Proceedings 52 (2025) 319-326

Materials Research Forum LLC
https://doi.org/10.21741/9781644903551-39

The assessment of heavy-duty laser cutting efficiency and environmental impact through different optical setup

Masoud Kardan[1,a] *, Brent Hendrickx[1,b] and Joost R. Duflou[1,c]

[1]KU Leuven, Department of Mechanical Engineering, Celestijnenlaan 300, Heverlee 3001, Belgium, Member of Flanders Make

[a]masoud.kardan@kuleuven.be, [b]brent.hendrickx@kuleuven.be, [c]joost.duflou@kuleuven.be

Keywords: Laser, Cutting, Energy Efficiency

Abstract. Improving energy efficiency has become a crucial priority, as the laser cutting industry is moving toward higher power levels and dealing with increasing energy costs. Recent advancements aim to enhance process performance by optimizing the intensity distribution of the laser beam. This paper evaluates the energy consumption of laser cutting using two different technologies: the conventional static beam strategy (SBS) and dynamic beam shaping (DBS). In addition to energy consumption, this study assesses the assist gas usage and environmental impact for both technologies. The findings reveal a 20% reduction in energy consumption and a 25% decrease in the assist gas consumption with DBS, along with improved cut quality. The study ultimately demonstrates a significant reduction in environmental impact when DBS cutting is used instead of SBS cutting. These results highlight DBS as a more efficient and environmentally friendly alternative to the traditional SBS technique.

Introduction

Laser cutting is widely used in industry because of its versatility, cost-effectiveness, and high precision. There are two primary types of lasers used for sheet metal cutting: fiber lasers and CO2 lasers. Fiber lasers are more productive compared to CO2 lasers due to their higher absorption when processing commonly cut metals. When comparing the energy consumption of CO2 and fiber laser cutting machines, CO2 systems use significantly more energy, particularly during active production. In contrast, fiber laser machines, which consume no energy when in standby mode, can save up to 16.6 MWh annually due to their superior efficiency [1]. Several studies have investigated the environmental impact of the laser cutting process. An overview of the environmental performance, including energy and resource efficiency of various laser cutting systems and strategies was provided in [2]. A life cycle assessment [3] has examined the environmental impact of laser cutting in terms of material waste. A comprehensive mathematical model for estimating CO2 laser cutting costs, focusing on the assist gas consumption has been proposed in [4]. A quantitative method to estimate the process efficiency, considering effectiveness and specific power consumption based on laser power and cutting speed, was introduced in [5]. A robust decision-making rule (RDMR), that integrates six multi-criteria decision-making (MCDM) methods with Taguchi's robust design principles to improve consistency in ranking laser cutting conditions, was presented in [6]. A study evaluated the energy consumption and carbon emissions of laser-based manufacturing, finding that auxiliary systems, rather than the laser itself, are the main contributors to energy use and that higher processing rates often lead to a greater specific energy consumption [7]. A stochastic simulation model to analyze the assist gas costs for various sheet thicknesses in CO2 laser oxygen cutting has been proposed [8]. Finally, a methodology for analyzing energy consumption in laser processing cells without disrupting normal operations was introduced [9].

Content from this work may be used under the terms of the Creative Commons Attribution 3.0 license. Any further distribution of this work must maintain attribution to the author(s) and the title of the work, journal citation and DOI. Published under license by Materials Research Forum LLC.

Sheet Metal 2025   Materials Research Forum LLC
Materials Research Proceedings 52 (2025) 319-326   https://doi.org/10.21741/9781644903551-39

The laser cutting industry is rapidly expanding due to recent advancements in fiber lasers and optical technologies. High-power laser sources are now available for heavy-duty cutting, enabling the processing of thicker materials with higher speeds and better quality. However, the use of higher laser power levels leads to increased energy and maintenance costs for companies. Consequently, there is a growing interest in smart manufacturing solutions aimed at enhancing cutting efficiency and reducing costs without increasing power levels. One approach within this trend is the focus on laser beam shaping strategies. Research has demonstrated that dynamic beam shaping (DBS) can significantly improve the cutting performance across various materials and thicknesses [10]. While the performance of dynamic beam shaping has been investigated and compared with the conventional static beam strategy (SBS) in terms of cutting speed and quality in many investigations, there are no studies specifically focused on comparing the energy consumption of these technologies. In this paper, the energy consumption of a cutting job performed using the two different strategies is evaluated and compared. Moreover, the assist gas consumption and environmental impact are included in this study. Therefore, this study involves experiments including laser cutting, power measurements, and quality evaluations assuring functional equivalence.

Experimental setup
Dynamic beam shaping. To enable the generation of a dynamic beam, the cutting head was equipped with a Newson Cyclops scanner (CYA-A38), chosen for its lightweight and compact design. This single-mirror system can produce a variety of dynamic beam shapes across a wide range of oscillation speeds, making it well-suited for industrial laser cutting of thick plates. In addition to the static beam, a dynamic beam with a longitudinal shape is utilized (Fig. 1a). The longitudinal dynamic beam shape improves the cutting performance through the linear oscillation of the beam focus point in the cutting direction, following a sinusoidal speed profile, as extensively explored in previous studies [11,12]. In a longitudinal dynamic beam, the amplitude represents the distance between the centers of the two extreme spots when the laser changes direction, measured at the laser beam focal plane (as shown in Fig. 1a). The oscillation frequency is defined as the number of full oscillations (back and forth) per second.

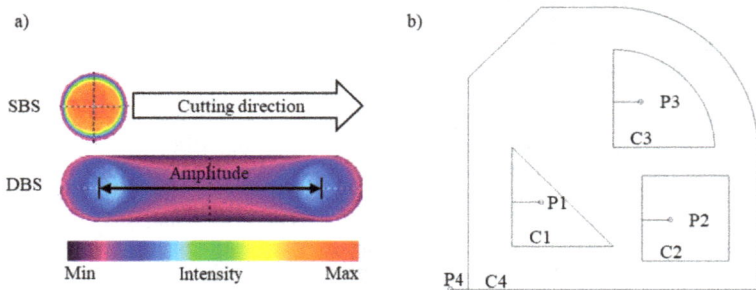

Fig. 1a: Laser beam intensity distribution for static beam strategy (SBS) and dynamic beam shaping (DBS). b: The cutting job used in this study.

Cutting job. To compare the energy consumption of both strategies the same cutting job was used. Fig. 1b illustrates the cutting job employed in this study, which includes various features representative of real-life laser cutting scenarios while maintaining simple geometries. This allows for correlating different states of the operating powers with their corresponding geometries. Labels

Sheet Metal 2025
Materials Research Proceedings 52 (2025) 319-326

Materials Research Forum LLC
https://doi.org/10.21741/9781644903551-39

P indicate the piercing points where the cutting is initiated, and labels C denote the cutting contours for each geometry. In total, there are four piercing points and four continuous cutting contours.

Laser cutting. Experiments have been conducted using an industrial 2D flat-bed laser cutting machine with a 6 kW laser power, equipped with a multimode fiber laser source that produces a top-hat intensity distribution. The laser beam is directed to the process zone through an optical fiber with a 100 µm core, a 100 mm collimation lens, and a 200 mm focusing lens. The beam waist diameter and beam parameter product were measured at about 225 µm and 3.1 mm mrad, respectively, using Ophir BeamGage and BeamWatch systems. This setup is used to perform tests on 10 mm thick mild steel S235JR. All process parameters are kept constant for both strategies, except for the focal point position (FPP), which significantly affects the cutting speed and cut surface quality depending on the applied strategy [13].

Power consumption measurement. To evaluate the efficiency of laser cutting using different strategies, the energy demand of the process has been measured using a Hobut C-Tran Power Transducer. The C-Tran is a programmable 3-phase multifunction AC power transducer and current transformer. To assess the electricity demand of the entire cutting process, power measurements were taken simultaneously at three different locations: the cooling system, exhaust, and laser source. These measurements have been collected using separate devices at each location and recorded using LabView. The total energy consumption of the process is calculated as the integral of the operating power of the entire system over the processing time.

Estimation of environmental impact reduction. The power consumption and cutting profile results are translated into an overall environmental impact assessment, taking into account the energy usage for cutting, nitrogen gas consumption, and the fiber laser's expected lifespan of 200,000 hours. A life cycle assessment (LCA) approach is employed, considering the differences in resource consumption and the anticipated emissions of the system. SimaPro (version 9) software, along with the EcoInvent 3.9 database, is used to convert the consumed resources and produced emissions into their equivalent CO_2 emissions based on the IPCC 2013 method (20 year horizon) and to determine the overall single score impact using the ReCiPe 2016 (H, endpoint) method. The existing process, "*Laser machining, metal, with CO2-Laser, 6000W power {RER}, APOS, U,*" is used to model air emissions (NOx, fine particulate matter), while the inputs of the process are replaced with those measured in this study. Nitrogen gas consumption is estimated at 115 m³ per hour of cutting, based on the nozzle diameter and assist gas pressure. Functional equivalence in the study is ensured by comparing data from different cutting methods applied to the same job (Fig. 1b). Initial sheet, produced parts, and scrap are omitted from the study to facilitate a direct comparison of processing conditions from an environmental impact perspective. European average data for production are considered for all system inputs.

Quality evaluation. To evaluate the quality of the cut edge, an optical microscope with focus variation capabilities (Keyence VHX-6000) has been used. The cut edge is scanned at 100x magnification, generating both 2D and 3D images. An image-based algorithm is applied to the 2D images to measure the dross area. The 3D images are utilized to assess the areal roughness and perpendicularity of the cut edge, as demonstrated in [14]. These quality factors are then combined into a quality index [14], providing a comprehensive indication of laser cut quality and enabling the comparison of different technologies.

Results and discussion

In fiber laser cutting, there is no need for a "warming up" phase for either the laser or the entire system. When the laser cutting machine is powered on, the cooling system activates instantly and operates based on the laser source temperature. In standby mode, the laser source consumes about 0.01 kWh, while the exhaust system does not use any energy during this period. The interval between powering on the machine and pressing the cycle start button to begin cutting is referred to as preparation time. This period includes tasks necessary to initiate cutting, such as changing

Sheet Metal 2025   Materials Research Forum LLC
Materials Research Proceedings 52 (2025) 319-326   https://doi.org/10.21741/9781644903551-39

the table, nozzle replacement, calibration, and job setup. The graph and resulting energy consumption may vary based on the specific preparation activities involved. If the preparation time is extended, additional cooling cycles may be required to cool down the laser source. Since this preparation time is unrelated to the technology used, it is excluded from the energy consumption comparison between the two technologies. To ensure an accurate comparison of energy consumption per part produced, the power consumption was measured from the moment that the cycle start button was pressed until the completion of the cutting process and cool-down phase. Fig. 2 presents these measurements for the two different technologies. The cooling down period refers to the time between the end of cutting and the reduction in energy consumption as the exhaust system deactivates and the cooling system reaches the standby state.

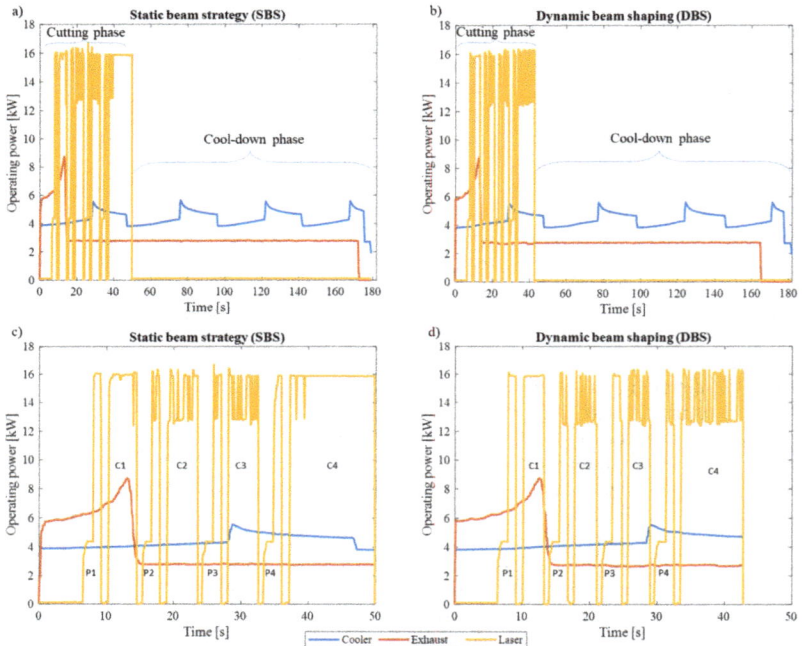

Fig. 2. The energy consumption of the laser cutting job: cutting and cooling down phases for SBS (a), cutting and cooling down phases for DBS (b), cutting phase for SBS (c), and cutting phase for DBS (d).

Fig. 2 illustrates the power demand for laser cutting of the sample (Fig. 1b) and the cool-down phase using both SBS and DBS techniques. The exhaust unit demonstrates a clear pattern that is directly tied to the cutting process. It activates shortly before the cutting is initiated and continues operating for a brief period after the cutting ends. As a result, the initial rise of the operating power for this unit can be used to define the actual start of the cutting process. A more detailed illustration of the cutting phase is shown in Fig. 2c and Fig. 2d. The total cutting times are 50 seconds for SBS and 43 seconds for DBS. As depicted, the peak power for each unit (cooling, exhaust, and laser) remains consistent between the two techniques, indicating that only the cutting time affects the total energy consumption. Additionally, the piercing times (P1, ..., P4) are identical for both

Sheet Metal 2025
Materials Research Proceedings 52 (2025) 319-326

Materials Research Forum LLC
https://doi.org/10.21741/9781644903551-39

strategies due to the same piercing approach being used. Therefore, the difference in cutting times is attributed to cutting the contours (C1, ..., C4). Based on power measurements, the machine (positioning system, control unit, ...) operates at an average power of approximately 0.5 kW for internal energy needs, including the movement of stages along different axes.

Fig. 3 illustrates the detailed power consumption for the two different strategies. Fig. 3a and Fig. 3b display the power consumption of SBS cutting across various electrical units during the cutting and cool-down phases respectively. Similarly, Fig. 3c and Fig. 3d illustrate the power consumption of DBS cutting for the cutting and cool-down phases. The total energy consumption during the cutting phase (Fig. 3a and Fig. 3c) decreased from 0.256 kWh for SBS to 0.207 kWh for DBS cutting, resulting in nearly a 20% reduction in energy use. However, energy consumption during the cooling phase increased from 0.270 kWh with SBS to 0.281 kWh with DBS cutting. As shown in Fig. 2a and Fig. 2b, the total process time (including both cutting and cool-down phases) remains 180 seconds for both techniques. Consequently, the reduced cutting time with DBS leads to a longer cool-down phase, an effect that is only present for short (<180 sec.) cutting jobs on this machine. When considering the electrical energy consumption per part and the average electricity cost of 0.2 EUR/kWh in Europe for non-household consumption, the electricity cost per part cutting is 0.05 EUR for SBS and 0.04 EUR for DBS, respectively. The piercing time for both cutting methods are equal, indicating higher potential savings for longer continuous cuts.

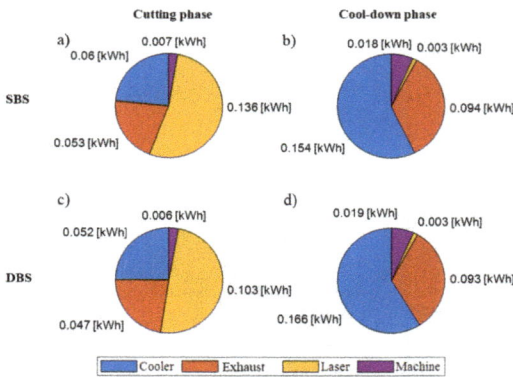

Fig. 3. The detailed energy consumption for laser cutting job using SBS (a,b) and DBS (c,d) technologies for both cutting (a,c) and cool-down (b,d) phases.

In addition to energy consumption, the assist gas usage is a crucial factor in determining cutting costs. Given that the same assist gas parameters are used for both techniques, time becomes a key factor. As a result, using DBS for the cutting job achieved a 25% reduction in gas consumption (net consumption during active cutting time only). Based on the estimated nitrogen consumption during cutting, the assist gas consumption and its cost per part cutting can be estimated based on the cutting time for each part. It is estimated that the assist gas cost per part cutting (Fig. 1b) is approximately 0.47 EUR for SBS cutting and 0.36 EUR for DBS cutting, assuming a nitrogen gas cost of 0.50 EUR/m3. As indicated, the assist gas cost constitutes the largest portion of the production cost per part [4]. A 25% reduction in cutting costs and increased productivity through DBS cutting can significantly enhance the efficiency of laser cutting machines.

Environmental impact values have been calculated for both cutting scenarios. A comparison of the results allows us to conclude that for the reference part, DBS cutting is expected to cause 20.5% less associated carbon emissions, and 20.3% less overall associated impact according to the ReCiPe 2016 (H) method (Fig. 4). The anticipated longer lifespan of the laser due to the reduced cutting time has a negligible contribution to the overall impact of the cutting process. Midpoint impact categories global warming potential, fine particulate matter formation, and human toxicity are the major contributors to the overall endpoint impact, and the electricity consumed for the laser cutting and the production of liquid nitrogen amount to >99% of the impact in these categories, with the exception of fine particulate matter formation. Within this impact category the emissions during processing contribute 5.6% to the weighted cumulative impact. Nitrogen gas consumption during cutting contributes to 46% of the overall emissions associated with processing. This ratio represents a significant difference compared to the results previously obtained by Kellens et al. [2], a shift that is attributed to the substantially higher energy demand of the CO_2 laser in their study, with power levels up to 40-50 kW for their system compared to a maximal 30 kW detected on the current set-up.

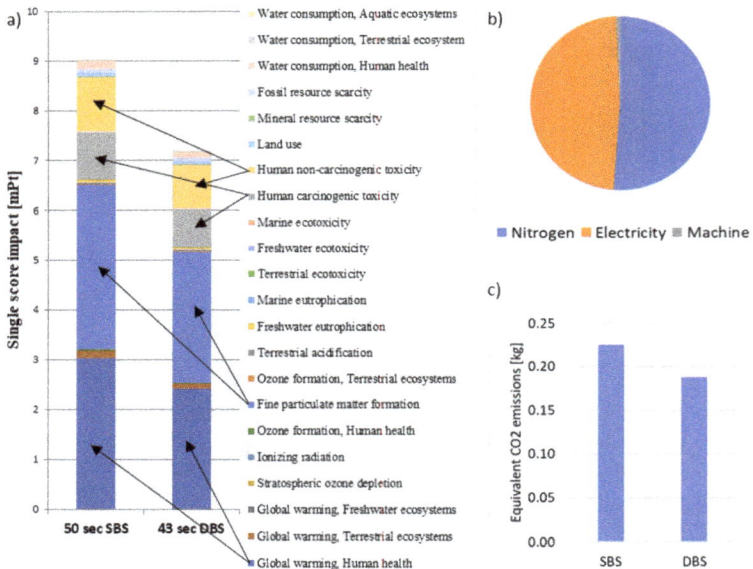

Fig. 4. ReCiPe endpoint (H) single score impact, subdivision per damage pathway (a), overall contribution to CO2 impact for laser cutting per input type for SBS cutting (b), and IPCC global warming potential per cutting type (c).

Another key factor affecting laser cutting efficiency is the quality of the cut surface. This quality can be measured through various factors, including roughness, dross, and, for thick plates, the perpendicularity of the cut edge. To quantify these quality factors, the methodology outlined in the previous section was employed, and the results have been summarized in Table 1. To evaluate the cut edge quality comprehensively, the quality index developed in [14] was utilized. To compare the improvement in quality from SBS cutting to DBS cutting, the quality numbers from SBS cutting were used as the reference values. In other words, the quality indexes have been normalized using the quality factors from SBS cutting, so the quality index for SBS cutting is equal to 1. A quality index greater than 1 signifies lower cut edge quality, while a value less than 1 indicates higher cut edge quality.

Table 1. The quality values of the cut edge for both SBS and DBS cutting.

Quality factors	SBS	DBS
Perpendicularity [µm]	196.78	130.74
Areal roughness, Sa [µm]	3.80	4.56
Dross area [mm²/mm]	0.745	0.083
Quality Index	1.00	0.93

The quality index demonstrates a notable improvement in cut edge quality when using DBS compared to SBS.

Summary
In this study, the energy consumption of a laser cutting job was measured and compared between two different cutting strategies: static beam strategy (SBS) and dynamic beam shaping (DBS). The results reveal that DBS reduces total energy consumption by 20% compared to SBS. Additionally, DBS achieves a significant reduction of approximately 25% in assist gas consumption, leading to a notable decrease in the cutting cost per part. However, it is important to consider that production costs are influenced by various factors beyond energy and gas consumption, including investment costs, maintenance expenses, and operator costs, which also need to be accounted for when evaluating overall production expenses.

The reduction of cutting time, and thus electricity and nitrogen gas consumption, also translates directly into a reduction of the environmental impact associated with the cutting process. As process emissions are assumed to remain the same, the environmental impact reduction is slightly lower (20%) compared to the combined effect of the reduction of electricity demand (20%) and nitrogen gas consumption (25%) on their own. Direct emissions from the cutting process itself are mainly found in the categories relating to human toxicity and fine particulate matter formation.

In addition to reductions in energy and assist gas consumption, the quality index reveals a significant improvement with DBS compared to SBS. This enhanced quality leads to further cost savings by reducing the need for post-processing. These advancements highlight DBS as a more efficient and cost-effective option in laser cutting technology, providing substantial advantages over traditional SBS methods.

References
[1] T. Devoldere, W. Dewulf, W. Deprez, J.R. Duflou, Energy Related Life Cycle Impact and Cost Reduction Opportunities in Machine Design: The Laser Cutting Case, in:15th CIRP International Conference on Life Cycle Engineering, Sydney, N.S.W., 2008.

[2] K. Kellens, G.C. Rodrigues, W. Dewulf, J.R. Duflou, Energy and Resource Efficiency of Laser Cutting Processes, Phys Procedia 56 (2014) 854-864. https://doi.org/10.1016/j.phpro.2014.08.104

[3] B.S. Yilbas, M.M. Shaukat, F. Ashraf, Laser cutting of various materials: Kerf width size analysis and life cycle assessment of cutting process, Opt Laser Technol 93 (2017) 67-73. https://doi.org/10.1016/j.optlastec.2017.02.014

Sheet Metal 2025   Materials Research Forum LLC
Materials Research Proceedings 52 (2025) 319-326   https://doi.org/10.21741/9781644903551-39

[4] M. Madić, M. Radovanović, B. Nedić, M. Gostimirović, CO2 Laser Cutting Cost Estimation: Mathematical Model and Application, Int. Journ. of Laser Science 1 (2018) 169-183.

[5] C.C. Girdu, L.A. Mihail, M. V Dragoi, Estimation of laser cutting process efficiency, IOP Conf Ser Mater Sci Eng 659 (2019) 12045. https://doi.org/10.1088/1757-899X/659/1/012045

[6] M. Madić, G. Petrović, D. Petković, J. Antucheviciene, D. Marinković, Application of a Robust Decision-Making Rule for Comprehensive Assessment of Laser Cutting Conditions and Performance, Machines 10 (2022) 153. https://doi.org/10.3390/machines10020153

[7] L.C.R. Jones, N. Goffin, J. Ouyang, N. Mirhossein, J. Xiong, Y. Li, L. Li, J. Tyrer, Z. Liu, E. Woolley, Y. He, G. Mi, P. Mativenga, Laser specific energy consumption: How do laser systems compare to other manufacturing processes, J Laser Appl 34 (2022) 042029. https://doi.org/10.2351/7.0000790

[8] Miloš Madić, Saša Ranđelović, Milan Trifunović, Stochastic simulation model for the analysis of assist gas costs in CO2 laser oxygen cutting, in: The 9th International Conference "Transport and Logistics," Niš, Serbia, 2023.

[9] N. Goffin, L.C.R. Jones, J.R. Tyrer, J. Ouyang, P. Mativenga, L. Li, E. Woolley, Industrial Energy Optimisation: A Laser Cutting Case Study, International Journal of Precision Engineering and Manufacturing-Green Technology 11 (2024) 765-779. https://doi.org/10.1007/s40684-023-00563-y

[10] M. Kardan, N. Levichev, S. Castagne, J.R. Duflou, Dynamic beam shaping requirements for fiber laser cutting of thick plates, J Manuf Process 103 (2023) 287-297. https://doi.org/10.1016/j.jmapro.2023.08.048

[11] N. Levichev, M.R. Vetrano, J.R. Duflou, Melt flow and cutting front evolution during laser cutting with dynamic beam shaping, Opt Lasers Eng 161 (2023) 107333. https://doi.org/10.1016/j.optlaseng.2022.107333

[12] C. Goppold, T. Pinder, S. Schulze, P. Herwig, A.F. Lasagni, Improvement of laser beam fusion cutting of mild and stainless steel due to longitudinal, linear beam oscillation, Applied Sciences 10 (2020) 11-13. https://doi.org/10.3390/app10093052

[13] M. Kardan, N. Levichev, J.R. Duflou, Experimental and numerical investigation of thick plate laser cutting using dynamic beam shaping., Procedia CIRP 111 (2022) 740-745. https://doi.org/10.1016/j.procir.2022.08.115

[14] M. Kardan, N. Levichev, A.T. García, Revisiting image-based quality evaluation of laser cut edges, 25 (2023) 363-370. https://doi.org/10.21741/9781644902417-45

Sheet Metal 2025
Materials Research Proceedings 52 (2025) 327-333

Materials Research Forum LLC
https://doi.org/10.21741/9781644903551-40

Sustainable steel production and application

Amalia KOLETTI[1,a] *, Fabian BOTZ[1,b] and Thomas FLOETH[1,c]

[1]Thyssenkrupp Steel Europe AG, Kaiser-Wilhelm-Straße 100, 47166 Duisburg, Germany

[a]amalia.koletti@thyssenkrupp-steel.com, [b]fabian.botz@thyssenkrupp-steel.com,
[c]thomas.floeth@thyssenkrupp-steel.com

Keywords: CO_2 Emissions, Hot Stamping, Life Cycle Assessment (LCA)

Abstract. thyssenkrupp Steel Europe is committed to lead the way to sustainable steel production by achieving net-zero emissions by 2045. To reach this goal, thyssenkrupp Steel is embarking on an ambitious transformation journey, focusing on the development of hydrogen-based steel production to significantly reduce CO_2 emissions. Since steel is one of the most important materials for automotive production, the decarbonization of the steel industry is crucial for the reduction of Scope 3 upstream emissions in the automotive supply chain. Besides the reduction of emissions during steel production, significant emission reductions can be achieved during its use phase by lightweight steel applications using advanced high-strength steels (AHSS). These types of steels enable weight reduction, which in turn lowers CO_2 emissions. By optimizing both the production and the application of steel, the industry can make substantial strides towards sustainability and environmental performance.

Introduction

The European Green Deal, as a strategic initiative, guides Europe toward climate neutrality and sustainability by 2050 [1]. thyssenkrupp Steel is actively supporting the decarbonization of the European steel industry and has committed to reduce direct CO_2-emissions by more than 30 % by 2030 compared to 2018 levels and achieve net-zero steel production by 2045. Through the funding from Germany's federal and state governments, the transformation of steel production at thyssenkrupp Steel is becoming reality; the production of premium steel with green electricity and hydrogen in direct reduction plants and no longer in coal-fired blast furnaces [2].

Steel accounts for around 65% of a vehicle's weight and is considered one of the most sustainable materials used in vehicles due to its high recyclability and relatively low CO_2 intensity compared to other metals like aluminum and magnesium [3]. Producing one metric ton primary steel emits about 2.3 metric tons of CO_2, which is significantly lower than the 15.1 metric tons of CO_2 emitted per metric ton of primary aluminum [4, 5]. Steel is essential for the automotive industry due to its strength, durability, and versatility, which are crucial for body-in-white, battery housings, and chassis solutions. At the same time, the automotive industry is facing significant challenges, due to stricter crash and safety requirements and the need to balance vehicle weight with sustainability goals [6, 7, 8, 9]. These demands make the role of steel even more critical, as it must meet both performance and environmental standards.

By adopting CO_2-reduced steel production methods that use hydrogen and green electricity instead of coal and coke, the automotive industry can significantly reduce its upstream emissions. Additionally, innovative steels like the Advanced High-Strength Steels (AHSS) are increasingly utilized to address challenges related to crash safety, vehicle weight, and sustainability. AHSS have a high strength-to-weight ratio, which not only reduces emissions during steel production but also enhances the energy efficiency of battery-electric vehicles (BEVs).

Content from this work may be used under the terms of the Creative Commons Attribution 3.0 license. Any further distribution of this work must maintain attribution to the author(s) and the title of the work, journal citation and DOI. Published under license by Materials Research Forum LLC.

Sheet Metal 2025   Materials Research Forum LLC
Materials Research Proceedings 52 (2025) 327-333   https://doi.org/10.21741/9781644903551-40

Transformation path to sustainable steel production

Currently, 72% of global steel production takes place in energy- and CO_2-intensive coal-fired blast furnaces [4]. An alternative production route for primary steel involves the use of direct reduction plants. These plants produce direct reduced iron (DRI), commonly referred to as sponge iron - a solid form of iron obtained by reducing iron ore without the need for melting (Fig. 1). While still hot, the DRI is further processed into liquid iron in electrically powered melters. Placing the melters directly next to the direct reduction plant enables the immediate conversion of the solid material into liquid iron, significantly enhancing the efficiency of the entire process.

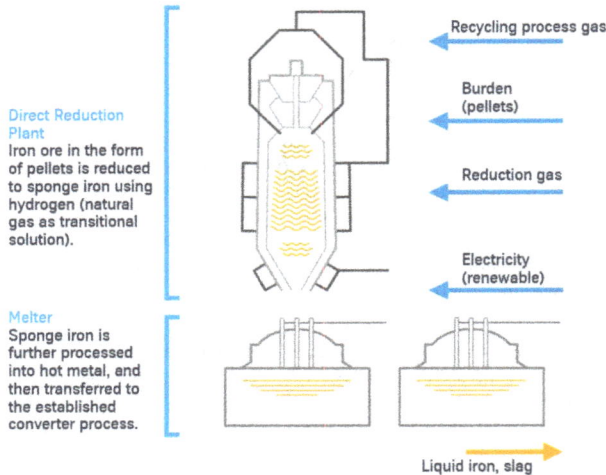

Fig. 1 Primary steel production via a direct reduction plant with melters [2]

A melter, also known as a submerged arc furnace (SAF), produces liquid iron that can be directly processed in the basic oxygen furnace (BOF), ensuring continuous operation and efficient use of existing infrastructure. This continuous operation enhances productivity and reduces downtime compared to the batch processes typical of electric arc furnaces (EAF). Additionally, the SAF allows for further reduction of iron oxides, leading to maximum iron yield. The slag produced is not a waste product but can be processed into granulated blast furnace sand, which can be used as a CO_2-saving clinker substitute in the cement industry, like the current blast furnace slag. Since the SAF process does not require scrap input, there is no dependency on scrap availability and scrap quality.

Since the quality and properties of this liquid iron will remain consistent with current standards, the direct reduction plant with the two melters can be seamlessly integrated into the surrounding steelworks infrastructure. This integration will ensure the availability of the complete current product portfolio and maintain the usual high quality in terms of formability, strength, surface, and electromagnetic properties [2]. By combining a direct reduction plant with two electric melters, the reduction and melting of the iron ores can be achieved by using hydrogen and green electricity instead of coke and coal, potentially reducing direct CO_2 emissions by more than 90% [10].

In spring 2023, thyssenkrupp Steel placed an order with SMS group for the engineering, delivery, and construction of the first direct reduction plant with two melting units in Duisburg. Once the plant is commissioned, 2.5 million tons of DRI will be produced. With the use of green hydrogen and green electricity, 3.5 million tons of CO_2 emissions will be avoided per year. This ambitious transformation towards sustainable steel production is receiving substantial support from both the federal government of Germany and the state government of North Rhine-Westphalia [11].

CO_2-reduced steel available today

For OEMs and suppliers who want to use CO_2-reduced steel products already today, thyssenkrupp Steel offers certified steel products under the brand bluemint® Steel. These products not only help reduce emissions but also promote circularity within the steel industry.

Recycling saves valuable resources and protects the environment. Hardly any other material used in the automotive industry is as recyclable as steel. This is why scrap steel is a sought-after raw material today. As steel is predominantly used in particularly durable consumer and capital goods, a large proportion of the world's steel is tied up in products and is only available for recycling years or even decades later. At the same time, the global demand for steel is continuing to rise sharply. Therefore, "fresh" steel must be continuously produced from iron ore, this being referred to as primary steel.

At thyssenkrupp Steel we have set ourselves the goal of making the previously CO_2-intensive production of primary steel more climate-friendly in the short term. We achieve this by adding a mixture of iron ore and specially prepared scrap to the blast furnace. In this way, we are reducing CO_2 emissions from steel production in real terms, and without requiring compensation measures (Fig. 2).

Conventional steel¹ [t CO_2eq/t steel]		bluemint recycled 25	CO_2 reduction
Hot strip	2.07	1.74	16 %
Hot-dip galvanized	2.32	1.98	15 %
NGO electrical steel	2.75	2.35	15 %

¹ Typical value as of March 2024 taking into account scope 1, 2 and partially 3 (only pre-material chain).
² Balanced recycled product with 25% scrap input in accordance with ISO 22095. Recyclate ratio blast furnace and converter > 50% without internal scrap.

Fig. 2 Carbon footprint of conventional and bluemint®
recycled 25 steel products

- ✓ CO_2-reduced crude steel via the blast furnace route
- ✓ Real CO_2 reduction using steel scrap in the blast furnace – no compensation certificate
- ✓ CO_2 reduction without quality loss
- ✓ Complete steel portfolio available, including electrical steel
- ✓ No adjustment to customer processes necessary

When it comes to recycling materials, especially in the automotive industry, the quality of scrap material is critical. It's the quality of scrap that counts, as not all scrap is the same – the purity and quality of the scrap is important if, for example, old cars are to be turned into new ones. To solve this problem, tkSE has developed a high-quality used scrap product, TSR40, in collaboration with its recycling partner TSR Recycling GmbH. This scrap product has a defined composition. TSR40 consists entirely of post-consumer material but is still characterized by the highest purity and an iron content of more than 98 percent. As a result, scrap can now be legally treated as a raw material and no longer as waste. TSR40 replaces part of the iron ore in the blast furnace, thereby reducing CO_2 emissions during hot metal production.

Sheet Metal 2025
Materials Research Proceedings 52 (2025) 327-333

Materials Research Forum LLC
https://doi.org/10.21741/9781644903551-40

Environmental impact of steel components
To evaluate the environmental impact of a product, a holistic approach is necessary. In this case, the emissions from raw material extraction, energy sources, and production of the steel, as well as the emissions resulting from the product manufacturing and application should be considered. For the calculation of the carbon footprint of the automotive parts in this paper, the LCA for Experts (GaBi) Software equipped with the GaBi Professional Database was used. The evaluation of the use phase is based on a battery-electric vehicle (BEV) for a mileage of 200,000 km. After the utilization phase the cut-off approach was applied, and thus recycling was not considered.

Sustainability of hot-formed components
Hot forming, also referred as hot stamping of steel sheets has been a success story the last few decades, offering a versatile and efficient solution to achieve stronger, safer, and lighter components. With strengths up to 2,000 MPa, hot-formed components offer a superior crash resistance, enabling weight reduction, which in turn results in CO_2 reduction. Additionally, hot forming allows manufacturers to create components having complex geometries that would be difficult or even impossible to manufacture utilizing cold forming.

Although initially hot-formed steels were used in areas where crash performance is crucial and thus any deformation during a crash should be kept minimal, the development of tailored tempering has allowed steel components to have varying levels of strength and ductility by controlling the cooling process during production. This makes it possible to produce components with ultra-high strength zones for structural integrity and more ductile regions for energy absorption during a crush. At the same time, hot forming can incorporate tailored blanks, in which case steel sheets can have varying thicknesses within the same component. This allows the optimization of material use, by reinforcing critical areas while minimizing weight in others.

Life Cycle Assessment of cold- and hot- formed B-pillars
When used according to requirements, hot forming offers sustainability advantages over cold forming, despite the additional heating process required. Due to its higher crash deformation resistance, the application of the press-hardened manganese-boron steel MBW®1500 can lead to weight savings compared to the cold-formed DP-K®700Y980T steel, enabling lightweight construction for

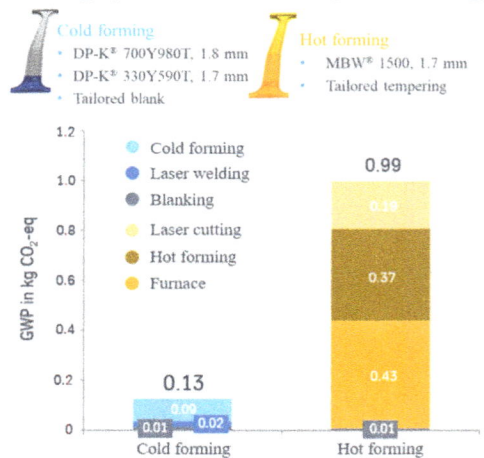

Fig. 3 Emissions from blank and component production.

the same crash performance. Additionally, since hot forming does not require any sheet holders, smaller blanks can be used for the final component, reducing material use.

Manufacturing of a hot-formed B-pillar, including reheating the blank and hot forming, results in significantly higher emissions compared to cold forming (Fig. 3). However, this disadvantage is mitigated when considering the overall emissions. Fig. 4 illustrates the emissions from steel production, part production, and the utilization phase. Comparing the two B-pillars reveals that the lightweight construction and reduced material usage enabled by hot forming are crucial for

Sheet Metal 2025   Materials Research Forum LLC
Materials Research Proceedings 52 (2025) 327-333   https://doi.org/10.21741/9781644903551-40

environmental performance, rather than the furnace process and manufacturing emissions. In this practical example, the hot-formed B-pillar has 7% less emissions compared to the cold formed B-pillar, and therefore, for this application the hot-formed concept is the most sustainable solution. It becomes clear from this example, that optimization of a part-specific carbon footprint is achieved through weight savings. By using higher strength steel grades for hot forming, up to 1,900 MPa, and Tailored Welded Blanks (TWB), a significant reduction in the carbon footprint can be achieved. These measures not only enhance the structural integrity and efficiency of automotive components but also contribute to a more sustainable and environmentally friendly manufacturing process.

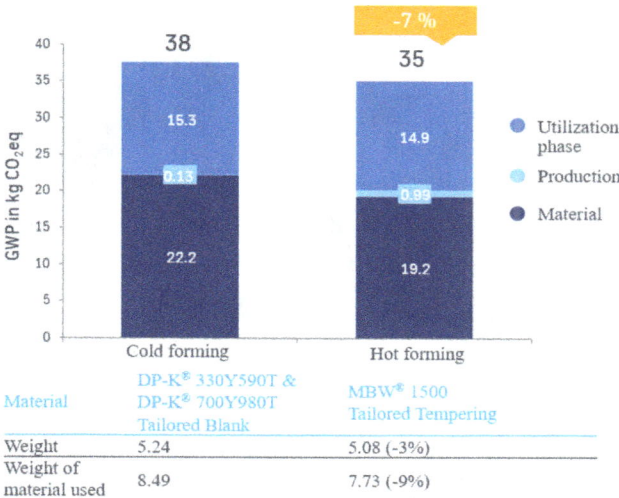

	Cold forming	Hot forming
Material	DP-K® 330Y590T & DP-K® 700Y980T Tailored Blank	MBW® 1500 Tailored Tempering
Weight	5.24	5.08 (-3%)
Weight of material used	8.49	7.73 (-9%)

Fig. 4 Total emissions of cold-formed and hot-formed B-pillars.

Life Cycle Assessment of seat cross members

Seat cross members consisting of AHSS must have high structural strength and stiffness to support the vehicle and enhance crash safety. Additionally, they should absorb energy during collisions, especially during side impacts, to reduce the forces transmitted to passengers. Often, additional components are attached to seat cross members to mount other parts, such as seats, and to facilitate load transfer and energy absorption. Crushable zones designed to bend and deform or alternatively consoles made of lower strength steels with high ductility, can be used to absorb energy during collisions.

Hot-formed tailored tempered manganese-boron steels can integrate all these functionalities into a single part (Fig. 5). This integration significantly reduces the component's weight, enhancing vehicle efficiency and contributing to overall sustainability. The environmental benefits of using high-strength, hot-formed steels for weight reduction become evident when comparing the emissions of roll-formed and hot-formed seat cross-members (Fig. 6).

Two designs were considered for this LCA comparison; the reference design consists of two roll-formed seat cross members made of martensitic steel (MS) with a strength of 1,700 MPa. Each roll-formed part includes deep drawn consoles made of dual phase steel DP-K®330Y590T (DP600) at the sides of the cross members, to ensure high ductility, while a console in the middle serves as seat mounting part. The hot-formed seat cross members are one-part components made of manganese-boron steel MBW®1900 with a strength of 1,900 MPa, featuring a soft zone on the sides adjusted through tailored tempering.

Fig. 5 Designs of reference roll-formed seat cross members with five deep drawn consoles and single part hot-formed seat cross members.

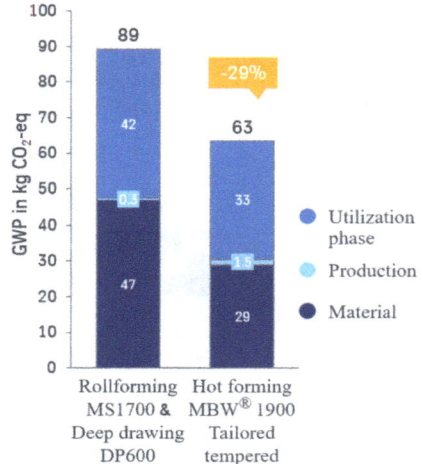

Fig. 6 Total emissions of rolled-formed and hot-formed seat cross members.

Fig. 6 shows the total emissions from steel production, component production, and use phase for both the reference and hot-formed seat cross members. Due to lower material usage, the emissions from steel production for the hot-formed part are lower. Additionally, the lower weight of the hot-formed component results in lower emissions during the use phase. The emissions from component production mainly originate from the furnace process before hot forming. Since these emissions constitute less than 2% of the total emissions, their impact on the carbon footprint of the hot-formed cross members is negligible.

Summary

Thyssenkrupp Steel Europe is committed to reduce its CO_2 emissions by more than 30% by 2030 and achieve net-zero steel production by 2045, supported by German government funding. Steel is essential for the automotive industry due to its strength, recyclability and versatility. Advanced High-Strength Steels (AHSS) play a crucial role in meeting safety and sustainability goals by reducing emissions originating from the steel production and enhancing the efficiency of battery-electric vehicles.

thyssenkrupp Steel's bluemint® Steel brand provides CO_2-reduced steel products for OEMs and suppliers, incorporating a high-quality scrap product to the current blast furnace-based steel production. The TSR40 scrap product, developed in collaboration with TSR Recycling GmbH, further reduces emissions, supporting the automotive industry's sustainability goals.

Evaluating the environmental impact of steel components requires a holistic approach, considering emissions from raw material extraction, energy sources, production processes and utilization. Hot forming of steel sheets has proven effective in creating stronger, safer, and lighter components, which help reduce CO_2 emissions. Life Cycle Assessments (LCA) show that hot-formed components, despite higher emissions during component production, offer overall sustainability benefits due to weight savings and reduced material use. For example, hot-formed

Sheet Metal 2025   Materials Research Forum LLC
Materials Research Proceedings 52 (2025) 327-333   https://doi.org/10.21741/9781644903551-40

B-pillars have 7% lower emissions compared to cold-formed ones, making them a more sustainable choice for automotive applications. Seat cross members made of Advanced High-Strength Steels (AHSS) must provide high structural strength and stiffness to support vehicles and enhance crash safety, while also absorbing energy during collisions. Hot-formed tailored tempered manganese-boron steels can integrate these functionalities into a single, lighter component, improving vehicle efficiency and sustainability.

References

[1] J. Somers, Technologies to decarbonise the EU steel industry, Publications Office of the European Union, Luxembourg, 2022

[2] Information on : https://www.thyssenkrupp-steel.com/de/unternehmen/nachhaltigkeit/klimastrategie/ klimastrategie.html

[3] Information on: https://www.worldautosteel.org/life-cycle-thinking/recycling/

[4] World Steel Association, "World Steel in Figures", Brussels, 2023

[5] Information on: Emissions-reduction-factsheet-v3.6-1.pdf (international-aluminium.org)

[6] Information on: https://www.euroncap.com/en/about-euro-ncap/timeline/euro-ncap-introduces-new-crash-tests-the-mobile-progressive-deformable-barrier-and-far-side-impact

[7] Information on: https://www.iihs.org/ratings/about-our-tests/test-protocols-and-technical-information

[8] Proposal for a regulation of the European Parliament and of the Council on circularity requirements for vehicle design and on management of end-of-life vehicles, amending Regulations (EU) 2018/858 and 2019/1020 and repealing Directives 2000/53/EC and 2005/64/EC

[9] Regulation (EU) 2023/1542 of the European Parliament and of the Council of 12 July 2023 concerning batteries and waste batteries, amending Directive 2008/98/EC and Regulation (EU) 2019/1020 and repealing Directive 2006/66/EC

[10] J. Suer, F. Ahrenhold, N. Jäger, Carbon Footprint and Energy Transformation Analysis of Steel Produced via a Direct Reduction Plant with an Integrated Electric Melting Unit, Journal of Sustainable Metallurgy, 8 (2022) 1532–1545. https://doi.org/10.1007/s40831-022-00585-x

[11] Information on: https://www.thyssenkrupp-steel.com/en/newsroom/press-releases/thyssenkrupp-steel-to-receive-federal-and-state-government-funding-totaling-around-two-billion-euros.html

Sheet Metal 2025
Materials Research Proceedings 52 (2025) 334-341

Materials Research Forum LLC
https://doi.org/10.21741/9781644903551-41

Experimental analyses of lubricant reduction in an industrial progressive tool

Eugen Stockburger[1,a,*], Leonard Kürbis[1] and Margarethe Nickel[1]

[1]Euscher GmbH & Co. KG, Johanneswerkstraße 22, 33611 Bielefeld, Germany

[a]eugen.stockburger@euscher.com

Keywords: Sustainable Development, CO2 Emission, Deep Drawing

Abstract. Due to political regulations and the general trend towards increased sustainability as well as efficient use of resources, companies are developing processes with a lower carbon footprint. Many small and medium-sized companies are creating their product based on years of experience without rethinking the production process. Hence, many processes contain high possibilities of improvement regarding the sustainability. Therefore, an experimental study at company Euscher was conducted. Six tests were performed using an industrial progressive die for deep drawing sleeves. The amount of lubrication was consequently reduced and parameters, such as the tool temperature, forming force and amount of lubrication, were tracked. Further, the part quality was checked during the tests regarding various properties, such as part dimensions and surface roughness. The amount of lubrication could be reduced significantly without any high impact on process parameters and part dimension. As expected, the roughness of the parts was increased.

Introduction

The process deep drawing is a well-known production method [1] and standardized in the DIN 8584-1 [2]. During deep drawing, a flat sheet metal is formed into a hollow body [3]. Prior to forming, a sheet metal blank is fixed between the drawing die and the blank holder, then the bottom of the sleeve is stretch-formed by the punch and afterwards the sides of the sleeve are deep drawn. If the part complexity exceeds the limits of a single deep-drawing operation, multi-stage processes are required to achieve a larger overall drawing ratio or a combination with other manufacturing processes [4]. For example, depending on the degree of complexity, multiple tool stages are used in the production of car body parts, which include bending, flanging, punching and embossing operations in addition to conventional deep drawing operations [5]. The implementation of a multi-stage tool on a common base plate creates a progressive tool and enables the economical and resource-saving production of challenging components in large batch sizes. Typical requirements for progressive tools are dimensional accuracy, complexity, mechanical properties and low costs [6]. The large number of influencing factors and forming requirements present a challenge, in particular the propagation of defects, hardening effects and the accumulation of heat over time [7]. Therefore, the design and construction of progressive tools require extensive expertise and many years of practical experience. Fig. 1 (A) shows the process chain for forming sleeves with a progressive tool.

Content from this work may be used under the terms of the Creative Commons Attribution 3.0 license. Any further distribution of this work must maintain attribution to the author(s) and the title of the work, journal citation and DOI. Published under license by Materials Research Forum LLC.

Sheet Metal 2025
Materials Research Proceedings 52 (2025) 334-341

Materials Research Forum LLC
https://doi.org/10.21741/9781644903551-41

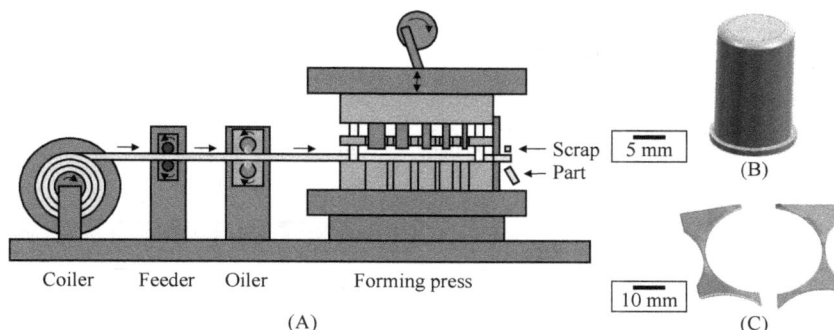

Fig. 1: Process chain for forming with a progressive tool (A), example of a produced sleeve (B) and resulting scrap (C)

First, material is transported by the feeder to the oiling station, where usually a coil lubrication supplies the material with oil. Then, the material is cut into a blank and formed into a sleeve, as shown in Fig. 1 (B), using the multiple deep drawing stations. Next to the part, a lot of scrap is produced, which is exemplarily displayed in Fig. 1 (C).

The use of lubricants for reducing friction and tool wear, as well as for the improvement of component quality, is a common practice. The lubricant, which is frequently derived from mineral oils, is applied to the blank prior to forming and subsequently removed by chemical cleaners after the forming process. In this context, both economic and ecological reasons motivate the reduction of lubrication use. In addition to the development of high quality and innovative products, success on international markets also depends on political requirements and restrictions, for example regarding sustainability. One example is the regulation on registration, evaluation, authorization and restriction of chemicals (REACH) issued by the European Union [8]. This decision increasingly obliges industry to refrain from using substances that are harmful to the environment and on human health. Against this background, all manufacturers had to rethink their production processes. Deep drawing is one of the most important sheet metal forming processes and therefore, a reduction in the environmental impact of this process would have a significant effect on the ecological balance of the entire production chain [9].

The lubricants themselves and the cleaning agents usually contain substances that are harmful to the environment and health. The environmental policy requirements of the REACH regulation and the environmental protection targets of Agenda 2020 for Europe [10] as well as the general trend towards increased sustainability and efficient use of resources are motivating processes with lower carbon footprint. Therefore, the objective of these industrial tests was to firstly investigate the potential of lubricant reduction on a commercial progressive deep drawing process at company Euscher.

Materials and Methods
In order to investigate a possibility for lubricant reduction, an existing progressive tool was used. The parts, which are produced, are sleeves and consist of the aluminium Al99.5 blank W7 in 1 mm thickness. The progressive tool was used in a forming machine from company Haulik&Ross having a lubrication system from company Raziol Zibulla & Sohn, where different lubrication stations can be connected to. Setting parameters of the oiling system are the tact and the pulse. For the tact, one spray signal is triggered after each number and the pulse indicates the duration of the spray signal. The progressive tool consists of six drawing steps and five more forming operations with a reduction ratio of about 1.6 for the first drawing step and 1.2 for the following steps. A

stroke rate of 150 min^{-1} resulting in a punch speed of about 200 mm/s was used for the tests. For lubrication the RENOFORM 19 B from company Fuchs Lubricants was tested, which is a high-performance forming oil for deep drawing of steel and aluminum.

A coil lubrication and three tool lubrications were used in the forming tool. For the coil lubrication, the parameters tact and pulse were fixed and not varied. In Table 1, the test matrix is shown with the variation of the lubrication parameters of the tool lubrications.

Table 1: Test matrix

Test	1. Start values		2. Increase tact		3. Decrease pulse		4. Increase tact 2		5. Turn off 1		6. Turn off 2	
Parameter	Tact	Pulse	Tact	Pulse	Tact	Pulse	Tact	Pulse	Tact	Pulse	Tact	Pulse
Unit	[-]	[s]	[-]	[s]	[-]	[s]	[-]	[s]	[-]	[s]	[-]	[s]
Lubrication 1	6	0.5	10	0.5	10	0.3	15	0.3	0	0	0	0
Lubrication 2									15	0.3		
Lubrication 3											15	0.3

For the first test, the conventional parameters of the lubrication system of the production were used. These values originate from the tool setup and were set high in order to an achieve optimum part quality. Here, the values were selected according to the experience of the workers, without determining a possible minimum quantity of lubrication. The tact was set to per 30 strokes for the coil lubrication and for the tool lubrications to per 6 strokes. The pulse was set to 0.06 s for the coil lubrication and to 0.5 s for the tool lubrications. In the second test, the tact was increased for the tool lubrication to a value of per 10 strokes or by 66.6 %. Further, the pulse of the tool lubrication was decreased by 40 % to 0.3 s in test number 3. Another increase of the tact was performed in test number 4. The tact for the tool lubrication was set to per 15 strokes or increased by 50 %. In test number 5, the first tool lubrication was completely turned off (100 % reduction) and in test number 6, the first and the second tool lubrication were switched off (100 % reduction).

To investigate the influence of the lubricant reduction on the forming process, the tests were analysed with various methods. First, a possible change in the distribution of the temperature in the progressive tool was monitored. Therefore, the thermographic camera E60 from the company Flir was used. To ensure a steady temperature distribution, the temperature measurements were carried out after 30 min of production. For an investigation of the continuous increase or decrease of the temperature in the forming tool, temperature sensors were used. The temperature sensors TMCx as well as the data logger UX120 from the company Onset were used and attached as close as possible to the active forming area. While testing, the forming force was observed five times from the force monitoring of the forming machine and evaluated five times per test to determine a possible influence of the lubricant reduction. Besides the forming force, the quality of the parts was tracked in accordance to the tests. The dimensions of the parts, such as different diameters and heights, were measured using the measuring device Surfcom from company Zeiss for all six tests five times during testing. For each specimen five parts were taken and measured. Further, the roughness of the parts was surveyed with the T8000 from company Hommel-Etamic in the same quantity as the dimensions. Finally, to quantify the reduction of lubrication in the tests, parts and scrap were weighted, degreased with cleaning petrol and then weight again. A precision scale XPR204 from company Mettler-Toledo was used. For each tests, 15 parts and 15 scraps, as shown in Fig. 1, were analysed.

Sheet Metal 2025
Materials Research Proceedings 52 (2025) 334-341

Materials Research Forum LLC
https://doi.org/10.21741/9781644903551-41

Results

Process parameters. For a first test, the continuous running of the forming machine was stopped after 30 min. The thermographic camera was used to search for the warmest point of the forming tool. As expected, the point with the highest temperature was located in the last deep-drawing stage of the tool. The drawing dies could not be analyzed in this case. Hence, the forming machine was run for 30 min to achieve a steady state of the process as far as possible. Then the machine was stopped to measure the temperature with the thermographic camera. In Fig. 2, the thermography measurements in the last deep-drawing stage are shown for each of the six tests.

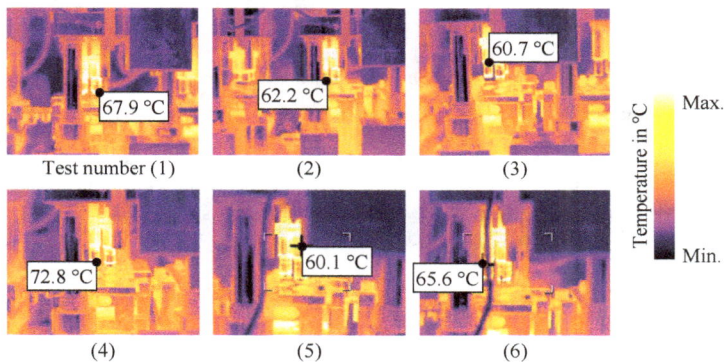

Fig. 2: Thermography measurements in the last deep-drawing stage

During the tests, attention was paid to whether new or bigger hot spots form with a reduction in the amount of lubricant in the forming tool. The temperature distribution remained uniform across all tests, with the maximum always being in the lower or upper part of the drawing die. The maximum temperature ranged from 60.7 °C to 72.8 °C, whereby the temperatures did not increase with a reduction of lubricant. Considering the different batches of material and other parameters such as the room temperature, the variation in temperature of around 12 °C can be well explained. Therefore, no new hot spots or a drastically increased temperature could be detected.

In order to not only observe a momentary view of the temperature behavior, temperature elements were also mounted to the last deep-drawing stage. To have an overview of the temperature development on the macro scale, the temperature elements were attached as close as possible to the forming zone. Two temperature elements (MP1 and MP2) were attached to the bottom of the forming tool close to the last drawing die and one temperature element (MP3) to the top of the forming tool near the last punch. The results are shown in Fig. 3, where the temperature is plotted over time for each test. For a better comparison, the tests are displayed as connected measurements. As before, no significant increase of the temperature with a reduction of the amount of lubricant could be detected. The temperature fluctuates strongly during the measurement. A peak temperature of 50 °C is reached during the tests and the lowest temperature is about 20 °C, which can be explained by the test procedure. Each test was carried out at the start of the production run. The parameters of the oiling system were set and a pallet of parts was produced. As a tool from actual production and not a test tool was used, the temperature decreased during downtime, while for example a failure was repaired. The minimum temperature was reached three times, as production was started and then the forming tool was shut down for a longer period of time due to production of spare parts, weekends or staff shortages. Nevertheless, the temperature development of the last test is similar to the first test.

Sheet Metal 2025   Materials Research Forum LLC
Materials Research Proceedings 52 (2025) 334-341   https://doi.org/10.21741/9781644903551-41

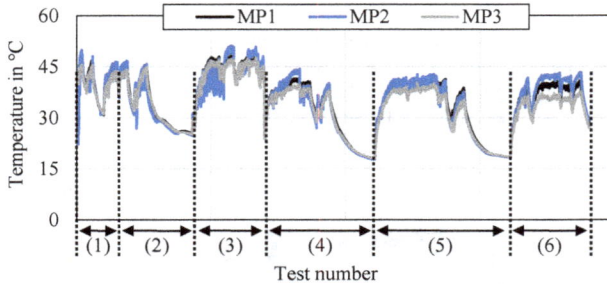

Fig. 3: Local temperature measurements during testing

In Fig. 4 (A), the maximum forming force for each tests is shown. Five measurements have been carried out five times during each test and analyzed as it was assumed that the reduction in lubricant would be reflected in an increase of the forming force. Firstly, a reduction in the forming force from circa 500 kN by 3.4 % to circa 483 kN can be determined for tests 1 to 3. Subsequently, the forming force increases again by 8.9 % to circa 526 kN, falls by 2.8 % to circa 511 kN and increases again by 4.3 % to circa 533 kN for tests 4 to 6.

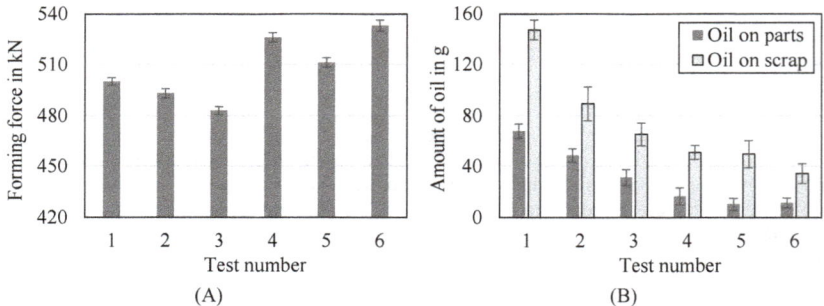

Fig. 4: Maximum forming forces (A) and oil consumption (B)

To evaluate this variation, the mechanical properties of the material used in each test were analyzed. Therefore, the test reports of the individual material batches of the manufacturer were checked to ensure identical material properties. The mechanical properties were determined by the manufacturer using tensile tests according to DIN 6892 and a tensile test specimen A50 [11]. It was found that the ultimate tensile strength was varying according to the forming forces. For example, the forming force was increasing 10.3 % from 483 MPa for test number 3 up to 533 MPa for test number 6, while the tensile strength was rising 16.4 % from 73 MPa to 85 MPa. Therefore, surprisingly no correlating influence of the reduction of the lubricant on the forming force could be determined.

In Fig. 4 (B), the amount of oil on 1000 pieces of parts and of scrap with standard deviation is depicted. A steady reduction of lubricant on the parts and scrap can be observed. The amount of oil on the parts was reduced by 83.2 % from 67.8 g per 1000 pieces to 11.4 g. For the scrap, the amount of oil was reduced from 147.2 g per 1000 pieces to 34.3 g or about 76.7 %.

Sheet Metal 2025
Materials Research Proceedings 52 (2025) 334-341

Materials Research Forum LLC
https://doi.org/10.21741/9781644903551-41

Component properties. In order to ensure a consistent quality of the components, five random samples of five components were taken and measured for each test. Fig. 5 shows the diameter, the height, the thickness of the walls and the roughness as a function of the tests as normalized values with standard deviation. Hereby, the normalization was carried out in relation to test number 1. This can be identified by test number 1 always being at 100 %. Furthermore, the tolerances were normalized and added to the diagram.

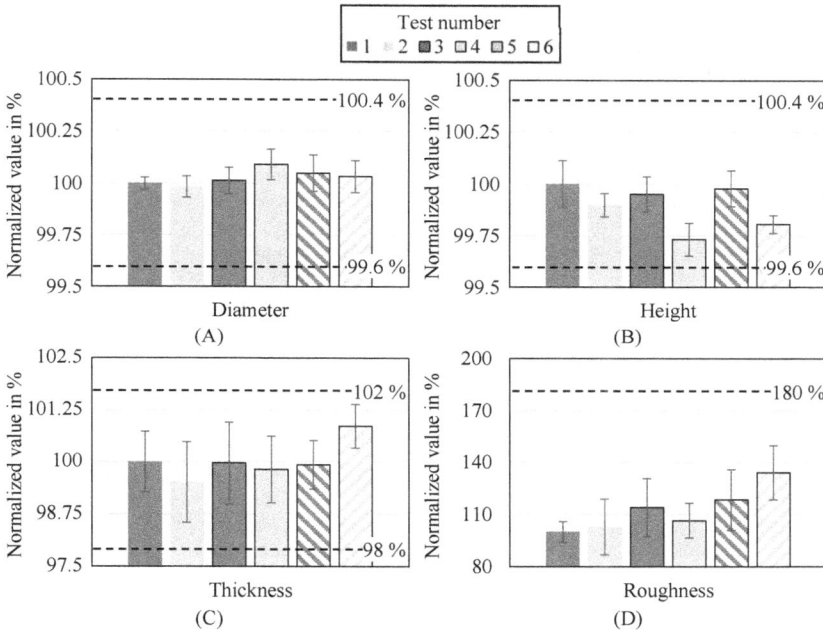

Fig. 5: Dimensional accuracy of the produced parts for the diameter (A), the height (B), wall thickness (C) and roughness Rz (D)

At first instance, the diameter of the part in Fig. 5 (A) seems to show an increase of its value to 100.1 % up to the test number 4. For the further tests, the diameter falls again, being 100.03 % for test number 6. Nevertheless, the diameter is never close to the tolerances of 99.6 % and 100.4 %. Considering the standard deviation of the diameter, it can be concluded that the diameter is nearly constant for the six tests. The height in Fig. 5 (B) shows no tendency regarding the oil reduction. The value decreases and increases randomly from maximum 100 % for test number 1 to minimum 99.7 % for test number 4 showing no clear correlation of the oil reduction on the height of the part. However, the height varies only slightly by around -0.35 %. The tolerance limits 99.6 % and 100.4 % are not reached. In Fig. 5 (C), the wall thickness of the parts is shown for the six tests. The same behavior can be observed here as with the height. The wall thickness fluctuate during production from maximum 99.5 % for test number 2 up to 100.9 % for test number 6 and no clear influence of the oil reduction is noticeable. Here, the given tolerances 98 % and 102 % are not met. As expected, an influence on the roughness can be observed by reducing the oil quantity, which is shown in Fig. 5 (D). The roughness value Rz, which was measured at four positions of each part, is displayed for the six tests. An increase of the roughness beginning with the second test is clearly visible. By the sixth test, the roughness increased from 0.72 μm to 0.96 μm or by 33.4 % compared to the first test. As the roughness is still at a relatively low value compared to the maximum tolerance of 180 % or 3.2 μm, the reduction in the amount of lubricant was classified as successful. However, due to the sharp increase in surface roughness, a further reduction is only conceivable with a slower decrease in the amount of lubrication in order to avoid damaging the forming tool and producing faulty parts.

Summary and Conclusion

To summarize, an industrial study was shown in which the amount of lubricant was reduced using the example of a progressive tool. The quantity of lubricant used in an existing production tool was successively reduced. For this purpose, the tact was increased, the pulse reduced and entire tool lubrications switched off. The temperature in the tool was monitored using thermographic images for one point in time and local temperature measurements for the entire test period. No significant increase in temperature with the reduction in the amount of lubricant was detected in the rather macroscopic tests. Furthermore, the forming force and the quality of the parts were analyzed as a function of the amount of lubricant. Surprisingly, no correlating influence of the lubricant quantity on the forming force was detected. The dimensional accuracy of the parts also remained almost constant. However, an increase in roughness was detected with a reduction of lubricant. Nevertheless, the reduction increased the roughness to a still acceptable value, so that the parts had a sufficiently high quality. In order to quantify the savings in lubricant quantity, the parts and scrap were weighed before and after degreasing. A total reduction of 78.8 % in the amount of lubricant was achieved, with only the roughness of the parts increasing slightly.

In addition to the improved sustainability and the lower direct costs for the lubricant, there are further savings effects. Parts with less lubricant are available, which have to be degreased for a shorter time with environmentally unfriendly agents. Furthermore, parts with less lubricant make it possible to use other, more environmentally friendly methods such as degreasing with modified alcohol or aqueous cleaning. In addition, fewer work steps are required to clean the parts with less lubricant. This saves additional machine and labor costs. A further positive effect is, that less lubricant is on the scrap resulting in input material for recycling with less lubricant and hence, a more sustainable recycling process. Other minor savings in terms of labor are achieved by the fact that less oil has to be ordered, transported and filled into the lubrication system. In the coming years, the greatest savings will be achieved due to the rising price of CO_2, meaning that many processes will have to be designed to be more CO_2-neutral. The challenge here is to maintain the quality of the parts at a high standard.

References

[1] E. Doege, B.-A. Behrens: Handbuch Umformtechnik - Grundlagen, Technologien, Maschinen. 3rd edition, Springer Vieweg, Berlin, 2018.

[2] DIN 8584-1:2003-09 - Manufacturing processes forming under combination of tensile and compressive conditions. DIN Media (2003)

[3] E. Stockburger, H. Wester, B.-A. Behrens: Fracture Characterisation and Modelling of AHSS Using Acoustic Emission Analysis for Deep Drawing. J. Manuf. Mater. Process. 7 (2023) 127. https://doi.org/10.3390/jmmp7040127

[4] A.S. Takalkar, L.B.M. Chinnapandi: Multi-stage deep drawing process of axis-symmetric extra deep drawing steel cylindrical cup. Eng. Res. Exp. 2 (2020) 025008. https://doi.org/10.1088/2631-8695/ab872a

Sheet Metal 2025
Materials Research Proceedings 52 (2025) 334-341

Materials Research Forum LLC
https://doi.org/10.21741/9781644903551-41

[5] J. Dietrich: Praxis der Umformtechnik - Umform- und Zerteilverfahren, Werkzeuge, Maschinen. 12th edition, Springer Vieweg, Dresden, 2018.

[6] C. Löbbe: Temperaturunterstütztes Biegen und Wärmebehandeln in mehrstufigen Werkzeugen. Dissertation, Dortmund, 2018.

[7] A.M. Engels: Beitrag zur Temperaturprognose in kombinierten Tiefzieh- und Abstreckgleitziehprozessen. Dissertation, Darmstadt, 2013.

[8] European Union: Regulation (EC) No 1907/2006 of the European Parliament and of the Council. https://eur-lex.europa.eu/legal-content/de/TXT/?uri=CELEX%3A32006R1907

[9] F. Klocke, W. König: Fertigungsverfahren 4 - Umformen. 5th edition, Springer-Verlag, Berlin, Heidelberg, 2006. https://doi.org/10.1007/978-3-540-39533-1

[10] J. Mischke: Europa 2020 Die Zukunftsstrategie der EU. Statistisches Bundesamt, Wiesbaden, 2013.

[11] DIN 6892-1:2020-06 - Metallic materials - Tensile testing - Part 1: Method of test at room temperature. DIN Media (2020)

Welding and additive manufacturing

Sheet Metal 2025
Materials Research Proceedings 52 (2025) 343-350

Materials Research Forum LLC
https://doi.org/10.21741/9781644903551-42

Influence of Liquid metal embrittlement on load-bearing capacity of resistance spot welds under crash loads:
A study based on S-Rail components

Keke Yang[1,a*], Max Biegler[2,b], Linus Happe[1,c], Marius Striewe[1,d], Viktoria Olfert[1,e], David Hein[1,f], Michael Rethmeier[4,3,2,g] and Gerson Meschut[1,h]

[1]Laboratory for Material and Joining Technology (LWF), 33098 Paderborn, Germany

[2]Fraunhofer Institute for Production Systems and Design Technology (IPK), 10587 Berlin, Germany

[3]Federal Institute for Materials Research and Testing (BAM), 12205 Berlin, Germany

[4]Institute for Machine Tools and Factory Management (IWF), Technical University of Berlin, 10623 Berlin, Germany

[a]keke.yang@lwf.upb.de, [b]max.biegler@ipk.fraunhofer.de, [c]linus.happe@lwf.upb.de, [d]marius.striewe@lwf.upb.de, [e]viktoria.olfert@lwf.upb.de, [f]david.hein@lwf.upb.de, [g]michael.rethmeier@ipk.fraunhofer.de, [h]meschut@lwf.upb.de

Keywords: Welding, Finite Element Method (FEM), Liquid Metal Embrittlement (LME)

Abstract. Liquid Metal Embrittlement (LME) cracking is a well-documented issue encountered during resistance spot welding (RSW) of zinc-coated advanced high-strength steels (AHSS) in automotive manufacturing. Given that existing research has predominantly focused on laboratory-scale samples and lacks investigation into the load-bearing capacity of joints under crash conditions, this study aims to fill these gaps by analyzing third-generation zinc-coated AHSS. S-Rail components were produced through stamping to replicate real-world manufacturing conditions and geometries of automotive parts. To account for the disturbances typically encountered in production, samples with LME cracks were intentionally fabricated. Subsequently, a modified three-point bending test, assisted by numerical simulations, was developed to effectively apply loads to the weld spots of the S-Rail components. Results from crash tests demonstrated that observed light crack severity does not significantly compromise the joint's load-bearing capacity or lead to earlier joint failure.

Introduction

Automobile manufacturers emphasize lightweight designs to improve both driving safety and to lower emissions simultaneously [1]. Advanced High-Strength Steels (AHSS) are highly regarded for their ideal balance between elongation and tensile strength, which supports these objectives [2]. In the production of body-in-white (BIW) components, Resistance Spot Welding (RSW) is the standard technique for joining these types of steel, which are frequently zinc-coated to enhance corrosion resistance [3]. Previous studies have revealed a potential risk of Liquid Metal Embrittlement (LME) during the RSW of zinc-coated steels [4]. The interaction of electrode force and the stresses produced during welding allows liquid zinc to infiltrate the grain boundaries of AHSS, weakening the structural integrity and resulting in LME-induced fractures [5].

Recent investigations have predominantly focused on assessing the influence of LME cracking on the load-bearing capacity of joints, which is largely attributed to the degradation of the base material induced by LME. Numerous studies have explored various steel grades and classified cracks based on their location, as defined by the [6] classification system. Currently, four types of cracks have been identified. Type A cracks occur in the contact area between the electrode cap's working surface and the material, Type B cracks appear in the shoulder region of the electrode

Content from this work may be used under the terms of the Creative Commons Attribution 3.0 license. Any further distribution of this work must maintain attribution to the author(s) and the title of the work, journal citation and DOI. Published under license by Materials Research Forum LLC.

cap, Type C cracks are found in the heat-affected zone (HAZ), and Type D cracks develop at the interface between the sheets. For instance, Rethmeier et al. employed CP800 and TRIP700 steel sheets, each with a thickness of 1.5 mm, as experimental materials. Cracks were categorized as Type B and Type C, with crack depths not exceeding 43% of the sheet thickness. This investigation, carried out under cyclic shear-tensile loading, concluded that the presence of cracks did not significantly affect fatigue life or structural stiffness [7].

Further research, such as those by [8], analyzed AHSS800, AHSS1050, and AHSS1180 steel grades, with thicknesses ranging from 1.0 mm to 1.4 mm. Type A and Type B cracks, which extended nearly through the full thickness of the sheets, were shown to adversely affect both strength and head tensile load-bearing capacity. These studies documented a 30% reduction in shear strength, a 20% decline in head tensile strength, and an 8% decrease in fatigue life under cyclic loading conditions. Another investigation by [9], focusing on TRIP1180 steel with a thickness of 1.6 mm, reported a 40% reduction in head tensile strength and energy absorption capacity when crack depth reached 40%, under both cyclic head tensile and quasi-static shear-tensile loads. Additionally, [10] assessed TRIP690 and DP980 steel, each with a thickness of 1.2 mm, and identified reductions in shear strength of 2.6% and 7.1%, respectively, in comparison to zinc-removed materials. The study further revealed that shear strength decreased by 43.5% when crack depth reached 68% in TRIP1100 steel. Research by [11] demonstrated that Type A and B cracks resulted in crack depths of 31% in TRIP1100 steel (1.6 mm thick) and 22% in TRIP1180 steel (1.2 mm thick). The conclusions emphasized that crack location plays a critical role in the primary load-bearing path, leading to a reduction in shear strength of up to 30% when cracks were present along this path.

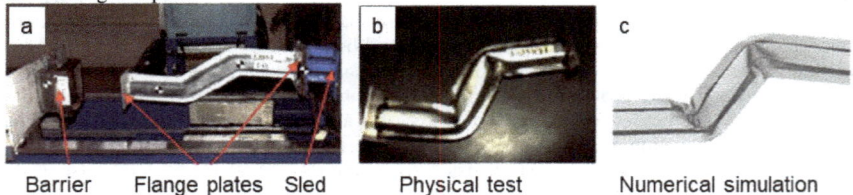

Barrier   Flange plates   Sled   Physical test   Numerical simulation

Fig. 1: Traditional crash test system of S-rail component (a) and Comparison of deformation pattern between the physical test (b) and simulation (c) according to [12]

Despite these research findings, studies on the impact of LME cracking on the load-bearing capacity of joints have been limited to single flat-point specimens, with the majority conducted under quasi-static loading conditions. As a result, the applicability of these findings to real-world conditions involving impact loads and the complex dimensions and structures of actual components is limited. This limitation is particularly evident in the S-Rail component, which is used for energy absorption in car bodies. As shown in Fig. 1, the asymmetry in the dimensions of the S-Rail component in conventional testing methods leads to bending at the corners, which hinders the effective transfer of loads to the joints at the flanges. Consequently, the load-bearing capacity of the joints under impact conditions cannot be fully assessed.

To address these limitations, the present study employs third-generation AHSS as the research material. Based on studies [13,14], the S-Rail component was selected as the specimen due to its representative geometry, which is commonly used in automotive structures. A novel testing method, specifically designed to account for the unique geometric complexities of the S-Rail, was developed using the finite element method (FEM). This approach facilitates a comprehensive evaluation of the effects of LME-cracking on the load-bearing capacity of RSW joints under impact loading conditions. The findings of this study are expected to provide insights into the behavior of LME in real-world applications.

Sheet Metal 2025
Materials Research Proceedings 52 (2025) 343-350

Materials Research Forum LLC
https://doi.org/10.21741/9781644903551-42

Material and welding equipment

In the experiments, a third-generation Advanced High-Strength Steel (AHSS) grade RA1180 EG, with a thickness of 1.4 mm and electrogalvanized (EG) coating on both sides, was used to manufacture the S-Rail component, positioned as the anode during welding. This component was paired with a bottom sheet made of hot-dip galvanized (HDG) mild steel, MS270 GI, with a thickness of 2.0 mm, which was laser-cut to match the shape of the S-Rail. The chemical composition and mechanical properties of the materials are summarized in Table 1 and Table 2, respectively. The selection of this material combination was based on the work of Benlatreche et al., in line with current industrial practices [8]. Resistance spot welding was performed using a 1000 Hz medium-frequency direct current (DC) and a pedestal-mounted servo-electric C-type welding gun. The electrodes were water-cooled at a rate of 4 liters per minute, and the electrode caps conformed to the F1-16-20-50-5.5 specifications outlined in ISO 5821. After welding, the zinc coating was removed with a 20% hydrochloric acid solution to enable better observation of potential LME cracks on the AHSS surface. A visual inspection of the sheets was then conducted using an optical microscope.

Table 1: Chemical composition of the utilized material (expressed in weight- %).

	C	Si	Mn	Al	Cr
RA1180	0.202	1,49	1.22	0.055	0.03
MS270	0.0019	0.01	0.09	0.045	0.02

Table 2: Mechanical properties of the utilized material combination

	$R_{p0.2}$ [MPA]	R_m [MPa]	A [%]
RA1180	888-995	1228-1241	14.9-16.8
MS270	262	342	39

Design of a novel testing method

Current testing methods have limitations, particularly in addressing the multiaxial nature of real-world loading conditions, where tensile shear and cross tensile stresses rarely act independently but rather combine to form peeling stresses. To overcome these challenges, this study developed a novel testing method for the S-Rail component, using a three-point bending experiment.

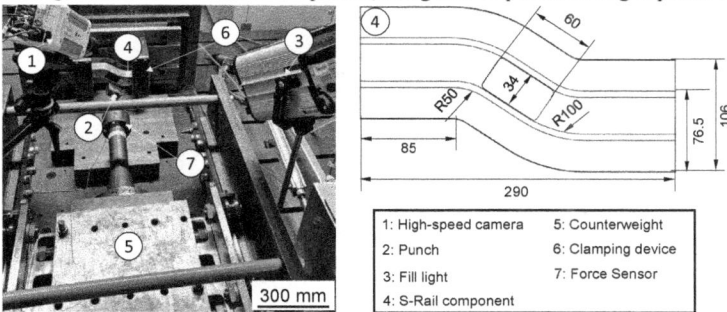

1: High-speed camera	5: Counterweight
2: Punch	6: Clamping device
3: Fill light	7: Force Sensor
4: S-Rail component	

Fig. 2: Experimental setup for the crash test and geometry of S-Rail-Component

The crash test was performed on a sled test rig, with the experimental setup and specimen dimensions shown in Fig. 2. Before the welding experiment, a window was cut in the center of the top section of the S-Rail component, which is made of AHSS (see label 4). The S-Rail was then joined to the bottom sheet via spot welds on the flanges. During the crash test, the punch (see label 2) directly applied the load to the bottom sheet through the window. As both ends of the combined structure (S-Rail component and bottom sheet) are fixed by a clamping device (see label 6), the

Sheet Metal 2025
Materials Research Proceedings 52 (2025) 343-350

Materials Research Forum LLC
https://doi.org/10.21741/9781644903551-42

bottom sheet will undergo plastic deformation in the direction of the applied load. In the same time, Due to the secure fixation of the weld spots on the flanges, a defined and repeatable peeling stress can be consistently applied to the weld spots.

Results and discussion

Upon establishing the testing method, the FE-Method was employed to validate its feasibility, with the primary objective of determining whether the spot welds were subjected to load under the applied conditions. Modeling was performed using the commercial simulation software Simufact Welding 2023, as shown in Fig. 3 (a). The S-Rail simulation model was constructed using 3D scan data obtained with a GOM ATOS patterned-light scanner. After cleaning, the data was imported into MSC APEX for finite element meshing. Hexahedral elements were employed in circular patterns around the weld zones to represent the RSW process's axisymmetric characteristics, while coarser elements were applied to regions farther from the welds. Thickness-wise, two elements were used, with additional refinement in a 6 mm radius near weld zones, resulting in eight layers through the thickness in these critical areas. The total element count was 17,628 for the S-Rail and 11,894 for the base plate pre-refinement. A fixed time step of 0.009 s was selected for welding and cooling simulations. Boundary conditions included central and edge supports for the bottom sheet, mimicking the experimental setup, while 2 kN clamps with 1 kN/mm stiffness were placed at the S-Rail's edges. These clamps were released 5 seconds after the final weld to record residual deformations and stresses. The calculations were performed on a workstation with an Intel XEON 2295 CPU and 64 GB RAM, completing in 93.1 hours. In this model, "F" denotes the weld spot, and the accompanying numbers indicate the welding sequence, which corresponds to the sequence used in the experimental procedure. The boundary conditions applied in the simulation were consistent with those used in the actual experiments.

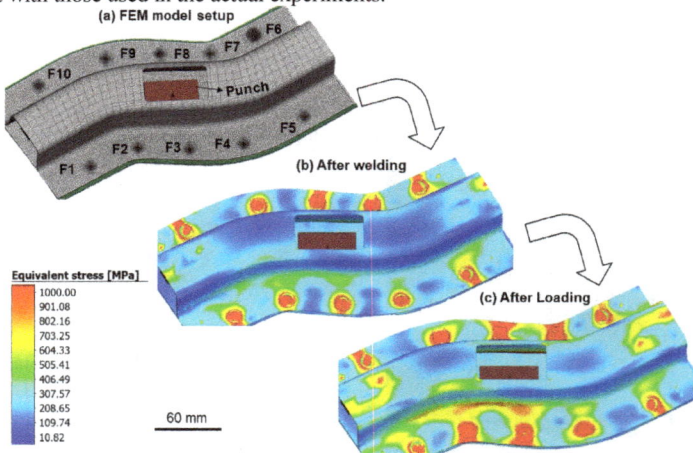

Fig. 3: Welding model setup (a), equivalent stress distribution after welding (b), and after applying the load (c)

Following the completion of the welding simulation, higher equivalent stress was observed in the weld spot regions, as shown in Fig. 3 (b), which can be attributed to residual stresses generated during welding. Subsequently, the punch was moved downward according to the predetermined settings, applying load to the bottom sheet. After the load was applied, Fig. 3 (c) revealed that the weld spots in the central area of the S-Rail component experienced significant loading. This

Sheet Metal 2025
Materials Research Proceedings 52 (2025) 343-350

Materials Research Forum LLC
https://doi.org/10.21741/9781644903551-42

observation confirms the effectiveness of the testing method. Since the primary aim of the simulation-based validation was to verify the effective loading of the weld spots, rather than to assess joint failure, the developed testing method was subjected to experimental validation prior to the initiation of the crash test.

The results of the experimental validation, as depicted in Fig. 4 (a), demonstrate substantial bending of the S-Rail component along the Z-axis, with no evidence of weld failure. Severe deformation occurred at the cut-out window, which nearly assumed a diamond shape. This deformation is attributed to the excessive proximity of the weld spots located on the convex-concave side (F2, F4, F7, and F9) to the central weld spots (F3 and F8). The close spacing of the welds resulted in an overconcentration of load near the cut-out window, leading to stress concentration. This, in turn, increased the material stiffness in the load-bearing region by reducing the effective force-distribution area, hindering the plastic deformation of the bottom sheet. Consequently, more of the absorbed kinetic energy was transferred to the weld spots on the convex-concave side, causing them to inwardly contract under the load in the Y direction, thereby exacerbating the deformation at the window. To minimize the impact of the unique dimensions of the S-Rail component on the crash test, adjustments were made to each welded S-Rail component sample prior to the crash test. The weld spots at positions F2, F4, F7, and F9 were drilled out, leaving only the weld spots in the central flange region (F3 and F8) intact. This adjustment aimed to increase the load-bearing area, prevent excessive deformation of the specimen, and ensure that failure occurs at the weld spots. After this adjustment, as shown in Fig. 4 (b), this ensures that there is no longer weld spot failure in the non-collinear bending areas of the S-Rail component. Consequently, it eliminates additional deformation in these regions, ultimately reducing the overall deformation of the S-Rail component during the crash test. The set kinetic energy was maximally utilized for loading the weld spots, effectively leading to weld spot failure. Therefore, subsequent experiments will continue using this adjusted test method.

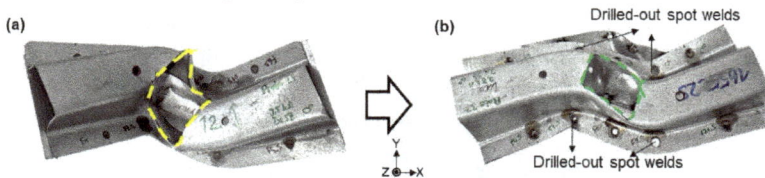

Fig. 4: Failure modes of S-Rail components before (a) and after (b) adjustment by drilling out the spot welds F2, F4, F7 and F9

In this study, the occurrence of LME was characterized under both normal conditions and conditions with electrode misalignment. To ensure sufficient heat input, the welding duration for both scenarios was set to twice the length recommended by the SEP1220-2 standard. This approach was adopted to cover the worst-case scenario, thereby enhancing the reliability of the results. As depicted in Fig. 5, the force-displacement curves illustrate that following the initial contact between the punch and the bottom sheet, subsequent plastic deformation leads to a gradual increase in displacement, reaching approximately 30 mm. A decline in the curve is observed thereafter, signifying the onset of spot weld failure. The subsequent pronounced fluctuations in the curve correspond to the complete failure of the two remaining weld spots in the central region of the S-Rail component. Notably, the presence of LME-induced cracks did not result in a statistically significant effect on the average F_{max} value, where F_{max} denotes the maximum load sustained immediately prior to the initiation of weld spot failure. To further analyze the impact of LME cracks on joint load capacity, the failure modes of weld spots in each sample will be examined with the help of a high-speed camera. The criterion for assessing whether LME affects the joint's

Sheet Metal 2025   Materials Research Forum LLC
Materials Research Proceedings 52 (2025) 343-350   https://doi.org/10.21741/9781644903551-42

load capacity is whether LME cracks participated in and facilitated the weld spot failure process. The specific criteria are as follows:

• Case 1: If LME cracks are present on the AHSS surface, but the joint fails on the mild steel side, it indicates that LME does not affect the joint's load capacity in this scenario.

• Case 2: If LME cracks are present on the AHSS surface and the joint fails on the AHSS side, but the failure does not follow the existing LME cracks, it indicates that LME cracks did not affect the joint strength.

• Case 3: If the conditions are the same as in Case 2, but the weld spot failure follows the LME cracks, it indicates that the presence of LME cracks facilitated the weld spot failure (since part of the material surface had already failed due to the cracks). This suggests that LME cracks negatively affect the joint's load capacity.

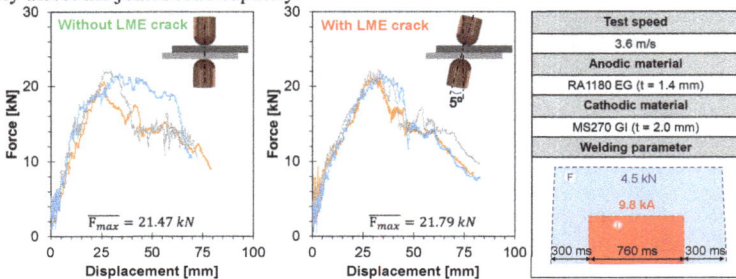

Fig. 5: Force-displacement curves under standard welding conditions (without LME) and with electrode misalignment (with LME)

In the control group, which adhered to standard welding conditions, no LME cracks were observed on the surface of the two central weld spots tested. Analysis of the high-speed camera footage revealed that weld spot failure initiated on the surface of the AHSS, specifically on the side closest to the applied load. The failure ultimately progressed, resulting in complete detachment on the AHSS side of the weld.

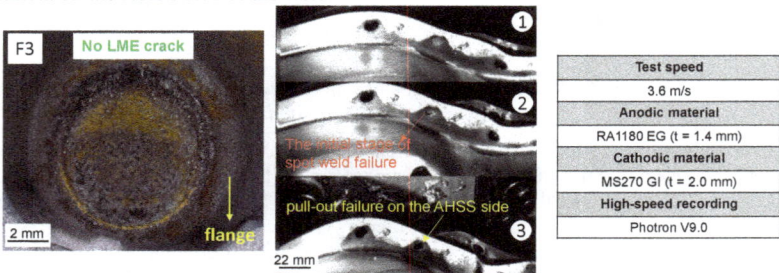

Fig. 6: Spot weld failure in standard welding condition (without LME cracking)

Subsequently, under standard conditions, electrode misalignment was introduced to further induce the occurrence of LME (see Fig. 7). Further investigation into the weld failure modes revealed that, although weld spot F3 exhibited slight Type A LME cracks, failure occurred in the form of a pull-out on the AHSS side. High-speed camera footage confirmed that the failure occurred entirely around the weld spot, indicating that the Type A crack did not contribute to the failure process. Additionally, in the same specimen, weld spot F8 exhibited both slight Type A and Type C cracks. However, the observed failure mode was a partial pull-out on the mild steel

Sheet Metal 2025   Materials Research Forum LLC
Materials Research Proceedings 52 (2025) 343-350   https://doi.org/10.21741/9781644903551-42

side, with no failure occurring on the AHSS side. Due to the size of the S-Rail component, it was challenging to perform non-destructive testing to assess crack depth prior to the crash test. Therefore, parallel experiments were conducted under identical boundary conditions to obtain samples with LME cracks similar to those in the crash test. The purpose was to determine the crack depth through metallographic examination, as shown in Fig. 7 (b) and (c). The observed depth of Type A and Type C cracks in these samples was approximately 700 µm and 140 µm. Based on these observations, it can be concluded that Type A and Type C cracks at this strength grade do not promote failure under impact loading. Furthermore, when combined with the force-displacement curve analysis, it can be inferred that such cracks do not negatively impact the maximum load-bearing capacity of the specimen.

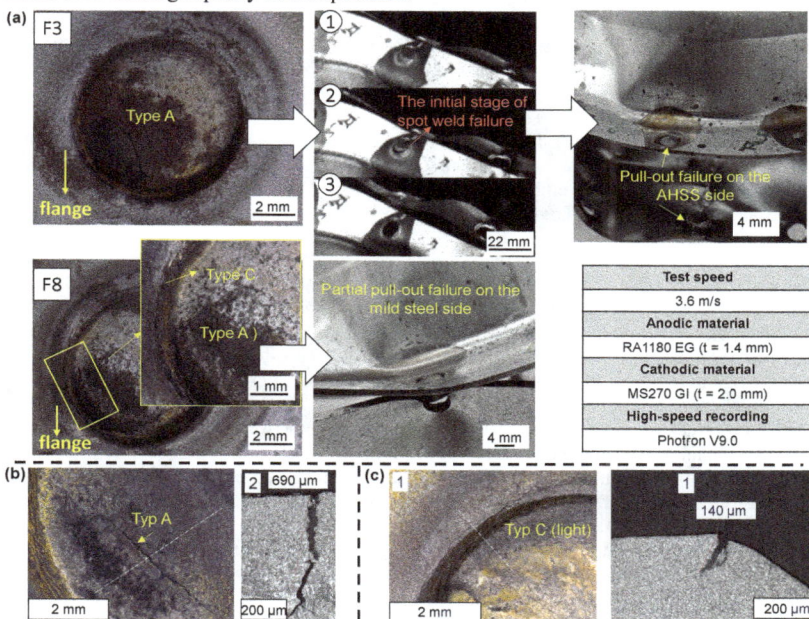

Fig. 7: Spot weld failure under welding condition with electrode misalignment

Summary and Outlook

In this study, numerical simulations were employed to develop and validate a novel testing method for evaluating the load-bearing capacity of joints in full-scale S-Rail components. The investigation revealed that third-generation AHSS shows relatively low sensitivity to LME, as no LME cracks were observed under standard welding conditions. It was only after introducing electrode misalignment, which induces LME, that Type A and Type C cracks appeared. However, these cracks did not significantly impact the load-bearing capacity of the RSW joints under crash loading conditions, nor did they contribute to the initiation of joint failure. The testing method developed in this study can also be adapted for future research to evaluate other joining techniques, such as Self-Piercing Riveting and clinching, at the component level in real-world applications.

Acknowledgement
This study was supported by the WorldAutoSteel consortium and conducted as part of a transnational research project, aiming to extend the current investigation of liquid metal embrittlement to the level of real components.

References
[1] G. Meschut, M. Matzke, R. Hoerhold, and T. Olfermann, "Hybrid technologies for joining ultra-high-strength boron steels with aluminum alloys for lightweight car body structures," *Procedia CIRP*, vol. 23, pp. 19–23, 2014. http://doi.org/10.1016/j.procir.2014.10.089

[2] Y. Ma, W. Jin, Y. Park, "Fracture modeling of resistance spot welded ultra-high-strength steel considering the effect of liquid metal embrittlement crack," *Materials & Design*, vol. 210, p. 110075, 2021. http://doi.org/10.1016/j.matdes.2021.110075

[3] W. Jin, A. Lalachan, S. P. Murugan, C. Ji, and Y. Park, "Effect of process parameters and nugget growth rate on liquid metal embrittlement (LME) cracking in the resistance spot welding of zinc-coated steels," *Journal of Welding and Joining*, vol. 40, no. 6, pp. 464–477, 2022. http://doi.org/10.5781/JWJ.2022.40.6.2

[4] W.-S. Jeon, A. Sharma, and J. P. Jung, "Liquid metal embrittlement of galvanized TRIP steels in resistance spot welding," *Metals*, vol. 10, no. 6, p. 787, 2020. http://doi.org/10.3390/met10060787.

[5] J. Mendala, "Liquid metal embrittlement of steel with galvanized coatings," *IOP Conference Series: Materials Science and Engineering*, vol. 35, p. 012002, 2012. http://doi.org/10.1088/1757-899X/35/1/012002

[6] S. P. Murugan, Y. Park, V. Vijayan, and C. Ji, "Four types of LME cracks in resistance spot welding of zinc-coated AHSS," *Welding Journal*, vol. 99, no. 3, pp. 75s–92s, 2020. http://doi.org/10.29391/2020.99.008

[7] M. Rethmeier, G. Weber, H. Gaul, and S. Brauser, Eds., *Untersuchungen zur Ermüdungsfestigkeit von Widerstandspunktschweißverbindungen aus hochfesten Mehrphasenstählen unter Berücksichtigung fertigungsspezifischer Einflüsse*, Düsseldorf: Verl. und Vertriebsges. mbH, 2011.

[8] Y. Benlatreche, M. Duchet, T. Dupuy, D. Cornette, and G. Carollo, "No effect of liquid metal embrittlement cracks on the mechanical performances of spot welds," presented at the Materials Science and Technology Conference, 2017.

[9] D. Choi, S. Uhm, C. Enloe, H. Lee, and G. Kim, "Liquid metal embrittlement of resistance spot welded 1180TRIP steel: Effects of crack geometry on weld mechanical performance," in *Contributed Papers from MS&T17*, 2017, pp. 454–462

[10] C. DiGiovanni, E. Biro, and N. Y. Zhou, "Impact of liquid metal embrittlement cracks on resistance spot weld static strength," *Science and Technology of Welding and Joining*, vol. 24, no. 3, pp. 218–224, 2019. http://doi.org/10.1080/13621718.2018.1518363

[11] C. DiGiovanni, X. Han, A. Powell, E. Biro, and N. Y. Zhou, "Experimental and numerical analysis of liquid metal embrittlement crack location," *Journal of Materials Engineering and Performance*, vol. 28, no. 4, pp. 2045–2052, 2019. http://doi.org/10.1007/s11665-019-04005-2

[12] K. Cai and D. Wang, "Optimizing the design of automotive S-rail using grey relational analysis coupled with grey entropy measurement to improve crashworthiness," *Structural Multidisciplinary Optimization*, vol. 56, no. 6, pp. 1539–1553, 2017. http://doi.org/10.1007/s00158-017-1728-y

[13] J. Choi, J. Lee, G. Bae, F. Barlat, and M.-G. Lee, "Evaluation of springback for DP980 S-rail using anisotropic hardening models," *JOM*, vol. 68, no. 7, pp. 1850–1857, 2016. http://doi.org/10.1007/s11837-016-1924-z

[14] R. Stocki, P. Tauzowski, and J. Knabel, "Reliability analysis of a crashed thin-walled S-rail accounting for random spot weld failures," *International Journal of Crashworthiness*, vol. 13, no. 6, pp. 693–706, 2008. http://doi.org/10.1080/13588260802055213

Sheet Metal 2025
Materials Research Proceedings 52 (2025) 351-356

Materials Research Forum LLC
https://doi.org/10.21741/9781644903551-43

A numerical model to study the temperature and residual stress profiles in hybrid additive manufacturing

Gaetano Pollara[1,a] *, Dina Palmeri[1,b], Gianluca Buffa[1,c] and Livan Fratini[1,d]

[1]1Dipartimento di Ingegneria, Università Degli Studi di Palermo, Viale delle Scienze, Palermo, 90128, Italy

[a]gaetano.pollara@unipa.it, [b]dina.palmeri@unipa.it, [c]gianluca.buffa@unipa.it, [d]livan.fratini@unipa.it

Keywords: Hybrid Manufacturing, Selective Laser Melting (SLM), Numerical Simulation

Abstract. Recently, there has been an increasing interest in hybrid additive manufacturing (HAM) technologies to overcome the limits of conventional and additive manufacturing (AM) technologies. In the case of metals, HAM can be used to combine AM with forming operations. This concept can be applied in both the production of bulk and sheet metal parts. When sheet metal parts are taken into consideration, usually AM technology such as laser powder bed fusion (L-PBF) and direct energy deposition (DED) can be combined with traditional forming operations. L-PBF is preferred when small details have to be applied to the metal sheet before undergoing the forming process. Thus, mass customization can be achieved by using the flexibility of the AM process, its ability to print complex geometries, and the speed of the sheet metal forming process. In this study, a numerical model was developed in order to analyze the influence of the L-PBF process on the metal sheet. The results show how the metal sheet is strongly influenced by the thermal input due to the deposition of the AM part. Moreover, the presence of residual stress can be observed within the metal sheet, which can result in distortion and create problems in the following forming step. The numerical model highlights also the more critical area, in which high-stress concentration is observed.

Introduction

Additive manufacturing (AM), thanks to its characteristic of fabricating parts in a layer-by-layer fashion, allows the production of complex structures with the possibility of creating lightweight structures and multi-functional components [1]. For this reason, AM is widely used in many industrial fields, such as biomedical and aerospace, where customization and complex geometries are usually required [2]. Nevertheless, the spreading of AM in other fields is limited to the large production time and poor geometrical accuracy. Recently, in order to overcome these disadvantages, a new trend of combining AM with conventional manufacturing processes is rising. The combination of AM with conventional manufacturing processes is called Hybrid Additive Manufacturing (HAM), where the goal is to overcome the challenges of both processes by taking advantage of their pros [3]. One strategy that can be used to improve the quality of AM products is combining AM technology with subtractive technology. Usually, when the production of metal parts is considered, Direct Energy Deposition (DED) is combined with the milling operation, which is performed for every layer after the material deposition to obtain a better surface roughness [4]. On the other hand, HAM can be used to shorten the processing time of AM technologies, improve its productivity, and overcome the limitations of traditional forming processes in geometrical complexity. In this way, HAM allows reaching the so-called mass customization [5]. This concept can be applied in both bulk and sheet metal forming by using different AM processes. In the case of bulk metal forming, DED is usually preferred with respect to other AM processes due to its high flexibility. In this case, the forming process is performed before the AM process [6]. When sheet metal forming is considered, instead, Laser Powder Bed Fusion (L-PBF) is usually

Content from this work may be used under the terms of the Creative Commons Attribution 3.0 license. Any further distribution of this work must maintain attribution to the author(s) and the title of the work, journal citation and DOI. Published under license by Materials Research Forum LLC.

Sheet Metal 2025   Materials Research Forum LLC
Materials Research Proceedings 52 (2025) 351-356   https://doi.org/10.21741/9781644903551-43

employed thanks to its superior geometrical resolution. Here, two different approaches can be adopted: 1) sheet metal forming followed by L-PBF, and 2) L-PBF followed by sheet metal forming. Most of the time, the second approach is used because the first one usually requires too complex clamping systems. For the second approach, the metal sheet has to be fixed on the build plate during the L-PBF process. After the end of the printing process, the sheet metal is removed from the platform, and the forming step is performed [7].

Papke et al. [8] investigated the bonding zone between the sheet metal and the AM part. In detail, a Ti-6Al-4V cylindrical geometry was printed on the metal sheet made of the same material as the printed part. The bonding zone was characterized through shear tests and Vickers microhardness. Some works can also be found with more complex geometries, as in [9], where a gear component geometry has been manufactured with the combination of L-PBF and sheet metal forming for 316L. As for the authors' knowledge, a numerical approach in order to investigate the bonding zone in the hybrid process is still missing. Such a numerical model can help in predicting the temperature and residual stress profiles within the AM part and the sheet metal. In this paper, a FEM model was developed in order to evaluate the temperature distribution and residual stress in hybrid additive manufacturing of Ti-6Al-4V. This study's result can help engineering understand the process better and facilitate the design phase by saving time and material.

Numerical model set-up

The FEA commercial software DEFORM-3D™ v12.0 (V12.0, SFC, Columbus, OH, USA) was used to perform the numerical simulation. In the frame of hybrid additive manufacturing, L-PBF was performed on a Ti-6Al-4V sheet metal fixed to the build platform. The thermal and mechanical properties of the metal sheet are presented in Table 1.

Table 1. Thermal and mechanical properties of the Ti-6Al-4V sheet

Thermal properties	Melting point	1648 °C
	Beta transus temperature	980 ± 4°C
	Thermal conductivity at 20°C	6,7 W/ m°C
Mechanical properties	Yield strength	870 MPa
	Tensile strength	920 MPa
	Elongation	10%

The goal is to simulate the printing process of a Ti-6Al-4V parallelepiped geometry of 15 mm x 10 mm x 5 mm on the metal sheet. A fillet of 1 mm in all the edges was applied to have a smoother transition between the AM part and the sheet metal. In order to shorten the simulation time, a layer-by-layer approach was adopted where the whole layer is deposited simultaneously. The elements affected by the heat source are activated through a search algorithm, which works thanks to the voxel mesh. In this way, according to the deposition strategy (the process parameters being selected), only the elements within the voxel mesh for that particular layer will be activated. Further details about the use of the voxel mesh can be found in [10]. In order to predict the temperature and residual stress profiles on the metal sheet due to the L-PBF process, an area of 60 mm x 60 mm was considered in the numerical simulation. Three geometries were modeled in DEFORM-3D™ v12.0 and designed with Autodesk Fusion 360 (Fig. 1).

The first body (I) is the AM part already described above; the second body (II) is the metal sheet with the dimensions of 60 mm x 60 mm x 2 mm, and the third body (III) is the built plate equal to 60 mm x 60 mm x 25 mm. For each body, a different number of mesh elements were used: 200000 elements for I, 100000 for II, and 30000 for III.

Materials Research Forum LLC

https://doi.org/10.21741/9781644903551-43

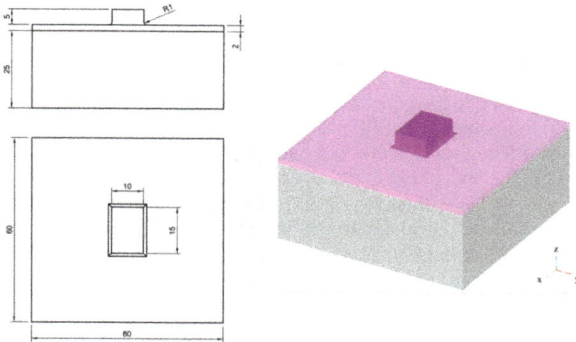

Fig 1. The designed geometries used in this study.

The geometries and the corresponding meshes are reported in Fig. 2. While body III was considered as a rigid material, bodies I and II were modeled as an elastoplastic material to observe the residual stress due to the thermal history they undergo during the L-PBF process.

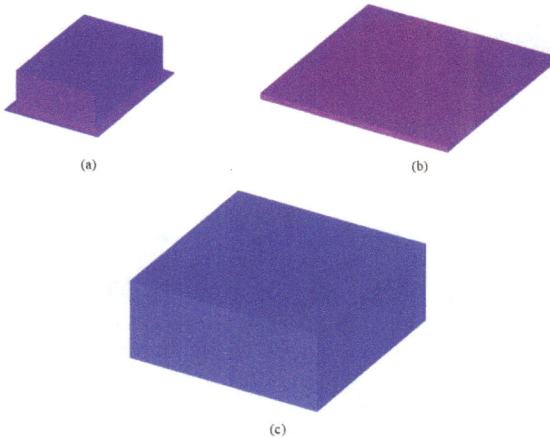

Fig 2. Geometries modeled within the FEA software:(a) meshed body I, (b) meshed body II, and (c) meshed body III.

A voxel mesh dimension of 0.5 mm x 0.5 mm x 0.5 mm was used for the element activation, considering a deposition strategy consisting of laser power of 250 W, scanning speed of 1400 mm/s, hatch distance of 120 μm, and layer thickness of 30 μm.

Sticking conditions were applied between I-II and II-III during the simulation of the L-PBF process and the cooling phase, where the residual stresses will develop. The build chamber temperature was set at 40 °C, while the temperature of II and III was set at 200°C.

Only convection heat exchange with argon was considered in the numerical model using h_{conv} = 1 W/m² °C, simplifying the computational complexity. A Newton-Rapson iteration coupled with a MUMPS (MUltifrontal Massively Parallel Sparse) solver was used to solve the

Sheet Metal 2025   Materials Research Forum LLC
Materials Research Proceedings 52 (2025) 351-356   https://doi.org/10.21741/9781644903551-43

thermomechanical problem. The simulation was carried out with a 12th Gen Intel (R) Core (TM) i9-12900 2.40 GHz processor.

Results and Discussions

The FEM model was used to analyze the temperature and residual stress profiles on the AM part and the metal sheet following the L-PBF process. In order to simulate the printing process, a voxel mesh dimension of 0.5 mm x 0.5 mm x 0.5 mm was used. This will result in a number of voxel mesh elements of 30, 20, and 10 along the x, y, and z directions, respectively. Since the height of the AM part is 5 mm, the layer thickness is 30 μm, and the voxel mesh element along z is 10, the computational layer will include about 16 real layers. In this way, it is possible to simulate the L-PBF process with a low computational cost. In detail, using a 12th Gen Intel (R) Core (TM) i9-12900 2.40 GHz processor results in a simulation time of 39 min and 15 sec (2355 total seconds). The temperature distribution predicted with the numerical simulation at the end of the L-PBF process is shown in Fig. 3.

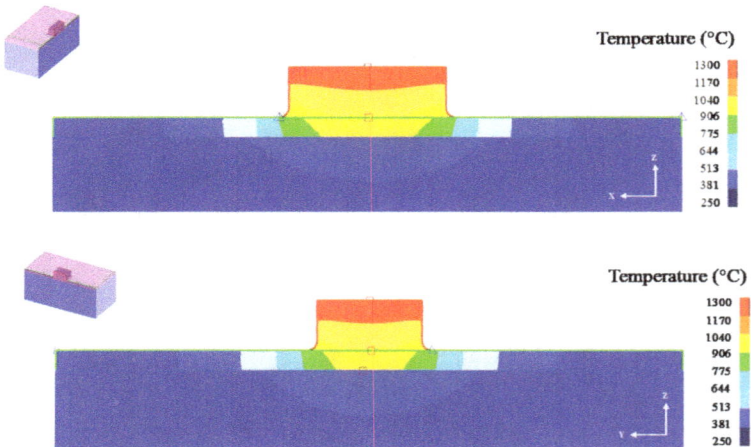

Fig 3. Temperature profiles along the longitudinal and transversal cross sections effectuated with two different planes, XZ and YZ, respectively.

It can be observed how the heat exchange between the AM part and the metal sheet is preferred with respect to the build platform. Near the bonding zone, the metal sheet is strongly affected by the thermal input, which melts the metal powder during the scanning of the new layer. In that area, it is possible to reach about 1000°C. The temperature along the metal sheet will decrease up to 400°C a few millimeters away from the printed part. This can be seen in both the longitudinal section (XZ plane) and the transversal section (YZ plane). The thermal history will be responsible for residual stresses during the cooling phase. In Fig. 4 the residual stresses σ_x and σ_y are presented. Strong residual stresses in the middle of the metal sheet can be observed.

The maximum value of the predicted residual stresses is around 900 MPa, such residual stresses can lead to distortion after removing the metal sheet from the build plate and can affect the mechanical response of the metal sheet. These residual stresses must be considered because, in the sheet metal HAM process, the metal sheet will undergo other metal forming operations.

Sheet Metal 2025
Materials Research Proceedings 52 (2025) 351-356

Materials Research Forum LLC
https://doi.org/10.21741/9781644903551-43

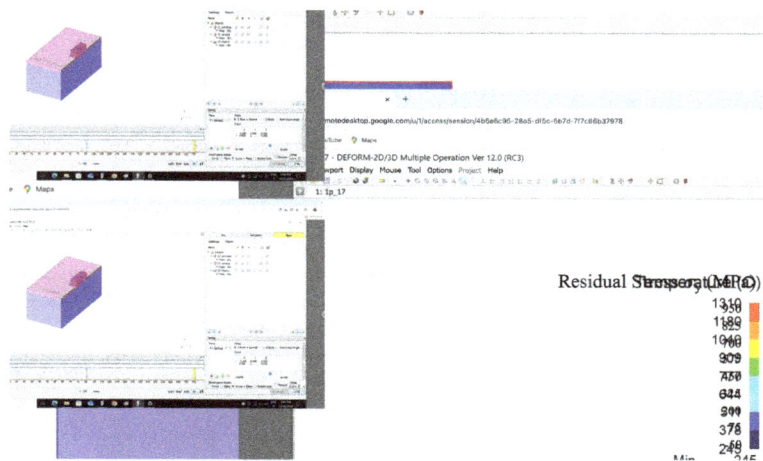

Fig 4. Residual stresses σ_x and σ_y along the XZ plane.

From the numerical simulation also, the distribution of the effective stress was analyzed (Fig. 5). It is worth noting that the most stressed area is the one near the edges of the AM part even if 1 mm fillet were applied in order to have a smoother transition and avoid stress concentration. This means particular attention must be paid to these areas where the AM part and the metal sheet can be detached.

Fig 5. Effective stresses due to the L-PBF process and the cooling phase.

Conclusions

A commercial software, DEFORM 3D v.12, based on the finite element method, was used to simulate the deposition of a parallelepiped geometry on a metal sheet during the L-PBF process of Ti-6Al-4V alloy. The main finding of this study can be summarized as follows:

- The deposition of the AM part strongly influences the metal sheet temperature profiles, a temperature of about 1000°C can be observed at the core of the metal sheet and decrease up to 400°C a few millimeters away from the AM part.
- The metal sheet's thermal history leads to large residual stress, especially below the AM geometry, where high temperatures were detected at the end of the L-PBF process.
- The residual stresses σ_x and σ_y can result in distortion in the XZ and YZ planes after the removal from the platform. This has to be considered in the case of HAM, where the AM processes will be followed by a forming process.

- Particular attention must be paid to the corners in which delamination between the AM part and the metal sheet can occur due to the high-stress concentration.

Acknowledgements

Part of this study was carried out within the activity of project using Italian MUR funds: "Aluminum Low Impact Circular Economy (ALICE)", funded by the European Union - Next Generation EU, Mission 4 Component 1, Project: P2022WN727, CUP: B53D23027330001 PRIN 2022-PNRR.

References

[1] T. Ngo, A. Kashani, G. Imbalzano, T. Nguyen, D. Hui, Additive manufacturing (3D printing): A review of materials, methods, applications and challenges, Compos B Eng 143 (2018) 25.

[2] B. Blakey-Milner, P. Gradl, G. Snedden, M. Brooks, J. Pitot, E. Lopez, M. Leary, F. Berto, A. du Plessis, Metal additive manufacturing in aerospace: A review, Mater Des 209 (2021) 110008. https://doi.org/10.1016/j.matdes.2021.110008.

[3] B. Ahuja, A. Schaub, M. Karg, R. Schmidt, M. Merklein, M. Schmidt, High power laser beam melting of Ti6Al4V on formed sheet metal to achieve hybrid structures, in: H. Helvajian, A. Piqué, M. Wegener, B. Gu (Eds.), 2015: p. 93530X. https://doi.org/10.1117/12.2082919.

[4] D. Svetlizky, M. Das, B. Zheng, A.L. Vyatskikh, S. Bose, A. Bandyopadhyay, J.M. Schoenung, E.J. Lavernia, N. Eliaz, Directed energy deposition (DED) additive manufacturing: Physical characteristics, defects, challenges and applications, Materials Today 49 (2021) 271–295. https://doi.org/10.1016/j.mattod.2021.03.020.

[5] A. Schaub, B. Ahuja, L. Butzhammer, J. Osterziel, M. Schmidt, M. Merklein, Additive Manufacturing of Functional Elements on Sheet Metal, Phys Procedia 83 (2016) 797–807. https://doi.org/10.1016/j.phpro.2016.08.082.

[6] D.-G. Ahn, Correction to: Directed Energy Deposition (DED) Process: State of the Art, International Journal of Precision Engineering and Manufacturing-Green Technology 9 (2022) 1215–1215. https://doi.org/10.1007/s40684-021-00401-z.

[7] J. Hafenecker, D. Bartels, C.-M. Kuball, M. Kreß, R. Rothfelder, M. Schmidt, M. Merklein, Hybrid process chains combining metal additive manufacturing and forming – A review, CIRP J Manuf Sci Technol 46 (2023) 98–115. https://doi.org/10.1016/j.cirpj.2023.08.002.

[8] T. Papke, P. Dubjella, L. Butzhammer, F. Huber, O. Petrunenko, D. Klose, M. Schmidt, M. Merklein, Influence of a bending operation on the bonding strength for hybrid parts made of Ti-6Al-4V, Procedia CIRP 74 (2018) 290–294. https://doi.org/10.1016/j.procir.2018.08.113.

[9] M. Merklein, R. Schulte, T. Papke, An innovative process combination of additive manufacturing and sheet bulk metal forming for manufacturing a functional hybrid part, J Mater Process Technol 291 (2021) 117032. https://doi.org/10.1016/j.jmatprotec.2020.117032.

[10] D. Palmeri, G. Pollara, R. Licari, F. Micari, Finite Element Method in L-PBF of Ti-6Al-4V: Influence of Laser Power and Scan Speed on Residual Stress and Part Distortion, Metals (Basel) 13 (2023) 1907. https://doi.org/10.3390/met13111907.

Sheet Metal 2025
Materials Research Proceedings 52 (2025) 357-364

Materials Research Forum LLC
https://doi.org/10.21741/9781644903551-44

Effect of process parameters on local thickening of Mg-Zn-Zr alloy sheets in TIG welding

Ecem Ozden[1,a*], Oleksandr Kurtov[2,b], Hans Vanhove[1,c] and Joost R. Duflou[1,d]

[1]KU Leuven, Department of Mechanical Engineering / Flanders Make, Celestijnenlaan 300B, B-3001 Leuven, Belgium

[2] KU Leuven, Jan Pieter De Nayerlaan 5, B-2860 Sint-Katlijne-Waver, Belgium

[a]ecem.ozden@kuleuven.be, [b]oleksandr.kurtov@kuleuven.be, [c]hans.vanhove@kuleuven.be, [d]joost.duflou@kuleuven.be

Keywords: Magnesium, TIG Welding, Local Thickening

Abstract. This study introduces a local thickening technique for thin Mg-Zn-Zr alloy sheets using Tungsten Inert Gas (TIG) welding deposition with pure argon shielding. The research examines how key welding parameters, including welding current, welding time, and filler diameter, affect the deposition morphology. Results show that higher heat inputs increase the deposition diameter, reduce height, and adjust contact angles. For dome-shaped deposition with consistent geometry, smooth deposition was achieved at contact angles below 70°. Excessive heat input, however, caused geometric inconsistencies and contamination at contact angles below 30°. By defining the impact of heat input parameters, process operating windows were identified for 4 mm and 5 mm filler materials. Importantly, no major distortion was observed in the 1.6 mm sheet substrate, supporting the potential of this technique for thin-sheet applications. Ultimately, this preliminary study provides valuable insights into the local thickening of magnesium sheets, with the particular aim of advancing implant design technology.

Introduction

Magnesium and its alloys stand out as leading candidates for biomedical applications, primarily due to their biocompatibility, biodegradability, potential to support bone formation, and mechanical properties that closely resemble those of bone structure [1,2]. Over the past two decades, the unique properties of magnesium have driven significant commercial use and research development efforts to optimize their use in various biomedical fields [3]. However, challenges remain in orthopedic implant applications, particularly in balancing optimal degradation rates with sufficient mechanical strength, making alloy composition and production techniques key aspects of ongoing investigation.

As a production technique for orthopedic implants, Incremental Sheet Forming (ISF) has emerged as a rapid and cost-effective method, enabling the production of patient-specific, thin-shelled implants with complex geometries, including cranial, facial, and clavicle implants, which have been extensively investigated [4,5]. While thin-shell implant designs are essential for minimizing patient discomfort, localized thickening may be required in certain cases to provide sufficient structural support [6]. Here, combining ISF with post-processing additive manufacturing (AM) techniques emerges as a promising solution.

Tungsten Inert Gas (TIG) welding, also known as Gas Tungsten Arc Welding (GTAW), offers a promising solution for reinforcing ISF-produced implants [6]. The use of non-consumable electrodes and inert gas shielding minimizes atmospheric contamination, resulting in clean, defect-free welds [7,8]. However, magnesium alloys pose significant challenges in welding due to their low melting point (650 °C) and high susceptibility to oxidation, which limits their high-temperature applications [7]. Moreover, an important aspect of the welding process involves

Content from this work may be used under the terms of the Creative Commons Attribution 3.0 license. Any further distribution of this work must maintain attribution to the author(s) and the title of the work, journal citation and DOI. Published under license by Materials Research Forum LLC.

Sheet Metal 2025   Materials Research Forum LLC
Materials Research Proceedings 52 (2025) 357-364   https://doi.org/10.21741/9781644903551-44

residual stresses in welded components, which arise from solidification shrinkage and thermal contraction [9]. These phenomena can lead to distortion, particularly in thin sheet substrates and high-energy input welding applications, as well as solidification cracking [9, 10]. Addressing the described welding defects relies on two core strategies: refining welding techniques and careful control of process parameters, particularly through the adjustment of heat input strategies [7]. Excessive heat input can degrade mechanical properties, promote weld defects, and increase the risk of distortion. Focusing on welding in thin sheets, Costanzi [11] proposes a novel method for robotic welding directly onto thin pre-bent sheets, extensively investigating tool-path strategies and geometric design. In a different approach, Wehren et al. [12] explore the use of WAAM for forming shell structures, enabling varying degrees of bending. Moreover, insufficient heat input can hinder weld formation due to the rapid solidification rate of magnesium alloys [13]. Zhang et al. [13] investigated peak current and peak time as the main heat input parameters, demonstrating that the optimal parameter range for regulated short-circuit welding of Mg-Gd-Y-Zr alloy is relatively narrow to achieve stable droplet transfer. Similarly, Zhou et al. [14] further explored the effects of welding current on the microstructure and mechanical properties of Mg-Zn-Cu alloy joints, demonstrating that excessive heat input can lead to grain coarsening, resulting in a decrease in strength.

Previous research highlights the importance of precise process control for high-quality welds, yet understanding how welding affects the quality of thin magnesium alloy sheets, particularly in applications involving welding onto complex geometries, remains limited. Integrating this knowledge with insights into various welding parameters is essential for optimizing process control and enhancing weld quality.

This paper presents a preliminary investigation into how welding parameters affect the local thickening of Mg-Zn-Zr alloy sheets during TIG welding deposition. The study examines deposition morphology and observes process defects, focusing on the influence of welding current, welding time, and filler material diameter. Based on these evaluations, a process window is established, providing critical insights for the optimization of these parameters.

Materials and Methods
This study investigated TIG welding of commercially available ZK61 magnesium alloy sheets, a promising material for orthopedic implants due to its controllable degradation, mechanical strength, and availability in sheet form. Accordingly, hot-rolled and annealed ZK61 sheets with a nominal composition of Mg-5.3Zn-0.5Zr (wt%) were used in this study.

For the TIG welding experiments, the substrate and filler materials were extracted from the same material and thickness: as-received 1.6 mm thick sheets. The substrates were cut to dimensions of 25 x 20 mm to ensure compatibility with subsequent mechanical testing. The filler material was prepared by cutting the sheet into cylindrical shapes with diameters of 4 mm and 5 mm using Wire Electrical Discharge Machining (EDM). Prior to each experiment, to improve deposition quality, all surfaces of the filler material, as well as both the front and back sides of the substrate, were lightly ground using 600 and 1200 grit abrasive paper respectively, and then cleaned with acetone.

Figure 1a illustrates the TIG welding setup, employing a MacGregor arc power source with a current range of 60-99 A (0.1 A increments). This setup enables precise arc pulsing, with main peak times (0-999 ms), and upslope/downslope periods (0-99 ms/ 0-999 ms), each with a 1 ms resolution [6]. To ensure an inert process environment, high-purity argon gas is directed through the TIG torch, with a gas lens surrounding the area. Additionally, a backing gas was used at a flow rate of 2 ℓ/min to further support the inert atmosphere. Notably, during the experiments, instead of feeding the filler material, it was positioned at the center top of the flat substrate, and the deposition was performed while maintaining a constant arc length (Fig. 1b). Consequently, dome-shaped geometries were produced on the substrates.

Sheet Metal 2025
Materials Research Proceedings 52 (2025) 357-364

Materials Research Forum LLC
https://doi.org/10.21741/9781644903551-44

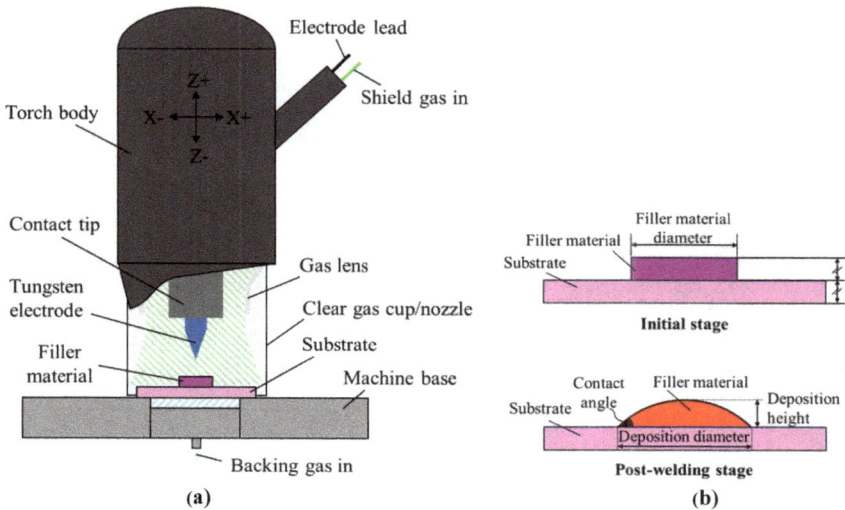

Fig. 1: (a) Schematic of the TIG welding setup, (b) pre- and post-welding stages of the filler material and substrate

Table 1 presents the welding process parameters investigated during the deposition process, including welding current, welding time and filler material diameter. For the 4 mm filler material diameter, welding times ranging from 500 to 900 ms were employed, with welding currents adjusted between 60 and 80 A. To enable a comparison of the domed-shaped geometries produced with the 4 mm filler material diameter, a narrower experimental range of welding times from 700 to 900 ms, and a current range of 65 to 80 A were applied to the 5 mm filler material diameter.

Table 1: Deposition process variables

Process parameter	Applied variable
Welding current [A]	60, 65, 70, 75, 80
Welding time [ms]	500, 600, 700, 800, 900
Filler material diameter [mm]	4, 5

Several parameters were held constant: the front argon flow was maintained at 8 ℓ/min, with upslope and downslope times of 10 ms and 600 ms, respectively. A tungsten electrode with a 60° tip angle was used, set at a constant arc length of 2.4 mm from the filler material surface. Additionally, the electrode was inspected after each process to mitigate potential contamination or material buildup.

Following the deposition process, the sheets were scanned using a Coord3 MC16 CMM equipped with an LC60Dx Laser Line Scanner. The scans were then processed with GOM Inspect

Sheet Metal 2025   Materials Research Forum LLC
Materials Research Proceedings 52 (2025) 357-364   https://doi.org/10.21741/9781644903551-44

2019 software to analyze the morphology of the deposition mainly in terms of diameter, height, and contact angle (Fig. 1b).

Results and Discussion

Cross-Sectional Deposition Profile of the Substrate. Figure 2 illustrates the effect of heat input parameters (welding current and welding time) on the deposition profile geometry, with a 4 mm filler diameter used for the experiment. Cross-sectional analyses were conducted along the X and Z axes through the deposition center, varying welding current and welding time. A limited number of parameter combinations were investigated to establish a baseline.

The results indicate that increasing heat input (longer welding times and/or higher current) leads to larger deposition diameters, reduced welding heights, and decreased contact angles. At the lowest heat input (60 A, 500 ms), the deposition height slightly exceeds the initial filler material thickness (1.6 mm), and the welding angle approaches 90°. Such a high angle can limit bonding due to reduced surface interaction between the molten metal and substrate. Consequently, angles below 90° are often preferred to facilitate smoother deposition and ensure adequate bonding [15]. Furthermore, reduced heat input may induce undesirable deposition geometry issues, such as higher roundness and uneven height distribution.

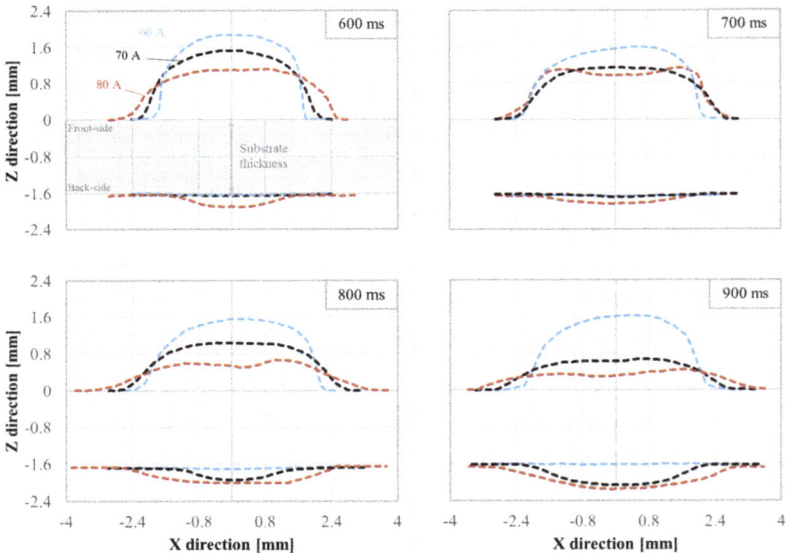

Fig. 2: Influence of process current and welding time on the deposition geometry for 4 mm filler material diameter

Conversely, increasing the heat input (e.g., 70 A, 700 ms) yields a more evenly distributed geometry with a thickness less than the filler material (see Fig. 2). However, further increasing the heat input (e.g., 70 A, 800 ms) leads to excess weld penetration and a visible back mark on the back-side of the substrate. Moreover, excessive heat input (e.g., 80 A, 700 ms) resulted in an improper deposition profile, characterized by a peak deposition height at the edges rather than the center. This phenomenon is attributed to a temperature-dependent surface tension gradient, which drives molten metal outward from the center to the edges [16]. Excessive heat input at this level

Sheet Metal 2025   Materials Research Forum LLC
Materials Research Proceedings 52 (2025) 357-364   https://doi.org/10.21741/9781644903551-44

leads to not only geometric inconsistency but also increased contamination and deposition/substrate surface bursts.

Deposition Geometry and Contact Angle. Figure 3 illustrates the deposition geometry, including mean diameter, height, and contact angle, for filler material with 4 mm and 5 mm diameters. The mean values and standard deviations, calculated from three cross-sectional measurements at 120° intervals, quantify the deposition roundness and height distribution.

For the 4 mm filler material (Fig. 3a), low heat input led to significant variations in height and diameter, as well as a high contact angle, attributed to the rapid solidification of magnesium. Conversely, high heat input resulted in increased geometrical variations due to the aforementioned surface tension gradients. Smooth deposition was achieved with a contact angle below 70°, corresponding to a deposition diameter exceeding the filler diameter and a thickness less than the sheet thickness. Similarly, for higher heat input, a contact angle above 30° is necessary for smooth deposition morphology.

Figure 3b presents deposition values for a 5 mm filler diameter. At the lowest applied heat input (65 A, 700 ms), the observed geometrical accuracy was comparable to that of the 4 mm filler material at its lowest heat input (60 A, 700 ms), indicating the minimum process limits for the 5 mm diameter. Additionally, at the maximum applied heat input (80 A, 900 ms), the domed-shaped deposition produced with the 5 mm filler material diameter exhibited superior geometrical accuracy compared to that of the 4 mm filler material diameter, without notable surface contamination. In contrast, the use of the 4 mm filler material diameter at 80 A, 900 ms exhibited excessive deposition with a high surface tension gradient, resulting in poor geometrical distribution and surface oxidation.

Process Operating Window. Figure 4a and Figure 4b illustrate the experimentally determined process windows for 4 mm and 5 mm filler diameters, respectively. Optimal deposition conditions, highlighted in green, were identified based on metrics such as deposition roundness, height consistency, and post-deposition evaluations for contamination and surface burst formation. Excessive heat input resulted in burst formation at the deposition center, which was classified as a process defect.

The process window was defined exclusively by the factors above, ensuring smooth, stable deposition by controlling key parameters. Specifically, for the 4 mm filler diameter, relatively smooth deposition was achieved within a contact angle ranging from 30° to 70°, and deposition height ranging from 0.6 mm to 1.6 mm, corresponding to the initial sheet thickness. It is important to note that applications requiring local thickening, such as achieving adequate thread length for locking screws in specific implant geometries [6], necessitate precise control over deposition diameter and height, potentially narrowing the acceptable parameter range within this process window.

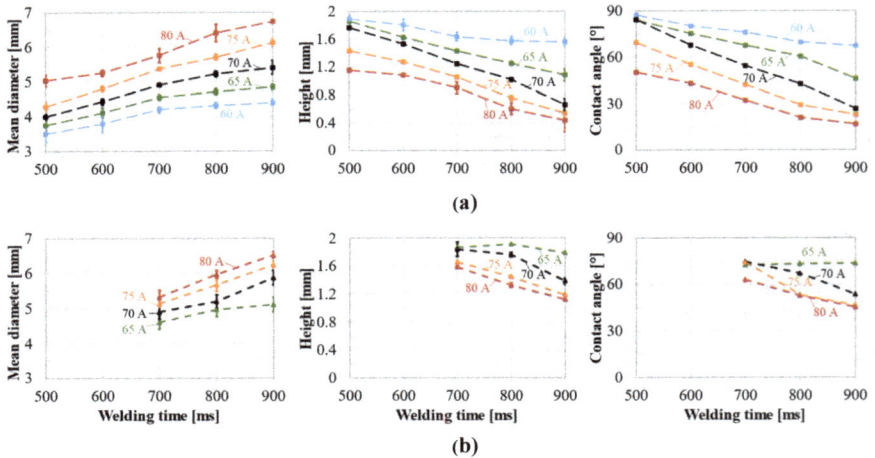

Fig. 3: Effect of TIG welding parameters on deposition geometry for different filler material diameters: (a) 4 mm, (b) 5 mm

A comparison of the two process windows (Fig. 4a and Fig. 4b) reveals a similar trend. However, smooth deposition with the 5 mm filler diameter requires higher heat inputs. This underscores the significant role of filler diameter alongside welding current and time. Furthermore, the upper limit for the 5 mm filler material remains undetermined, suggesting potential for smoother deposition profiles.

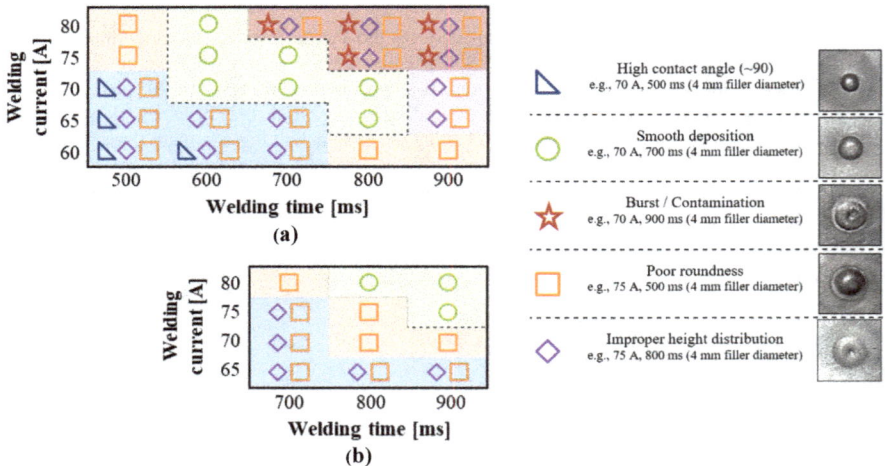

Fig. 4: Process operating window for different filler material diameters: (a) 4 mm, (b) 5 mm

Conclusions and Outlook

This study introduces a local reinforcement technique for Mg-Zn-Zr alloy thin flat sheets using TIG welding deposition, with pure argon providing an inert atmosphere on both the front and back surfaces during the process. The research examines the effects of welding input parameters—namely, welding current, welding time, and filler material diameter—on localized thickening during TIG deposition. In this method, the filler material is strategically positioned at the top of the substrate with a defined arc length. The key findings are summarized as follows:

- Remarkably, even at the highest applied heat input, no major distortion was detected in the 1.6mm-thick substrate geometry, which is encouraging for applications that involve thinner and more complex geometries.
- Increased heat input, achieved through prolonged welding times and elevated currents, results in larger deposition diameters, reduced heights, and varied contact angles. A dome-shaped deposition with reduced roundness, balanced height distribution, and minimal contamination was achieved at a contact angle below 70°. However, excessive heat input can introduce adverse effects, such as reduced geometric consistency and increased contamination, particularly evident at contact angles below 30°.
- The operating window is established for 4 mm filler diameter conditions to achieve smooth and consistent deposition. The effective deposition was achieved within specific ranges for diameter and height consistency, while also considering contamination caused by high energy input. In contrast, the 5 mm diameter required higher energy inputs to establish the upper process limit for geometric accuracy. These findings indicate a positive correlation between filler diameter and the required energy input for smooth deposition geometry.

As a preliminary investigation, this research provides a foundation for future studies focused on the microstructural and mechanical properties of the deposits, as well as their impact on overall geometrical stability and penetration characteristics. Gaining a deeper understanding of these microstructural and mechanical aspects is essential for advancing deposition processes on thinner, more complex geometries, while also evaluating the distortion limits. Ultimately, the insights from this work are critical steps toward refining ISF processes and contributing to post-forming additive manufacturing techniques that offer local thickening or coating solutions for high-quality, patient-specific implants with enhanced performance characteristics.

Acknowledgment
This research was conducted within the framework of the STIFF project, facilitated by the KU Leuven C2 research fund.

References
[1] H. Ibrahim, S.N. Esfahani, B. Poorganji, D. Dean, M. Elahinia, Resorbable bone fixation alloys, forming, and post-fabrication treatments, Mater. Sci. Eng. C 70 (2017) https://doi.org/10.1016/j.msec.2016.09.069

[2] A. Torroni, C. Xiang, L. Witek, E.D. Rodriguez, P.G. Coelho, N. Gupta, Biocompatibility and degradation properties of WE43 Mg alloys with and without heat treatment: In vivo evaluation and comparison in a cranial bone sheep model, J. Cranio-Maxillofac. Surg. 45 (2017) https://doi.org/10.1016/j.jcms.2017.09.016

[3] P. Sekar, S. Narendranath, V. Desai, Recent progress in in vivo studies and clinical applications of magnesium-based biodegradable implants - A review, J. Magnes. Alloys 9 (2021) https://doi.org/10.1016/j.jma.2020.11.001

[4] H. Vanhove, Y. Carette, S. Vancleef, J.R. Duflou, Production of thin shell clavicle implants through Single Point Incremental Forming, Procedia Eng. 183 (2017) https://doi.org/10.1016/j.proeng.2017.04.058

[5] J.R. Duflou, A.M. Habraken, J. Cao, R. Malhotra, M. Bambach, D. Adams, H. Vanhove, A. Mohammadi, J. Jeswiet, Single point incremental forming: state-of-the-art and prospects, Int. J. Mater. Form. 11 (2018) https://doi.org/10.1007/s12289-017-1387-y

[6] H. Vanhove, O. Kurtov, A. Dejans, E. Ozden, J.R. Duflou, Local reinforcement of titanium sheet by means of GTAW droplet deposition for threaded connections, Mater. Res. Proc. 41 (2024) 363-370. https://doi.org/10.21741/9781644903131-41

[7] D.E.P. Klenam, G.S. Ogunwande, T. Omotosho, B. Ozah, N.B. Maledi, S.I. Hango, A.A. Fabuyide, L. Mohlala, J.W. Van Der Merwe, M.O. Bodunrin, Welding of magnesium and its alloys: An overview of methods and process parameters and their effects on mechanical behaviour and structural integrity of the welds, Manuf. Rev. 8 (2021) https://doi.org/10.1051/mfreview/2021028

[8] W. Zhou, Q. Le, L. Ren, Y. Shi, Y. Jiang, Q. Liao, Effect of welding current and Laves phase on the microstructures and mechanical properties of GTAW joints for ZC63 magnesium alloy, Mater. Sci. Eng. A 872 (2023) https://doi.org/10.1016/j.msea.2023.144954

[9] S. Kou, Welding Metallurgy, 3rd ed., Book, 2021.

[10] K. Liu, S. Kou, Susceptibility of magnesium alloys to solidification cracking, Sci. Technol. Weld. Join. 25 (2020) 3. https://doi.org/10.1080/13621718.2019.1681160

[11] B. Costanzi, Proposed Hybrid WAAM and Thin Sheet Metal Welding, in: Reinforcing and Detailing of Thin Sheet Metal Using Wire Arc Additive Manufacturing as an Application in Facades, Mechanik, Werkstoffe und Konstruktion im Bauwesen, vol. 68, Springer Vieweg, Wiesbaden, 2023. https://doi.org/10.1007/978-3-658-41540-2

[12] C. Wehren, J. Kellerwessel, M. Trautz, R. Sharma, U. Reisgen, Using WAAM Metal Distortion for Sheet Metal Forming, in: Proceedings of the IASS 2024 Symposium Redefining the Art of Structural Design, Zurich, Switzerland, August 26-30, 2024.

[13] T. Zhang, C. Xu, J. Cheng, Y. Huang, Y. Peng, L. Wang, K. Wang, Study on droplet transfer behavior and spattering mechanism in Mg-Gd-Y-Zr alloy regulated short circuit (Fronius CMT) welding, J. Manuf. Process. 126 (2024) 35-47. https://doi.org/10.1016/j.jmapro.2024.07.075

[14] W. Zhou, Q. Le, L. Ren, Y. Shi, Y. Jiang, Q. Liao, Effect of welding current and Laves phase on the microstructures and mechanical properties of GTAW joints for ZC63 magnesium alloy, Mater. Sci. Eng. A 872 (2023) https://doi.org/10.1016/j.msea.2023.144954

[15] Kapil, A., Sharma, V., Pauw, J. De, & Sharma, A. Exploring impact, spreading, and bonding dynamics in molten metal deposition for novel drop-on-demand printing, Mater. Des. 238 (2024) 112633. https://doi.org/10.1016/j.matdes.2024.112633

[16] F. Wu, K. V. Falch, D. Guo, P. English, M. Drakopoulos, W. Mirihanage, Time evolved force domination in arc weld pools, Mater. Des. 190 (2020) 108534. https://doi.org/10.1016/j.matdes.2020.108534

Sheet Metal 2025
Materials Research Proceedings 52 (2025) 365-373

Materials Research Forum LLC
https://doi.org/10.21741/9781644903551-45

Local adaptation of aluminum blanks through laser de-alloying and wire alloying

Marcel Stephan[1,2,a,*], Henrik Zieroth[3], Simona Samland[1], Dominic Bartels[1,2,4], Marion Merklein[3] and Michael Schmidt[1,2,4]

[1]Bayerisches Laserzentrum GmbH (BLZ), Konrad-Zuse-Straße 2-6, 91052 Erlangen, Germany

[2]Erlangen Graduate School in Advanced Optical Technologies (SAOT), Friedrich-Alexander-Universität Erlangen-Nürnberg (FAU), Paul-Gordan-Str. 6, 91052 Erlangen, Germany

[3]Institute of Manufacturing Technology (LFT), FAU, Egerlandstraße 13, 91058 Erlangen, Germany

[4]Institute of Photonic Technologies (LPT), FAU, Konrad-Zuse-Straße 3-5, 91052 Erlangen, Germany

[a]m.stephan@blz.org

Keywords: Laser Welding, Aluminum, Hybrid Manufacturing

Abstract. The 7xxx series aluminum alloys, renowned for their exceptional specific strength, are promising materials for automotive manufacturing. Despite their advantages, the widespread application of these alloys is inhibited by their low formability. The high strength but low formability is caused by Zn and Mg precipitations, which hinder dislocation motion. In this study, the local adaption of the elemental composition of aluminum blanks through selective evaporation and alloying via auxiliary wire is investigated. A high-power laser is used for melting and evaporating low boiling elements such as Mg and Zn while adding AlSi5 in parallel to improve formability in the forming zone. The resulting element distribution is studied and correlated with the applied process conditions, to gain an insight on the possibilities and limitations of the laser induced alloy adaption.

Introduction

Aluminum alloys are important materials for applications in various industries, mainly due to their high specific strength. Ternary 6xxx (Al-Mg-Si) series alloys, like AA6082, are used in the automotive industry for structural components (i.e. body panels), because of their good formability and considerably high strength [1]. In comparison, quaternary 7xxx series alloys (Al-Zn-Mg-Cu), like AA7075, possess more strength. Among other things, their poor formability and ductility, due to Zn and Mg precipitations, is a major hurdle. This hinders their application in lightweighting of crash relevant parts in automotive applications.

Vollertsen et al. [2] previously investigated the improvement of formability by means of laser application. They showed that by heat treating plate material with a laser the formability can be increased by dissolution of precipitation. This process is time sensitive as the heat treatment creates a metastable super-saturated solid solution and with enough time precipitation will restart [3].

Adjusting the composition locally in areas that require high formability offers the possibility to avoid this time-sensitive procedure. Two main approaches could be followed in this instance: On the one hand, the chemical composition could be adjusted by removing, in this case evaporating low-boiling elements. This *de-alloying* could be used specifically to remove the main strengthening elements Zn and Mg. These elements possess a significantly lower boiling point compared to the other elements, as shown in Table 1. The combination of the lower boiling point and high vapor pressure of these elements enables the concept of *de-alloying*. Correspondingly, the local energy input during laser welding in combination with the low evaporation temperatures

Content from this work may be used under the terms of the Creative Commons Attribution 3.0 license. Any further distribution of this work must maintain attribution to the author(s) and the title of the work, journal citation and DOI. Published under license by Materials Research Forum LLC.

of these elements could be exploited to achieve a selective adjustment of the chemical composition. Accordingly, this method can be used to reduce the formation of Zn or Mg precipitations. However, this approach is only feasible for alloys in which low-boiling elements are responsible for the final strength of the material. On the other hand, the overall concentration of specific elements could be tackled by adding auxiliary material and thus generating a new alloy. Through in-situ alloying an adapted material that consists of the blank and the auxiliary material could be generated. The element composition of the material can be adapted locally by the elements removed and added in this way.

Table 1 Elemental composition of AA6082 and AA7075 according EN-573-3 [4] and boiling points of elements [5]

Element	Al	Zn	Mg	Si	Cu	Fe	Cr	Ti	Mn
AA6082 EN-573-3 [wt.-%]	Bal.	0.2	0.6 – 1.2	0.7 – 1.3	0.1	0.5	0.25	0.1	0.4 – 1.0
AA7075 EN-573-3 [wt.-%]	Bal.	5.1 – 6.1	2.1 – 2.9	0.4	1.2 – 2.0	0.5	0.18 – 0.28	0.2	0.3
Boiling Point [°C]	2520	907	1090	3270	2560	2860	2672	3285	2061

Goal of this work is to investigate the potentials and limitations of *laser-induced alloy adaptation* (German: Laser-induzierte Legierungsadaption, LILA) when processing hardly formable alloys from the 7xxx series. First, the influence of a processing strategy from preliminary experiments on the resulting properties in the field of de-alloying is studied. In the next step, this alloy is modified through wire addition. For this purpose, an AlSi5 wire is supplied to improve the formability of the material by increasing the Si concentration, which in theory should result in a higher ductility and also reduces hot crack susceptibility [6]. Following these thoughts, this study aims to investigate the possibilities of localized alloying by applying an AlSi5 wire on previously evaporated aluminum banks and studies the influence of process parameters on process stability and element distribution.

Materials and Methods

The base material was AA7075-F (Novelis Deutschland GmbH, Göttingen, Germany) and was delivered as sheets with the dimension 50x50x2 mm³. A schematic illustration of the concept of LILA is shown in Fig. 1.

In the first step, volatile elements such as Zn and Mg were de-alloyed using a set of parameters developed in preliminary experiments (see Table 2). In the next step, AlSi5 wire (MTC GmbH, Meerbusch, Germany) with a thickness of 0.8 mm was fixed in the previously evaporated gap using a pressing device and then molten to apply it on the de-alloyed area in order to reach the desired elemental composition similar to AA6082.

Sheet Metal 2025   Materials Research Forum LLC
Materials Research Proceedings 52 (2025) 365-373   https://doi.org/10.21741/9781644903551-45

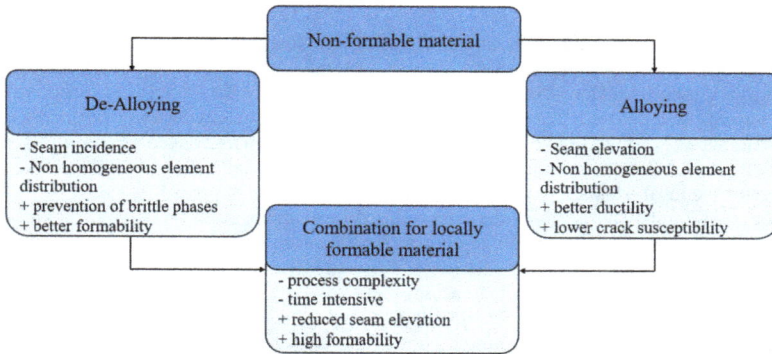

Fig. 1 Concept of Laser Induced Alloy Adaption (LILA)

Table 2 Parameter sets used for wire-alloying

Parameter set	Laser Power P_L [W]	Feed Speed v [mm/s]	Beam Size d_L [µm]	Number of Passes
P0	1500	5	307	2
P1	800	25	307	1
P2	1200	25	307	1
P3	1600	25	307	1
P4	1400 (1st pass) 1600 (2nd pass)	25	307	2

Joining the wire with the blank was carried out using 6 kW disk laser of the type TruDisk 6001 (TRUMPF SE + Co. KG, Germany) and focusing optics of the type BEO D70 (TRUMPF SE + Co. KG, Germany) with a beam diameter of 196.02 µm in the focal plane. To achieve a larger spot diameter, the laser beam was defocused.

In preliminary experiments different sets were tested in order to achieve depth of at least half the thickness of the blanks, higher process stability and more homogeneous elemental distribution. The parameter sets investigated in this work are listed in Table 2. The parameter sets P1 to P4 are used for melting and applying the wire on the de-alloyed area. De-alloying was conducted through evaporating volatile elements, Zn and Mg, using the P0 parameter set. This parameter set was identified in preliminary experiments. The aim of P1 to P3 was to improve the penetration depth and process stability by increasing the laser power in several steps. P4 follows the thought of improving and homogenizing element distribution by remelting the alloyed area. For applying the wire and remelting the modified seam different laser power, 1400 W and 1600 W respectively, was used in consideration of the absorption, heat conduction and changed process conditions, like a new oxide layer or defects. For P0 and P4 a waiting time of 180 s was implemented between the passes to prevent excessive heat accumulation.

For comprehensive measurement across the entire adapted area, Large Area Mapping (LAM) was employed. LAM acquires data sequentially over adjacent fields, achieving high spatial resolution at low magnifications, thereby enabling detailed compositional analysis of extensive regions.

Sheet Metal 2025
Materials Research Proceedings 52 (2025) 365-373

Materials Research Forum LLC
https://doi.org/10.21741/9781644903551-45

Results and Discussion

Analyses of depth, density and element concentration and distribution were carried out. The depth determines the size of the functionalized area. In turn, the relative material density related to the total cross-sectional area along the seam as a measure of the defect density has significant influence on the forming and failure behavior of the blank. Any defect, such as pores or cracks, is a weak point in the sample during forming, e.g. during bending. The element concentrations of Mg, Si and Zn and their distribution within the adapted seam are the main focus of this work, which investigates the local adaptation of aluminum blanks.

Penetration depth. The depth of the wire-alloyed seam defines the functionalized area of the blank. Consequently, the penetration depth has significant influence on further processing of the blank, e.g. forming. A depth of at least 1 mm is targeted since this resembles half the thickness of the blanks. Thus, it would be possible to achieve an overall penetration depth of 2 mm by turning the sheets around. Since not only the overall depth but also the evaporation depth is of importance, both characteristics need to be determined. The respective average depths for the different process parameters are shown in Fig. 2. Each measurement was conducted in the middle of the seam and repeated for 3 samples respectively.

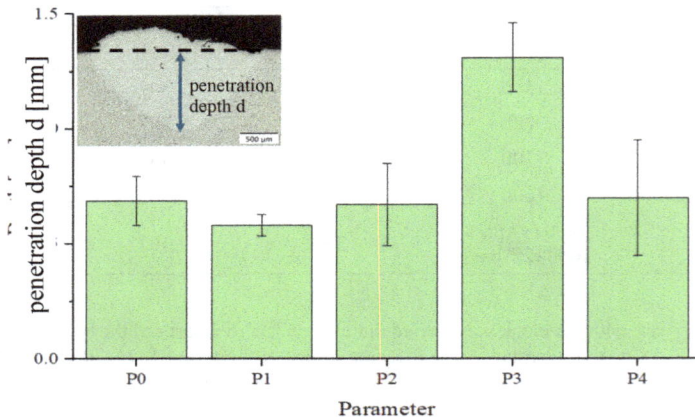

Fig. 2 Plot of the penetration depth depending on the parameter set used. P0 stands for the parameter set used for de-alloying and P1 – P4 are standing for parameter sets used for wire-alloying

Processing the material using P1 resulted in an insufficient penetration depth even lower than the depth generated by P0. The limited depth is most likely caused by a too low energy input. Thus, the laser power was increased in the following (P2 and P3). The penetration depth does not increase linearly with the laser power (see Fig. 2). A potential explanation for this can be the transition from a heat conduction welding regime (P1 and P2) towards deep penetration welding (P3). This assumption is underlined by the cross-sections, which reveal an increased penetration depth for P3 while the width of the adapted seam appears rather comparable (see Fig. 3). Double-illumination (P4) of the specimen results in a comparable penetration depth as for P2, even though the laser power was comparable to P3 in the second pass. Several findings can be concluded from this. First, this indicates that the threshold for the onset of deep penetration welding is somewhere between 1,400 W (0.67 ± 0.18 mm) and 1,600 W (1.31 ± 0.15 mm). Furthermore, it needs to be assumed

Sheet Metal 2025
Materials Research Proceedings 52 (2025) 365-373

Materials Research Forum LLC
https://doi.org/10.21741/9781644903551-45

that an oxide layer is formed between the first and second illumination of the surface. This oxide layer is characterized by different optical and physical properties. The absorption of the laser wavelength and thus the energy coupling into the specimens should be improved due to the better absorptivity of the oxide layer [7]. However, the higher melting point of the newly formed oxide layer (2052 °C) counteracts an increase in penetration depth [8].

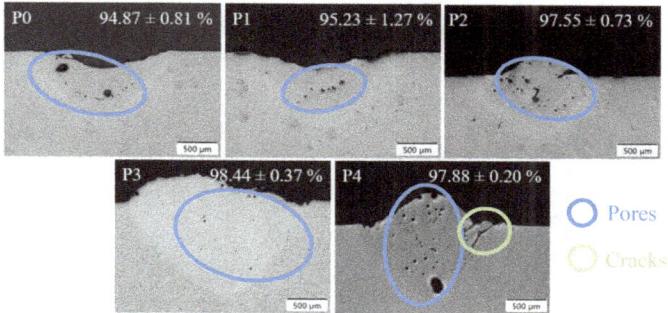

Fig. 3 Exemplary cross-sections of the adapted seams generated using parameters P0 to P4 as well as respective density along the whole seam relative to the longitudinal cross-section of the blank.

Regarding the resulting penetration depth, only P3 assures a sufficient depth of above 1 mm, which is needed to modify the entire 2 mm sheet metals in z-direction. Nevertheless, the lower penetration depths of the other parameter sets are not necessarily detrimental. Further investigations are needed on this topic to assess e.g. the formability for only selectively modified specimen (in z-direction). In this instance, the relative density of the generated tracks is of higher importance, since internal defects such as cracks or pores promote a premature failure during forming operations. The lowest part density is obtained for P1. It is most likely that this laser power was not sufficient to re-fill the defects that were obtained during the initial evaporation process. When applying laser powers comparable to the one used for de-alloying with P0 (1500 W), a relative density of the adapted tracks in the range of 97 to 99 % was obtained.

The cross-section for P1 shows no seam elevation. This can be explained with Balling effects of the wire material. Balling effects occur when molten material loses its bond to the surface and forms spherical structures [9]. In this case, this is probably due to the fact that the base material under the wire is not melting sufficiently and is not fusing with the molten wire.

The results show that further work is needed to improve the quality of the adapted seams. Potential approaches to improve the part density could be the application of beam shaping to tailor the energy input and the thermal field within the melt pool [10]. These adjusted temperature fields could be helpful to avoid peak temperatures, which would result in undesired element evaporation of higher boiling elements (keyhole porosity), and also provide the melt pool with sufficient time to re-fill emerging gaps in the liquid state. Further potential for improvement is also recognizable for the parameter set (P0), which was used for de-alloying and served as the starting point for wire-alloying. However, further investigations into de-alloying are part of independent work that was not in the focus here. Despite the lower relative part density compared to the initial aluminum blank, the adapted alloying composition might be helpful to improve the formability of the materials. Consequently, the chemical composition is investigated in the next step.

Elemental distribution. The overall concentration of the different key alloying elements Mg, Zn, and Si in the entire adapted seam is shown for the different parameter combinations in Fig. 4.

Sheet Metal 2025   Materials Research Forum LLC
Materials Research Proceedings 52 (2025) 365-373   https://doi.org/10.21741/9781644903551-45

After evaporation, an Mg and Zn content of 1.9 and 0.02 wt.-% was obtained. Slight fluctuations of the values occur due to reflections during LAM measurement. These values are the reference state for the upcoming investigations regarding the alloying with the auxiliary AlSi5 wire material.

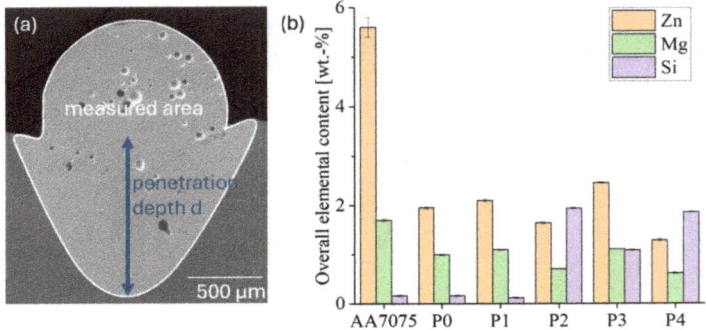

Fig. 4 (a) Exemplary schematic illustration of the measured area on a cross-section for P2 and (b) Overall concentration of Zn, Mg of the modified seams depending on the parameters used. The overall concentrations were measured by LAM except for the reference values for AA7075 which were measured using a small rectangular area in the middle of the unmodified blank by EDS

During alloying, the average Mg and Zn concentration trend to remain rather constant or decrease slightly for the different parameter combinations investigated. This trend can be explained by the fact that the penetration depth of P1, P2 and P4 is below the evaporation depth of P0. Consequently, the evaporation of low-boiling elements is further promoted. However, specimens processed using parameter combination P3 show an increase in Mg and Zn concentration. Here, the penetration depth of P3 surpasses the evaporation depth of P0. It is possible that the depth hinders the rise of low-boiling elements to the surface of the melt pool and thus their evaporation from the liquid material. Comparing the concentrations of key alloying elements Mg, Si and Zn of the AA6082 shown in Table 1 with the P0 to P4, several conclusions can be drawn. First, concentration of Mg is in the range of the target concentration for all parameter sets. Second, concentration of Zn is multiple times too high for all parameter sets. Third, concentration of Si is either too low or too high compared to the target concentration. Accordingly, the potential for further investigations and improvement in both de-alloying and alloying remains. Looking at the EDS mappings (Fig. 5) reveals that the wire-alloyed region is almost free of Mg and Zn, while higher concentrations can be found when moving towards the unmodified aluminum blanks. To validate this assumption, EDS line scans along the penetration depth direction were performed. Fig. 5 shows an exemplary EDS mapping for the alloying elements Zn when processing the material with different parameter combinations.

It can be seen that the Zn content is varying along the depth of the adapted seams. An increase of the Zn content can be determined once the depth of the adapted area is exceeded. This jump can be explained by the higher concentration of the base material compared to the de-alloyed one. For example, P3 is characterized by the lowest Zn concentration of all parameters at a depth of around 0.8 mm. All other parameter combinations possess higher Zn concentrations at this depth. The depth at which an increase in Zn content takes place is predefined by first the evaporation depth (de-alloying) and second the alloying depth during wire-alloying. Following this thought, the

Sheet Metal 2025   Materials Research Forum LLC
Materials Research Proceedings 52 (2025) 365-373   https://doi.org/10.21741/9781644903551-45

previous increase in P2 can be explained by an insufficient alloying depth. Through wire-alloying, an increase in Si concentration can be obtained for most parameter combinations. The highest Si content was obtained for P2, followed by P4. Here, the wire was mixed rather homogeneously with the previously de-alloyed material. When increasing the laser power (P3), a lower overall Si content was determined. This effect is logical since the higher penetration depth (see Fig. 6) results in a larger dilution zone. Since the supplied wire material is the same in all cases, a larger depth result in a lower average element concentration of the auxiliary element Si. This finding also correlates well with the previously described increase in Mg and Zn.

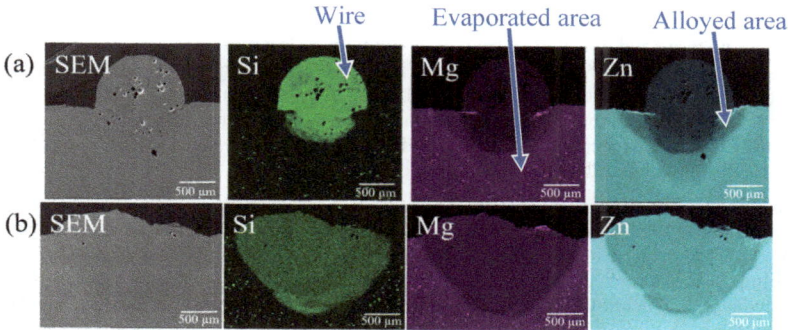

Fig. 5 SEM images and large area mappings of the elements Zn, Mg and Si of parameters (a) P2 and (b) P3

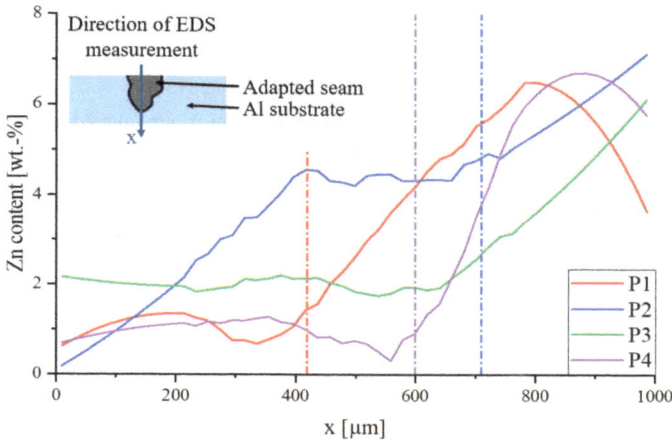

Fig. 6 Plot of Zn content along the depth of the adapted seams. The end of the adjusted areas is marked with a vertical line respectively except for P3 where the depth exceeded the measured area.

Sheet Metal 2025

Materials Research Proceedings 52 (2025) 365-373

Materials Research Forum LLC

https://doi.org/10.21741/9781644903551-45

Summary

In this work, a novel approach consisting of a targeted de-alloying and a wire-based alloying process using laser welding is proposed. The combination of these two process steps supports the generation of Tailor Alloy Blanks. Through de-alloying, it was possible to reduce the Mg and Zn concentration to low levels, likely hindering the formation of precipitations that promote strengthening effects. A decrease in Mg and Zn concentration was achieved in initial investigations. However, the evaporated regions only possess low relative part density, revealing the need for optimized processing strategies.

Applying additional wire material helped to introduce Si into the material. Si was distributed homogeneously within the adapted seams. Thus, it was possible to adapt the chemical composition of the base material and move away from 7xxx alloy towards a more ductile 6xxx alloys. Further investigations will focus on the development of improved processing strategies that result in lower defect tendency. Applying beam shaping or oscillating illumination patterns could help to improve element evaporation and avoid crack formation by controlling the thermal fields within the melt pool. Furthermore, the mechanical properties, especially formability, need to be studied to provide insight on the potential of the proposed LILA concept.

Acknowledgement

We would like to thank the German Research Foundation (Deutsche Forschungsgemeinschaft, DFG) for funding the research project "Tailor Alloyed Blanks" (project number 521490180). Furthermore, the authors gratefully acknowledge funding of the Erlangen Graduate School in Advanced Optical Technologies (SAOT) by the Bavarian State Ministry for Science and Art.

References

[1] M. Tisza, I. Czinege, Comparative study of the application of steels and aluminium in lightweight production of automotive parts, International Journal of Lightweight Materials and Manufacture 1 (2018) 229-238. https://doi.org/10.1016/j.ijlmm.2018.09.001

[2] F. Vollertsen, K. Lange, Enhancement of Drawability by Local Heat Treatment, CIRP Annals 47 (1998) 181-184. https://doi.org/10.1016/S0007-8506(07)62813-3

[3] A. Poznak, D. Freiberg, P. Sanders, Automotive Wrought Aluminium Alloys, in: Fundamentals of Aluminium Metallurgy, Elsevier, 2018, pp. 333-386. https://doi.org/10.1016/B978-0-08-102063-0.00010-2

[4] Deutsches Institut für Normung, DIN EN 573-3, Aluminium and aluminium alloys - chemical composition and form of wrought products. Part 3, Chemical composition and form of products, Deutsche Fassung, Beuth Verlag GmbH, Berlin, 2024.

[5] W.M. Haynes, CRC Handbook of Chemistry and Physics, CRC Press, 2014. https://doi.org/10.1201/b17118

[6] D. Fabrègue, A. Deschamps, M. Suéry, Influence of the silicon content on the mechanical properties of AA6xxx laser welds, Materials Science and Engineering: A 506 (2009) 157-164. https://doi.org/10.1016/j.msea.2008.11.033

[7] R. Indhu, V. Vivek, L. Sarathkumar, A. Bharatish, S. Soundarapandian, Overview of Laser Absorptivity Measurement Techniques for Material Processing, Lasers Manuf. Mater. Process. 5 (2018) 458-481. https://doi.org/10.1007/s40516-018-0075-1

[8] O.O. Oladimeji, Trend and innovations in laser beam welding of wrought aluminum alloys, Welding in the world 60 (2016) 415-457. https://doi.org/10.1007/s40194-016-0317-9

Sheet Metal 2025
Materials Research Proceedings 52 (2025) 365-373

Materials Research Forum LLC
https://doi.org/10.21741/9781644903551-45

[9] C. Teng, D. Pal, H. Gong, K. Zeng, K. Briggs, N. Patil, B. Stucker, A review of defect modeling in laser material processing, Additive Manufacturing 14 (2017) 137-147. https://doi.org/10.1016/j.addma.2016.10.009

[10] F. Nahr, D. Bartels, R. Rothfelder, M. Schmidt, Influence of Novel Beam Shapes on Laser-Based Processing of High-Strength Aluminium Alloys on the Basis of EN AW-5083 Single Weld Tracks, JMMP 7 (2023) 93. https://doi.org/10.3390/jmmp7030093

Keyword Index

About the Editor

www.ingramcontent.com/pod-product-compliance
Lightning Source LLC
Chambersburg PA
CBHW071318210326
41597CB00015B/1267

* 9 7 8 1 6 4 4 9 0 3 5 4 4 *